General Motors

Buick Regal
Chevrolet Lumina
Olds Cutlass Supreme
Pontiac Grand Prix

Automotive Repair Manual

by Robert Maddox and John H Haynes

Member of the Guild of Motoring Writers

Models covered:
Buick, Chevrolet (up to 1994), Oldsmobile and Pontiac
Full-size front-wheel drive models (W body style)
1988 through 1999

(11C10 - 38010)
(1671)

Haynes Publishing Group
Sparkford Nr Yeovil
Somerset BA22 7JJ England

Haynes North America, Inc
861 Lawrence Drive
Newbury Park
California 91320 USA

Acknowledgements

Wiring diagrams provided exclusively for Haynes North America, Inc. by Valley Forge Technical Communications. Technical writers who contributed to this project include Larry Warren and Brian Styve

A book in the Haynes Automotive Repair Manual Series

Printed in the U.S.A.

ISBN 1 56392 371 8

Library of Congress Catalog Card Number 00-100318

Contents

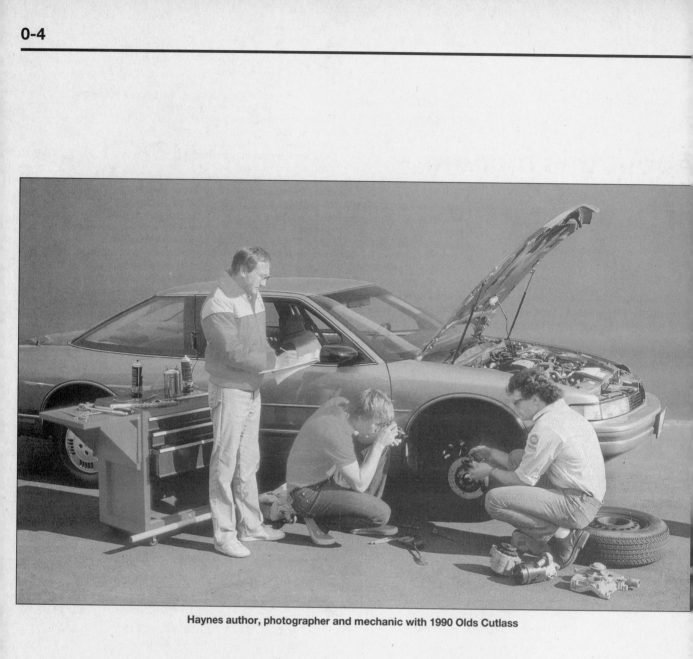

Haynes author, photographer and mechanic with 1990 Olds Cutlass

About this manual

Its purpose

The purpose of this manual is to help you get the best value from your vehicle. It can do so in several ways. It can help you decide what work must be done, even if you choose to have it done by a dealer service department or a repair shop; it provides information and procedures for routine maintenance and servicing; and it offers diagnostic and repair procedures to follow when trouble occurs.

We hope you use the manual to tackle the work yourself. For many simpler jobs, doing it yourself may be quicker than arranging an appointment to get the vehicle into a shop and making the trips to leave it and pick it up. More importantly, a lot of money can be saved by avoiding the expense the shop must pass on to you to cover its labor and overhead costs. An added benefit is the sense of satisfaction and accomplishment that you feel after doing the job yourself.

Using the manual

The manual is divided into Chapters. Each Chapter is divided into numbered Sections, which are headed in bold type between horizontal lines. Each Section consists of consecutively numbered paragraphs.

At the beginning of each numbered Section you will be referred to any illustrations which apply to the procedures in that Section. The reference numbers used in illustration captions pinpoint the pertinent Section and the Step within that Section. That is, illustration 3.2 means the illustration refers to Section 3 and Step (or paragraph) 2 within that Section.

Procedures, once described in the text, are not normally repeated. When it's necessary to refer to another Chapter, the reference will be given as Chapter and Section number. Cross references given without use of the word "Chapter" apply to Sections and/or paragraphs in the same Chapter. For example, "see Section 8" means in the same Chapter.

References to the left or right side of the vehicle assume you are sitting in the driver's seat, facing forward.

Even though we have prepared this manual with extreme care, neither the publisher nor the author can accept responsibility for any errors in, or omissions from, the information given.

NOTE

A **Note** provides information necessary to properly complete a procedure or information which will make the procedure easier to understand.

CAUTION

A **Caution** provides a special procedure or special steps which must be taken while completing the procedure where the Caution is found. Not heeding a Caution can result in damage to the assembly being worked on.

WARNING

A **Warning** provides a special procedure or special steps which must be taken while completing the procedure where the Warning is found. Not heeding a Warning can result in personal injury.

Introduction to the Chevrolet Lumina, Oldsmobile Cutlass, Buick Regal and Pontiac Grand Prix

These front-wheel drive full-size General Motors vehicles, some-times referred to as the "W" body style or 10 series, are available in two and four-door sedan body styles.

Engines used in these vehicles include the 2.2 liter and 2.5 liter overhead valve (OHV) four-cylinder, the 2.3 liter overhead cam (OHC) Quad-4 cylinder, the 2.8 liter V6, the 3.1 liter and 3100 V6, the 3.4 liter double overhead cam V6, and the 3800 V6.

Throttle body injection (TBI) is used on the 2.5 liter four-cylinder engine. All other engines are equipped with multi-port fuel injection.

The engine drives the front wheels through either a manual or automatic transaxle via driveaxles equipped with Constant Velocity (CV) joints. The power assisted rack and pinion steering is mounted behind the engine.

The front suspension is composed of MacPherson struts, three point control arms and a stabilizer bar. The rear suspension is in-dependent, with trailing arms, strut/shock absorber units, a transversely mounted leaf spring, and in some cases coil springs.

The brakes are disc at all four wheels on most models, although some are equipped with drums at the rear wheels. Power assist is standard equipment.

Vehicle identification numbers

Modifications are a continuing and unpublicized part of vehicle manufacturing. Since spare parts manuals and lists are compiled on a numerical basis, the individual vehicle numbers are essential to correctly identify the component required.

Vehicle Identification Number (VIN)

This very important identification number is stamped on a plate attached to the left side of the dashboard and is visible through the driver's side of the windshield (see illustration). The VIN also appears on the Vehicle Certificate of Title and Registration. It contains valuable information such as where and when the vehicle was manufactured, the model year and the body style.

Body identification plate

This metal plate is usually located on the top side of the radiator support or on the upper surface of the fan shroud. Like the VIN, it contains important information concerning the production of the vehicle as well as information about how the vehicle came equipped from the factory. It's especially useful for matching the color and type of paint during repair work.

Service parts identification label

This label is located in the trunk, usually either on the inside of the trunk lid or on the spare tire cover. It lists the VIN number, wheelbase, paint number, options and other information specific to the vehicle it's attached to. Always refer to this label when ordering parts.

Engine identification numbers

The engine code number on the 2.2 liter and 2.5 liter four-cylinder engine is stamped on a pad on the radiator side, at the rear of the block. The 2.3 liter OHC (Quad-4) engine VIN number is stamped into the rear of the block, near the starter motor. The Quad-4 engine also has a code label attached to the rear edge of the timing belt housing. On V6 engines, the code number is found on a pad on the rear of the block, just above the starter motor. On the 3800, the number is also stamped on a pad adjacent to the water pump.

Manual transaxle number

The Muncie transaxle has an adhesive-backed identification label attached to the rear of the case and an ID number stamped into the front. On the Getrag five-speed transaxle, the VIN is on a pad at the front (radiator) side of the case edge, near the bell-housing. If the transaxle label is missing or unreadable, use the Service parts identification label to determine which transaxle was installed at the factory.

Automatic transaxle number

The nameplate/ID number on the THM 125C/3T40 transaxle is attached to the upper surface of the case, near the rear (see illustration). On the 440-T4/4T60-E and 4T65-E transaxle, the transaxle VIN number is stamped into the right rear of the housing; a unit number label is attached above it.

Vehicle Emissions Control Information label

This label is found in the engine compartment. See Chapter 6 for more information on this label.

The Vehicle Identification Number (VIN) is on a plate attached to the top of the dashboard on the driver's side of the vehicle - it can be seen from outside the vehicle, looking through the windshield

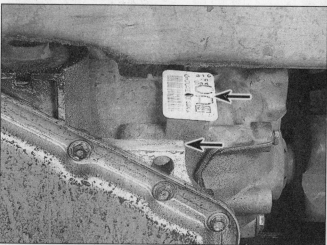

On the THM 125 transaxle, the ID number is on a plate attached to the upper surface of the transmission case

Buying parts

Replacement parts are available from many sources, which generally fall into one of two categories - authorized dealer parts departments and independent retail auto parts stores. Our advice concerning these parts is as follows:

Retail auto parts stores: Good auto parts stores will stock frequently needed components which wear out relatively fast, such as clutch components, exhaust systems, brake parts, tune-up parts, etc. These stores often supply new or reconditioned parts on an exchange basis, which can save a considerable amount of money. Discount auto parts stores are often very good places to buy materials and parts needed for general vehicle maintenance such as oil, grease, filters, spark plugs, belts, touch-up paint, bulbs, etc. They also usually sell tools and general accessories, have convenient hours, charge lower prices and can often be found not far from home.

Authorized dealer parts department: This is the best source for parts which are unique to the vehicle and not generally available elsewhere (such as major engine parts, transmission parts, trim pieces, etc.).

Warranty information: If the vehicle is still covered under warranty, be sure that any replacement parts purchased - regardless of the source - do not invalidate the warranty!

To be sure of obtaining the correct parts, have engine and chassis numbers available and, if possible, take the old parts along for positive identification.

Maintenance techniques, tools and working facilities

Maintenance techniques

There are a number of techniques involved in maintenance and repair that will be referred to throughout this manual. Application of these techniques will enable the home mechanic to be more efficient, better organized and capable of performing the various tasks properly, which will ensure that the repair job is thorough and complete.

Fasteners

Fasteners are nuts, bolts, studs and screws used to hold two or more parts together. There are a few things to keep in mind when working with fasteners. Almost all of them use a locking device of some type, either a lockwasher, locknut, locking tab or thread adhesive. All threaded fasteners should be clean and straight, with undamaged threads and undamaged corners on the hex head where the wrench fits. Develop the habit of replacing all damaged nuts and bolts with new ones. Special locknuts with nylon or fiber inserts can only be used once. If they are removed, they lose their locking ability and must be replaced with new ones.

Rusted nuts and bolts should be treated with a penetrating fluid to ease removal and prevent breakage. Some mechanics use turpentine in a spout-type oil can, which works quite well. After applying the rust penetrant, let it work for a few minutes before trying to loosen the nut or bolt. Badly rusted fasteners may have to be chiseled or sawed off or removed with a special nut breaker, available at tool stores.

If a bolt or stud breaks off in an assembly, it can be drilled and removed with a special tool commonly available for this purpose. Most automotive machine shops can perform this task, as well as other repair procedures, such as the repair of threaded holes that have been stripped out.

Flat washers and lockwashers, when removed from an assembly, should always be replaced exactly as removed. Replace any damaged washers with new ones. Never use a lockwasher on any soft metal surface (such as aluminum), thin sheet metal or plastic.

Fastener sizes

For a number of reasons, automobile manufacturers are making wider and wider use of metric fasteners. Therefore, it is important to be able to tell the difference between standard (sometimes called U.S. or SAE) and metric hardware, since they cannot be interchanged.

All bolts, whether standard or metric, are sized according to diameter, thread pitch and length. For example, a standard 1/2 - 13 x 1 bolt is 1/2 inch in diameter, has 13 threads per inch and is 1 inch long. An M12 - 1.75 x 25 metric bolt is 12 mm in diameter, has a thread pitch of 1.75 mm (the distance between threads) and is 25 mm long. The two bolts are nearly identical, and easily confused, but they are not interchangeable.

In addition to the differences in diameter, thread pitch and length, metric and standard bolts can also be distinguished by examining the bolt heads. To begin with, the distance across the flats on a standard bolt head is measured in inches, while the same dimension on a metric bolt is sized in millimeters (the same is true for nuts). As a result, a standard wrench should not be used on a metric bolt and a metric wrench should not be used on a standard bolt. Also, most stan-

dard bolts have slashes radiating out from the center of the head to denote the grade or strength of the bolt, which is an indication of the amount of torque that can be applied to it. The greater the number of slashes, the greater the strength of the bolt. Grades 0 through 5 are commonly used on automobiles. Metric bolts have a property class (grade) number, rather than a slash, molded into their heads to indicate bolt strength. In this case, the higher the number, the stronger the bolt. Property class numbers 8.8, 9.8 and 10.9 are commonly used on automobiles.

Strength markings can also be used to distinguish standard hex nuts from metric hex nuts. Many standard nuts have dots stamped into one side, while metric nuts are marked with a number. The greater the number of dots, or the higher the number, the

greater the strength of the nut.

Metric studs are also marked on their ends according to property class (grade). Larger studs are numbered (the same as metric bolts), while smaller studs carry a geometric code to denote grade.

It should be noted that many fasteners, especially Grades 0 through 2, have no distinguishing marks on them. When such is the case, the only way to determine whether it is standard or metric is to measure the thread pitch or compare it to a known fastener of the same size.

Standard fasteners are often referred to as SAE, as opposed to metric. However, it should be noted that SAE technically refers to a non-metric fine thread fastener only. Coarse thread non-metric fasteners are referred to as USS sizes.

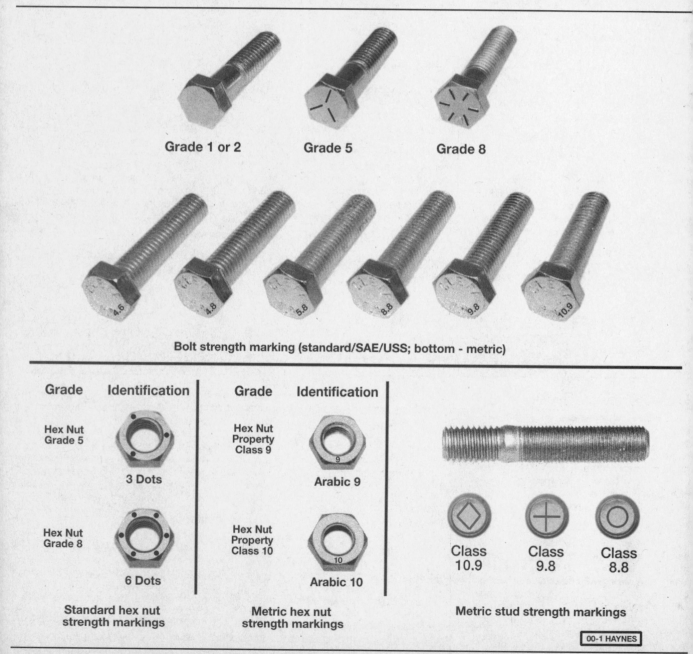

Grade 1 or 2   Grade 5   Grade 8

Bolt strength marking (standard/SAE/USS; bottom - metric)

Grade   Identification   Grade   Identification

Hex Nut Grade 5   3 Dots   Hex Nut Property Class 9   Arabic 9

Hex Nut Grade 8   6 Dots   Hex Nut Property Class 10   Arabic 10

Class 10.9   Class 9.8   Class 8.8

Standard hex nut strength markings   **Metric hex nut strength markings**   **Metric stud strength markings**

00-1 HAYNES

Since fasteners of the same size (both standard and metric) may have different strength ratings, be sure to reinstall any bolts, studs or nuts removed from your vehicle in their original locations. Also, when replacing a fastener with a new one, make sure that the new one has a strength rating equal to or greater than the original.

Tightening sequences and procedures

Most threaded fasteners should be tightened to a specific torque value (torque is the twisting force applied to a threaded component such as a nut or bolt). Overtightening the fastener can weaken it and cause it to break, while undertightening can cause it to eventually come loose. Bolts, screws and studs, depending on the material they are made of and their thread diameters, have specific torque values, many of which are noted in the Specifications at the beginning of each Chapter. Be sure to follow the torque recommendations closely. For fasteners not assigned a specific torque, a general torque value chart is presented here as a guide. These torque values are for dry (unlubricated) fasteners threaded into steel or cast iron (not aluminum). As was previously mentioned, the size and grade of a fastener determine the amount of torque that can safely be applied to it. The figures listed here are approximate for Grade 2 and Grade 3 fasteners. Higher grades can tolerate higher torque values.

Fasteners laid out in a pattern, such as cylinder head bolts, oil pan bolts, differential cover bolts, etc., must be loosened or tightened in sequence to avoid warping the component. This sequence will normally be shown in the appropriate Chapter. If a specific pattern is not given, the following procedures can be used to prevent warping.

Metric thread sizes	Ft-lbs	Nm
M-6	6 to 9	9 to 12
M-8	14 to 21	19 to 28
M-10	28 to 40	38 to 54
M-12	50 to 71	68 to 96
M-14	80 to 140	109 to 154
Pipe thread sizes		
1/8	5 to 8	7 to 10
1/4	12 to 18	17 to 24
3/8	22 to 33	30 to 44
1/2	25 to 35	34 to 47
U.S. thread sizes		
1/4 - 20	6 to 9	9 to 12
5/16 - 18	12 to 18	17 to 24
5/16 - 24	14 to 20	19 to 27
3/8 - 16	22 to 32	30 to 43
3/8 - 24	27 to 38	37 to 51
7/16 - 14	40 to 55	55 to 74
7/16 - 20	40 to 60	55 to 81
1/2 - 13	55 to 80	75 to 108

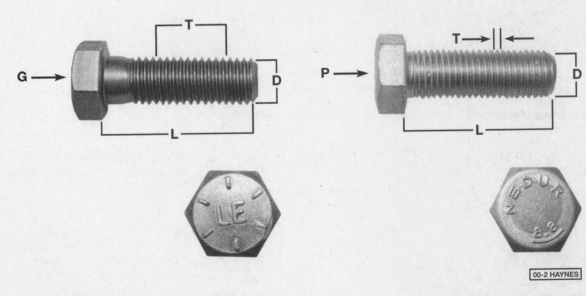

00-2 HAYNES

Standard (SAE and USS) bolt dimensions/grade marks

G   Grade marks (bolt strength)
L   Length (in inches)
T   Thread pitch (number of threads per inch)
D   Nominal diameter (in inches)

Metric bolt dimensions/grade marks

P   Property class (bolt strength)
L   Length (in millimeters)
T   Thread pitch (distance between threads in millimeters)
D   Diameter

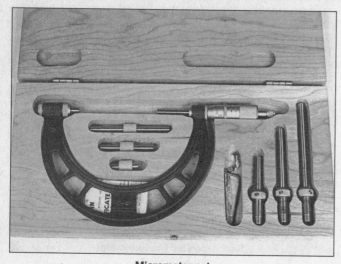

Micrometer set

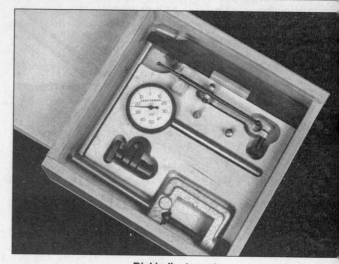

Dial indicator set

Initially, the bolts or nuts should be assembled finger-tight only. Next, they should be tightened one full turn each, in a criss-cross or diagonal pattern. After each one has been tightened one full turn, return to the first one and tighten them all one-half turn, following the same pattern. Finally, tighten each of them one-quarter turn at a time until each fastener has been tightened to the proper torque. To loosen and remove the fasteners, the procedure would be reversed.

Component disassembly

Component disassembly should be done with care and purpose to help ensure that the parts go back together properly. Always keep track of the sequence in which parts are removed. Make note of special characteristics or marks on parts that can be installed more than one way, such as a grooved thrust washer on a shaft. It is a good idea to lay the disassembled parts out on a clean surface in the order that they were removed. It may also be helpful to make sketches or take instant photos of components before removal.

When removing fasteners from a component, keep track of their locations. Sometimes threading a bolt back in a part, or putting the washers and nut back on a stud, can prevent mix-ups later. If nuts and bolts cannot be returned to their original locations, they should be kept in a compartmented box or a series of small boxes. A cupcake or muffin tin is ideal for this purpose, since each cavity can hold the bolts and nuts from a particular area (i.e. oil pan bolts, valve cover bolts, engine mount bolts, etc.). A pan of this type is especially helpful when working on assemblies with very small parts, such as the carburetor, alternator, valve train or interior dash and trim pieces. The cavities can be marked with paint or tape to identify the contents.

Whenever wiring looms, harnesses or connectors are separated, it is a good idea to identify the two halves with numbered pieces of masking tape so they can be easily reconnected.

Gasket sealing surfaces

Throughout any vehicle, gaskets are used to seal the mating surfaces between two parts and keep lubricants, fluids, vacuum or pressure contained in an assembly.

Many times these gaskets are coated with a liquid or paste-type gasket sealing compound before assembly. Age, heat and pressure can sometimes cause the two parts to stick together so tightly that they are very difficult to separate. Often, the assembly can be loosened by striking it with a soft-face hammer near the mating surfaces. A regular hammer can be used if a block of wood is placed between the hammer and the part. Do not hammer on cast parts or parts that could be easily damaged. With any particularly stubborn part, always recheck to make sure that every fastener has been removed.

Avoid using a screwdriver or bar to pry apart an assembly, as they can easily mar the gasket sealing surfaces of the parts, which must remain smooth. If prying is absolutely necessary, use an old broom handle, but keep in mind that extra clean up will be necessary if the wood splinters.

After the parts are separated, the old gasket must be carefully scraped off and the gasket surfaces cleaned. Stubborn gasket material can be soaked with rust penetrant or treated with a special chemical to soften it so it can be easily scraped off. A scraper can be fashioned from a piece of copper tubing by flattening and sharpening one end. Copper is recommended because it is usually softer than the surfaces to be scraped, which reduces the chance of gouging the part. Some gaskets can be removed with a wire brush, but regardless of the method used, the mating surfaces must be left clean and smooth. If for some reason the gasket surface is gouged, then a gasket sealer thick enough to fill scratches will have to be used during reassembly of the components. For most applications, a non-drying (or semi-drying) gasket sealer should be used.

Hose removal tips

Warning: *If the vehicle is equipped with air conditioning, do not disconnect any of the A/C hoses without first having the system depressurized by a dealer service department or a service station.*

Hose removal precautions closely parallel gasket removal precautions. Avoid scratching or gouging the surface that the hose mates against or the connection may leak. This is especially true for radiator hoses. Because of various chemical reactions, the rubber in hoses can bond itself to the metal spigot that the hose fits over. To remove a hose, first loosen the hose clamps that secure it to the spigot. Then, with slip-joint pliers, grab the hose at the clamp and rotate it around the spigot. Work it back and forth until it is completely free, then pull it off. Silicone or other lubricants will ease removal if they can be applied between the hose and the outside of the spigot. Apply the same lubricant to the inside of the hose and the outside of the spigot to simplify installation.

As a last resort (and if the hose is to be replaced with a new one anyway), the rubber can be slit with a knife and the hose peeled from the spigot. If this must be done, be careful that the metal connection is not damaged.

If a hose clamp is broken or damaged, do not reuse it. Wire-type clamps usually weaken with age, so it is a good idea to replace them with screw-type clamps whenever a hose is removed.

Tools

A selection of good tools is a basic requirement for anyone who plans to maintain and repair his or her own vehicle. For the owner who has few tools, the initial investment might seem high, but when compared to the spiraling costs of professional auto maintenance and repair, it is a wise one.

To help the owner decide which tools are needed to perform the tasks detailed in this manual, the following tool lists are offered: *Maintenance and minor repair,*

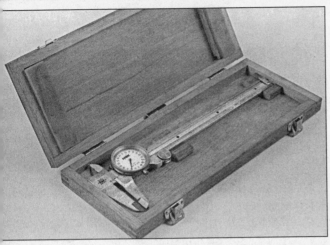

Dial caliper

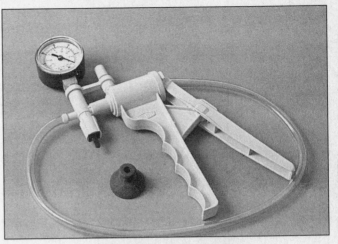

Hand-operated vacuum pump

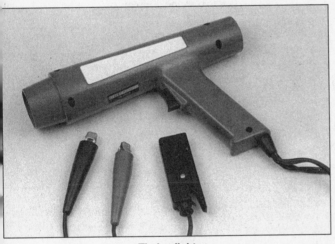

Timing light

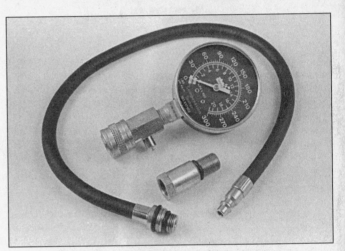

Compression gauge with spark plug hole adapter

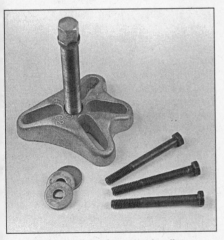

Damper/steering wheel puller

General purpose puller

Hydraulic lifter removal tool

Repair/overhaul and *Special.*

The newcomer to practical mechanics should start off with the *maintenance and minor repair* tool kit, which is adequate for the simpler jobs performed on a vehicle. Then, as confidence and experience grow, the owner can tackle more difficult tasks, buying additional tools as they are needed.

Eventually the basic kit will be expanded into the *repair and overhaul* tool set. Over a period of time, the experienced do-it-yourselfer will assemble a tool set complete enough for most repair and overhaul procedures and will add tools from the special category when it is felt that the expense is justified by the frequency of use.

Maintenance and minor repair tool kit

The tools in this list should be considered the minimum required for performance of routine maintenance, servicing and minor repair work. We recommend the purchase of combination wrenches (box-end and open-

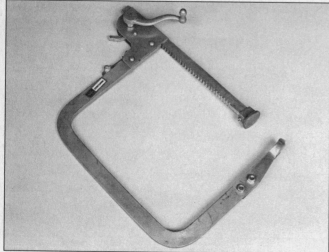

Valve spring compressor

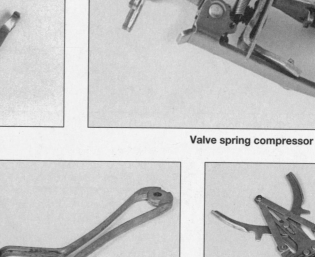

Valve spring compressor

Ridge reamer

Piston ring groove cleaning tool

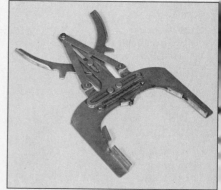

Ring removal/installation tool

end combined in one wrench). While more expensive than open end wrenches, they offer the advantages of both types of wrench.

> *Combination wrench set (1/4-inch to 1 inch or 6 mm to 19 mm)*
> *Adjustable wrench, 8 inch*
> *Spark plug wrench with rubber insert*
> *Spark plug gap adjusting tool*
> *Feeler gauge set*
> *Brake bleeder wrench*
> *Standard screwdriver (5/16-inch x 6 inch)*
> *Phillips screwdriver (No. 2 x 6 inch)*
> *Combination pliers - 6 inch*
> *Hacksaw and assortment of blades*
> *Tire pressure gauge*
> *Grease gun*
> *Oil can*
> *Fine emery cloth*
> *Wire brush*
> *Battery post and cable cleaning tool*
> *Oil filter wrench*
> *Funnel (medium size)*
> *Safety goggles*
> *Jackstands (2)*
> *Drain pan*

Note: *If basic tune-ups are going to be part of routine maintenance, it will be necessary to purchase a good quality stroboscopic timing*

light and combination tachometer/dwell meter. Although they are included in the list of special tools, it is mentioned here because they are absolutely necessary for tuning most vehicles properly.

Repair and overhaul tool set

These tools are essential for anyone who plans to perform major repairs and are in addition to those in the maintenance and minor repair tool kit. Included is a comprehensive set of sockets which, though expensive, are invaluable because of their versatility, especially when various extensions and drives are available. We recommend the 1/2-inch drive over the 3/8-inch drive. Although the larger drive is bulky and more expensive, it has the capacity of accepting a very wide range of large sockets. Ideally, however, the mechanic should have a 3/8-inch drive set and a 1/2-inch drive set.

> *Socket set(s)*
> *Reversible ratchet*
> *Extension - 10 inch*
> *Universal joint*
> *Torque wrench (same size drive as sockets)*
> *Ball peen hammer - 8 ounce*
> *Soft-face hammer (plastic/rubber)*

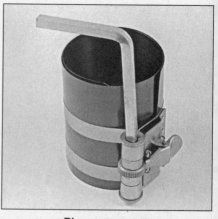

Ring compressor

> *Standard screwdriver (1/4-inch x 6 inch)*
> *Standard screwdriver (stubby - 5/16-inch)*
> *Phillips screwdriver (No. 3 x 8 inch)*
> *Phillips screwdriver (stubby - No. 2)*
> *Pliers - vise grip*
> *Pliers - lineman's*
> *Pliers - needle nose*
> *Pliers - snap-ring (internal and external)*
> *Cold chisel - 1/2-inch*

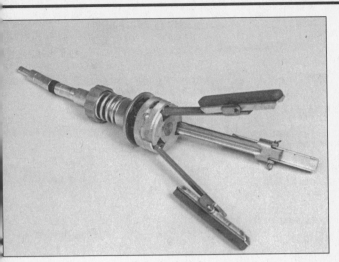

Cylinder hone

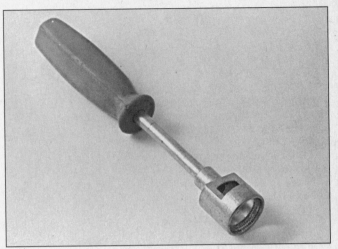

Brake hold-down spring tool

Scribe
Scraper (made from flattened copper
    tubing)
Centerpunch
Pin punches (1/16, 1/8, 3/16-inch)
Steel rule/straightedge - 12 inch
Allen wrench set (1/8 to 3/8-inch or
    4 mm to 10 mm)
A selection of files

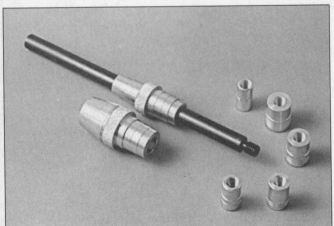

Brake cylinder hone

Wire brush (large)
Jackstands (second set)
Jack (scissor or hydraulic type)

Note: *Another tool which is often useful is an electric drill with a chuck capacity of 3/8-inch and a set of good quality drill bits.*

Special tools

The tools in this list include those which are not used regularly, are expensive to buy, or which need to be used in accordance with their manufacturer's instructions. Unless these tools will be used frequently, it is not very economical to purchase many of them. A consideration would be to split the cost and use between yourself and a friend or friends. In addition, most of these tools can be obtained from a tool rental shop on a temporary basis.

This list primarily contains only those tools and instruments widely available to the public, and not those special tools produced by the vehicle manufacturer for distribution to dealer service departments. Occasionally, references to the manufacturer's special tools are included in the text of this manual. Generally, an alternative method of doing the job without the special tool is offered. How-

ever, sometimes there is no alternative to their use. Where this is the case, and the tool cannot be purchased or borrowed, the work should be turned over to the dealer service department or an automotive repair shop.

Valve spring compressor
Piston ring groove cleaning tool
Piston ring compressor
Piston ring installation tool
Cylinder compression gauge
Cylinder ridge reamer
Cylinder surfacing hone
Cylinder bore gauge
Micrometers and/or dial calipers
Hydraulic lifter removal tool
Balljoint separator
Universal-type puller
Impact screwdriver
Dial indicator set
Stroboscopic timing light (inductive
    pick-up)
Hand operated vacuum/pressure pump
Tachometer/dwell meter
Universal electrical multimeter
Cable hoist
Brake spring removal and installation
    tools
Floor jack

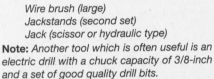

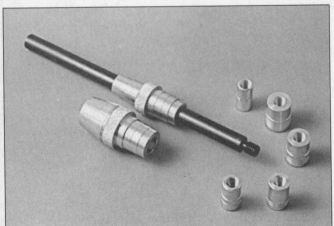

Clutch plate alignment tool

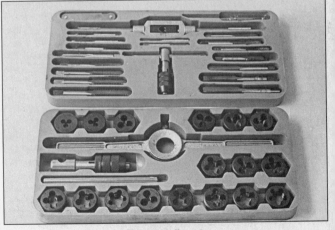

Tap and die set

Buying tools

For the do-it-yourselfer who is just starting to get involved in vehicle maintenance and repair, there are a number of options available when purchasing tools. If maintenance and minor repair is the extent of the work to be done, the purchase of individual tools is satisfactory. If, on the other hand, extensive work is planned, it would be a good idea to purchase a modest tool set from one of the large retail chain stores. A set can usually be bought at a substantial savings over the individual tool prices, and they often come with a tool box. As additional tools are needed, add-on sets, individual tools and a larger tool box can be purchased to expand the tool selection. Building a tool set gradually allows the cost of the tools to be spread over a longer period of time and gives the mechanic the freedom to choose only those tools that will actually be used.

Tool stores will often be the only source of some of the special tools that are needed, but regardless of where tools are bought, try to avoid cheap ones, especially when buying screwdrivers and sockets, because they won't last very long. The expense involved in replacing cheap tools will eventually be greater than the initial cost of quality tools.

Care and maintenance of tools

Good tools are expensive, so it makes sense to treat them with respect. Keep them clean and in usable condition and store them properly when not in use. Always wipe off any dirt, grease or metal chips before putting them away. Never leave tools lying around in the work area. Upon completion of a job, always check closely under the hood for tools that may have been left there so they won't get lost during a test drive.

Some tools, such as screwdrivers, pliers, wrenches and sockets, can be hung on a panel mounted on the garage or workshop wall, while others should be kept in a tool box or tray. Measuring instruments, gauges, meters, etc. must be carefully stored where they cannot be damaged by weather or impact from other tools.

When tools are used with care and stored properly, they will last a very long time. Even with the best of care, though, tools will wear out if used frequently. When a tool is damaged or worn out, replace it. Subsequent jobs will be safer and more enjoyable if you do.

How to repair damaged threads

Sometimes, the internal threads of a nut or bolt hole can become stripped, usually from overtightening. Stripping threads is an all-too-common occurrence, especially when working with aluminum parts, because aluminum is so soft that it easily strips out.

Usually, external or internal threads are only partially stripped. After they've been cleaned up with a tap or die, they'll still work. Sometimes, however, threads are badly damaged. When this happens, you've got three choices:

1) *Drill and tap the hole to the next suitable oversize and install a larger diameter bolt, screw or stud.*

2) *Drill and tap the hole to accept a threaded plug, then drill and tap the plug to the original screw size. You can also buy a plug already threaded to the original size. Then you simply drill a hole to the specified size, then run the threaded plug into the hole with a bolt and jam nut. Once the plug is fully seated, remove the jam nut and bolt.*

3) *The third method uses a patented thread repair kit like Heli-Coil or Slimsert. These easy-to-use kits are designed to repair damaged threads in straight-through holes and blind holes. Both are available as kits which can handle a variety of sizes and thread patterns. Drill the hole, then tap it with the special included tap. Install the Heli-Coil and the hole is back to its original diameter and thread pitch.*

Regardless of which method you use, be sure to proceed calmly and carefully. A little impatience or carelessness during one of these relatively simple procedures can ruin your whole day's work and cost you a bundle if you wreck an expensive part.

Working facilities

Not to be overlooked when discussing tools is the workshop. If anything more than routine maintenance is to be carried out, some sort of suitable work area is essential.

It is understood, and appreciated, that many home mechanics do not have a good workshop or garage available, and end up removing an engine or doing major repairs outside. It is recommended, however, that the overhaul or repair be completed under the cover of a roof.

A clean, flat workbench or table of comfortable working height is an absolute necessity. The workbench should be equipped with a vise that has a jaw opening of at least four inches.

As mentioned previously, some clean, dry storage space is also required for tools, as well as the lubricants, fluids, cleaning solvents, etc. which soon become necessary.

Sometimes waste oil and fluids, drained from the engine or cooling system during normal maintenance or repairs, present a disposal problem. To avoid pouring them on the ground or into a sewage system, pour the used fluids into large containers, seal them with caps and take them to an authorized disposal site or recycling center. Plastic jugs, such as old antifreeze containers, are ideal for this purpose.

Always keep a supply of old newspapers and clean rags available. Old towels are excellent for mopping up spills. Many mechanics use rolls of paper towels for most work because they are readily available and disposable. To help keep the area under the vehicle clean, a large cardboard box can be cut open and flattened to protect the garage or shop floor.

Whenever working over a painted surface, such as when leaning over a fender to service something under the hood, always cover it with an old blanket or bedspread to protect the finish. Vinyl covered pads, made especially for this purpose, are available at auto parts stores.

Booster battery (jump) starting

Observe the following precautions when using a booster battery to start a vehicle:

a) *Before connecting the booster battery, make sure the ignition switch is in the Off position.*

b) *Turn off the lights, heater and other electrical loads.*

c) *Your eyes should be shielded. Safety goggles are a good idea.*

d) *Make sure the booster battery is the same voltage as the dead one in the vehicle.*

e) *The two vehicles MUST NOT TOUCH each other.*

f) *Make sure the transmission is in Neutral (manual transaxle) or Park (automatic transaxle).*

g) *If the booster battery is not a maintenance-free type, remove the vent caps and lay a cloth over the vent holes.*

Connect the red jumper cable to the positive (+) terminals of each battery.

Connect one end of the black cable to the negative (-) terminal of the booster battery. The other end of this cable should be connected to a good ground on the engine block **(see illustration)**. Make sure the cable will not come into contact with the fan, drivebelts or other moving parts of the engine.

Start the engine using the booster battery, then, with the engine running at idle speed, disconnect the jumper cables in the reverse order of connection.

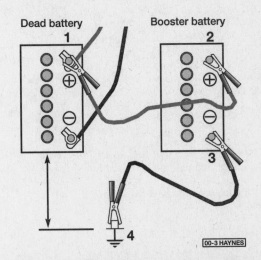

Make the booster battery cable connections in the numerical order shown (note that the negative cable of the booster battery is NOT attached to the negative terminal of the dead battery)

Jacking and towing

Jacking

Warning: *The jack supplied with the vehicle should only be used for changing a tire or placing jackstands under the frame. Never work under the vehicle or start the engine while this jack is being used as the only means of support.*

The vehicle should be on level ground. Place the shift lever in Park, if you have an automatic, or Reverse if you have a manual transaxle. Block the wheel diagonally opposite the wheel being changed. Set the parking brake.

Remove the spare tire and jack from stowage. Remove the wheel cover and trim ring (if so equipped) with the tapered end of the lug nut wrench by inserting and twisting the handle and then prying against the back of the wheel cover. On aluminum wheels, tap the back side of the wheel hub cover after removing the wheel (do not attempt to pull off the wheel hub cover by hand). Loosen, but do not remove, the lug nuts (one-half turn is sufficient).

Place the scissors-type jack under the side of the vehicle and adjust the jack height until it fits between the notches in the vertical rocker panel flange nearest the wheel to be changed. There is a front and rear jacking point on each side of the vehicle **(see illustration)**.

Turn the jack handle clockwise until the tire clears the ground. Remove the lug nuts and pull the wheel off. Replace it with the spare.

Install the lug nuts with the beveled edges facing in. Tighten them snugly. Don't attempt to tighten them completely until the vehicle is lowered or it could slip off the jack. Turn the jack handle counterclockwise to lower the vehicle. Remove the jack and tighten the lug nuts in a diagonal pattern.

Install the cover (and trim ring, if used) and be sure it's snapped into place all the way around.

Stow the tire, jack and wrench. Unblock the wheels.

Towing

As a general rule, the vehicle should be towed with the front (drive) wheels off the ground. If they can't be raised, place them on a dolly. The ignition key must be in the ACC position, since the steering lock mechanism isn't strong enough to hold the front wheels straight while towing.

On 1990 and earlier models, vehicles equipped with an automatic transaxle can be towed from the front only with all four wheels on the ground, provided that speeds don't exceed 30 mph and the distance is not over 50 miles. Before towing, check the transmission fluid level (see Chapter 1). If the level is below the HOT line on the dipstick, add fluid or use a towing dolly.

When towing a vehicle equipped with a manual transaxle with all four wheels on the ground, be sure to place the shift lever in neutral and release the parking brake.

Equipment specifically designed for towing should be used. It should be attached to the main structural members of the vehicle, not the bumpers or brackets.

Safety is a major consideration when towing and all applicable state and local laws must be obeyed. A safety chain system must be used at all times.

On 1991 and later models, the manufacturer does not recommend towing except with a towing dolly under the front wheels. In an emergency the vehicle can be towed a short distance with a cable or chain attached to one of the towing eyelets located under the front or rear bumpers following the precautions above. The driver must remain in the vehicle to operate the steering and brakes (remember that power steering and power brakes will not work with the engine off).

Jacking points

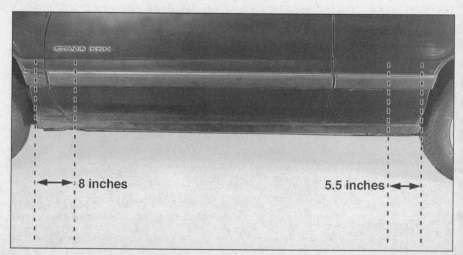

8 inches   5.5 inches

Automotive chemicals and lubricants

A number of automotive chemicals and lubricants are available for use during vehicle maintenance and repair. They include a wide variety of products ranging from cleaning solvents and degreasers to lubricants and protective sprays for rubber, plastic and vinyl.

Cleaners

Carburetor cleaner and choke cleaner is a strong solvent for gum, varnish and carbon. Most carburetor cleaners leave a dry-type lubricant film which will not harden or gum up. Because of this film it is not recommended for use on electrical components.

Brake system cleaner is used to remove grease and brake fluid from the brake system, where clean surfaces are absolutely necessary. It leaves no residue and often eliminates brake squeal caused by contaminants.

Electrical cleaner removes oxidation, corrosion and carbon deposits from electrical contacts, restoring full current flow. It can also be used to clean spark plugs, carburetor jets, voltage regulators and other parts where an oil-free surface is desired.

Demoisturants remove water and moisture from electrical components such as alternators, voltage regulators, electrical connectors and fuse blocks. They are non-conductive, non-corrosive and non-flammable.

Degreasers are heavy-duty solvents used to remove grease from the outside of the engine and from chassis components. They can be sprayed or brushed on and, depending on the type, are rinsed off either with water or solvent.

Lubricants

Motor oil is the lubricant formulated for use in engines. It normally contains a wide variety of additives to prevent corrosion and reduce foaming and wear. Motor oil comes in various weights (viscosity ratings) from 0 to 50. The recommended weight of the oil depends on the season, temperature and the demands on the engine. Light oil is used in cold climates and under light load conditions. Heavy oil is used in hot climates and where high loads are encountered. Multi-viscosity oils are designed to have characteristics of both light and heavy oils and are available in a number of weights from 5W-20 to 20W-50.

Gear oil is designed to be used in differentials, manual transmissions and other areas where high-temperature lubrication is required.

Chassis and wheel bearing grease is a heavy grease used where increased loads and friction are encountered, such as for wheel bearings, balljoints, tie-rod ends and universal joints.

High-temperature wheel bearing grease is designed to withstand the extreme temperatures encountered by wheel bearings in disc brake equipped vehicles. It usually contains molybdenum disulfide (moly), which is a dry-type lubricant.

White grease is a heavy grease for metal-to-metal applications where water is a problem. White grease stays soft under both low and high temperatures (usually from -100 to +190-degrees F), and will not wash off or dilute in the presence of water.

Assembly lube is a special extreme pressure lubricant, usually containing moly, used to lubricate high-load parts (such as main and rod bearings and cam lobes) for initial start-up of a new engine. The assembly lube lubricates the parts without being squeezed out or washed away until the engine oiling system begins to function.

Silicone lubricants are used to protect rubber, plastic, vinyl and nylon parts.

Graphite lubricants are used where oils cannot be used due to contamination problems, such as in locks. The dry graphite will lubricate metal parts while remaining uncontaminated by dirt, water, oil or acids. It is electrically conductive and will not foul electrical contacts in locks such as the ignition switch.

Moly penetrants loosen and lubricate frozen, rusted and corroded fasteners and prevent future rusting or freezing.

Heat-sink grease is a special electrically non-conductive grease that is used for mounting electronic ignition modules where it is essential that heat is transferred away from the module.

Sealants

RTV sealant is one of the most widely used gasket compounds. Made from silicone, RTV is air curing, it seals, bonds, waterproofs, fills surface irregularities, remains flexible, doesn't shrink, is relatively easy to remove, and is used as a supplementary sealer with almost all low and medium temperature gaskets.

Anaerobic sealant is much like RTV in that it can be used either to seal gaskets or to form gaskets by itself. It remains flexible, is solvent resistant and fills surface imperfections. The difference between an anaerobic sealant and an RTV-type sealant is in the curing. RTV cures when exposed to air, while an anaerobic sealant cures only in the absence of air. This means that an anaerobic sealant cures only after the assembly of parts, sealing them together.

Thread and pipe sealant is used for sealing hydraulic and pneumatic fittings and vacuum lines. It is usually made from a Teflon compound, and comes in a spray, a paint-on liquid and as a wrap-around tape.

Chemicals

Anti-seize compound prevents seizing, galling, cold welding, rust and corrosion in fasteners. High-temperature anti-seize, usually made with copper and graphite lubricants, is used for exhaust system and exhaust manifold bolts.

Anaerobic locking compounds are used to keep fasteners from vibrating or working loose and cure only after installation, in the absence of air. Medium strength locking compound is used for small nuts, bolts and screws that may be removed later. High-strength locking compound is for large nuts, bolts and studs which aren't removed on a regular basis.

Oil additives range from viscosity index improvers to chemical treatments that claim to reduce internal engine friction. It should be noted that most oil manufacturers caution against using additives with their oils.

Gas additives perform several functions, depending on their chemical makeup. They usually contain solvents that help dissolve gum and varnish that build up on carburetor, fuel injection and intake parts. They also serve to break down carbon deposits that form on the inside surfaces of the combustion chambers. Some additives contain upper cylinder lubricants for valves and piston rings, and others contain chemicals to remove condensation from the gas tank.

Miscellaneous

Brake fluid is specially formulated hydraulic fluid that can withstand the heat and pressure encountered in brake systems. Care must be taken so this fluid does not come in contact with painted surfaces or plastics. An opened container should always be resealed to prevent contamination by water or dirt.

Weatherstrip adhesive is used to bond weatherstripping around doors, windows and trunk lids. It is sometimes used to attach trim pieces.

Undercoating is a petroleum-based, tar-like substance that is designed to protect metal surfaces on the underside of the vehicle from corrosion. It also acts as a sound-deadening agent by insulating the bottom of the vehicle.

Waxes and polishes are used to help protect painted and plated surfaces from the weather. Different types of paint may require the use of different types of wax and polish. Some polishes utilize a chemical or abrasive cleaner to help remove the top layer of oxidized (dull) paint on older vehicles. In recent years many non-wax polishes that contain a wide variety of chemicals such as polymers and silicones have been introduced. These non-wax polishes are usually easier to apply and last longer than conventional waxes and polishes.

Conversion factors

Length (distance)

Inches (in)	X	25.4	= Millimetres (mm)	X 0.0394	= Inches (in)
Feet (ft)	X	0.305	= Metres (m)	X 3.281	= Feet (ft)
Miles	X	1.609	= Kilometres (km)	X 0.621	= Miles

Volume (capacity)

Cubic inches (cu in; in³)	X	16.387	= Cubic centimetres (cc; cm³)	X 0.061	= Cubic inches (cu in; in³)
Imperial pints (Imp pt)	X	0.568	= Litres (l)	X 1.76	= Imperial pints (Imp pt)
Imperial quarts (Imp qt)	X	1.137	= Litres (l)	X 0.88	= Imperial quarts (Imp qt)
Imperial quarts (Imp qt)	X	1.201	= US quarts (US qt)	X 0.833	= Imperial quarts (Imp qt)
US quarts (US qt)	X	0.946	= Litres (l)	X 1.057	= US quarts (US qt)
Imperial gallons (Imp gal)	X	4.546	= Litres (l)	X 0.22	= Imperial gallons (Imp gal)
Imperial gallons (Imp gal)	X	1.201	= US gallons (US gal)	X 0.833	= Imperial gallons (Imp gal)
US gallons (US gal)	X	3.785	= Litres (l)	X 0.264	= US gallons (US gal)

Mass (weight)

Ounces (oz)	X	28.35	= Grams (g)	X 0.035	= Ounces (oz)
Pounds (lb)	X	0.454	= Kilograms (kg)	X 2.205	= Pounds (lb)

Force

Ounces-force (ozf; oz)	X	0.278	= Newtons (N)	X 3.6	= Ounces-force (ozf; oz)
Pounds-force (lbf; lb)	X	4.448	= Newtons (N)	X 0.225	= Pounds-force (lbf; lb)
Newtons (N)	X	0.1	= Kilograms-force (kgf; kg)	X 9.81	= Newtons (N)

Pressure

Pounds-force per square inch (psi; lbf/in²; lb/in²)	X	0.070	= Kilograms-force per square centimetre (kgf/cm²; kg/cm²)	X 14.223	= Pounds-force per square inch (psi; lbf/in²; lb/in²)
Pounds-force per square inch (psi; lbf/in²; lb/in²)	X	0.068	= Atmospheres (atm)	X 14.696	= Pounds-force per square inch (psi; lbf/in²; lb/in²)
Pounds-force per square inch (psi; lbf/in²; lb/in²)	X	0.069	= Bars	X 14.5	= Pounds-force per square inch (psi; lbf/in²; lb/in²)
Pounds-force per square inch (psi; lbf/in²; lb/in²)	X	6.895	= Kilopascals (kPa)	X 0.145	= Pounds-force per square inch (psi; lbf/in²; lb/in²)
Kilopascals (kPa)	X	0.01	= Kilograms-force per square centimetre (kgf/cm²; kg/cm²)	X 98.1	= Kilopascals (kPa)

Torque (moment of force)

Pounds-force inches (lbf in; lb in)	X	1.152	= Kilograms-force centimetre (kgf cm; kg cm)	X 0.868	= Pounds-force inches (lbf in; lb in)
Pounds-force inches (lbf in; lb in)	X	0.113	= Newton metres (Nm)	X 8.85	= Pounds-force inches (lbf in; lb in)
Pounds-force inches (lbf in; lb in)	X	0.083	= Pounds-force feet (lbf ft; lb ft)	X 12	= Pounds-force inches (lbf in; lb in)
Pounds-force feet (lbf ft; lb ft)	X	0.138	= Kilograms-force metres (kgf m; kg m)	X 7.233	= Pounds-force feet (lbf ft; lb ft)
Pounds-force feet (lbf ft; lb ft)	X	1.356	= Newton metres (Nm)	X 0.738	= Pounds-force feet (lbf ft; lb ft)
Newton metres (Nm)	X	0.102	= Kilograms-force metres (kgf m; kg m)	X 9.804	= Newton metres (Nm)

Vacuum

Inches mercury (in. Hg)	X	3.377	= Kilopascals (kPa)	X 0.2961	= Inches mercury
Inches mercury (in. Hg)	X	25.4	= Millimeters mercury (mm Hg)	X 0.0394	= Inches mercury

Power

Horsepower (hp)	X	745.7	= Watts (W)	X 0.0013	= Horsepower (hp)

Velocity (speed)

Miles per hour (miles/hr; mph)	X	1.609	= Kilometres per hour (km/hr; kph)	X 0.621	= Miles per hour (miles/hr; mph)

Fuel consumption*

Miles per gallon, Imperial (mpg)	X	0.354	= Kilometres per litre (km/l)	X 2.825	= Miles per gallon, Imperial (mpg)
Miles per gallon, US (mpg)	X	0.425	= Kilometres per litre (km/l)	X 2.352	= Miles per gallon, US (mpg)

Temperature

Degrees Fahrenheit = (°C x 1.8) + 32

Degrees Celsius (Degrees Centigrade; °C) = (°F - 32) x 0.56

*It is common practice to convert from miles per gallon (mpg) to litres/100 kilometres (l/100km), where mpg (Imperial) x l/100 km = 282 and mpg (US) x l/100 km = 235

Safety first!

Regardless of how enthusiastic you may be about getting on with the job at hand, take the time to ensure that your safety is not jeopardized. A moment's lack of attention can result in an accident, as can failure to observe certain simple safety precautions. The possibility of an accident will always exist, and the following points should not be considered a comprehensive list of all dangers. Rather, they are intended to make you aware of the risks and to encourage a safety conscious approach to all work you carry out on your vehicle.

Essential DOs and DON'Ts

DON'T rely on a jack when working under the vehicle. Always use approved jackstands to support the weight of the vehicle and place them under the recommended lift or support points.

DON'T attempt to loosen extremely tight fasteners (i.e. wheel lug nuts) while the vehicle is on a jack - it may fall.

DON'T start the engine without first making sure that the transmission is in Neutral (or Park where applicable) and the parking brake is set.

DON'T remove the radiator cap from a hot cooling system - let it cool or cover it with a cloth and release the pressure gradually.

DON'T attempt to drain the engine oil until you are sure it has cooled to the point that it will not burn you.

DON'T touch any part of the engine or exhaust system until it has cooled sufficiently to avoid burns.

DON'T siphon toxic liquids such as gasoline, antifreeze and brake fluid by mouth, or allow them to remain on your skin.

DON'T inhale brake lining dust - it is potentially hazardous (see *Asbestos* below).

DON'T allow spilled oil or grease to remain on the floor - wipe it up before someone slips on it.

DON'T use loose fitting wrenches or other tools which may slip and cause injury.

DON'T push on wrenches when loosening or tightening nuts or bolts. Always try to pull the wrench toward you. If the situation calls for pushing the wrench away, push with an open hand to avoid scraped knuckles if the wrench should slip.

DON'T attempt to lift a heavy component alone - get someone to help you.

DON'T rush or take unsafe shortcuts to finish a job.

DON'T allow children or animals in or around the vehicle while you are working on it.

DO wear eye protection when using power tools such as a drill, sander, bench grinder, etc. and when working under a vehicle.

DO keep loose clothing and long hair well out of the way of moving parts.

DO make sure that any hoist used has a safe working load rating adequate for the job.

DO get someone to check on you periodically when working alone on a vehicle.

DO carry out work in a logical sequence and make sure that everything is correctly assembled and tightened.

DO keep chemicals and fluids tightly capped and out of the reach of children and pets.

DO remember that your vehicle's safety affects that of yourself and others. If in doubt on any point, get professional advice.

Asbestos

Certain friction, insulating, sealing, and other products - such as brake linings, brake bands, clutch linings, torque converters, gaskets, etc. - may contain asbestos. Extreme care must be taken to avoid inhalation of dust from such products, since it is hazardous to health. If in doubt, assume that they do contain asbestos.

Fire

Remember at all times that gasoline is highly flammable. Never smoke or have any kind of open flame around when working on a vehicle. But the risk does not end there. A spark caused by an electrical short circuit, by two metal surfaces contacting each other, or even by static electricity built up in your body under certain conditions, can ignite gasoline vapors, which in a confined space are highly explosive. Do not, under any circumstances, use gasoline for cleaning parts. Use an approved safety solvent.

Always disconnect the battery ground (-) cable at the battery before working on any part of the fuel system or electrical system. Never risk spilling fuel on a hot engine or exhaust component. It is strongly recommended that a fire extinguisher suitable for use on fuel and electrical fires be kept handy in the garage or workshop at all times. Never try to extinguish a fuel or electrical fire with water.

Fumes

Certain fumes are highly toxic and can quickly cause unconsciousness and even death if inhaled to any extent. Gasoline vapor falls into this category, as do the vapors from some cleaning solvents. Any draining or pouring of such volatile fluids should be done in a well ventilated area.

When using cleaning fluids and solvents, read the instructions on the container carefully. Never use materials from unmarked containers.

Never run the engine in an enclosed space, such as a garage. Exhaust fumes contain carbon monoxide, which is extremely poisonous. If you need to run the engine, always do so in the open air, or at least have the rear of the vehicle outside the work area.

If you are fortunate enough to have the use of an inspection pit, never drain or pour gasoline and never run the engine while the vehicle is over the pit. The fumes, being heavier than air, will concentrate in the pit with possibly lethal results.

The battery

Never create a spark or allow a bare light bulb near a battery. They normally give off a certain amount of hydrogen gas, which is highly explosive.

Always disconnect the battery ground (-) cable at the battery before working on the fuel or electrical systems.

If possible, loosen the filler caps or cover when charging the battery from an external source (this does not apply to sealed or maintenance-free batteries). Do not charge at an excessive rate or the battery may burst.

Take care when adding water to a non maintenance-free battery and when carrying a battery. The electrolyte, even when diluted, is very corrosive and should not be allowed to contact clothing or skin.

Always wear eye protection when cleaning the battery to prevent the caustic deposits from entering your eyes.

Household current

When using an electric power tool, inspection light, etc., which operates on household current, always make sure that the tool is correctly connected to its plug and that, where necessary, it is properly grounded. Do not use such items in damp conditions and, again, do not create a spark or apply excessive heat in the vicinity of fuel or fuel vapor.

Secondary ignition system voltage

A severe electric shock can result from touching certain parts of the ignition system (such as the spark plug wires) when the engine is running or being cranked, particularly if components are damp or the insulation is defective. In the case of an electronic ignition system, the secondary system voltage is much higher and could prove fatal.

Troubleshooting

Contents

This section provides an easy reference guide to the more common problems which may occur during the operation of your vehicle. Various symptoms and their possible causes are grouped under headings denoting components or systems, such as Engine, Cooling system, etc. They also refer to the Chapter and/or Section that deals with the problem.

Remember that successful troubleshooting isn't a mysterious "black art" practiced only by professional mechanics. It's simply the result of knowledge combined with an intelligent, systematic approach to a problem. Always use a process of elimination, starting with the simplest solution and working through to the most complex - and never overlook the obvious. Anyone can run the gas tank dry or leave the lights on overnight, so don't assume that you're exempt from such oversights.

Finally, always establish a clear idea why a problem has occurred and take steps to ensure that it doesn't happen again. If the electrical system fails because of a poor connection, check all other connections in the system to make sure they don't fail as well. If a particular fuse continues to blow, find out why - don't just go on replacing fuses. Remember, failure of a small component can often be indicative of potential failure or incorrect functioning of a more important component or system.

Engine and performance

1 Engine will not rotate when attempting to start

1 Battery terminal connections loose or corroded (Chapter 1).
2 Battery discharged or faulty (Chapter 1).
3 Automatic transaxle not completely engaged in Park (Chapter 7) or clutch not completely depressed (Chapter 8). Also, the neutral start switch (automatic transaxle) or starter/clutch interlock switch (manual transaxle) could be faulty (Chapter 7 or 8).
4 Broken, loose or disconnected wiring in the starting circuit (Chapters 5 and 12).
5 Starter motor pinion jammed in flywheel ring gear (Chapter 5).
6 Starter solenoid faulty (Chapter 5).
7 Starter motor faulty (Chapter 5).
8 Ignition switch faulty (Chapter 12).
9 Starter pinion or flywheel teeth worn or broken (Chapter 5).

2 Engine rotates but will not start

1 Fuel tank empty.
2 Battery discharged (engine rotates slowly) (Chapter 5).
3 Battery terminal connections loose or corroded (Chapter 1).
4 Leaking fuel injector(s), fuel pump, pressure regulator, etc. (Chapter 4).
5 Fuel not reaching fuel injection system (Chapter 4).
6 Ignition components damp or damaged (Chapter 5).
7 Worn, faulty or incorrectly gapped spark plugs (Chapter 1).
8 Broken, loose or disconnected wiring in the starting circuit (Chapter 5).
9 Broken, loose or disconnected wires at the ignition coil(s) or faulty coil(s) (Chapter 5).

3 Engine hard to start when cold

1 Battery discharged or low (Chapter 1).
2 Fuel system malfunctioning (Chapter 4).
3 Injector(s) leaking (Chapter 4).

4 Engine hard to start when hot

1 Air filter clogged (Chapter 1).
2 Fuel not reaching the fuel injection system (Chapter 4).
3 Corroded battery connections, especially ground (Chapter 1).

5 Starter motor noisy or excessively rough in engagement

1 Pinion or flywheel gear teeth worn or broken (Chapter 5).
2 Starter motor mounting bolts loose or missing (Chapter 5).

6 Engine starts but stops immediately

1 Loose or faulty electrical connections at coil pack or alternator (Chapter 5).
2 Insufficient fuel reaching the fuel injectors (Chapter 4).
3 Vacuum leak at the gasket between the intake manifold/plenum and throttle body (Chapters 1 and 4).

7 Oil puddle under engine

1 Oil pan gasket and/or oil pan drain bolt seal leaking (Chapters 1 and 2).
2 Oil pressure sending unit leaking (Chapter 2).
3 Rocker arm cover gaskets leaking (Chapter 2).
4 Engine oil seals leaking (Chapter 2).
5 Timing cover sealant or sealing flange leaking (Chapter 2).

8 Engine lopes while idling or idles erratically

1 Vacuum leaks (Chapter 4).
2 Leaking EGR valve or plugged PCV valve (Chapters 1 and 6).
3 Air filter clogged (Chapter 1).
4 Fuel pump not delivering sufficient fuel to the fuel injection system (Chapter 4).
5 Leaking head gasket (Chapter 2).
6 Timing chain and/or gears worn (Chapter 2).
7 Camshaft lobes worn (Chapter 2).

9 Engine misses at idle speed

1 Spark plugs worn or not gapped properly (Chapter 1).
2 Faulty spark plug wires (Chapter 1).
3 Vacuum leaks (Chapters 1 and 4).
4 Ignition system malfunctioning (Chapter 5).
5 Uneven or low compression (Chapter 2).

10 Engine misses throughout driving speed range

1 Fuel filter clogged and/or impurities in the fuel system (Chapters 1 and 4).
2 Low fuel output at the injector (Chapter 4).
3 Faulty or incorrectly gapped spark plugs (Chapter 1).
4 Leaking spark plug wires (Chapter 1).
5 Other fault in the ignition system (Chapter 5).
6 Faulty emission system components (Chapter 6).
7 Low or uneven cylinder compression pressures (Chapter 2).
8 Weak or faulty ignition system (Chapter 5).
9 Vacuum leak in fuel injection system, intake manifold or vacuum hoses (Chapter 4).

11 Engine stumbles on acceleration

1 Spark plugs fouled (Chapter 1).
2 Fuel injection system needs adjustment or repair (Chapter 4).
3 Fuel filter clogged (Chapter 1).
4 Faulty spark plug wires (Chapter 1).
5 Intake manifold air leak (Chapter 4).

12 Engine surges while holding accelerator steady

1 Intake air leak (Chapter 4).
2 Fuel pump faulty (Chapter 4).
3 Loose fuel injector harness connections (Chapter 4).
4 Defective ECM (Chapter 6).

13 Engine stalls

1 Fuel filter clogged and/or water and

impurities in the fuel system (Chapters 1 and 4).
2   Ignition components damp or damaged (Chapter 5).
3   Faulty emissions system components (Chapter 6).
4   Faulty or incorrectly gapped spark plugs (Chapter 1).
5   Faulty spark plug wires (Chapter 1).
6   Vacuum leak in the fuel injection system, intake manifold or vacuum hoses (Chapter 4).

14 Engine lacks power

1   Faulty or incorrectly gapped spark plugs (Chapter 1).
2   Fuel injection system out of adjustment or malfunctioning (Chapter 4).
3   Faulty coil(s) (Chapter 5).
4   Brakes binding (Chapter 1).
5   Automatic transaxle fluid level incorrect (Chapter 1).
6   Clutch slipping (Chapter 8).
7   Fuel filter clogged and/or impurities in the fuel system (Chapter 1).
8   Emission control system not functioning properly (Chapter 6).
9   Low or uneven cylinder compression pressures (Chapter 2).

15 Engine backfires

1   Emissions system not functioning properly (Chapter 6).
2   Spark plug wires not routed correctly (Chapter 1).
3   Faulty secondary ignition system (Chapter 5).
4   Fuel injection system in need of adjustment or worn excessively (Chapter 4).
5   Vacuum leak at fuel injectors, intake manifold or vacuum hoses (Chapter 4).
6   Valves sticking (Chapter 2).

16 Pinging or knocking engine sounds during acceleration or uphill

1   Incorrect grade of fuel.
2   Excessive carbon build-up in combustion chambers.
3   Fuel injection system in need of repair (Chapter 4).
4   Improper or damaged spark plugs or wires (Chapter 1).
5   Worn or damaged ignition components (Chapter 5).
6   Faulty emissions system (Chapter 6).
7   Vacuum leak (Chapter 4).

17 Engine runs with oil pressure light on

1   Low oil level (Chapter 1).

2   Short in wiring circuit (Chapter 12).
3   Faulty oil pressure sender (Chapter 2).
4   Worn engine bearings and/or oil pump (Chapter 2).

18 Engine diesels (continues to run) after switching off

1   Excessive engine operating temperature (Chapter 3).
2   Excessive carbon build-up in combustion chambers (Chapter 2).
3   Fuel injection system malfunctioning (Chapter 4).
4   Ignition system malfunctioning (Chapter 5).

Engine electrical system

19 Battery will not hold a charge

1   Alternator drivebelt defective or not adjusted properly (Chapter 1).
2   Battery terminals loose or corroded (Chapter 1).
3   Alternator not charging properly (Chapter 5).
4   Loose, broken or faulty wiring in the charging circuit (Chapter 5).
5   Short in vehicle wiring (Chapters 5 and 12).
6   Internally defective battery (Chapters 1 and 5).

20 Voltage warning light fails to go out

1   Faulty alternator or charging circuit (Chapter 5).
2   Alternator drivebelt defective or out of adjustment (Chapter 1).
3   Alternator voltage regulator inoperative (Chapter 5).

21 Voltage warning light fails to come on when key is turned on

1   Warning light bulb defective (Chapter 12).
2   Fault in the printed circuit, dash wiring or bulb holder (Chapter 12).

Fuel system

22 Excessive fuel consumption

1   Dirty or clogged air filter element (Chapter 1).
2   Fuel leak (usually accompanied by a fuel

odor) (see the next Section).
3   Emissions system not functioning properly (Chapter 6).
4   Fuel injection internal parts excessively worn or damaged (Chapter 4).
5   Low tire pressure or incorrect tire size (Chapter 1).
6   Clutch (Chapter 8) or automatic transaxle (Chapter 7) slipping (normally you will also hear excessive engine revving during acceleration).

23 Fuel leakage and/or fuel odor

1   Leak in a fuel feed or vent line (Chapter 4).
2   Tank overfilled.
3   Evaporative canister filter clogged (Chapters 1 and 6).
4   Fuel injector internal parts excessively worn (Chapter 4).

Cooling system

24 Overheating

1   Insufficient coolant in system (Chapter 1).
2   Water pump drivebelt defective or out of adjustment (Chapter 1).
3   Radiator core blocked or grille restricted (Chapter 3).
4   Thermostat faulty (Chapter 3).
5   Electric cooling fan blades broken or cracked (Chapter 3).
6   Radiator cap not maintaining proper pressure (Chapter 3).

25 Overcooling

Faulty thermostat (Chapter 3).

26 External coolant leakage

1   Deteriorated/damaged hoses or loose clamps (Chapters 1 and 3).
2   Water pump seal defective (Chapters 1 and 3).
3   Leakage from radiator core or header tank (Chapter 3).
4   Engine drain or water jacket core plugs leaking (Chapter 2).

27 Internal coolant leakage

1   Leaking cylinder head gasket (Chapter 2).
2   Cracked cylinder bore or cylinder head (Chapter 2).

28 Coolant loss

1 Too much coolant in system (Chapter 1).
2 Coolant boiling away because of over-heating (Chapter 3).
3 Internal or external leakage (Chapter 3).
4 Faulty radiator cap (Chapter 3).

29 Poor coolant circulation

1 Inoperative water pump (Chapter 3).
2 Restriction in cooling system (Chapters 1 and 3).
3 Water pump drivebelt defective or out of adjustment (Chapter 1).
4 Thermostat sticking (Chapter 3).

Clutch

30 Pedal travels to floor - no pressure or very little resistance

1 Master or release cylinder faulty (Chapter 8).
2 Hose/pipe burst or leaking (Chapter 8).
3 Connections leaking (Chapter 8).
4 No fluid in reservoir (Chapter 8).
5 If fluid is present in master cylinder dust cover, rear master cylinder seal has failed (Chapter 8).
6 If fluid level in reservoir rises as pedal is depressed, master cylinder center valve seal is faulty (Chapter 8).
7 Broken release bearing or fork (Chapter 8).

31 Fluid in area of master cylinder dust cover and on pedal

Rear seal failure in master cylinder (Chapter 8).

32 Fluid on release cylinder

Release cylinder seal faulty (Chapter 8).

33 Pedal feels spongy when depressed

Air in system (Chapter 8).

34 Unable to select gears

1 Faulty transaxle (Chapter 7).
2 Faulty clutch disc (Chapter 8).
3 Fork and bearing not assembled properly (Chapter 8).
4 Faulty pressure plate (Chapter 8).
5 Pressure plate-to-flywheel bolts loose (Chapter 8).

35 Clutch slips (engine speed increases with no increase in vehicle speed)

1 Clutch plate worn (Chapter 8).
2 Clutch plate is oil soaked by leaking rear main seal (Chapter 8).
3 Clutch plate not seated. It may take 30 or 40 normal starts for a new one to seat.
4 Warped pressure plate or flywheel (Chapter 8).
5 Weak diaphragm spring (Chapter 8).
6 Clutch plate overheated. Allow to cool.

36 Grabbing (chattering) as clutch is engaged

1 Oil soaked, burned or glazed linings (Chapter 8).
2 Worn or loose engine or transaxle mounts (Chapters 2 and 7).
3 Worn splines on clutch plate hub (Chapter 8).
4 Warped pressure plate or flywheel (Chapter 8).

37 Noise in clutch area

1 Fork shaft improperly installed (Chapter 8).
2 Faulty release bearing (Chapter 8).

38 Clutch pedal stays on floor

1 Fork shaft binding in housing (Chapter 8).
2 Broken release bearing or fork (Chapter 8).

39 High pedal effort

1 Fork shaft binding in housing (Chapter 8).
2 Pressure plate faulty (Chapter 8).

Manual transaxle

40 Vibration

1 Rough wheel bearing (Chapter 10).
2 Damaged driveaxle (Chapter 8).
3 Out-of-round tires (Chapter 1).
4 Tire out-of-balance (Chapter 10).
5 Worn or damaged CV joint (Chapter 8).

41 Noisy in Neutral with engine running

Damaged clutch release bearing (Chapter 8).

42 Noisy in one particular gear

1 Damaged or worn constant mesh gears (Chapter 7).
2 Damaged or worn synchronizers (Chapter 7).

43 Noisy in all gears

1 Insufficient lubricant (Chapter 1).
2 Damaged or worn bearings (Chapter 7).
3 Worn or damaged input gear shaft and/or output gear shaft (Chapter 7).

44 Slips out of gear

1 Worn or improperly adjusted linkage (Chapter 7).
2 Transaxle loose on engine (Chapter 7).
3 Shift linkage does not work freely, binds (Chapter 7).
4 Input shaft bearing retainer broken or loose (Chapter 7).
5 Dirt between clutch cover and engine housing (Chapter 7).
6 Worn shift fork (Chapter 7).

45 Leaks lubricant

1 Excessive amount of lubricant in transaxle (Chapter 1).
2 Loose or broken input shaft bearing retainer (Chapter 7).
3 Input shaft bearing retainer O-ring and/or lip seal damaged (Chapter 7).

Automatic transaxle

Note: *Due to the complexity of the automatic transaxle, it's difficult for the home mechanic to properly diagnose and service this component. For problems other than the following, the vehicle should be taken to a dealer service department or a transmission shop.*

46 Fluid leakage

1 Automatic transmission fluid is a deep red color. Fluid leaks should not be confused with engine oil, which can easily be blown by air flow to the transaxle.
2 To pinpoint a leak, first remove all built-up dirt and grime from the transaxle housing with degreasing agents and/or steam cleaning. Drive the vehicle at low speeds so air flow will not blow the leak far from its source. Raise the vehicle and determine where the leak is coming from. Common areas of leakage are:

a) *Pan (Chapters 1 and 7)*
b) *Filler pipe (Chapter 7)*
c) *Transaxle oil lines (Chapter 7)*
d) *Speedometer gear or sensor (Chapter 7)*
e) *Modulator*

47 Transaxle fluid brown or has a burned smell

Transaxle overheated. Change fluid (Chapter 1).

48 General shift mechanism problems

1   Chapter 7 Part B deals with checking and adjusting the shift linkage on automatic transaxles. Common problems which may be attributed to poorly adjusted linkage are:
a) *Engine starting in gears other than Park or Neutral.*
b) *Indicator on shifter pointing to a gear other than the one actually being used.*
c) *Vehicle moves when in Park.*
2   Refer to Chapter 7 Part B for the shift linkage adjustment procedure.

49 Transaxle will not downshift with accelerator pedal pressed to the floor

Throttle valve (TV) cable out of adjustment (Chapter 7).

50 Engine will start in gears other than Park or Neutral

Starter safety switch malfunctioning (Chapter 7).

51 Transaxle slips, shifts roughly, is noisy or has no drive in forward or reverse gears

There are many probable causes for the above problems, but the home mechanic should be concerned with only one possibility - fluid level. Before taking the vehicle to a repair shop, check the level and condition of the fluid as described in Chapter 1.

Correct the fluid level as necessary or change the fluid and filter if needed. If the problem persists, have a professional diagnose the probable cause.

Driveaxles

52 Clicking noise in turns

Worn or damaged outer CV joint. Check for cut or damaged boots (Chapter 1). Repair as necessary (Chapter 8).

53 Knock or clunk when accelerating after coasting

Worn or damaged outer CV joint. Check for cut or damaged boots (Chapter 1). Repair as necessary (Chapter 8).

54 Shudder or vibration during acceleration

1   Excessive inner CV joint angle. Check and correct as necessary (Chapter 8).
2   Worn or damaged CV joints. Repair or replace as necessary (Chapter 8).
3   Sticking inboard joint assembly. Correct or replace as necessary (Chapter 8).

Brakes

Note: *Before assuming that a brake problem exists, make sure . . .*
a) *The tires are in good condition and properly inflated (Chapter 1).*
b) *The front end alignment is correct (Chapter 10).*
c) *The vehicle isn't loaded with weight in an unequal manner.*

55 Vehicle pulls to one side during braking

1   Incorrect tire pressures (Chapter 1).
2   Front end out of line (have the front end aligned).
3   Unmatched tires on same axle.
4   Restricted brake lines or hoses (Chapter 9).
5   Malfunctioning brake assembly (Chapter 9).
6   Loose suspension parts (Chapter 10).
7   Loose brake calipers (Chapter 9).

56 Noise (high-pitched squeal when the brakes are applied)

Front disc brake pads worn out. The noise comes from the wear sensor rubbing against the disc. Replace pads with new ones immediately (Chapter 9).

57 Brake roughness or chatter (pedal pulsates)

1   Excessive front brake disc lateral runout (Chapter 9).
2   Parallelism not within specifications (Chapter 9).
3   Uneven pad wear caused by caliper not sliding due to improper clearance or dirt (Chapter 9).
4   Defective brake disc (Chapter 9).

58 Excessive pedal effort required to stop vehicle

1   Malfunctioning power brake booster (Chapter 9).
2   Partial system failure (Chapter 9).
3   Excessively worn pads (Chapter 9).
4   One or more caliper pistons or wheel cylinders seized or sticking (Chapter 9).
5   Brake pads contaminated with oil or grease (Chapter 9).
6   New pads installed and not yet seated. It will take a while for the new material to seat.

59 Excessive brake pedal travel

1   Partial brake system failure (Chapter 9).
2   Insufficient fluid in master cylinder (Chapters 1 and 9).
3   Air trapped in system (Chapters 1 and 9).

60 Dragging brakes

1   Master cylinder pistons not returning correctly (Chapter 9).
2   Restricted brakes lines or hoses (Chapters 1 and 9).
3   Incorrect parking brake adjustment (Chapter 9).

61 Grabbing or uneven braking action

1   Malfunction of proportioner valves (Chapter 9).
2   Malfunction of power brake booster unit (Chapter 9).
3   Binding brake pedal mechanism (Chapter 9).

62 Brake pedal feels spongy when depressed

1   Air in hydraulic lines (Chapter 9).
2   Master cylinder mounting bolts loose (Chapter 9).
3   Master cylinder defective (Chapter 9).

63 Brake pedal travels to the floor with little resistance

Little or no fluid in the master cylinder reservoir caused by leaking caliper or wheel cylinder pistons, loose, damaged or disconnected brake lines (Chapter 9).

64 Parking brake does not hold

Check the parking brake (Chapter 9).

Suspension and steering systems

Note: *Before attempting to diagnose the suspension and steering systems, perform the following preliminary checks:*

a) *Check the tire pressures and look for uneven wear.*
b) *Check the steering universal joints or coupling from the column to the steering gear for loose fasteners and wear.*
c) *Check the front and rear suspension and the steering gear assembly for loose and damaged parts.*
d) *Look for out-of-round or out-of-balance tires, bent rims and loose and/or rough wheel bearings.*

65 Vehicle pulls to one side

1 Mismatched or uneven tires (Chapter 10).
2 Broken or sagging springs (Chapter 10).
3 Front wheel alignment incorrect (Chapter 10).
4 Front brakes dragging (Chapter 9).

66 Abnormal or excessive tire wear

1 Front wheel alignment incorrect (Chapter 10).
2 Sagging or broken springs (Chapter 10).
3 Tire out-of-balance (Chapter 10).
4 Worn shock absorber (Chapter 10).
5 Overloaded vehicle.
6 Tires not rotated regularly.

67 Wheel makes a "thumping" noise

1 Blister or bump on tire (Chapter 1).
2 Improper shock absorber action (Chapter 10).

68 Shimmy, shake or vibration

1 Tire or wheel out-of-balance or out-of-round (Chapter 10).
2 Loose or worn wheel bearings (Chapter 10).
3 Worn tie-rod ends (Chapter 10).
4 Worn balljoints (Chapter 10).
5 Excessive wheel runout (Chapter 10).
6 Blister or bump on tire (Chapter 1).

69 Hard steering

1 Lack of lubrication at balljoints, tie-rod ends and steering gear assembly (Chapter 10).
2 Front wheel alignment incorrect (Chapter 10).
3 Low tire pressure (Chapter 1).

70 Steering wheel does not return to center position correctly

1 Lack of lubrication at balljoints and tie-rod ends (Chapter 10).
2 Binding in steering column (Chapter 10).
3 Defective rack-and-pinion assembly (Chapter 10).
4 Front wheel alignment problem (Chapter 10).

71 Abnormal noise at the front end

1 Lack of lubrication at balljoints and tie-rod ends (Chapter 1).
2 Loose upper strut mount (Chapter 10).
3 Worn tie-rod ends (Chapter 10).
4 Loose stabilizer bar (Chapter 10).
5 Loose wheel lug nuts (Chapter 1).
6 Loose suspension bolts (Chapter 10).

72 Wander or poor steering stability

1 Mismatched or uneven tires (Chapter 10).
2 Lack of lubrication at balljoints or tie-rod ends (Chapters 1 and 10).
3 Worn shock absorbers (Chapter 10).
4 Loose stabilizer bar (Chapter 10).
5 Broken or sagging springs (Chapter 10).
6 Front wheel alignment incorrect (Chapter 10).
7 Worn steering gear clamp bushings (Chapter 10).

73 Erratic steering when braking

1 Wheel bearings worn (Chapters 8 and 10).
2 Broken or sagging springs (Chapter 10).
3 Leaking wheel cylinder or caliper (Chapter 9).
4 Warped brake discs (rotors) (Chapter 9).
5 Worn steering gear clamp bushings (Chapter 10).

74 Excessive pitching and/or rolling around corners or during braking

1 Loose stabilizer bar (Chapter 10).
2 Worn shock absorbers or mounts (Chapter 10).
3 Broken or sagging springs (Chapter 10).
4 Overloaded vehicle.

75 Suspension bottoms

1 Overloaded vehicle.

2 Worn shock absorbers (Chapter 10).
3 Incorrect, broken or sagging springs (Chapter 10).

76 Cupped tires

1 Front wheel alignment incorrect (Chapter 10).
2 Worn shock absorbers (Chapter 10).
3 Wheel bearings worn (Chapters 8 and 10).
4 Excessive tire or wheel runout (Chapter 10).
5 Worn balljoints (Chapter 10).

77 Excessive tire wear on outside edge

1 Inflation pressures incorrect (Chapter 1).
2 Excessive speed in turns.
3 Front end alignment incorrect (excessive toe-in or positive camber). Have professionally aligned.
4 Suspension arm bent or twisted (Chapter 10).

78 Excessive tire wear on inside edge

1 Inflation pressures incorrect (Chapter 1).
2 Front end alignment incorrect (toe-out or excessive negative camber). Have professionally aligned.
3 Loose or damaged steering components (Chapter 10).

79 Tire tread worn in one place

1 Tires out-of-balance.
2 Damaged or buckled wheel. Inspect and replace if necessary.
3 Defective tire (Chapter 1).

80 Excessive play or looseness in steering system

1 Wheel bearings worn (Chapter 10).
2 Tie-rod end loose or worn (Chapter 10).
3 Steering gear loose (Chapter 10).

81 Rattling or clicking noise in rack and pinion

Steering gear clamps loose (Chapter 10).

Notes

Chapter 1   Tune-up and routine maintenance

Contents

Specifications

Recommended lubricants and fluids

Note: *Listed here are manufacturer recommendations at the time this manual was written. Manufacturers occasionally upgrade their fluid and lubricant specifications, so check with your local auto parts store for current recommendations.*

Engine oil
    Type .... API grade SG, SG/CC or SG/CD multigrade and fuel-efficient oil
    Viscosity .... See accompanying chart
Automatic transaxle fluid .... Dexron III Automatic Transmission Fluid (ATF)
Manual transaxle lubricant .... See your owner's manual or consult a dealer service department
Engine coolant .... 50/50 mixture of water and the specified ethylene glycol-based
    (green color) antifreeze or "DEX-COOL", silicate-free (orange-color)
    coolant - DO NOT mix the two types (refer to Sections 4, 9 and 30)
Brake fluid .... Delco Supreme II or DOT 3 fluid
Clutch fluid .... Delco Supreme II or DOT 3 fluid
Power steering fluid .... GM power steering fluid or equivalent
Chassis grease .... SAE NLGI no. 2 chassis grease

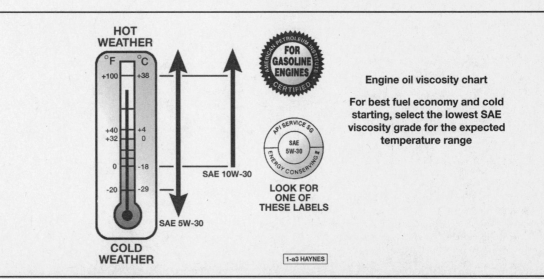

Engine oil viscosity chart

For best fuel economy and cold starting, select the lowest SAE viscosity grade for the expected temperature range

HOT WEATHER

COLD WEATHER

FOR GASOLINE ENGINES

AMERICAN PETROLEUM INSTITUTE CERTIFIED

API SERVICE SG
SAE 5W-30
ENERGY CONSERVING

LOOK FOR ONE OF THESE LABELS

SAE 10W-30

SAE 5W-30

1-a3 HAYNES

Capacities

Engine oil (with filter change)
- 2.5L four cylinder .... 3.0 qts
- 3.4L V6 .... 5.0 qts
- All others
  - 1996 and earlier .... 4.0 qts
  - 1997 and later .... 4.5 qts

Fuel tank
- 2.2L four-cylinder .... 17.1 gals
- All others
  - 1988 through 1990 .... 16 gals
  - 1991 through 1996 .... 16.5 gals
  - 1997 and later .... 17.0 qts
- Cooling system................................. Depends on engine, transaxle and cooling system options. Consult your owner's manual for exact capacity

Automatic transaxle (fluid and filter change)
- 3T40 transaxle................................. 4.0 qts
- All others
  - 1995 and earlier .... 6.0 qts
  - 1996 and 1997 .... 7.0 qts
  - 1998 and later .... 8.0 qts
- Manual transaxle (approximate) .... 2.1 qts

All capacities approximate. Add as necessary to bring to appropriate level.

Ignition system

Spark plug type and gap
- Four cylinder engines
  - 2.2L... AC type R44LTSMA or equivalent @ 0.045 inch
  - 2.3L... AC type FR3L or equivalent @ 0.035 inch
  - 2.5L... AC type R43CTS6 or equivalent @ 0.060 inch
- V6 engines
  - 1990 and earlier .... AC type R43LTSE or equivalent @ 0.045 inch
  - 1991 through 1995
    - 3.1L .... AC type R44LTSM or equivalent @ 0.045 inch
    - 3100 .... AC type R44LTSM or equivalent @ 0.060 inch
    - 3.4L .... AC type R42LTSM or equivalent @ 0.045 inch
    - 3800 .... AC type R44LTS6 or equivalent @ 0.060 inch
  - 1996 and later
    - 3100 .... AC type 41-940 or equivalent @ 0.060 inch
    - 3.4L .... AC type 41-919 or equivalent @ 0.045 inch
    - 3800 .... AC type 41-921 or equivalent @ 0.060 inch
- Firing order
  - Four-cylinder engines 1-3-4-2
  - V6 engines...................................... 1-2-3-4-5-6
  - 3800 engines 1-6-5-4-3-2

General

- Radiator cap pressure rating............. 15 psi
- Brake pad lining wear limit .... 1/8 inch
- Brake shoe lining wear limit.............. 1/16 inch from rivets

Torque specifications

Ft-lbs (unless otherwise indicated)

Automatic transaxle oil pan bolts
- 3T40 .... 97 to 120 in-lbs
- 4T60-E and 4T65-E .... 156 in-lbs
- Engine oil drain plug........................ 15 to 20
- Throttle body nuts/bolts................... 144 in-lbs
- Torque strut bolts 50
- Spark plugs
  - Four cylinder engines
    - 2.2L... 132 in-lbs
    - 2.3L and 2.5L 15
  - V6 engines
    - 2.8L and 3.1L............................... 20
    - 3100, 3.4L and 3800 15
- Wheel lug nuts................................. 100

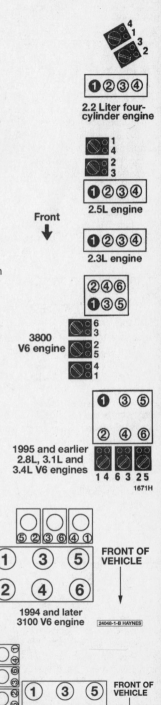

2.2 Liter four-cylinder engine

Front ↓

2.5L engine

2.3L engine

3800 V6 engine

1995 and earlier 2.8L, 3.1L and 3.4L V6 engines

1671H

FRONT OF VEHICLE

1994 and later 3100 V6 engine

24048-1-B HAYNES

FRONT OF VEHICLE

1996 3.4L V6 engine

24048-1-C HAYNES

Cylinder and coil terminal locations

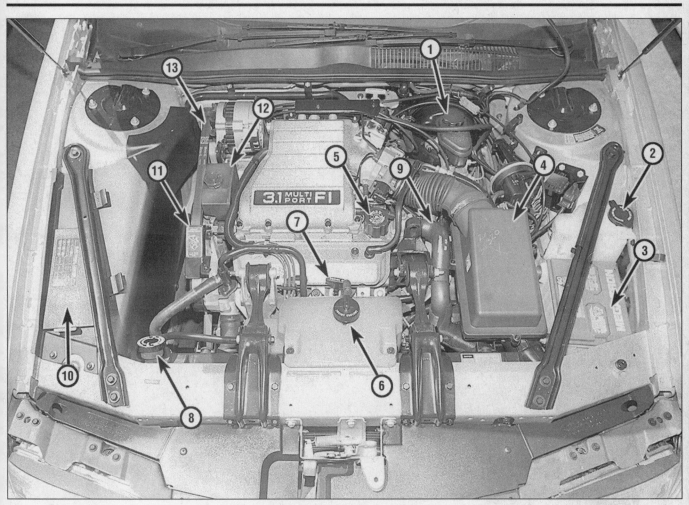

A typical V6 engine compartment

1	Brake fluid reservoir	6	Engine coolant reservoir
2	Windshield washer fluid reservoir	7	Engine oil dipstick
3	Battery	8	Radiator cap
4	Air cleaner housing	9	Radiator hose
5	Engine oil filler cap		

10	Engine compartment fuse block
11	Drivebelt routing decal
12	Power steering fluid reservoir
13	Serpentine drivebelt

1 Introduction

This Chapter is designed to help the home mechanic maintain the Chevrolet Lumina, Pontiac Grand Prix, Buick Regal and Oldsmobile Cutlass W-body models with the goals of maximum performance, economy, safety and reliability in mind.

Included is a master maintenance schedule (page 1-7), followed by procedures dealing specifically with each item on the schedule. Visual checks, adjustments, component replacement and other helpful items are included. Refer to the accompanying illustrations of the engine compartment and the underside of the vehicle for the locations of various components.

Servicing your vehicle in accordance with the mileage/time maintenance schedule and the step-by-step procedures will result in a planned maintenance program that should produce a long and reliable service life. Keep in mind that it's a comprehensive plan, so maintaining some items but not others at the specified intervals will not produce the same results.

As you service your vehicle, you'll discover that many of the procedures can - and should - be grouped together because of the nature of the particular procedure you're performing or because of the close proximity of two otherwise unrelated components to one another.

For example, if the vehicle is raised, you should inspect the exhaust, suspension, steering and fuel system, while you're under the vehicle. When you're rotating the tires, it makes good sense to check the brakes since the wheels are already removed. Finally, let's suppose you have to borrow or rent a torque wrench. Even if you only need it to tighten the spark plugs, you might as well check the torque of as many critical fasteners as time allows.

The first step in this maintenance program is to prepare yourself before the actual work begins. Read through all the procedures you're planning to do, then gather up all the parts and tools needed. If it looks like you might run into problems during a particular job, seek advice from a mechanic or an experienced do-it-yourselfer.

A typical Quad-4 engine compartment

1	Automatic transaxle fluid dipstick	6	Air cleaner housing
2	Brake fluid reservoir	7	Engine coolant reservoir
3	Power steering fluid reservoir	8	Engine oil filler cap/dipstick
4	Windshield washer fluid reservoir	9	Radiator cap
5	Battery	10	Engine compartment fuse block

Typical engine compartment underside view

1	Engine oil filter	3	Automatic transaxle fluid pan	5	Front disc brake caliper
2	Driveaxle boot	4	Exhaust system		

A typical rear underside view

1 Muffler	2 Fuel tank	3 Fuel filter	4 Rear disc brake caliper

2 Buick Regal, Chevrolet Lumina, Pontiac Grand Prix, Olds Cutlass Supreme Maintenance schedule

The following maintenance intervals are based on the assumption that the vehicle owner will be doing the maintenance or service work, as opposed to having a dealer service department do the work. Although the time/mileage intervals are loosely based on factory recommendations, most have been shortened to ensure, for example, that such items as lubricants and fluids are checked/changed at intervals that promote maximum engine/driveline service life. Also, subject to the preference of the individual owner interested in keeping his or her vehicle in peak condition at all times, and with the vehicle's ultimate resale in mind, many of the maintenance procedures may be performed more often than recommended in the following schedule. We encourage such owner initiative.

When the vehicle is new it should be serviced initially by a factory authorized dealer service department to protect the factory warranty. In many cases the initial maintenance check is done at no cost to the owner (check with your dealer service department for more information).

Every 250 miles or weekly, whichever comes first

Check the engine oil level (Section 4)
Check the engine coolant level (Section 4)
Check the windshield washer fluid level (Section 4)
Check the brake and clutch fluid levels (Section 4)
Check the tires and tire pressures (Section 5)

Every 3000 miles or 3 months, whichever comes first

All items listed above plus:
Check the automatic transmission fluid level (Section 6)
Check the power steering fluid level (Section 7)
Check the cooling system (Section 9)
Change the engine oil and filter (Section 12)

Every 7500 miles or 6 months, whichever comes first

All items listed above plus:
Check and service the battery (Section 8)
Inspect and replace, if necessary, all underhood hoses (Section 10)
Inspect and replace, if necessary, the windshield wiper blades (Section 11)
Check the driveaxle boots (Section 13)
Inspect the suspension and steering components (Section 14)
Inspect the exhaust system (Section 15)
Check the manual transaxle lubricant level (Section 16)
Rotate the tires (Section 17)
Check the brakes (Section 18)*
Lubricate the chassis (Section 36)
Check the throttle linkage (Section 37)

Every 15,000 miles or 12 months, whichever comes first

Inspect the fuel system (Section 19)
Replace the air filter and the PCV filter (Section 20)
Check the throttle body nut/bolt torque - 2.5L four cylinder engine only (Section 21)
Check the engine drivebelt(s) (Section 22)
Inspect the seat belts (Section 23)
Check the starter safety switch (Section 24)
Check the seat back latch (Section 25)
Check the spare tire and jack (Section 26)

Every 30,000 miles or 24 months, whichever comes first

All items listed above plus:
Replace the fuel filter (Section 27)***
Change the automatic transaxle fluid (Section 28)**
Change the manual transaxle lubricant (Section 29)***
Service the cooling system (drain, flush and refill) (green-colored ethylene glycol anti-freeze only) (Section 30)
Inspect and replace, if necessary, the PCV valve (Section 31)
Inspect the evaporative emissions control system (Section 32)
Replace the spark plugs (conventional [non-platinum] spark plugs) (Section 33)
Inspect the spark plug wires (Section 34)

Every 60,000 miles or 48 months, whichever comes first

Inspect and replace, if necessary the timing belt (3.4L V6 engines) (Chapter 2E). **Note:** *After 60,000 miles, the timing belt should be inspected every 15,000 miles.*

Every 100,000 miles or 5 years, whichever comes first

Replace the spark plugs (platinum-tipped spark plugs) (Section 33)
Service the cooling system (drain, flush and refill) (orange-colored "DEX-COOL" silicate-free coolant only) (Section 30)
* *If the vehicle frequently tows a trailer, is operated primarily in stop-and-go conditions or its brakes receive severe usage for any other reason, check the brakes every 3000 miles or three months.*
** *If operated under one or more of the following conditions, change the automatic transaxle fluid every 15,000 miles:*

In heavy city traffic where the outside temperature regularly reaches 90-degrees F (32-degrees C) or higher
In hilly or mountainous terrain
Frequent trailer pulling

*** *On later models these may not be required maintenance items. Consult your owner's manual.*

4.2 The engine oil dipstick is clearly marked - "ENGINE OIL", as is the oil filler cap, which threads into the engine - on the Quad-4 shown here, the filler cap incorporates the dipstick

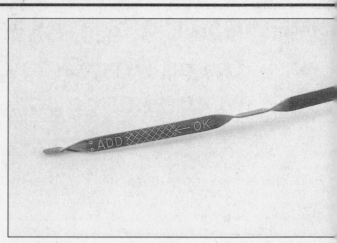

4.4 The oil level should be in the cross-hatched area - if it's below the ADD line, add enough oil to bring the level into the cross-hatched area

3   Tune-up general information

The term tune-up is used in this manual to represent a combination of individual operations rather than one specific procedure.

If, from the time the vehicle is new, the routine maintenance schedule is followed closely and frequent checks are made of fluid levels and high wear items, as suggested throughout this manual, the engine will be kept in relatively good running condition and the need for additional work will be minimized.

More likely than not, however, there will be times when the engine is running poorly due to lack of regular maintenance. This is even more likely if a used vehicle, which has not received regular and frequent maintenance checks, is purchased. In such cases, an engine tune-up will be needed outside of the regular routine maintenance intervals.

The first step in any tune-up or diagnostic procedure to help correct a poor running engine is a cylinder compression check. A compression check (see Chapter 2, Part G) will help determine the condition of internal engine components and should be used as a guide for tune-up and repair procedures. If, for instance, a compression check indicates serious internal engine wear, a conventional tune-up won't improve the performance of the engine and would be a waste of time and money. Because of its importance, the compression check should be done by someone with the right equipment and the knowledge to use it properly.

The following procedures are those most often needed to bring a generally poor running engine back into a proper state of tune.

Minor tune-up

Check all engine related fluids (Section 4)
Clean, inspect and test the battery
    (Section 8)
Check and adjust the drivebelts
    (Section 22)

Replace the spark plugs (Section 33)
Inspect the spark plug wires (Section 34)
Check the PCV valve (Section 31)
Check the air and PCV filters (Section 20)
Check the cooling system (Section 9)
Check all underhood hoses (Section 10)

Major tune-up

All items listed under Minor tune-up
    plus . . .
Check the EGR system (Chapter 6)
Check the ignition system (Section 33 and
    Chapter 5)
Check the charging system (Chapter 5)
Check the fuel system (Section 19)
Replace the air and PCV filters (Section 20)
Replace the spark plug wires (Section 34)

4   Fluid level checks

Note: *The following are fluid level checks to be done on a 250 mile or weekly basis. Additional fluid level checks can be found in specific maintenance procedures, which follow. Regardless of intervals, be alert to fluid leaks under the vehicle, which would indicate a problem to be corrected immediately.*

1   Fluids are an essential part of the lubrication, cooling, brake and windshield washer systems. Because the fluids gradually become depleted and/or contaminated during normal operation of the vehicle, they must be periodically replenished. See *Recommended lubricants and fluids* at the beginning of this Chapter before adding fluid to any of the following components. **Note:** *The vehicle must be on level ground when fluid levels are checked.*

Engine oil

Refer to illustrations 4.2 and 4.4

2   The engine oil level is checked with a dipstick **(see illustration)**. The dipstick extends through a metal tube down into the oil pan.

3   The oil level should be checked before the vehicle has been driven, or about 15 minutes after the engine has been shut off. If the oil is checked immediately after driving the vehicle, some of the oil will remain in the upper part of the engine, resulting in an inaccurate reading on the dipstick.

4   Pull the dipstick from the tube and wipe all the oil from the end with a clean rag or paper towel. Insert the clean dipstick all the way back into the tube and pull it out again. Note the oil at the end of the dipstick. Add oil as necessary to keep the level above the ADD mark in the cross-hatched area of the dipstick **(see illustration)**.

5   Do not overfill the engine by adding too much oil since this may result in oil fouled spark plugs, oil leaks or oil seal failures.

6   Oil is added to the engine after removing a twist-off cap located on the engine. On the Quad-4 engine, add the oil through the tube where the dipstick was removed. A funnel may help to reduce spills.

7   Checking the oil level is an important preventive maintenance step. A consistently low oil level indicates oil leakage through damaged seals, defective gaskets or past worn rings or valve guides. If the oil looks milky in color or has water droplets in it, the cylinder head gasket may be blown or the head or block may be cracked. The engine should be checked immediately. The condition of the oil should also be checked. Whenever you check the oil level, slide your thumb and index finger up the dipstick before wiping off the oil. If you see small dirt or metal particles clinging to the dipstick, the oil should be changed (Section 12).

Engine coolant

Refer to illustration 4.9

Warning: *Do not allow antifreeze to come in contact with your skin or painted surfaces of the vehicle. Rinse off spills immediately with plenty of water. Antifreeze is highly toxic if ingested. Never leave antifreeze lying around in an open container or in puddles on the floor; children and pets are attracted by its*

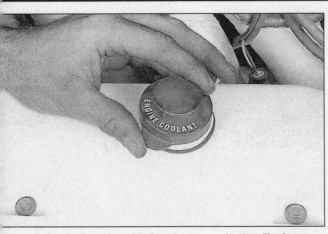

4.9 Coolant can be added to the reservoir after flipping the cap up

4.14 Flip the windshield washer fluid cap up to add fluid

weet smell and may drink it. Check with local authorities about disposing of used antifreeze. Many communities have collection centers which will see that antifreeze is disposed of safely. **Caution:** *Never mix green-colored ethylene glycol anti-freeze and orange-colored "DEX-COOL" silicate-free coolant because doing so will destroy the efficiency of the "DEX-COOL" coolant which is designed to last for 100,000 miles or five years.*

All vehicles covered by this manual are equipped with a pressurized coolant recovery system. A white plastic coolant reservoir located at the front of the engine compartment is connected by a hose to the radiator filler neck. As the engine warms up and the coolant expands, it escapes through a valve in the radiator cap and travels through the hose into the reservoir. As the engine cools, the coolant is automatically drawn back into the cooling system to maintain the correct level.

The coolant level in the reservoir should be checked regularly. **Warning:** *Do not remove the radiator cap to check the coolant level when the engine is warm. The level in the reservoir varies with the temperature of the engine.* When the engine is cold, the coolant level should be at or slightly above the FULL COLD mark on the reservoir. Once the engine has warmed up, the level should be at or near the FULL HOT mark. If it isn't, add coolant to the reservoir. To add coolant simply flip up the cap and add a 50/50 mixture of ethylene glycol based green-colored antifreeze or orange-colored "DEX-COOL" silicate-free coolant and water (see **Caution** above). The coolant and windshield washer reservoirs look similar, so be sure to add the correct fluids; the caps are clearly marked **(see illustration)**.

10 Drive the vehicle and recheck the coolant level. If only a small amount of coolant is required to bring the system up to the proper level, water can be used. However, repeated additions of water will dilute the antifreeze and water solution. In order to maintain the proper ratio of antifreeze and water, always top up the coolant level with the correct mixture. An empty plastic milk jug or bleach bottle makes an excellent container for mixing coolant. Do not use rust inhibitors or additives.

11 If the coolant level drops consistently, there may be a leak in the system. Inspect the radiator, hoses, filler cap, drain plugs and water pump (see Section 9). If no leaks are noted, have the radiator cap pressure tested by a service station.

12 If you have to remove the radiator cap, wait until the engine has cooled completely, then wrap a thick cloth around the cap and turn it to the first stop. If coolant or steam escapes, let the engine cool down longer, then remove the cap.

13 Check the condition of the coolant as well. It should be relatively clear. If it is brown or rust colored, the system should be drained, flushed and refilled. Even if the coolant appears to be normal, the corrosion inhibitors wear out, so it must be replaced at the specified intervals.

Windshield washer fluid

Refer to illustration 4.14

14 Fluid for the windshield washer system is located in a plastic reservoir in the engine compartment **(see illustration)**. In milder climates, plain water can be used in the reservoir, but it should be kept no more than two-thirds full to allow for expansion if the water freezes. In colder climates, use windshield washer system antifreeze, available at any auto parts store, to lower the freezing point of the fluid. Mix the antifreeze with water in accordance with the manufacturer's directions on the container. **Caution:** *Do not use cooling system antifreeze - it will damage the vehicle's paint.*

15 To help prevent icing in cold weather, warm the windshield with the defroster before using the washer.

Battery electrolyte

16 All vehicles covered by this manual are equipped with a battery which is permanently sealed (except for vent holes) and has no filler caps. Water does not have to be added to

4.18 Unscrew the reservoir cap and make sure the brake fluid level is up to the bottom of the slot in the neck

these batteries at any time; however, if a maintenance-type battery has been installed on the vehicle since it was new, remove all the cell caps on top of the battery (usually there are two caps that cover three cells each). If the electrolyte level is low, add distilled water until the level is above the plates. There is usually a split-ring indicator in each cell to help you judge when enough water has been added. Add water until the electrolyte level is just up to the bottom of the split ring indicator. Do not overfill the battery or it will spew out electrolyte when it is charging.

Brake and clutch fluid

Refer to illustration 4.18

17 The brake master cylinder is mounted on the front of the power booster unit in the engine compartment. The clutch master cylinder used with manual transaxles is mounted adjacent to it on the firewall. On early models equipped with ABS, apply the brake 40 times before checking the fluid level.

18 Unscrew the cap and make sure the fluid level is even with the bottom of the filler neck slot **(see illustration)**. On early models equipped with ABS, release the clips and lift the reservoir cover off.

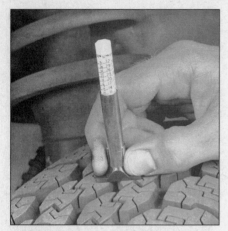

5.2 Use a tire tread depth indicator to monitor tire wear - they are available at auto parts stores and service stations and cost very little

19   When adding fluid, pour it carefully into the reservoir to avoid spilling it on surrounding painted surfaces. Be sure the specified fluid is used, since mixing different types of brake fluid can cause damage to the system. See Recommended lubricants and fluids at the front of this Chapter or your owner's manual. **Warning:** *Brake fluid can harm your eyes and damage painted surfaces, so use extreme caution when handling or pouring it. Do not use brake fluid that has been standing open or is more than one year old. Brake fluid absorbs moisture from the air. Excess moisture can cause a dangerous loss of braking effectiveness.*

20   At this time the fluid and master cylinder can be inspected for contamination. The system should be drained and refilled if deposits, dirt particles or water droplets are seen in the fluid.

21   After filling the reservoir to the proper level, make sure the cap is on tight to prevent fluid leakage.

22   The brake fluid level in the master cylinder will drop slightly as the pads at each wheel wear down during normal operation. If the master cylinder requires repeated replenishing to keep it at the proper level, this is an indication of leakage in the brake system, which should be corrected immediately. Check all brake lines and connections (see Section 18 for more information).

23   If, when checking the master cylinder fluid level, you discover one or both reservoirs empty or nearly empty, the brake system should be bled (Chapter 9).

5   Tire and tire pressure checks

Refer to illustrations 5.2, 5.3, 5.4a, 5.4b and 5.8

1   Periodic inspection of the tires ma spare you the inconvenience of beir stranded with a flat tire. It can also provic you with vital information regarding possib problems in the steering and suspension sy tems before major damage occurs.

2   The original tires on this vehicle a equipped with 1/2-inch side bands th appear when tread depth reaches 1/16-inc but they don't appear until the tires are wo out. Tread wear can be monitored with a sin ple, inexpensive device known as a trea depth indicator **(see illustration)**.

3   Note any abnormal tread wear **(se illustration)**. Tread pattern irregularities suc as cupping, flat spots and more wear on or side than the other are indications of fro: end alignment and/or balance problems. any of these conditions are noted, take th vehicle to a tire shop or service station to cc rect the problem.

4   Look closely for cuts, punctures ar embedded nails or tacks. Sometimes a ti will hold air pressure for short time or lea down very slowly after a nail has embedde

UNDERINFLATION

CUPPING

Cupping may be caused by:
- **Underinflation and/or mechanical irregularities such as out-of-balance condition of wheel and/or tire, and bent or damaged wheel.**
- **Loose or worn steering tie-rod or steering idler arm.**
- **Loose, damaged or worn front suspension parts.**

OVERINFLATION

INCORRECT TOE-IN OR EXTREME CAMBER

FEATHERING DUE TO MISALIGNMENT

5.3 This chart will help you determine the condition of the tires, the probable cause(s) of abnormal wear and the corrective action necessary

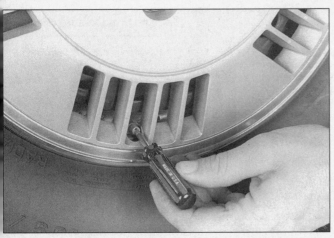

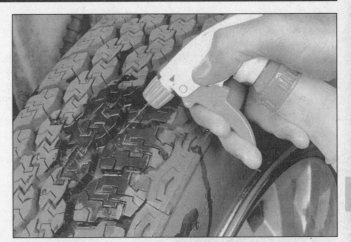

5.4a If a tire loses air on a steady basis, check the valve core first to make sure it's snug (special inexpensive wrenches are commonly available at auto parts stores)

5.4b If the valve core is tight, raise the corner of the vehicle with the low tire and spray a soapy water solution onto the tread as the tire is turned slowly - leaks will cause small bubbles to appear

5.8 To extend the life of the tires, check the air pressure at least once a week with an accurate gauge (don't forget the spare!)

6.3a On earlier models the transaxle dipstick is clearly marked "TRANS FLUID" . . .

6.3b . . . while on later models the transaxle dipstick is indicated by a RED handle (arrow) - In either case the dipstick is located at the rear of the engine compartment between the engine and the firewall

itself in the tread. If a slow leak persists, check the valve stem core to make sure it's tight **(see illustration)**. Examine the tread for an object that may have embedded itself in the tire or for a "plug" that may have begun to leak (radial tire punctures are repaired with a plug that's installed in a puncture). If a puncture is suspected, it can be easily verified by spraying a solution of soapy water onto the suspected area **(see illustration)**. The soapy solution will bubble if there's a leak. Unless the puncture is unusually large, a tire shop or service station can usually repair the tire.

5   Carefully inspect the inner sidewall of each tire for evidence of brake fluid. If you see any, inspect the brakes immediately.

6   Correct air pressure adds miles to the lifespan of the tires, improves mileage and enhances overall ride quality. Tire pressure cannot be accurately estimated by looking at a tire, especially if it's a radial. A tire pressure gauge is essential. Keep an accurate gauge in the vehicle. The pressure gauges attached to the nozzles of air hoses at gas stations are often inaccurate.

7   Always check tire pressure when the

tires are cold. Cold, in this case, means the vehicle has not been driven over a mile in the three hours preceding a tire pressure check. A pressure rise of four to eight pounds is not uncommon once the tires are warm.

8   Unscrew the valve cap protruding from the wheel or hubcap and push the gauge firmly onto the valve stem **(see illustration)**. Note the reading on the gauge and compare the figure to the recommended tire pressure shown on the label attached to the inside of the glove compartment door. Be sure to reinstall the valve cap to keep dirt and moisture out of the valve stem mechanism. Check all four tires and, if necessary, add enough air to bring them up to the recommended pressure.

9   Don't forget to keep the spare tire inflated to the specified pressure (refer to your owner's manual or the tire sidewall).

6   Automatic transaxle fluid level check

Refer to illustrations 6.3a, 6.3b and 6.6

1   The automatic transaxle fluid level

should be carefully maintained. Low fluid level can lead to slipping or loss of drive, while overfilling can cause foaming and loss of fluid.

2   With the parking brake set, start the engine, then move the shift lever through all the gear ranges, ending in Park. The fluid level must be checked with the vehicle level and the engine running at idle. **Note:** *Incorrect fluid level readings will result if the vehicle has just been driven at high speeds for an extended period, in hot weather in city traffic, or if it has been pulling a trailer. If any of these conditions apply, wait until the fluid has cooled (about 30 minutes).*

3   With the transaxle at normal operating temperature, remove the dipstick from the filler tube. The dipstick is located at the rear of the engine compartment **(see illustration)**.

4   Carefully touch the fluid at the end of the dipstick to determine if the fluid is cool, warm or hot. Wipe the fluid from the dipstick with a clean rag and push it back into the filler tube until the cap seats.

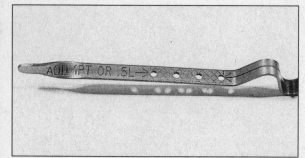

6.6 The automatic transaxle fluid level must be maintained within the cross-hatched area on the dipstick

7.2 The power steering fluid reservoir is located near the front (drivebelt end) of the engine; turn the cap clockwise for removal (V6 engine shown)

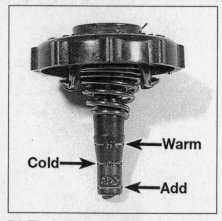

7.6 The marks on the dipstick indicate the safe fluid level range

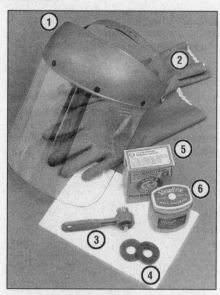

8.1 Tools and materials required for battery maintenance

level. The level should be at the HOT mark if the fluid was hot to the touch **(see illustration)**. It should be at the COLD mark if the fluid was cool to the touch. Note that on some models the marks (FULL HOT and COLD) are on opposite sides of the dipstick. At no time should the fluid level drop below the ADD mark.

7   If additional fluid is required, pour the specified type directly into the reservoir using a funnel to prevent spills.

8   If the reservoir requires frequent fluid additions, all power steering hoses, hose connections, the power steering pump and the rack and pinion assembly should be carefully checked for leaks.

8   Battery check and maintenance

Refer to illustrations 8.1, 8.4, 8.5a, 8.5b, 8.5c and 8.5d

Warning: *Certain precautions must be followed when checking and servicing the battery. Hydrogen gas, which is highly flammable, is always present in the battery cells, so keep lighted tobacco and all other open flames and sparks away from the battery. The electrolyte inside the battery is actually dilute sulfuric acid, which will cause injury if splashed on your skin or in your eyes. It will also ruin clothes and painted surfaces. When removing the battery cables, always detach the negative cable first and hook it up last.*
Caution: *On models equipped with the Theft-lock audio system, be sure the lockout feature is turned off before performing any procedure which requires disconnecting the battery.*

1   Battery maintenance is an important procedure which will help ensure you aren't stranded because of a dead battery. Several tools are required for this procedure (see illustration).

2   A sealed battery is standard equipment on all vehicles covered by this manual. Although this type of battery has many advantages over the older, capped cell type, and never requires the addition of water, it

5   Pull the dipstick out again and note the fluid level.

6   If the fluid felt cool, the level should be about 1/8-to-3/8 inch below the "ADD 1 PT" mark **(see illustration)**. If it felt warm, the level should be close to the "ADD 1 PT" mark. If the fluid was hot, the level should be within the cross-hatched area. If additional fluid is required, pour it directly into the tube using a funnel. It takes about one pint to raise the level from the ADD mark to the upper edge of the cross-hatched area with a hot transaxle, so add the fluid a little at a time and keep checking the level until it's correct.

7   The condition of the fluid should also be checked along with the level. If the fluid at the end of the dipstick is a dark reddish-brown color, or if the fluid has a burned smell, the fluid should be changed. If you're in doubt about the condition of the fluid, purchase some new fluid and compare the two for color and smell.

7   Power steering fluid level check

Refer to illustrations 7.2 and 7.6

1   Unlike manual steering, the power steering system relies on fluid which may, over a period of time, require replenishing.

2   The fluid reservoir for the power steering pump is located behind the radiator near the front (drivebelt end) of the engine **(see illustration)**.

3   For the check, the front wheels should be pointed straight ahead and the engine should be off.

1   *Face shield/safety goggles - When removing corrosion with a brush, the acidic particles can easily fly up into your eyes*

2   *Rubber gloves - Another safety item to consider when servicing the battery - remember that's acid inside the battery!*

3   *Battery terminal/cable cleaner - This wire brush cleaning tool will remove all traces of corrosion from the battery and cable*

4   *Treated felt washers - Placing one of these on each terminal, directly under the cable end, will help prevent corrosion (be sure to get the correct type for side-terminal batteries)*

5   *Baking soda - A solution of baking soda and water can be used to neutralize corrosion*

6   *Petroleum jelly - A layer of this on the battery terminal bolts will help prevent corrosion*

4   Use a clean rag to wipe off the reservoir cap and the area around the cap. This will help prevent any foreign matter from entering the reservoir during the check.

5   Twist off the cap and check the temperature of the fluid at the end of the dipstick with your finger.

6   Wipe off the fluid with a clean rag, reinsert it, then withdraw it and read the fluid

8.4 Check the tightness of the battery cable terminal bolts

8.5a A tool like this one (available at auto parts stores) is used to clean the side terminal type battery contact area

8.5b Use the brush to finish the cleaning job

8.5c The result should be a clean, shiny terminal area

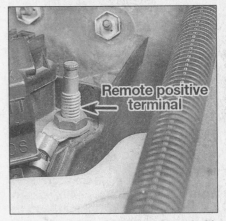

8.5d Some models are also equipped with a remote positive battery terminal - be sure to remove the cover and check for corrosion and a that the cable is securely fastened

should still be routinely maintained according to the procedures which follow.

Check

3   The battery is located on the left side of the engine compartment. The exterior of the battery should be inspected periodically for damage such as a cracked case or cover.

4   Check the tightness of the battery cable terminals and connections (see illustration) to ensure good electrical connections and check the entire length of each cable for cracks and frayed conductors.

5   If corrosion (visible as white, fluffy deposits) is evident, remove the cables from the terminals, clean them with a battery brush and reinstall the cables (see illustrations). Corrosion can be kept to a minimum by using special treated fiber washers available at auto parts stores or by applying a layer of petroleum jelly to the terminals and cables after they are assembled.

6   Make sure that the battery tray is in good condition and the hold-down clamp bolt is tight. If the battery is removed from the tray, make sure no parts remain in the bottom of the tray when the battery is reinstalled. When reinstalling the hold-down clamp bolt, do not overtighten it.

7   Information on removing and installing the battery can be found in Chapter 5. Information on jump starting can be found at the front of this manual. For more detailed battery checking procedures, refer to the *Haynes Automotive Electrical Manual.*

Cleaning

8   Corrosion on the hold-down components, battery case and surrounding areas can be removed with a solution of water and baking soda. Thoroughly rinse all cleaned areas with plain water.

9   Any metal parts of the vehicle damaged by corrosion should be covered with a zinc-based primer, then painted.

Charging

Warning: *When batteries are being charged, hydrogen gas, which is very explosive and flammable, is produced. Do not smoke or allow open flames near a charging or a recently charged battery. Wear eye protection when near the battery during charging. Also, make sure the charger is unplugged before connecting or disconnecting the battery from the charger.*

10   Slow-rate charging is the best way to restore a battery that's discharged to the point where it will not start the engine. It's also a good way to maintain the battery charge in a vehicle that's only driven a few miles between starts. Maintaining the battery charge is particularly important in the winter when the battery must work harder to start the engine and electrical accessories that drain the battery are in greater use.

11   It's best to use a one or two-amp battery charger (sometimes called a "trickle" charger). They are the safest and put the least strain on the battery. They are also the least expensive. For a faster charge, you can use a higher amperage charger, but don't use one rated more than 1/10th the amp/hour rating of the battery. Rapid boost charges that claim to restore the power of the battery in one to two hours are hardest on the battery and can damage batteries not in good condition. This type of charging should only be used in emergency situations.

12   The average time necessary to charge a battery should be listed in the instructions that come with the charger. As a general rule, a trickle charger will charge a battery in 12 to 16 hours.

13   Remove all of the cell caps (if equipped) and cover the holes with a clean cloth to prevent spattering electrolyte. Disconnect the negative battery cable and hook the battery charger leads to the battery posts (positive to positive, negative to negative), then plug in the charger. Make sure it is set at 12 volts if it has a selector switch. **Caution:** *On models equipped with the Theftlock audio system, be sure the lockout feature is turned off before performing any procedure which requires disconnecting the battery.*

14   If you're using a charger with a rate higher than two amps, check the battery regularly during charging to make sure it doesn't overheat. If you're using a trickle charger, you can safely let the battery charge overnight after you've checked it regularly for the first couple of hours.

15   If the battery has removable cell caps, measure the specific gravity with a hydrome-

ter every hour during the last few hours of the charging cycle. Hydrometers are available inexpensively from auto parts stores - follow the instructions that come with the hydrometer. Consider the battery charged when there's no change in the specific gravity reading for two hours and the electrolyte in the cells is gassing (bubbling) freely. The specific gravity reading from each cell should be very close to the others. If not, the battery probably has a bad cell(s).

16   Some batteries with sealed tops have built-in hydrometers on the top that indicate the state of charge by the color displayed in the hydrometer window. Normally, a bright-colored hydrometer indicates a full charge and a dark hydrometer indicates the battery still needs charging. Check the battery manufacturer's instructions to be sure you know what the colors mean.

17   If the battery has a sealed top and no built-in hydrometer, you can hook up a digital voltmeter across the battery terminals to check the charge. A fully charged battery should read 12.5 volts or higher.

9   Cooling system check

Refer to illustration 9.4
Caution: *Never mix green-colored ethylene glycol anti-freeze and orange-colored "DEX-COOL" silicate-free coolant because doing so will destroy the efficiency of the "DEX-COOL" coolant which is designed to last for 100,000 miles or five years.*

1   Many major engine failures can be attributed to a faulty cooling system. If the vehicle is equipped with an automatic transaxle, the cooling system also cools the transaxle fluid and plays an important role in prolonging transaxle life.

2   The cooling system should be checked with the engine cold. Do this before the vehicle is driven for the day or after the engine has been shut off for at least three hours.

3   Remove the radiator cap by turning it to the left until it reaches a stop. If you hear any hissing sounds (indicating there is still pressure in the system), wait until it stops. Now press down on the cap with the palm of your hand and continue turning to the left until the cap can be removed. Thoroughly clean the cap, inside and out, with clean water. Also clean the filler neck on the radiator. All traces of corrosion should be removed. The coolant inside the radiator should be relatively transparent. If it is rust colored, the system should be drained and refilled (Section 30). If the coolant level is not up to the top, add additional antifreeze/coolant mixture (see Section 4).

4   Carefully check the large upper and lower radiator hoses along with any smaller diameter heater hoses which run from the engine to the firewall. Inspect each hose along its entire length, replacing any hose which is cracked, swollen or shows signs of deterioration. Cracks may become more apparent if the hose is squeezed **(see illustration)**.

5   Make sure all hose connections are

Check for a chafed area that could fail prematurely.

Check for a soft area indicating the hose has deteriorated inside.

Overtightening the clamp on a hardened hose will damage the hose and cause a leak.

Check each hose for swelling and oil-soaked ends. Cracks and breaks can be located by squeezing the hose

9.4 Hoses, like drivebelts, have a habit of failing at the worst possible time - to prevent the inconvenience of a blown radiator or heater hose, inspect them carefully as shown here

tight. A leak in the cooling system will usually show up as white or rust colored deposits on the areas adjoining the leak. If wire-type clamps are used at the ends of the hoses, it may be wise to replace them with more secure screw-type clamps.

6   Use compressed air or a soft brush to remove bugs, leaves, etc. from the front of the radiator or air conditioning condenser. Be careful not to damage the delicate cooling fins or cut yourself on them.

7   Every other inspection, or at the first indication of cooling system problems, have the cap and system pressure tested. If you don't have a pressure tester, most gas stations and repair shops will do this for a minimal charge.

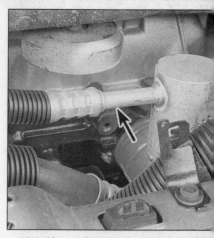

10.1 Air conditioning hoses are easily identified by the metal tubes used at all bends (arrow) - DO NOT disconnect or accidentally damage the air conditioning hoses (the system is under high pressure

10   Underhood hose check and replacement

General

Refer to illustration 10.1
Caution: *Replacement of air conditioning hoses must be left to a dealer service department or air conditioning shop that has the equipment to depressurize the system safely. Never remove air conditioning components or hoses* **(see illustration)** *until the system has been depressurized.*

2   High temperatures under the hood can cause the deterioration of the rubber and plastic hoses used for engine, accessory and emission systems operation. Periodic inspection should be made for cracks, loose clamps, material hardening and leaks. Information specific to the cooling system hoses can be found in Section 9.

3   Some, but not all, hoses are secured to the fittings with clamps. Where clamps are used, check to be sure they haven't lost their tension, allowing the hose to leak. If clamps aren't used, make sure the hose hasn't expanded and/or hardened where it slips over the fitting, allowing it to leak.

Vacuum hoses

4   It's quite common for vacuum hoses, especially those in the emissions system, to be color coded or identified by colored stripes molded into each hose. Various systems require hoses with different wall thicknesses, collapse resistance and temperature resistance. When replacing hoses, be sure the new ones are made of the same material.

5   Often the only effective way to check a hose is to remove it completely from the vehicle. If more than one hose is removed, be sure to label the hoses and fittings to ensure correct installation.

6   When checking vacuum hoses, be sure

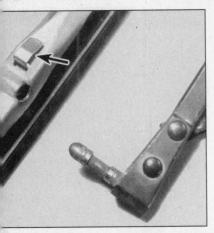

11.5a On 1995 and earlier models, lift the release lever (arrow), then slide the blade assembly off the wiper arm

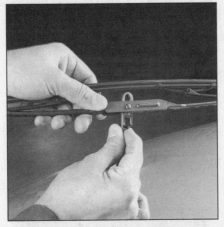

11.5b On 1996 and later models, press down the release tab, then slide the blade assembly down and out of the hook in the arm

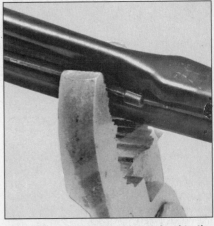

11.7 The rubber element is retained to the blade by small clips - the metal backing of the rubber element can be compressed at one end with pliers, allowing the element to slide out of the clips

to include any plastic T-fittings in the check. Inspect the fittings for cracks and the hose where it fits over the fitting for distortion, which could cause leakage.

7   A small piece of vacuum hose (1/4-inch inside diameter) can be used as a stethoscope to detect vacuum leaks. Hold one end of the hose to your ear and probe around vacuum hoses and fittings, listening for the "hissing" sound characteristic of a vacuum leak. **Warning:** *When probing with the vacuum hose stethoscope, be careful not to allow your body or the hose to come into contact with moving engine components such as the drivebelt, cooling fan, etc.*

Fuel hose

Warning: *Gasoline is extremely flammable, so take extra precautions when you work on any part of the fuel system. Don't smoke or allow open flames or bare light bulbs near the work area, and don't work in a garage where a natural gas-type appliance (such as a water heater or a clothes dryer) with a pilot light is present. Since gasoline is carcinogenic, wear latex gloves when there's a possibility of being exposed to fuel, and, if you spill any fuel on your skin, rinse it off immediately with soap and water. Mop up any spills immediately and do not store fuel-soaked rags where they could ignite. When you perform any kind of work on the fuel system, wear safety glasses and have a Class B type fire extinguisher on hand.*

8   Check all rubber fuel lines for deterioration and chafing. Check especially for cracks in areas where the hose bends and just before fittings, such as where a hose attaches to the fuel filter and fuel injection unit.

9   High quality fuel line, usually identified by the word Fluroelastomer printed on the hose, should be used for fuel line replacement. Never, under any circumstances, use unreinforced vacuum line, clear plastic tubing

or water hose for fuel lines.

10   Spring-type clamps are commonly used on fuel lines. These clamps often lose their tension over a period of time, and can be "sprung" during the removal process. As a result spring-type clamps be replaced with screw-type clamps whenever a hose is replaced.

Metal lines

11   Sections of steel tubing often used for fuel line between the fuel pump and fuel injection unit. Check carefully for cracks, kinks and flat spots in the line.

12   If a section of metal fuel line must be replaced, only seamless steel tubing should be used, since copper and aluminum tubing do not have the strength necessary to withstand normal engine vibration.

13   Check the metal brake lines where they enter the master cylinder and brake proportioning unit (if used) for cracks in the lines and loose fittings. Any sign of brake fluid leakage calls for an immediate thorough inspection of the brake system.

11 Windshield wiper blade inspection and replacement

Refer to illustrations 11.5a, 11.5b and 11.7

1   The windshield wiper and blade assembly should be inspected periodically for damage, loose components and cracked or worn blade elements.

2   Road film can build up on the wiper blades and affect their efficiency, so they should be washed regularly with a mild detergent solution.

3   The action of the wiping mechanism can loosen the bolts, nuts and fasteners, so they should be checked and tightened, as necessary, at the same time the wiper blades are checked.

4   If the wiper blade elements (sometimes called inserts) are cracked, worn or warped, they should be replaced with new ones.

5   On 1995 and earlier models remove the wiper blade assembly from the wiper arm by using a small screwdriver to lift the release lever while pulling on the blade to release it **(see illustration)**. On 1996 and later models lift the arm assembly away from the glass for clearance, press on the release lever, then slide the wiper blade assembly out of the hook in the end of the arm **(see illustration)**.

6   With the blade removed from the vehicle, you can remove the rubber element from the blade.

7   Using pliers, pinch the metal backing of the element **(see illustration)**, then slide the element out of the blade assembly.

8   Compare the new element with the old for length, design, etc.

9   Slide the new element into place. It will automatically lock at the correct location.

10   Reinstall the blade assembly on the arm, wet the windshield glass and test for proper operation.

12 Engine oil and filter change

Refer to illustrations 12.3, 12.9, 12.14 and 12.18

1   Frequent oil changes are the most important preventive maintenance procedures that can be done by the home mechanic. As engine oil ages, it becomes diluted and contaminated, which leads to premature engine wear.

2   Although some sources recommend oil filter changes every other oil change, we feel that the minimal cost of an oil filter and the relative ease with which it is installed dictate that a new filter be used every time the oil is changed.

3   Gather together all necessary tools and

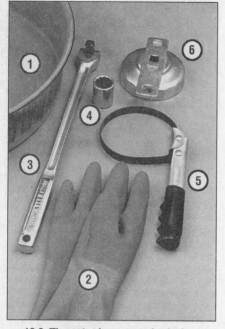

12.3 These tools are required when changing the engine oil and filter

1   *Drain pan* - It should be fairly shallow in depth, but wide to prevent spills
2   *Rubber gloves* - When removing the drain plug and filter, you will get oil on your hands (the gloves will prevent burns)
3   *Breaker bar* - Sometimes the oil drain plug is tight, and a long breaker bar is needed to loosen it
4   *Socket* - To be used with the breaker bar or a ratchet (must be the correct size to fit the drain plug - six-point preferred)
5   *Filter wrench* - This is a metal band-type wrench, which requires clearance around the filter to be effective
6   *Filter wrench* - This type fits on the bottom of the filter and can be turned with a ratchet or breaker bar (different-size wrenches are available for different types of filters)

materials before beginning the procedure **(see illustration)**.
4   In addition, you should have plenty of clean rags and newspapers handy to mop up any spills. Access to the underside of the vehicle is greatly improved if the vehicle can be lifted on a hoist, driven onto ramps or supported by jackstands. **Warning:** *Do not work under a vehicle which is supported only by a bumper, hydraulic or scissors-type jack.*
5   If this is your first oil change, get under the vehicle and familiarize yourself with the locations of the oil drain plug and the oil filter. The engine and exhaust components will be warm during the actual work, so note how they are situated to avoid touching them when working under the vehicle.
6   Warm the engine to normal operating temperature. If the new oil or any tools are

12.9 The engine oil drain plug is located at the rear of the oil pan (Quad-4 engine shown) - it is usually very tight, so use a box-end wrench to avoid rounding off the hex

needed, use this warm-up time to gather everything necessary for the job. The correct type of oil for your application can be found in *Recommended lubricants and fluids* at the beginning of this Chapter.
7   With the engine oil warm (warm engine oil will drain better and more built-up sludge will be removed with the oil), raise and support the vehicle. Make sure it's safely supported.
8   Move all necessary tools, rags and newspapers under the vehicle. Position the drain pan under the drain plug. Keep in mind that the oil will initially flow from the pan with some force, so place the pan accordingly.
9   Being careful not to touch any of the hot exhaust components, remove the drain plug at the bottom of the oil pan **(see illustration)**. Depending on how hot the oil is, you may want to wear gloves while unscrewing the plug the final few turns.
10   Allow the old oil to drain into the pan. It may be necessary to move the pan farther under the engine as the oil flow slows to a trickle.
11   After all the oil has drained, wipe off the drain plug with a clean rag. Small metal particles may cling to the plug which would immediately contaminate the new oil.
12   Clean the area around the drain plug opening and reinstall the plug. Tighten the plug securely with the wrench. If a torque wrench is available, use it to tighten the plug.
13   Move the drain pan into position under the oil filter.
14   Use the filter wrench to loosen the oil filter **(see illustration)**. Chain or metal band filter wrenches may distort the filter canister, but this is of no concern as the filter will be discarded anyway.
15   Completely unscrew the old filter. Be careful - it's full of oil. Empty the oil inside the filter into the drain pan.
16   Compare the old filter with the new one to make sure they are the same type.
17   Use a clean rag to remove all oil, dirt and sludge from the area where the oil filter

12.14 The oil filter is usually on very tight as well and will require a special wrench for removal - DO NOT use the wrench to tighten the new filter!

mounts on the engine. Check the old filter to make sure the rubber gasket isn't stuck to the engine. If the gasket is stuck to the engine (use a flashlight if necessary), remove it.
18   Apply a light coat of oil to the rubber gasket on the new oil filter **(see illustration)**. Open a can of oil and partially fill the oil filter with fresh oil. Oil pressure will not build in the engine until the oil pump has filled the filter with oil, so partially filling it at this time will reduce the amount of time the engine runs with no oil pressure.
19   Attach the new filter to the engine, following the tightening directions printed on the filter canister or box. Most filter manufacturers recommend against using a filter wrench due to the possibility of overtightening and damage to the seal.
20   Remove all tools, rags, etc. from under the vehicle, being careful not to spill the oil in the drain pan, then lower the vehicle.
21   Move to the engine compartment and locate the oil filler cap.
22   Pour the fresh oil through the filler opening. A funnel can be used.
23   Pour three quarts of fresh oil into the

12.18 Lubricate the oil filter gasket with clean engine oil before installing the filter on the engine

13.2 Push on the driveaxle boots to check for cracks

16.2 The manual transaxle dipstick (arrow) is located adjacent to the brake master cylinder

16.3 Follow the manual transaxle lubricant checking procedure printed on the dipstick

engine. Wait a few minutes to allow the oil to drain into the pan, then check the level on the dipstick (see Section 4 if necessary). If the oil level is above the ADD mark, start the engine and allow the new oil to circulate.

24   Run the engine for only about a minute and then shut it off. Immediately look under the vehicle and check for leaks at the oil pan drain plug and around the oil filter. If either is leaking, tighten with a bit more force.

25   With the new oil circulated and the filter now completely full, recheck the level on the dipstick and add more oil as necessary.

26   During the first few trips after an oil change, make it a point to check frequently for leaks and proper oil level.

27   The old oil drained from the engine cannot be reused in its present state and should be disposed of. Oil reclamation centers, auto repair shops and gas stations will normally accept the oil, which can be refined and used again. After the oil has cooled it can be drained into a container (capped plastic jugs, topped bottles, milk cartons, etc.) for transport to a disposal site.

13   Driveaxle boot check

Refer to illustration 13.2

1   The driveaxle boots are very important because they prevent dirt, water and foreign material from entering and damaging the constant velocity (CV) joints.

2   Inspect the boots for tears and cracks as well as loose clamps **(see illustration)**. If there is any evidence of cracks or leaking lubricant, they must be replaced as described in Chapter 8.

14   Suspension and steering check

1   Raise the front of the vehicle periodically and visually check the suspension and steering components for wear.

2   Be alert for excessive play in the steering wheel before the front wheels react,

excessive sway around corners, body movement over rough roads and binding at some point as the steering wheel is turned. If you notice any of the above symptoms, the steering and suspension systems should be checked.

3   Support the vehicle on jackstands placed under the frame rails. Because of the work to be done, make sure the vehicle cannot fall off the stands.

4   Check the front wheel hub nuts and make sure they are securely locked in place.

5   Working under the vehicle, check for loose bolts, broken or disconnected parts and deteriorated rubber bushings on all suspension and steering components. Look for grease or fluid leaking from the steering assembly. Check the power steering hoses and connections for leaks.

6   Have an assistant turn the steering wheel from side-to-side and check the steering components for free movement, chafing and binding. If the steering doesn't react with the movement of the steering wheel, try to determine where the slack is located.

15   Exhaust system check

1   With the engine cold (at least three hours after the vehicle has been driven), check the complete exhaust system from the engine to the end of the tailpipe. Ideally, the inspection should be done with the vehicle on a hoist to permit unrestricted access. If a hoist is not available, raise the vehicle and support it securely on jackstands.

2   Check the exhaust pipes and connections for evidence of leaks, severe corrosion and damage. Make sure that all brackets and hangers are in good condition and tight.

3   At the same time, inspect the underside of the body for holes, corrosion, open seams, etc. which may allow exhaust gases to enter the interior. Seal all body openings with silicone or body putty.

4   Rattles and other noises can often be traced to the exhaust system, especially the

mounts and hangers. Try to move the pipes, muffler and catalytic converter. If the components can come in contact with the body or suspension parts, secure the exhaust system with new mounts.

5   Check the running condition of the engine by inspecting inside the end of the tailpipe. The exhaust deposits here are an indication of engine state-of-tune. If the pipe is black and sooty or coated with white deposits, the engine is in need of a tune-up, including a thorough fuel system inspection and adjustment.

16   Manual transaxle lubricant level check

Refer to illustrations 16.2 and 16.3

1   A dipstick is used for checking the lubricant level in the manual transaxles used on these models.

2   With the transaxle cold (cool to the touch) and the vehicle parked on a level surface, remove the dipstick from the filler tube located at the left rear side of the engine compartment, adjacent to the brake master cylinder **(see illustration)**.

3   The level must be even with or slightly above the FULL COLD mark on the dipstick **(see illustration)**. Make sure the level is at the FULL COLD mark because lubricant may appear on the end of the dipstick even when the transaxle is several pints low.

4   If the level is low, add the specified lubricant through the filler tube, using a funnel.

5   Insert the dipstick into the filler tube and seat it securely.

17   Tire rotation

Refer to illustration 17.2

1   The tires should be rotated at the specified intervals and whenever uneven wear is noticed.

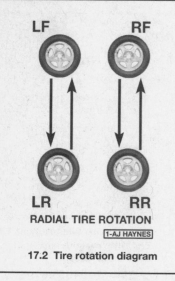

RADIAL TIRE ROTATION

1-AJ HAYNES

17.2 Tire rotation diagram

18.5 Look through the opening in the front of the caliper to check the brake pads (arrows) - the pad lining, which rubs against the disc, can also be inspected by looking through each end of the caliper

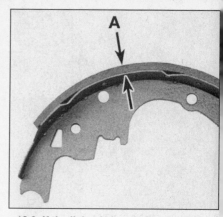

18.9 If the lining is bonded to the brake shoe, measure the lining thickness from the outer surface to the metal shoe, as shown here; if the lining is riveted to the shoe, measure from the lining outer surface to the rivet head

2   Front wheel drive vehicles require a special tire rotation pattern **(see illustration)**.

3   Refer to the information in Jacking and towing at the front of this manual for the proper procedures to follow when raising the vehicle and changing a tire. If the brakes are going to be checked, don't apply the parking brake as stated. Make sure the tires are blocked to prevent the vehicle from rolling as it's raised.

4   The entire vehicle should be raised at the same time. This can be done on a hoist or by jacking up each corner and then lowering the vehicle onto jackstands placed under the frame rails. Always use four jackstands and make sure the vehicle is safely supported.

5   After rotation, check and adjust the tire pressures as necessary and be sure to check the lug nut tightness.

18   Brake check

Refer to illustrations 18.5 and 18.9

Warning: *Brake system dust contains asbestos, which is hazardous to your health. DO NOT blow it out with compressed air or inhale it. DO NOT use gasoline or solvents to remove the dust. Use brake system cleaner or denatured alcohol only.*

Note: *For detailed photographs of the brake system, refer to Chapter 9.*

1   In addition to the specified intervals, the brakes should be inspected every time the wheels are removed or whenever a defect is suspected. Raise the vehicle and place it securely on jackstands. Remove the wheels (see Jacking and towing at the front of this manual, if necessary).

Disc brake pads

2   Disc brakes are used on the all four wheels of this vehicle. Extensive rotor damage can occur if the pads are not replaced when needed.

3   The disc brake pads have built-in wear indicators which make a high-pitched squealing or cricket-like warning sound when the pads are worn. **Caution:** *Expensive rotor dam-*

age can result if the pads are not replaced soon after the wear indicators start squealing.

4   The disc brake calipers, which contain the pads, are now visible. There is an outer pad and an inner pad in each caliper. All pads should be inspected.

5   Each caliper has one or two "windows" to inspect the pads **(see illustration)**. If the pad material has worn to about 1/8-inch thick or less, the pads should be replaced.

6   If you're unsure about the exact thickness of the remaining lining material, remove the pads for further inspection or replacement (refer to Chapter 9).

7   Before installing the wheels, check for leakage and/or damage at the brake hoses and connections. Replace the hose or fittings as necessary, referring to Chapter 9.

8   Check the condition of the brake rotor. Look for score marks, deep scratches and overheated areas (they will appear blue or discolored). If damage or wear is noted, the rotor can be removed and resurfaced by an automotive machine shop or replaced with a new one. Refer to Chapter 9 for more detailed inspection and repair procedures.

Drum brake shoes

9   Remove the brake drum and note the thickness of the lining material on brake shoes **(see illustration)**. If the material has worn away to within 1/16-inch of the recessed rivets or metal backing, the shoes should be replaced. If the linings look worn, but you are unable to determine their exact thickness, compare them with a new set at an auto parts store. The shoes should also be replaced if they are cracked, glazed (shiny surface) or contaminated with brake fluid.

10   Check to see that all the brake assembly springs are connected and in good condition.

11   Check the brake components for signs of fluid leakage. With your finger, carefully pry back the rubber cups on the wheel cylinder located at the top of the brake shoes. Any leakage is an indication that the wheel cylinders should be overhauled immediately (see Chapter 9). Also check the hoses and con-

nections for signs of leakage.

12   Clean the inside of the drum with brake system cleaner. Again, be careful not to breathe the dust.

13   Check the inside of the drum for cracks, scores, deep scratches and hard spots which will appear as small discolored areas. If imperfections cannot be removed with fine emery cloth, the drum must be taken to a machine shop for resurfacing.

14   After the inspection process, if all parts are found to be in good condition, reinstall the brake drum. Install the wheel and lower the vehicle to the ground, then tighten the lug nuts to the torque listed in this Chapter's Specifications.

Parking brake

15   The parking brake is operated by a foot pedal and locks the rear brake system. The easiest, and perhaps most obvious, method of periodically checking the operation of the parking brake assembly is to park the vehicle on a steep hill with the parking brake set and the transaxle in Neutral. If the parking brake cannot prevent the vehicle from rolling, it needs service (see Chapter 9).

19   Fuel system check

Warning: *Gasoline is extremely flammable, so take extra precautions when you work on any part of the fuel system. Don't smoke or allow open flames or bare light bulbs near the work area, and don't work in a garage where a natural gas-type appliance (such as a water heater or a clothes dryer) with a pilot light is present. Since gasoline is carcinogenic, wear latex gloves when there's a possibility of being exposed to fuel, and, if you spill any fuel on your skin, rinse it off immediately with soap and water. Mop up any spills immediately and do not store fuel-soaked rags where they could ignite. When you perform any kind of work on the fuel system, wear safety glasses and have a Class B type fire extinguisher on hand.*

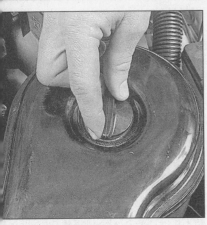

20.2a Remove the wingnut . . .

20.2b . . . lift off the cover and spark arrestor (if equipped) . . .

20.2c . . . and pull the air filter element out of the housing

The fuel system is most easily checked with the vehicle raised on a hoist so the components underneath the vehicle are readily visible and accessible.

If the smell of gasoline is noticed while driving or after the vehicle has been in the sun, the system should be thoroughly inspected immediately.

Remove the gas tank cap and check for damage, corrosion and an unbroken sealing imprint on the gasket. Replace the cap with a new one if necessary.

With the vehicle raised, inspect the gas tank and filler neck for punctures, cracks and other damage. The connection between the filler neck and tank is especially critical. Sometimes a rubber filler neck will leak due to loose clamps or deteriorated rubber, problems a home mechanic can usually rectify. **Warning:** *Do not, under any circumstances, try to repair a fuel tank yourself (except rubber components). A welding torch or any open flame can easily cause the fuel vapors to explode if the proper precautions are not taken.*

Carefully check all rubber hoses and metal lines leading away from the fuel tank. Check for loose connections, deteriorated hoses, crimped lines and other damage. Follow the lines to the front of the vehicle, carefully inspecting them all the way. Repair or replace damaged sections as necessary.

20 Air filter and PCV filter replacement

1   At the specified intervals, the air filter should be replaced with a new one. The filter should be inspected between changes.

Air filter replacement

2.5 liter four-cylinder and 2.8 liter V6 engine

Refer to illustrations 20.2a, 20.2b and 20.2c

2   The air filter is located inside the air cleaner housing located at the front of the engine, near the battery. Remove the wing nut, lift off the cover and spark arrestor (if equipped), then lift the filter out **(see illustrations)**.

3.1 liter V6 engine

Turbocharged models

3   Detach the clips, remove the PCV breather from the valve cover and lift the air cleaner housing off, then remove the filter from the housing.

Non-turbocharged models

4   Remove the bolts, lift the upper air cleaner housing off, then lift the filter element out.

3.4 liter V6, 3800 V6, 3100 V6 and 2.3 liter four-cylinder (Quad-4) engines

Refer to illustrations 20.5a and 20.5b

5   The air cleaner housing is located on the left (driver's) side of the engine compartment. Loosen or remove the nuts or retainer clips, lift the top cover off and withdraw the filter **(see illustrations)**.

All models

6   While the filter housing cover is off, be careful not to drop anything down into the throttle body or air cleaner assembly.
7   Wipe out the inside of the air cleaner housing with a clean rag.
8   Except on TBI models with a PCV filter, place the new filter in the air cleaner housing. Make sure it seats properly in the bottom of the housing. Install the top plate or cover.

PCV filter replacement

Refer to illustrations 20.9 and 20.10

9   Some 2.5L models are equipped with a PCV filter. It is usually located in a holder in the side of the filter housing and can be removed after disconnecting the hose and removing the clip **(see illustration)**.

20.5a On the Quad-4 engine, remove the two wingnuts . . .

20.5b . . . then lift up the cover and remove the filter element

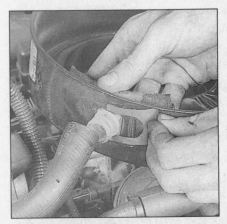

20.9 Remove the clip and withdraw the PCV filter housing from the air cleaner

20.10 Pull the PCV filter element out of the holder

10   Pull the PCV filter out of the holder (see illustration).
11   Install the PCV filter in the holder and attach it to the air cleaner housing. Install the air filter.

21   Throttle Body Injection (TBI) mounting bolt torque check (2.5L engine only)

1   The TBI throttle body is attached to the top of the intake manifold by two bolts. They can sometimes work loose from vibration and temperature changes during normal engine operation and cause a vacuum leak.
2   If you suspect a vacuum leak exists at the bottom of the throttle body, use a rubber hose as a stethoscope. Start the engine and place one end of the hose next to your ear as you probe around the base of the throttle body with the other end. You will hear a hissing sound if a leak exists (be careful of hot and moving engine components).
3   Remove the air cleaner assembly (see Chapter 4).
4   Locate the throttle body mounting bolts. Decide what special tools or adapters will be necessary, if any, to tighten them.
5   Tighten the bolts securely and evenly. Do not overtighten them, as the manifold threads could strip.
6   If, after the bolts are properly tightened, a vacuum leak still exists, the throttle body must be removed and a new gasket installed. See Chapter 4 for more information.
7   After tightening the fasteners, reinstall the air cleaner and return all hoses to their original positions.

22   Drivebelt check, adjustment and replacement

Refer to illustrations 22.2, 22.4, 22.5 and 22.7
1   A single serpentine drivebelt is located at the front of the engine and plays an impor-

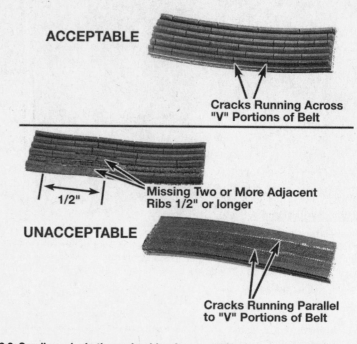

ACCEPTABLE

Cracks Running Across "V" Portions of Belt

1/2"

Missing Two or More Adjacent Ribs 1/2" or longer

UNACCEPTABLE

Cracks Running Parallel to "V" Portions of Belt

22.2 Small cracks in the underside of a serpentine belt are acceptable - lengthwise cracks, or missing pieces are cause for replacement

tant role in the overall operation of the engine and its components. Due to its function and material make up, the belt is prone to wear and should be periodically inspected. The serpentine belt drives the alternator, power steering pump, water pump and air conditioning compressor.
2   With the engine off, open the hood and use your fingers (and a flashlight, if necessary), to move along the belt checking for cracks and separation of the belt plies. Also check for fraying and glazing, which gives the belt a shiny appearance (see illustration). Both sides of the belt should be inspected, which means you will have to twist the belt to check the underside.
3   Check the ribs on the underside of the belt. They should all be the same depth, with none of the surface uneven.
4   The tension of the belt is checked visually. Locate the belt tensioner at the front of the engine under the air conditioning compressor on the right (passenger's) side, then find the tensioner operating marks (see illustration) located on the side of the tensioner. If the indicator mark is outside of the operating range, the belt should be replaced.
5   To replace the belt, rotate the tensioner to release belt tension (see illustration). Note: *On some models it may be necessary to remove the coolant recovery bottle in order to access the drive belt tensioner.*
6   Remove the belt from the auxiliary components and carefully release the tensioner.
7   Route the new belt over the various pulleys, again rotating the tensioner to allow the belt to be installed, then release the belt tensioner. Note: *Most models have a drivebelt*

routing decal on the belt guard to help during drivebelt installation (see illustration).

23   Seat belt check

1   Check the seat belts, buckles, latch plates and guide loops for obvious damage and signs of wear.
2   See if the seat belt reminder light comes on when the key is turned to the Run or Start position. A chime should also sound.
3   The seat belts are designed to lock up during a sudden stop or impact, yet allow free movement during normal driving. Make sure the retractors return the belt against your chest while driving and rewind the belt fully when the buckle is unlatched.
4   If any of the above checks reveal problems with the seat belt system, replace parts as necessary.

24   Starter safety switch check

Warning: *During the following checks there's a chance the vehicle could lunge forward, possibly causing damage or injuries. Allow plenty of room around the vehicle, apply the parking brake and hold down the regular brake pedal during the checks.*
1   On automatic transaxle equipped vehicles, try to start the engine in each gear. The engine should crank only in Park or Neutral.
2   If equipped with a manual transaxle, place the shift lever in Neutral and push the clutch pedal down about halfway. The engine

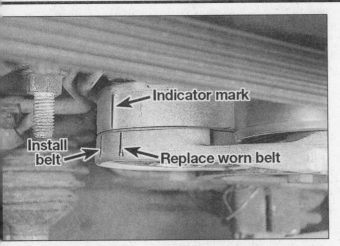

22.4 The drivebelt tensioner automatically keeps proper tension on the drivebelt, but does have limits - the indicator mark should remain in the nominal range, between the "INSTALL" and "REPLACE BELT" marks

22.5 On some models it will be necessary to insert a breaker bar into the notch on the tensioner arm, on other models it will be necessary to use a box end wrench on the pulley bolt to rotate the tensioner

should crank only with the clutch pedal fully depressed.

3   Make sure the steering column lock allows the key to go into the Lock position only when the shift lever is in Park (automatic transaxle) or Reverse (manual transaxle).

4   The ignition key should come out only in the Lock position.

25   Seatback latch check

1   It's important to periodically check the seatback latch mechanism on two-door models to prevent the seatback from moving forward during a sudden stop or an accident.

2   Grasping the top of the seat, attempt to tilt it forward. It should tilt only when the latch on the rear of the seat is pulled up. Note that there is a certain amount of free play built into the latch mechanism.

3   When returned to the upright position, the seatback should latch securely.

26   Spare tire and jack check

1   Periodically checking the security and condition of the spare tire and jack will help to familiarize you with the procedures necessary for emergency tire replacement and also ensure that no components work loose during normal vehicle operation.

2   Following the instructions in your owner's manual or under Jacking and towing near the front of this manual, remove the spare tire and jack.

3   Using a reliable air pressure gauge, check the pressure in the spare tire. It should be kept at the pressure marked on the tire sidewall.

4   Make sure the jack operates freely and all components are undamaged.

5   When finished, make sure the wing nuts hold the jack and tire securely in place.

27   Fuel filter replacement

Refer to illustration 27.3

Warning: *Gasoline is extremely flammable, so take extra precautions when you work on any part of the fuel system. Don't smoke or allow open flames or bare light bulbs near the work area, and don't work in a garage where a natural gas-type appliance (such as a water heater or a clothes dryer) with a pilot light is present. Since gasoline is carcinogenic, wear latex gloves when there's a possibility of being exposed to fuel, and, if you spill any fuel on your skin, rinse it off immediately with soap and water. Mop up any spills immediately and do not store fuel-soaked rags where they could ignite. When you perform any kind of work on the fuel system, wear safety glasses and have a Class B type fire extinguisher on hand.*

1   Relieve the fuel system pressure (see Chapter 4).

2   Raise the vehicle and support it securely on jackstands.

3   Unscrew the fuel line-to-fuel filter fittings **(see illustration)**. Use a back-up wrench to keep from twisting the line.

4   If the car is equipped with plastic "quick-connect" fittings, squeeze the tabs on the fitting and pull it apart. If the car is equipped with metal quick-connect fittings, pull back the rubber boot and twist the fitting one-quarter turn to loosen any dirt, then, if compressed air is available, blow the dirt away from the connection. Use a special fuel line separator tool to disconnect the fitting. (See Chapter 4 for more information on the quick-connect fittings.)

5   Remove the filter from the clip.

6   Snap the new filter securely into the clip. Make sure the arrow on the filter points toward the engine.

7   Using new O-rings, reattach the fuel lines to the filter. Apply a few drops of clean engine oil to the O-rings to ease reassembly.

22.7 The serpentine drivebelt routing diagram is located on the belt guard on most models

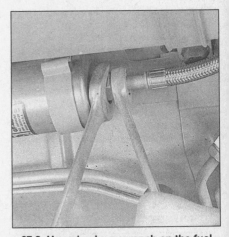

27.3 Use a back-up wrench on the fuel filter side of the fitting; use a flare-nut wrench to unscrew the outer nut

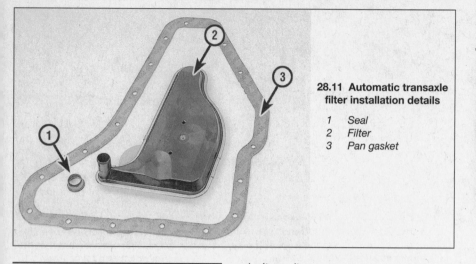

28.11 Automatic transaxle filter installation details

1 Seal
2 Filter
3 Pan gasket

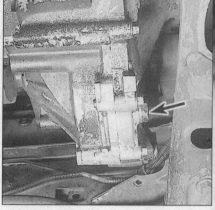

29.3 The manual transaxle drain plug (arrow) is located on the side of the case

28 Automatic transaxle fluid and filter change

Refer to illustration 28.11

1   At the specified time intervals, the transaxle fluid should be drained and replaced. Since the fluid will remain hot long after driving, perform this procedure only after everything has cooled down completely.
2   Before beginning work, purchase the specified transaxle fluid (see Recommended lubricants and fluids at the front of this Chapter) and a new filter.
3   Other tools necessary for this job include jackstands to support the vehicle in a raised position, a drain pan capable of holding several quarts, newspapers and clean rags.
4   Raise and support the vehicle on jackstands.
5   With a drain pan in place, remove the front and side transaxle pan mounting bolts.
6   Loosen the rear pan bolts one turn.
7   Carefully pry the transaxle pan loose with a screwdriver, allowing the fluid to drain.
8   Remove the remaining bolts, pan and gasket. Carefully clean the gasket surface of the transaxle to remove all traces of the old gasket and sealant.
9   Drain the fluid from the transaxle pan, clean the pan with solvent and dry it with compressed air. Be careful not to lose the magnet.
10   Remove the filter from the mount inside the transaxle.
11   Install a new filter and seal **(see illustration)**.
12   Make sure the gasket surface on the transaxle pan is clean, then install a new gasket. Put the pan in place against the transaxle and install the bolts. **Note:** *If the bolts are equipped with captive crush washers, they should be replaced.* Working around the pan, tighten each bolt a little at a time until the final torque figure is reached.
13   Lower the vehicle and add the specified amount of automatic transmission fluid through the filler tube (see Section 6).
14   With the shift lever in Park and the parking brake set, run the engine at a fast idle, but

don't race it.
15   Move the shift lever through each gear and back to Park. Check the fluid level.
16   Check under the vehicle for leaks during the first few trips.

29 Manual transaxle lubricant change

Refer to illustration 29.3

1   Raise the vehicle and support it securely on jackstands.
2   Move a drain pan, rags, newspapers and wrenches under the transaxle.
3   Remove the transaxle drain plug and allow the lubricant to drain into the pan **(see illustration)**.
4   After the lubricant has drained completely, reinstall the plug and tighten it securely.
5   Remove the transaxle dipstick. Using a hand pump, syringe or funnel, fill the transaxle with the correct amount of specified lubricant.
6   Lower the vehicle. Check the lubricant level as described in Section 16. Add more lubricant as necessary.

30 Cooling system servicing (draining, flushing and refilling)

Warning: *Make sure the engine is completely cool before performing this procedure.*
Caution: *Never mix green-colored ethylene glycol anti-freeze and orange-colored "DEX-COOL" silicate-free coolant because doing so will destroy the efficiency of the "DEX-COOL" coolant which is designed to last for 100,000 miles or five years.*
1   Periodically, the cooling system should be drained, flushed and refilled to replenish the antifreeze mixture and prevent formation of rust and corrosion, which can impair the performance of the cooling system and cause engine damage.
2   At the same time the cooling system is serviced, all hoses and the radiator cap

should be inspected and replaced if defective (see Section 9).
3   Since antifreeze is a corrosive and poisonous solution, be careful not to spill any of the coolant mixture on the vehicle's paint or your skin. If this happens, rinse it off immediately with plenty of clean water. Consult local authorities about the recycling of antifreeze before draining the cooling system. In many areas, reclamation centers have been set up to collect automobile oil and drained antifreeze/water mixtures, rather than allowing them to be added to the sewage system.
4   With the engine cold, remove the radiator cap.
5   On 2.5 liter four-cylinder engines, remove the thermostat.
6   On the 3800 V6, open the air bleed vents on the thermostat housing.
7   On the 3100 V6, open the air bleed vents on the thermostat housing and coolant pump.
8   On the 3.4L V6, open the air bleed vents on the thermostat housing and the heater coolant inlet pipe (near the brake master cylinder).
9   On the Quad-4 and on other V6 engines, open the vent plugs in the thermostat housing and (if equipped) bypass pipe.
10   If accessible, remove the engine block drains. **Note:** *On 3800 engines it will be necessary to remove the knock sensors from each side of the engine block to allow the block to drain.*
11   On the 3.4L engine, remove the inlet hose at the left side of the engine oil cooler.
12   Move a large container under the radiator to catch the coolant as it's drained.
13   Drain the radiator by removing (or opening) the plug at the bottom. If the plug is corroded and can't be turned easily, or if the radiator isn't equipped with a plug, disconnect the lower radiator hose to allow the coolant to drain. Be careful not to get antifreeze on your skin or in your eyes.
14   Disconnect the hose from the coolant reservoir and remove the reservoir. Flush it out with clean water.
15   Place a garden hose in the radiator filler neck and flush the system until the water

31.1 The PCV valve (arrow) on most engines is located in the valve cover

31.5 The PCV valve on 1995 and later 3800 engines is located at the front of the intake manifold below the MAP sensor (arrow)

31.13 Disconnect the electrical connector and unclip the MAP sensor from the PCV valve access cover

runs clear at all drain points.

16 In severe cases of contamination or clogging of the radiator, remove it (see Chapter 3) and reverse flush it. This involves inserting the hose in the bottom radiator outlet to allow the water to run against the normal flow, draining through the top. A radiator repair shop should be consulted if further cleaning or repair is necessary.

17 When the coolant is regularly drained and the system refilled with the correct antifreeze/water mixture, there should be no need to use chemical cleaners or descalers.

18 To refill the system, reconnect all hoses and the coolant reservoir.

19 Reinstall (or close) the radiator drain plug, and reinstall the engine block drains.

20 On V6 and Quad-4 models, fill the radiator with the recommended mixture of antifreeze and water (see Section 4) to the base of the filler neck, tighten the vent plugs securely but leave the radiator cap off.

21 On 2.5 liter four-cylinder engines, fill the cooling system through the thermostat housing until the level reaches the base of the radiator filler neck and install the cap. Fill the thermostat housing to one half inch from the top and install the thermostat and cap.

22 On all models add more coolant to the reservoir until it reaches the lower mark.

23 On 3.4L, 3100 and 3800 V6 engines, install the radiator cap and close all the air bleeds.

24 On other V6 engines and on the Quad 4, start the engine and run it until normal operating temperature is reached. With the engine idling, add additional coolant to the radiator and the reservoir. Install the radiator and reservoir caps.

25 Keep a close watch on the coolant level and the cooling system hoses during the first few miles of driving. Tighten the hose clamps and/or add more coolant as necessary. The coolant level should be a little above the HOT mark on the reservoir with the engine at normal operating temperature.

26 On 3.4L, 3100 and 3800 V6 engines, after flushing the system add two engine coolant supplement sealant pellets when refilling.

31 Positive Crankcase Ventilation (PCV) valve check and replacement

Check

All engines (except 1995 and later 3800 engines)

Refer to illustration 31.1

1 On most of the engines covered by this manual the PCV valve is located in the valve cover **(see illustration)**. On 3.4L engines the PCV valve is located in the intake manifold just behind the throttle body assembly.

2 With the engine idling at normal operating temperature, pull the valve (with hose attached) out of the rubber grommet in the intake plenum or valve cover.

3 Place your finger over the end of the valve. If there is no vacuum at the valve, check for a plugged hose, manifold port, or the valve itself. Replace any plugged or deteriorated hoses.

4 Turn off the engine and shake the PCV valve, listening for a rattle. If the valve doesn't rattle, replace it with a new one.

3800 engine (1995 and later)

Refer to illustration 31.5

5 On 3800 engines the PCV valve is located in the intake manifold under the MAP sensor at the front (passenger side) of the engine **(see illustration)**.

6 To check the valve it must first be removed (see Step 12). Then shake the PCV valve, listening for a rattle. If the valve doesn't rattle, replace it with a new one.

Replacement

All engines (except 1995 and later 3800 engines)

7 To replace the valve, pull it out of the end of the hose, noting its installed position and direction.

8 When purchasing a replacement PCV valve, make sure it's for your particular vehicle, model year and engine size. Compare the old valve with the new one to make sure they are the same.

9 Push the valve into the end of the hose until it's seated.

10 Inspect the rubber grommet for damage and replace it with a new one if necessary.

11 Push the PCV valve and hose securely into position.

3800 engine (1995 and later)

Refer to illustration 31.13, 31.14 and 31.15

12 Detach the fuel injector cover (see Chapter 4).

13 Disconnect the electrical connector from the MAP sensor and unclip the MAP sensor from the PCV valve access cover **(see illustration)**.

14 Press downward on the PCV valve access cover and rotate it counterclockwise to remove it **(see illustration)**.

15 To replace the valve, pull the PCV valve and O-ring assembly from the intake mani-

31.14 Press down on the PCV valve access cover, rotate it counterclockwise then pull up to remove it (be sure to check the O-ring and replace it if necessary)

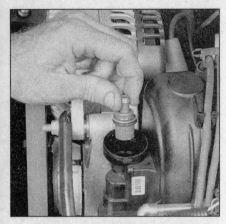

31.15 Remove the PCV valve and the O-ring from the intake manifold

fold noting its installed position and direction **(see illustration)**.

16   When purchasing a replacement PCV valve, make sure it's for your particular vehicle, model year and engine size. Compare the old valve with the new one to make sure they are the same.

17   Installation of the valve is the reverse of removal.

32 Evaporative emissions control system check

1   The function of the evaporative emissions control system is to draw fuel vapors from the gas tank and fuel system, store them in a charcoal canister and then burn them during normal engine operation.

2   The most common symptom of a fault in the evaporative emissions system is a strong fuel odor in the engine compartment. If a fuel odor is detected, inspect the charcoal canister, located at the front of the engine compartment. Check the canister and all hoses for damage and deterioration.

3   The evaporative emissions control system is explained in more detail in Chapter 6.

33 Spark plug replacement

Refer to illustrations 33.2, 33.5a, 33.5b, 33.6a, 33.6b, 33.6c, 33.6d, 33.8 and 33.10
Note: *On many V6 engines, it will be necessary to rotate the engine for access to the rear spark plugs, as outlined in Section 34.*

1   The spark plugs are located on the front (radiator) side of the engine on 2.2L and 2.5 liter four-cylinder models. The spark plugs on Quad-4 engines are located under a cover on the top of the engine, in the valley between the camshaft covers. On 3.4L V6 engines, the spark plugs are located in the valleys between the camshafts, three on each cylinder head. On other V6 engines, three plugs are located at the front and three at the rear (firewall) side of the engine. With the exception of the 3.4L and 3800 V6 engines, the

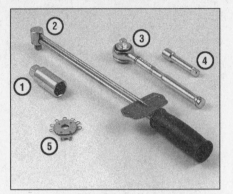

33.2 Tools required for changing spark plugs

1   **Spark plug socket** - *This will have special padding inside to protect the spark plug's porcelain insulator*
2   **Torque wrench** - *Although not mandatory, using this tool is the best way to ensure the plugs are tightened properly*
3   **Ratchet** - *Standard hand tool to fit the spark plug socket*
4   **Extension** - *Depending on model and accessories, you may need special extensions and universal joints to reach one or more of the plugs*
5   **Spark plug gap gauge** - *This gauge for checking the gap comes in a variety of styles. Make sure the gap for your engine is included*

engine must be rotated forward to gain access to rear spark plugs (see Section 34).
Note 1: *On 3.4L engines the upper intake manifold must be removed first to allow access to the right side (rear) spark plugs or plug wires. See Chapter 4 for the upper intake plenum removal procedure.*
Note 2: *On 1996 and later 3800 engines the acoustic engine cover must be removed first to access the right side (rear) spark plugs or plug wires (see Chapter 2F).*

2   In most cases, the tools necessary for spark plug replacement include a spark plug socket which fits onto a ratchet (spark plug sockets are padded inside to prevent damage to the porcelain insulators on the new plugs), various extensions and a gap gauge to check and adjust the gaps on the new plugs **(see illustration)**. A special plug wire removal tool is available for separating the wire boots from the spark plugs, and is a good idea on these models because the boots fit very tightly. On the 3800 V6 engine, a special tool is required to remove the spark plug heat shields from the right rear plugs. A torque wrench should be used to tighten the new plugs. Because the aluminum cylinder heads used on some of these models are easily damaged, allow the engine to cool before removing or installing the spark plugs.

3   The best approach when replacing the spark plugs is to purchase the new ones in advance, adjust them to the proper gap and replace the plugs one at a time. When buying the new spark plugs, be sure to obtain the

33.5a Spark plug manufacturers recommend using a wire-type gauge when checking the gap - if the wire does not slide between the electrodes with a slight drag, adjustment is required

33.5b To change the gap, bend the side electrode only, as indicated by the arrows, and be very careful not to crack or chip the porcelain insulator surrounding the center electrode

correct plug type for your particular engine. The plug type can be found in the Specifications at the front of this Chapter and on the Emission Control Information label located under the hood. If these two sources list different plug types, consider the emission control label correct.

4   Allow the engine to cool completely before attempting to remove any of the plugs. While you are waiting for the engine to cool, check the new plugs for defects and adjust the gaps.

5   Check the gap by inserting the proper thickness gauge between the electrodes at the tip of the plug **(see illustration)**. The gap between the electrodes should be the same as the one specified on the Emissions Control Information label. The wire should slide between the electrodes with a slight amount of drag. If the gap is incorrect, use the adjuster on the gauge body to bend the curved side electrode slightly until the proper gap is obtained **(see illustration)**. If the side electrode is not exactly over the center elec-

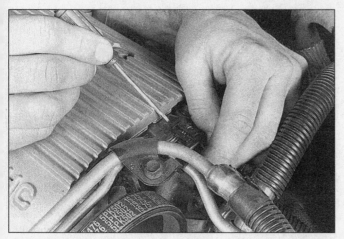

33.6a On Quad-4 engines, use a small screwdriver to detach the electrical connector

33.6b Remove the ignition cover bolts (arrows) (Quad-4 engine)

33.6c Lift the ignition cover straight up to detach the spark plug connectors

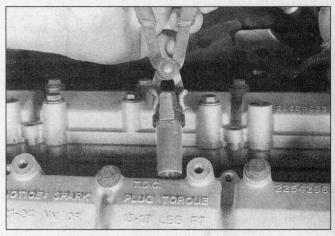

33.6d A spark plug boot removal tool will be required to remove the tight-fitting boots used on the Quad-4

trode, bend it with the adjuster until it is. Check for cracks in the porcelain insulator (if any are found, the plug should not be used).

6   On Quad-4 engines, unplug the electrical connector and remove the ignition cover for access to the spark plugs **(see illustrations)**. On all models, with the engine cool, remove the spark plug wire from one spark plug. Pull only on the boot at the end of the wire - do not pull on the wire **(see illustration)**. A plug wire removal tool should be used if available.

7   If compressed air is available, use it to blow any dirt or foreign material away from the spark plug hole. A common bicycle pump will also work. The idea here is to eliminate the possibility of debris falling into the cylinder as the spark plug is removed.

8   Place the spark plug socket over the plug and remove it from the engine by turning it in a counterclockwise direction **(see illustration)**.

9   Compare the spark plug to those shown in the accompanying photos to get an indication of the general running condition of the engine.

10   Thread one of the new plugs into the

33.8 Various sockets will be required to remove the spark plug on these models, because of access - here on the Quad-4, a long extension is needed to reach down into the camshaft valley

33.10 A length of 3/16-inch ID rubber hose will save time and prevent damaged threads when installing the spark plugs

hole until you can no longer turn it with your fingers, then tighten it with a torque wrench (if available) or the ratchet. It might be a good idea to slip a short length of rubber hose over

the end of the plug to use as a tool to thread it into place **(see illustration)**. The hose will grip the plug well enough to turn it, but will start to slip if the plug begins to cross-thread in the hole - this will prevent damaged

34.4 To reach the rear spark plugs, remove the strut bolts (arrows) and rotate the engine forward

34.5 Rotate the engine forward with a prybar and secure it in place

threads and the accompanying repair costs.

11   Before pushing the spark plug wire onto the end of the plug, inspect it following the procedures outlined in Section 35.

12   Attach the plug wire to the new spark plug, again using a twisting motion on the boot until it's seated on the spark plug.

13   Repeat the procedure for the remaining spark plugs, replacing them one at a time to prevent mixing up the spark plug wires.

34   Rotating the engine

Refer to illustration 34.4 and 34.5

1   Block the wheels, place the transaxle in neutral, and disconnect the negative battery cable. **Caution:** *On models equipped with the Theftlock audio system, be sure the lock-out feature is turned off before performing any procedure which requires disconnecting the battery.*

2   Disconnect the air duct.

3   Without disconnecting any hoses, remove the coolant recovery bottle and set it aside.

4   Remove the torque strut to engine bracket bolts and swing the struts aside. Place the bolt back into passenger side bracket **(see illustration)**.

5   Using a pry bar in the passenger side bracket (2.5L four-cylinder and V6 engines) or exhaust manifold (2.2L four-cylinder), rotate the engine forward and secure it in this position. To secure the engine in the forward (rotated) position on earlier models it will require placing a bolt through the slave hole on the driver's side torque strut and bracket. **(see Chapter 2B illustration 6.9).** To secure the engine in the forward (rotated) position on later models it will require connecting a ratchet strap to the engine and to the frame or radiator core support **(see illustration)**.

6   Reverse the procedure when you have completed the task that required rotating the engine. Torque the strut to engine bracket bolts to 50 ft-lbs.

35   Spark plug wire check and replacement

Note 1: *These models are equipped with distributorless ignition systems. The spark plug wires are connected directly to the ignition coils.*

Note 2: *On many V6 engines, it will be necessary to rotate the engine for access to the rear spark plugs or wires, as outlined in Section 34.*

Note 3: *Quad-4 engines are not equipped with spark plug wires.*

Note 4: *On 3.4L engines the upper intake manifold must be removed first to allow access to the right side (rear) spark plugs or plug wires. See Chapter 4 for the upper intake plenum removal procedure.*

Note 5: *On 1996 and later 3800 engines the acoustic engine cover must be removed first to access the right side (rear) spark plugs or plug wires (see Chapter 2F).*

1   The spark plug wires should be checked at the recommended intervals and whenever new spark plugs are installed in the engine.

2   The wires should be inspected one at a time to prevent mixing up the order, which is essential for proper engine operation.

3   Disconnect the plug wire from the spark plug. To do this, grab the rubber boot, twist slightly and pull the wire off. Do not pull on the wire itself, only on the rubber boot.

4   Check inside the boot for corrosion, which will look like a white crusty powder. Push the wire and boot back onto the end of the spark plug. It should be a tight fit on the plug. If it isn't, remove the wire and use pliers to carefully crimp the metal connector inside the boot until it fits securely on the end of the spark plug.

5   Using a clean rag, wipe the entire length of the wire to remove any built-up dirt and grease. Once the wire is clean, check for burns, cracks and other damage. Do not bend the wire excessively or pull the wire lengthwise - the conductor inside might break.

6   Disconnect the wire from the coil. Again, pull only on the rubber boot. Check for corrosion and a tight fit in the same manner as the spark plug end. Replace the wire at the coil.

7   Check the remaining spark plug wires one at a time, making sure they are securely fastened at the ignition coil and the spark plug when the check is complete.

8   If new spark plug wires are required, purchase a set for your specific engine model. Wire sets are available pre-cut, with the rubber boots already installed. Remove and replace the wires one at a time to avoid mix-ups in the firing order.

36   Chassis lubrication

Refer to illustrations 36.1, 36.3 and 36.10

1   Most suspension and steering joints are factory sealed and DO NOT require lubrication, although some model years do require the tie rod ends to be lubricated. Other various chassis components such as hood and door hinges, door locks, parking brake cable guides and transaxle shift linkage also require regularly scheduled lubrication. Specific materials and equipment are needed to perform these tasks **(see illustration)**.

2   To determine whether your vehicle will require tie rod lubrication, raise it with a jack and place jackstands under the frame. Make sure it is safely supported by the stands. If the wheels are to be removed at this interval for tire rotation or brake inspection, loosen the lug nuts slightly while the vehicle is still on the ground.

3   Look under the vehicle for grease fittings or plugs on the steering components **(see illustration)**. They are normally found on the tie-rod ends. If there are plugs, remove them and install grease fittings, which will thread into the component. An automotive parts store will be able to supply the correct fittings. Straight, as well as angled, fittings are available. If no grease fittings are found proceed to Step 9.

4   Refer to *Recommended lubricants and*

36.1 Materials required for chassis and body lubrication

1   **Engine oil** - *Light engine oil in a can like this can be used for door and hood hinges*
2   **Graphite spray** - *Used to lubricate lock cylinders*
3   **Grease** - *Grease, in a variety of types and weights, is available for use in a grease gun. Check the Specification for your requirements*
4   **Grease gun** - *A common grease gun, shown here with a detachable hose and nozzle, is needed for chassis lubrication. After use, clean it thoroughly!*

fluids at the front of this Chapter to obtain the necessary grease, etc.
5   Before beginning, force a little grease out of the nozzle to remove any dirt from the end of the gun. Wipe the nozzle clean with a rag.
6   With the grease gun and plenty of clean rags, crawl under the vehicle and begin lubricating the components.
7   Wipe the area around the grease fitting free of dirt, then squeeze the trigger on the grease gun to force grease into the component. Continue pumping grease into the fitting until it oozes out of the joint between the two components. If it escapes around the grease gun nozzle, the fitting is clogged or the nozzle is not completely seated on the fitting. Resecure the gun nozzle to the fitting and try again. If necessary, replace the fitting with a new one.
8   Wipe the excess grease from the components and the grease fitting. Repeat the procedure for the remaining fittings.

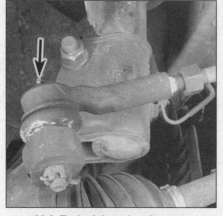

36.3 Typical tie rod end grease fitting (arrow)

9   While under the vehicle clean and lubricate the parking brake cable guides and levers. This can be done by using a grease gun. **Note:** *DO NOT lubricate the parking brake cables, as this will deteriorate the plastic coating used to prevent the cables from rusting.*
10   Clean and lubricate the transaxle shift linkage with engine oil **(see illustration)**.
11   Open the hood and smear a little chassis grease on the hood latch mechanism. Have an assistant pull the hood release lever from inside the vehicle as you lubricate the cable at the latch.
12   Lubricate all the hinges (door, hood, etc.) with engine oil to keep them in proper working order.
13   The key lock cylinders can be lubricated with spray graphite or silicone lubricant, which is available at auto parts stores.
14   Lubricate the door weatherstripping with silicone spray. This will reduce chafing and retard wear.

---

37   Throttle linkage check

Refer to illustration 37.2
1   At the specified intervals the throttle linkage should be inspected for kinks, binding and misalignment.
2   Starting at the throttle body in the engine compartment, with the engine OFF, grasp the throttle lever and open it to the full throttle position **(see illustration)**. Quickly release your hand from the throttle lever and note the amount of time it takes the throttle lever to return to the idle position. If the lever returns quickly to the idle position the throttle

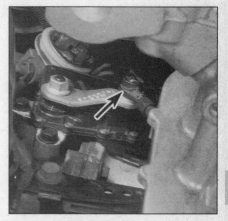

36.10 Lubricate the transaxle shift linkage (arrow) with clean engine oil

37.2 Check the throttle linkage for binding

linkage is in proper working order. If the lever returns slowly to the idle position, inspect the accelerator cable or cruise control cable for kinks or signs of binding. Also check the cable retaining brackets for missing retaining clips. **Note:** *Never lubricate the accelerator or cruise control cable as this will destroy the protective plastic coating on the outside of the cable.* If signs of kinks or binding exist in the cable assembly, replace the cable as described in Chapter 4.
3   If there's no evidence of binding in the cable assembly, remove the cables from the throttle lever and repeat the test described in Step 2. If the throttle lever returns slowly to the idle position with the cables detached, the problem lies in the throttle body assembly. See Chapter 4 for further inspection of the throttle body assembly.

Notes

Chapter 2 Part A
2.2 liter four-cylinder engine

Contents

Specifications

General

Cylinder numbers (drivebelt end-to-transaxle end)	1-2-3-4
Firing order	1-3-4-2

2.2 Liter four-cylinder engine

Cylinder and coil terminal location

Torque specifications

	Ft-lbs (unless otherwise indicated)
Camshaft sprocket bolt	77
Crankshaft pulley-to-hub bolts	37
Crankshaft pulley center bolt	77
Cylinder head bolts	
Step 1	
Short bolts	43
Long bolts	46
Step 2	Rotate all bolts an additional 90-degrees
Exhaust manifold fasteners	115 in-lbs
Flywheel bolts	54
Driveplate bolts	52
Intake manifold fasteners	
Upper	22
Lower	24
Oil pan nuts	89 in-lbs
Oil pump mounting bolt	32
Valve cover bolts	89 in-lbs
Rocker arm nuts	22
Timing chain cover bolts	97 in-lbs
Timing chain tensioner bolts	18

1 General information

This Part of Chapter 2 is devoted to in-vehicle repair procedures for the 2.2 liter four-cylinder overhead valve engine. The 2.2L engine became available in 1993, in the Lumina model only. These engines have cast iron blocks and cast aluminum pistons. The aluminum cylinder head has replaceable valve seats and guides. Stamped steel rocker arms and tubular pushrods actuate the valves. To positively identify this engine, locate the vehicle identification number on the left front corner of the instrument panel. This plate is visible from outside the vehicle (**see illustration 1.1** in Chapter 2, Part B). The eighth character in the sequence is the engine designation:

4 = 2.2L four-cylinder, overhead valve engine

All information concerning engine removal and installation and engine block and cylinder head overhaul can be found in Part G of this Chapter.

The following repair procedures are based on the assumption the engine is in the

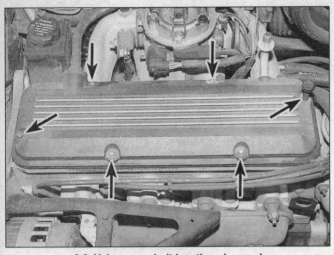

3.6 Valve cover bolt locations (arrows)

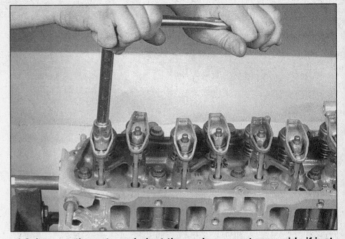

4.2 Loosen the nuts and pivot the rocker arms to one side if just the pushrods are being removed (otherwise, remove the nuts and lift off the pivot balls and rocker arms)

vehicle. If the engine has been removed from the vehicle and mounted on a stand, many of the steps outlined in this Part of Chapter 2 will not apply.

The Specifications included in this Part of Chapter 2 apply only to the procedures contained in this Part. Part G of Chapter 2 contains the Specifications necessary for cylinder head and engine block rebuilding.

2   Repair operations possible with the engine in the vehicle

Many major repair operations can be accomplished without removing the engine from the vehicle.

Clean the engine compartment and the exterior of the engine with some type of degreaser before any work is done. It'll make the job easier and help keep dirt out of the internal areas of the engine.

Depending on the components involved, it may be helpful to remove the hood to improve access to the engine as repairs are performed (refer to Chapter 11 if necessary). Cover the fenders to prevent damage to the paint. Special pads are available, but an old bedspread or blanket will also work.

If vacuum, exhaust, oil or coolant leaks develop, indicating a need for gasket or seal replacement, the repairs can generally be made with the engine in the vehicle. The intake and exhaust manifold gaskets, timing chain cover gasket, oil pan gasket, crankshaft oil seals and cylinder head gasket are all accessible with the engine in place.

Exterior engine components, such as the intake and exhaust manifolds, the oil pan (and the oil pump), the water pump, the starter motor, the alternator and the fuel system components can be removed for repair with the engine in place.

Since the cylinder head can be removed without pulling the engine, valve component servicing can also be accomplished with the engine in the vehicle. Replacement of the

timing chain and sprockets is also possible with the engine in the vehicle.

In extreme cases caused by a lack of necessary equipment, repair or replacement of piston rings, pistons, connecting rods and rod bearings is possible with the engine in the vehicle. However, this practice is not recommended because of the cleaning and preparation work that must be done to the components involved.

3   Valve cover - removal and installation

Refer to illustration 3.6

Removal

1   Remove the air cleaner assembly, tagging each hose to be disconnected with a piece of numbered tape to simplify installation.
2   Remove the breather hose from the valve cover.
3   Remove the spark plug wires from the spark plugs and from the valve cover clips.
4   Detach the control cable bracket at the intake plenum and valve cover.
5   Disconnect the cables at the throttle body.
6   Remove the valve cover bolts **(see illustration).**
7   Detach the valve cover from the head.
Note: *If the cover is stuck to the cylinder head, use a block of wood and hammer to dislodge it. If that doesn't work, try to slip a flexible putty knife between the head and cover to break the gasket seal. Don't pry at the cover-to-head joint or damage to the sealing surfaces may occur (leading to oil leaks in the future).*

Installation

8   The mating surfaces of the cylinder head and valve cover must be perfectly clean when the cover is installed. Use a gasket scraper to remove all traces of sealant or old

gasket, then clean the mating surfaces with lacquer thinner or acetone (if there's sealant or oil on the mating surfaces when the cover is installed, oil leaks may develop). The head and cover are made of aluminum, so be extra careful not to nick or gouge the mating surfaces with the scraper.
9   Clean the mounting bolt threads with a die if necessary to remove any corrosion and restore damaged threads. Make sure the threaded holes in the head are clean - run a tap into them if necessary to remove corrosion and restore damaged threads.
10   Apply a thin coat of RTV-type sealant to the sealing flange on the cover and install a new gasket.
11   Place the valve cover on the cylinder head and install the mounting bolts. Tighten the bolts a little at a time to the torque listed in this Chapter's Specifications. Work from the center out in a spiral pattern.
12   Complete the installation by reversing the removal procedure.

4   Rocker arms and pushrods - removal, inspection and installation

Refer to illustrations 4.2, 4.4a and 4.4b

Removal

1   Refer to Section 3 and detach the valve cover from the cylinder head.
2   Beginning at the front of the cylinder head, loosen the rocker arm nuts **(see illustration). Note:** *If the pushrods are the only items being removed, rotate the rocker arms to one side so the pushrods can be lifted out.*
3   Remove the nuts, the rocker arms and the pivot balls and store them in marked containers (they must be reinstalled in their original locations).
4   Remove the pushrods and store them separately to make sure they don't get mixed up during installation **(see illustrations).**
5   If the pushrod guides must be removed

4.4a The pushrods can be lifted straight out

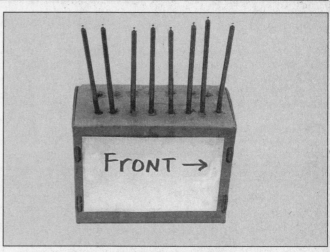

4.4b A perforated cardboard box can be used to store the pushrods to ensure they're reinstalled in their original locations - note the label indicating the front (drivebelt end) of the engine

for any reason, make sure they're marked so they can be reinstalled in their original locations.

Inspection

6   Check each rocker arm for wear, cracks and other damage, especially where the pushrods and valve stems contact the rocker arm faces.
7   Make sure the hole at the pushrod end of each rocker arm is open.
8   Check each rocker arm pivot area for wear, cracks and galling. If the rocker arms are worn or damaged, replace them with new ones and use new pivot balls as well.
9   Inspect the pushrods for cracks and excessive wear at the ends. Roll each pushrod across a piece of plate glass to see if it's bent (if it wobbles, it's bent).

Installation

10   Lubricate the lower ends of the pushrods with clean engine oil or moly-base grease and install them in their original locations. Make sure each pushrod seats completely in the lifter socket.
11   Apply moly-base grease to the ends of the valve stems and the upper ends of the pushrods before positioning the rocker arms and installing the nuts.
12   Set the rocker arms in place, then install the pivot balls and nuts. Apply moly-base grease to the pivot balls to prevent damage to the mating surfaces before engine oil pressure builds up. Tighten the nuts to the torque listed in this Chapter's Specifications.

5   Valve springs, retainers and seals - replacement

Refer to illustrations 5.4, 5.9 and 5.17
Note: Broken valve springs and defective valve stem seals can be replaced without removing the cylinder head. Two special tools

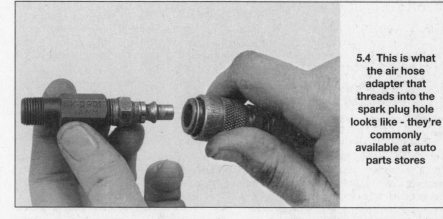

5.4 This is what the air hose adapter that threads into the spark plug hole looks like - they're commonly available at auto parts stores

and a compressed air source are normally required to perform this operation, so read through this Section carefully and rent or buy the tools before beginning the job. If compressed air isn't available, a length of nylon rope can be used to keep the valves from falling into the cylinder during this procedure.
1   Refer to Section 3 and remove the valve cover.
2   Remove the spark plug from the cylinder which has the defective component. If all of the valve stem seals are being replaced, all of the spark plugs should be removed.
3   Turn the crankshaft until the piston in the affected cylinder is at top dead center on the compression stroke (refer to Chapter 2, Part G, for instructions). If you're replacing all of the valve stem seals, begin with cylinder number one and work on the valves for one cylinder at a time. Move from cylinder-to-cylinder following the firing order sequence (see this Chapter's Specifications).
4   Thread an adapter into the spark plug hole (see illustration) and connect an air hose from a compressed air source to it. Most auto parts stores can supply the air hose adapter. Note: Many cylinder compression gauges utilize a screw-in fitting that may work with your air hose quick-disconnect fitting.

5   Remove the nut, pivot ball and rocker arm for the valve with the defective part and pull out the pushrod. If all of the valve stem seals are being replaced, all of the rocker arms and pushrods should be removed (refer to Section 4).
6   Apply compressed air to the cylinder. Warning: The piston may be forced down by compressed air, causing the crankshaft to turn suddenly. If the wrench used when positioning the number one piston at TDC is still attached to the bolt in the crankshaft nose, it could cause damage or injury when the crankshaft moves.
7   The valves should be held in place by the air pressure. If the valve faces or seats are in poor condition, leaks may prevent air pressure from retaining the valves - a "valve job" is necessary to correct this problem.
8   If you don't have access to compressed air, an alternative method can be used. Position the piston at a point just before TDC on the compression stroke, then feed a long piece of nylon rope through the spark plug hole until it fills the combustion chamber. Be sure to leave the end of the rope hanging out of the engine so it can be removed easily. Use a large ratchet and socket to rotate the crankshaft in the normal direction of rotation (clockwise) until slight resistance is felt.

5.9 Once the spring is compressed, remove the keepers with a magnet or needle-nose pliers

5.17 Keepers don't always stay in place, so apply a small dab of grease to each one, as shown here, before installation - it'll hold them in place on the valve stem as the spring is released

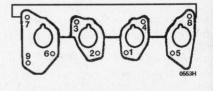

6.8 Intake manifold bolt tightening sequence

9   Stuff shop rags into the cylinder head holes above and below the valves to prevent parts and tools from falling into the engine, then use a valve spring compressor to compress the spring. Remove the keepers with small needle-nose pliers or a magnet **(see illustrations)**. **Note:** *A couple of different types of tools are available for compressing the valve springs with the head in place. One type utilizes the rocker arm stud and nut for leverage, while the other type grips the lower spring coils and presses on the retainer as the knob is turned. Both types work very well, although the lever type is usually less expensive.*

10   Remove the spring retainer and valve spring, then remove the valve guide seal. **Note:** *If air pressure fails to hold the valve in the closed position during this operation, the valve face or seat is probably damaged. If so, the cylinder head will have to be removed for additional repair operations.*

11   Wrap a rubber band or tape around the top of the valve stem so the valve won't fall into the combustion chamber, then release the air pressure. **Note:** *If a rope was used instead of air pressure, turn the crankshaft slightly in the direction opposite normal rotation.*

12   Inspect the valve stem for damage. Rotate the valve in the guide and check the end for eccentric movement, which would indicate the valve stem is bent.

13   Move the valve up-and-down in the guide and make sure it doesn't bind. If the valve stem binds, either the valve is bent or the guide is damaged. In either case, the head will have to be removed for repair.

14   Reapply air pressure to the cylinder to retain the valve in the closed position, then remove the tape or rubber band from the valve stem. If a rope was used instead of air pressure, rotate the crankshaft in the normal direction of rotation until slight resistance is felt.

15   Lubricate the valve stem with engine oil and install a new valve guide seal. **Note:** *Intake and exhaust valve seals are different.*

16   Install the spring in position over the valve.

17   Install the valve spring retainer. Com-

press the valve spring and carefully install the keepers in the groove. Apply a small dab of grease to the inside of each keeper to hold it in place if necessary **(see illustration)**. Remove the pressure from the spring tool and make sure the keepers are seated.

18   Disconnect the air hose and remove the adapter from the spark plug hole. If a rope was used in place of air pressure, pull it out of the cylinder.

19   Refer to Section 4 and install the rocker arm(s) and pushrod(s).

20   Install the spark plug(s) and hook up the wire(s).

21   Refer to Section 3 and install the valve cover.

22   Start and run the engine, then check for oil leaks and unusual sounds coming from the valve cover area.

6   Intake manifold - removal and installation

Refer to illustration 6.8
Note: *This procedure may be easier with the engine rotated forward (see Chapter 1).*
1   Relieve the fuel pressure (see Chapter 4), then disconnect the negative battery cable from the battery. **Caution:** *On models equipped with the Theftlock audio system, be sure the lockout feature is turned off before performing any procedure which requires disconnecting the battery.*
2   Remove the upper intake manifold assembly (see Chapter 4).
3   Raise the front of the vehicle and support it securely on jackstands. Drain the coolant (refer to Chapter 1).
4   If necessary for clearance, remove the power steering pump (if equipped) and tie it aside in an upright position (see Chapter 10).
5   Remove the heater hose and coolant line retaining nut near the bottom of the intake manifold.
6   Remove the mounting nuts from the intake manifold.
7   Separate the intake manifold and gasket from the engine. Scrape all traces of gasket material off the intake manifold and head

gasket mating surfaces. When scraping, be careful not to scratch or gouge the delicate aluminum gasket surfaces on the head and manifold. Clean the surfaces with a rag soaked in lacquer thinner or acetone.

8   Installation is the reverse of removal. Be sure to use a new gasket. Tighten the nuts/bolts to the torque listed in this Chapter's Specifications **(see illustration)**.

9   Add coolant, run the engine and check for leaks and proper operation.

7   Exhaust manifold - removal and installation

Refer to illustration 7.11
Caution: *Allow the engine to cool completely before beginning this procedure.*

1   Disconnect the negative cable from the battery. **Caution:** *On models equipped with the Theftlock audio system, be sure the lockout feature is turned off before performing any procedure which requires disconnecting the battery.*

2   Unplug the oxygen sensor lead.

3   Remove the air cleaner and air duct assembly.

4   Remove the air-inlet resonator from the upper tie bar.

5   Remove the lower air duct.

6   Disconnect the upper engine mounting torque strut from the engine.

7   Remove the nuts and bolts and discon-

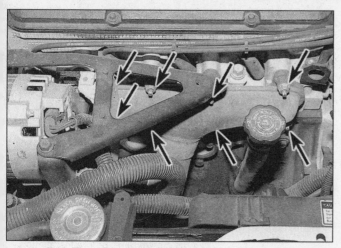

7.11 The exhaust manifold bolt locations (arrows)

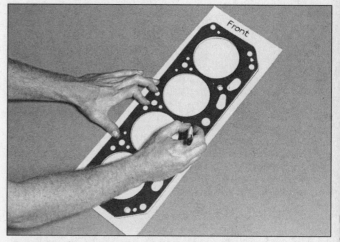

8.11a To avoid mixing up the head bolts, use the new gasket to transfer the bolt hole pattern to a piece of cardboard, then punch holes in the cardboard to accept the bolts

nect the engine strut bracket from the cylinder head.

8   Remove the drivebelt, the alternator and the alternator rear support bracket (see Chapters 1 and 5).

9   Remove the oil level dipstick and dipstick holder.

10   Raise the front of the vehicle, support it securely on jackstands and apply the parking brake. Block the rear wheels to keep the vehicle from rolling off the jackstands. Unbolt the exhaust pipe from the manifold. Lower the vehicle.

11   Remove the exhaust manifold-to-cylinder head nuts/bolts **(see illustration)**, pull the manifold off the engine and lift it out of the exhaust pipe flange. Remove and discard the gasket.

12   Scrape all traces of gasket material off the exhaust manifold and cylinder head mating surfaces. Be careful not to scratch or gouge the delicate aluminum cylinder head or exhaust leaks will develop. Wipe the surfaces clean with a rag soaked in lacquer thinner or acetone.

13   Clean all bolt and stud threads before installation. A wire brush can be used on the manifold mounting studs, while a tap works well when cleaning the cylinder head bolt holes.

14   If a new manifold is being installed, transfer the oxygen sensor from the old manifold to the new one.

15   Installation is the reverse of removal. Be sure to use a new gasket and tighten the nuts/bolts to the torque listed in this Chapter's Specifications. Work in a spiral pattern from the center out.

8   Cylinder head - removal and installation

Caution: *Allow the engine to cool completely before beginning this procedure.*

Note: *On vehicles with high mileage and dur-*

8.11b Loosen the cylinder head bolts in 1/4-turn increments to avoid warping the head

ing an engine overhaul, camshaft lobe height should be checked prior to cylinder head removal (see Chapter 2, Part G for instructions).

Removal

Refer to illustrations 8.11a and 8.11b

1   Relieve the fuel pressure (see Chapter 4). Disconnect the cable from the negative battery terminal. **Caution:** *On models equipped with the Theftlock audio system, be sure the lockout feature is turned off before performing any procedure which requires disconnecting the battery.*

2   Remove the alternator and brackets as described in Chapter 5.

3   Remove the intake manifold as described in Section 6.

4   Remove the exhaust manifold as described in Section 7.

5   Unbolt the drivebelt tensioner bracket.

6   Unbolt the power steering pump (if equipped) and set it aside without disconnecting the hoses (see Chapter 10).

7   Disconnect any remaining wires, hoses, fuel and vacuum lines from the cylinder head. Be sure to label them to simplify reinstallation.

8   Disconnect the spark plug wires and remove the spark plugs. Be sure the plug wires are labeled to simplify reinstallation.

9   Remove the valve cover (see Section 3).

10   Remove the rocker arms and pushrods (see Section 4).

11   Using the new head gasket, outline the cylinders and bolt pattern on a piece of cardboard **(see illustration)**. Be sure to indicate the front of the engine for reference. Punch holes at the bolt locations. Loosen each of the cylinder head mounting bolts 1/4-turn at a time until they can be removed by hand **(see illustration)**. Store the bolts in the cardboard holder as they're removed - this will ensure they are reinstalled in their original locations, which is absolutely essential.

12   Lift the head off the engine. If it's stuck, don't attempt to pry it off - you could damage the sealing surfaces. Instead, use a hammer and block of wood to tap the head and break the gasket seal. Place the head on a block of wood to prevent damage to the gasket surface.

13   Remove the cylinder head gasket.

14   Refer to Chapter 2, Part G, for cylinder head disassembly and valve service procedures.

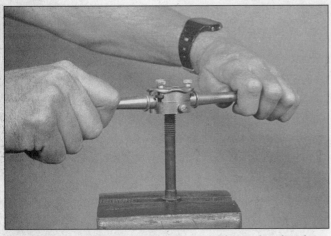

8.17 A die should be used to remove corrosion and sealant from the head bolt threads prior to installation

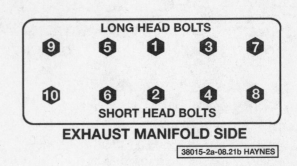

LONG HEAD BOLTS

9   5   1   3   7

10   6   2   4   8

SHORT HEAD BOLTS

EXHAUST MANIFOLD SIDE

38015-2a-08.21b HAYNES

8.21 Cylinder head bolt tightening sequence – note that the rear head bolts (long bolts) have a higher specified torque than the front bolts

10.4a Crankshaft pulley installation details

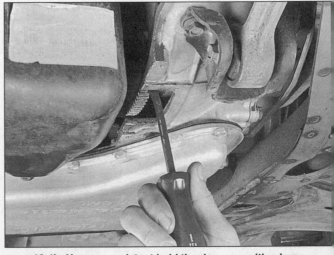

10.4b Have an assistant hold the ring gear with a large screwdriver as the pulley-to-crankshaft bolt is loosened/tightened

Installation

Refer to illustrations 8.17 and 8.21

15   If a new cylinder head is being installed, transfer all external parts from the old cylinder head to the new one.

16   If not already done, thoroughly clean the gasket surfaces on the cylinder head and the engine block. Do not gouge or otherwise damage the soft aluminum gasket surfaces.

17   To get the proper torque readings, the threads of the head bolts must be clean **(see illustration)**. This also applies to the threaded holes in the engine block. Run a tap through the holes to ensure they are clean.

18   Place the gasket in position over the engine block dowel pins. Note any marks like "THIS SIDE UP" and install the gasket accordingly.

19   Carefully lower the cylinder head onto the engine, over the dowel pins and the gasket.

20   Install the bolts finger tight. Don't tighten any of the bolts at this time.

21   Tighten each of the bolts in 1/4-turn increments in the recommended sequence

(see illustration). A special tool must be used on the head bolts. Note that the rear bolts have a different torque specification than the front bolts. Continue tightening in the recommended sequence until the torque (and angle of rotation) specified in this Chapter is reached. Mark each bolt with a felt-tip marker each time you tighten it to make sure none of the bolts have been left out of the sequence.

22   The remaining installation steps are the reverse of removal.

23   Be sure to refill the cooling system and change the oil and filter (see Chapter 1).

9   Hydraulic lifters - removal, inspection and installation

This procedure is essentially the same as for the 2.5L four-cylinder engine (see Chapter 2, Part B), except there's no pushrod cover to remove. You'll need a long magnetic tool, scribe or lifter removal tool, since you'll have to reach down into the cylinder block to remove the lifters.

10   Crankshaft pulley - removal and installation

Refer to illustrations 10.4a, 10.4b and 10.6

Removal

1   Remove the cable from the negative battery terminal. **Caution:** *On models equipped with the Theftlock audio system, be sure the lockout feature is turned off before performing any procedure which requires disconnecting the battery.*

2   Remove the drivebelt (see Chapter 1).

3   Raise the front of the vehicle and support it securely on jackstands. Remove the right front tire and the inner fender splash shield.

4   Remove the crankshaft pulley-to-crankshaft bolt **(see illustration)**. A breaker bar will probably be needed, since the bolt is very tight. If necessary, remove the lower bellhousing cover and insert a large screwdriver into the teeth of the flywheel/driveplate ring gear to prevent the crankshaft from turning **(see illustration)**.

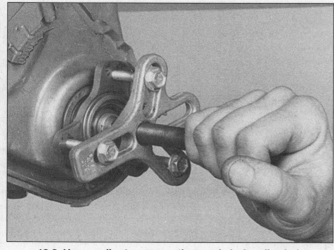

10.6 Use a puller to remove the crankshaft pulley hub

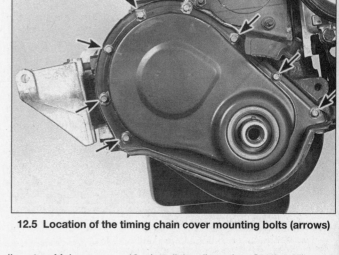

12.5 Location of the timing chain cover mounting bolts (arrows)

5   Remove the three bolts that attach the pulley to the hub.
6   Using a puller, remove the crankshaft pulley or hub from the crankshaft **(see illustration)**.

Installation

7   Refer to Section 11 for the oil seal replacement procedure.
8   Apply a thin layer of clean multi-purpose grease to the seal contact surface of the hub.
9   Position the hub on the crankshaft and slide it through the seal until it bottoms against the crankshaft gear. Note that the slot (keyway) in the hub must be aligned with the Woodruff key in the end of the crankshaft. The pulley-to-crankshaft bolt can be used to press the hub into position.
10   Tighten the pulley-to-crankshaft bolt to the torque listed in this Chapter's Specifications.
11   The remaining installation steps are the reverse of removal.

11   Crankshaft front oil seal - replacement

1   Remove the crankshaft pulley (see Section 10).
2   Pry the old oil seal out with a seal removal tool or a screwdriver. Be very careful not to nick or otherwise damage the crankshaft in the process. Wrap the screwdriver tip with vinyl tape to protect the crankshaft.
3   Apply a thin coat of RTV-type sealant to the outer edge of the new seal. Lubricate the seal lip with moly-base grease or clean engine oil.
4   Place the seal squarely in position in the bore and drive it into place with special tool J-35468 (or equivalent).
5   If you don't have the special, carefully tap the seal into place with a large socket or piece of pipe and a hammer. The outer diameter of the socket or pipe should be the same

size as the seal outer diameter. Make sure the seal is seated completely in the bore.
6   Install the crankshaft pulley (see Section 10).
7   Reinstall the remaining parts in the reverse order of removal.
8   Start the engine and check for oil leaks at the seal.

12   Timing chain cover - removal and installation

Refer to illustration 12.5

1   Remove the crankshaft pulley (and hub, if equipped) as described in Section 10.
2   Remove the oil pan (see Section 15).
3   Remove the bolts holding the power steering pump and position it aside.
4   Remove the alternator bolts and position it aside.
5   Remove the drivebelt tensioner and bracket and the timing chain cover-to-block bolts, then detach the cover **(see illustration)**.
6   Using a scraper and degreaser, remove all old gasket material from the sealing surfaces of the timing chain cover, engine block and oil pan.
7   If necessary, replace the front oil seal by carefully prying it out of the cover with a large screwdriver. Don't distort the cover.
8   Install the new seal with the spring side toward the inside of the cover. Drive the seal into place using a seal installation tool or a large socket and hammer. A block of wood will also work.
9   Use a thin coat of RTV-type sealant to position a new gasket on the timing chain cover.
10   Place the cover in position over the dowel pins on the block.
11   Install the bolts that secure the cover to the block, then tighten all of them to the torque listed in this Chapter's Specifications. Follow a criss-cross pattern to avoid distorting the cover.

12   Install the oil pan (see Section 15).
13   Complete the installation by reversing the removal procedure.

13   Timing chain and sprockets - inspection, removal and installation

Refer to illustrations 13.9 and 13.18

Inspection

1   Disconnect the cable from the negative battery terminal. **Caution:** *On models equipped with the Theftlock audio system, be sure the lockout feature is turned off before performing any procedure which requires disconnecting the battery.*
2   Remove the crankshaft pulley (see Section 10).
3   Remove the timing chain cover (see Section 12).
4   Before removing the chain and sprockets, visually inspect the teeth on the sprockets for wear and the chain for excessive slack. Also check the condition of the timing chain tensioner.
5   If either or both sprockets show any signs of wear (edges on the teeth of the camshaft sprocket not "square," bright or blue areas on the teeth of either sprocket, chipping, pitting, etc.), they should be replaced with new ones. Wear in these areas is very common.
6   Failure to replace a worn timing chain may result in erratic engine performance, loss of power and lowered fuel mileage.
7   If any one component requires replacement, all related components, including the tensioner, should be replaced as well.
8   If the chain and sprockets must be replaced, proceed as follows.

Removal

9   Temporarily install the bolt in the end of the crankshaft. Turn the engine over using

13.9 Align the camshaft and crankshaft sprocket marks as shown here - note that the marks will be positioned differently depending on which piston (number 1 or number 4) is at TDC on the compression stroke

13.18 Compress the timing chain tensioner and insert a drill bit or nail into the hole to retain the tensioner in its retracted position

the bolt until the marks on the camshaft and crankshaft line up (see illustration). Note: *Do not attempt to remove the timing chain until this is done and do not turn the crankshaft or camshaft while the sprockets/chain are off.*

10   Remove the timing chain tensioner upper bolt. Loosen the timing chain tensioner Torx bolt as far as possible but don't remove it.

11   Remove the camshaft sprocket retaining bolt and detach the camshaft sprocket and timing chain. It may be necessary to tap the sprocket with a soft-face hammer to dislodge it.

12   If the crankshaft sprocket must be removed, it can be drawn off the crankshaft with a puller.

Installation

13   Lubricate the thrust side with moly-base grease and install the crankshaft sprocket (if removed) on the crankshaft using a special tool. Be sure the Woodruff key is aligned with the keyway as the sprocket is installed. If the special tool is unavailable, drive the sprocket into place using a section of pipe just large enough to fit over the nose of the crankshaft.

14   Slip the timing chain onto the camshaft sprocket.

15   With the timing marks aligned, slip the chain over the crankshaft sprocket. Align the dowel in the camshaft with the dowel hole in the camshaft sprocket and install the sprocket on the camshaft. Draw the camshaft sprocket into place with the retaining bolt and tighten it to the torque listed in this Chapter's Specifications. Caution: *Do not hammer or attempt to drive the camshaft sprocket into place - it could dislodge the welch plug at the rear of the engine.*

16   With the chain and both sprockets in place, check again to make sure the timing marks on the two sprockets are properly aligned (see illustration 13.9). If not, remove the camshaft sprocket and move the chain until they are.

17   Lubricate the chain with clean engine oil.

18   The timing chain tensioner spring must be compressed using a special tool, prior to installation. Insert a nail or cotter pin into hole A (see illustration) to hold the spring in place during installation. Remove the nail or pin after installation.

19   Install the remaining components in the reverse order of removal.

14   Camshaft and bearings - removal, inspection and installation

Due to the fact the engine is mounted transversely in the vehicle, there isn't enough room to remove the camshaft with the engine in place. Therefore, the procedure is covered in Chapter 2, Part G.

15   Oil pan - removal and installation

1   Warm up the engine, then drain the oil and remove the oil filter (see Chapter 1).

2   Detach the cable from the negative battery terminal. Caution: *On models equipped with the Theftlock audio system, be sure the lockout feature is turned off before performing any procedure which requires disconnecting the battery.*

3   Remove the air cleaner and air-duct assembly.

4   Remove the drivebelt (see Chapter 1).

5   Disconnect the upper engine mounting strut from the engine bracket.

6   Install engine support fixture or, by other suitable means, ensure that the engine is properly supported.

7   Loosen the right front wheel lug nuts, then raise the vehicle and support it securely on jackstands.

8   Remove the right front wheel and tire.

9   Remove the right-side engine splash shield.

10   Remove the exhaust pipe and catalytic converter.

11   On air conditioned models, remove the air conditioner brace at the starter and compressor bracket.

12   Remove the starter and bracket (see Chapter 5).

13   Remove the lower bellhousing cover.

14   On air conditioned models, remove the air conditioner brace.

15   Remove the four right support bolts. Lower the support slightly to gain clearance for oil pan removal.

16   Remove the oil filter extension (automatic transaxle equipped models only).

17   Remove the bolts and nuts securing the oil pan to the engine block.

18   Tap on the pan with a soft-face hammer to break the gasket seal, then detach the oil pan from the engine.

19   Using a gasket scraper, remove all traces of old gasket and/or sealant from the engine block and oil pan. Make sure the threaded bolt holes in the block are clean. Wash the oil pan with solvent and dry it thoroughly.

20   Check the gasket flanges for distortion, particularly around the bolt holes. If necessary, place the pan on a block of wood and use a hammer to flatten and restore the gasket surfaces. Clean the mating surfaces with lacquer thinner or acetone.

21   Place a 2 mm diameter bead of RTV sealant on the oil pan-to-block sealing flanges and the oil pan-to-front cover surface.

22   Apply a thin coat of RTV sealant to the ends of the rear oil pan seal down to the ears. Press the oil pan seal into position.

23   Carefully place the oil pan against the block.

24   Install the bolts/nuts and tighten them in 1/4-turn increments to the torque listed in this Chapter's Specifications. Start with the bolts closest to the center of the pan and

17.3 Most flywheels and driveplates have locating dowels (arrow) - if the one you're working on doesn't, make some marks to ensure correct installation

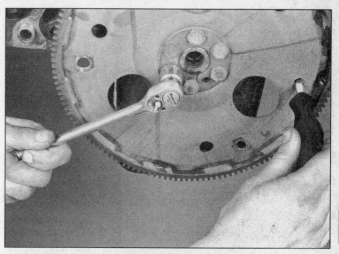

17.4 A large screwdriver wedged in the starter ring gear teeth or one of the holes in the driveplate can be used to keep the flywheel/driveplate from turning as the mounting bolts are removed

work out in a spiral pattern. Don't overtighten them or leakage may occur.

25   Reinstall components removed for access to the oil pan.

26   Add oil and install a new filter, run the engine and check for oil leaks.

16   Oil pump - removal and installation

1   Remove the oil pan (see Section 15).
2   Place a large drain pan under the engine.
3   Unbolt the pump from the rear main bearing cap.
4   Lower the pump and extension shaft from the engine.
5   Before installation, prime the pump with engine oil. Pour oil into the pick-up while the pump extension shaft is turned.
6   Attach the pump, extension shaft and retainer to the main bearing cap. While aligning the pump with the dowel pins at the bottom of the main bearing cap, align the top end of the extension shaft with the lower end of the oil pump drive. When aligned properly, it should slip into place easily.
7   Install the pump mounting bolt and tighten it to the torque listed in this Chapter's Specifications.
8   Install the oil pan and add oil (see Section 15).

17   Flywheel/driveplate - removal and installation

Refer to illustrations 17.3 and 17.4

1   Raise the vehicle and support it securely on jackstands, then refer to Chapter 7 and remove the transaxle. If it's leaking, now would be a very good time to replace the front pump seal/O-ring (automatic transaxle only).
2   Remove the pressure plate and clutch

disc (Chapter 8 - manual transaxle equipped vehicles). Now is a good time to check/replace the clutch components and pilot bearing.
3   If there is no dowel pin, make some marks on the flywheel/driveplate and crankshaft to ensure correct alignment during reinstallation **(see illustration)**.
4   Remove the bolts that secure the flywheel/driveplate to the crankshaft **(see illustration)**. If the crankshaft turns, wedge a screwdriver through the openings in the driveplate (automatic transaxle) or against the flywheel ring gear teeth (manual transaxle). Since the flywheel is fairly heavy, be sure to support it while removing the last bolt.
5   Remove the flywheel/driveplate from the crankshaft.
6   Clean the flywheel to remove grease and oil. Inspect the friction surface for cracks, rivet grooves, burned areas and score marks. Light scoring can be removed with emery cloth. Check for cracked and broken ring gear teeth. Lay the flywheel on a flat surface and use a straightedge to check for warpage.
7   Clean and inspect the mating surfaces of the flywheel/driveplate and the crankshaft. If the crankshaft rear seal is leaking, replace it before reinstalling the flywheel/driveplate.
8   Position the flywheel/driveplate against the crankshaft. Be sure to align the dowel or marks made during removal. Before installing the bolts, apply thread locking compound to the threads.
9   Keep the flywheel/driveplate from turning as described above while you tighten the bolts to the torque listed in this Chapter's Specifications.
10   The remainder of installation is the reverse of the removal procedure.

18   Rear main oil seal - replacement

1   Remove the flywheel/driveplate (see Section 17).

2   Using a thin screwdriver or seal removal tool, carefully remove the oil seal from the engine block. Be very careful not to damage the crankshaft surface while prying the seal out.
3   Clean the bore in the block and the seal contact surface on the crankshaft. Check the seal contact surface on the crankshaft for scratches and nicks that could damage the new seal lip and cause oil leaks - if the crankshaft is damaged, the only alternative is a new or different crankshaft. Inspect the seal bore for nicks and scratches. Carefully smooth it with a fine file if necessary, but don't nick the crankshaft in the process.
4   A special tool is recommended to install the new oil seal. Lubricate the oil seal lips. Slide the seal onto the mandril until the dust lip bottoms squarely against the collar of the tool. **Note:** *If the special tool isn't available, carefully work the seal lip over the crankshaft and tap it into place with a hammer and blunt punch.*
5   Align the dowel pin on the tool with the dowel pin hole in the crankshaft and attach the tool to the crankshaft by hand-tightening the bolts.
6   Turn the tool handle until the collar bottoms against the case, seating the seal.
7   Loosen the tool handle and remove the bolts. Remove the tool.
8   Check the seal and make sure it's seated squarely in the bore.
9   Install the flywheel/driveplate (see Section 17).
10   Install the transaxle.

19   Engine mounts - check and replacement

1   Engine mounts seldom require attention, but broken or deteriorated mounts should be replaced immediately or the added strain placed on the driveline components may cause damage or wear.

Check

2   During the check, the engine must be raised slightly to remove the weight from the mounts.

3   Raise the vehicle and support it securely on jackstands, then position a jack under the engine oil pan. Place a large block of wood between the jack head and the oil pan, then carefully raise the engine just enough to take the weight off the mounts. **Warning:** *DO NOT place any part of your body under the engine when it's supported only by a jack!*

4   Check the mounts to see if the rubber is cracked, hardened or separated from the metal plates. Sometimes the rubber will split right down the center.

5   Check for relative movement between the mount plates and the engine or frame (use a large screwdriver or prybar to attempt to move the mounts). If movement is noted, lower the engine and tighten the mount fasteners.

6   Rubber preservative should be applied to the mounts to slow deterioration.

Replacement

7   Disconnect the negative battery cable from the battery. **Caution:** *On models equipped with the Theftlock audio system, be sure the lockout feature is turned off before performing any procedure which requires disconnecting the battery.*

Lower mount

8   Remove the through-bolt from the engine bracket end of the upper mount strut and rotate the strut out of the way.

9   Raise the engine slightly with a jack or hoist. Remove the fasteners and detach the mount from the frame bracket.

10   Remove the mount-to-block bracket bolts/nuts and detach the mount.

Upper mount strut

11   Disconnect the negative battery cable from the battery and remove the bolts and nuts from the strut and the brace. **Caution:** *On models equipped with the Theftlock audio system, be sure the lockout feature is turned off before performing any procedure which requires disconnecting the battery.*

12   Installation is the reverse of removal. Use thread locking compound on the threads and be sure to tighten everything securely.

Chapter 2 Part B
2.5 liter four-cylinder engine

Contents

Specifications

General

Cylinder numbers (drivebelt end-to-transaxle end)	1-2-3-4
Firing order	1-3-4-2

Torque specifications

	Ft-lbs (unless otherwise indicated)
Crankshaft pulley center bolt	162
Cylinder head bolts (refer to illustration 8.21)	
First step	18
Second step	
All bolts except no. 9	26
Bolt no. 9	18
Fourth step (all bolts)	Rotate an additional 90-degrees (1/4 turn)
Force balancer assembly	
Step 1 - All bolts	107 in-lbs
Step 2	
Short bolts	11 plus 75-degrees additional rotation
Long bolts	11 plus 90-degrees additional rotation

Front

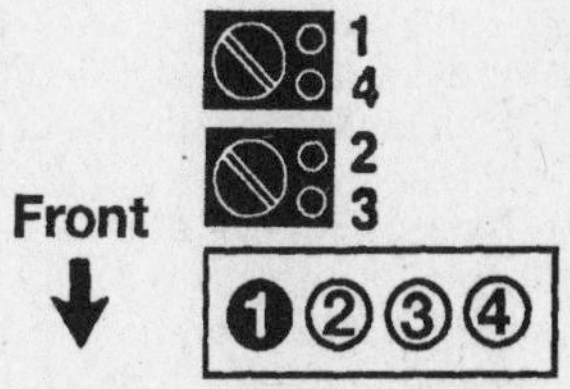

2.5L engine

Cylinder and coil terminal location

Torque specifications   Ft-lbs (unless otherwise indicated)

Exhaust manifold fasteners
    Inner bolts 37
    Outer bolts... 26
Driveplate bolts.. 55
Intake manifold fasteners 25
Oil pan bolts... 89 in-lbs
Oil pan drain plug 25
Oil pump cover bolts .... 89 in-lbs
Valve cover bolts .... 80 in-lbs
Rocker arm bolts 20
Pushrod cover nuts .... 89 in-lbs

1   General information

Refer to illustration 1.1

This Part of Chapter 2 is devoted to in-vehicle repair procedures for the 2.5 liter four-cylinder, overhead valve engine. The 2.5L engine became available in 1990, in the Lumina model only. These engines have cast iron blocks and cast aluminum pistons. The cast iron cylinder head has valve guides that are cast integrally. Steel rocker arms and tubular pushrods actuate the valves. This engine is also equipped with a force balancer assembly. This unit is attached to the crankshaft and it dampens the engine vibration using two eccentrically weighted shafts and gears complemented by a concentric gear on the crankshaft. The engine is transversely mounted and can be recognized by the black steel valve cover with the oil fill cap in the left rear corner (as viewed from the front of the engine compartment). To positively identify this engine, locate the vehicle identification number on the left front corner of the instrument panel. This plate is visible from outside the vehicle **(see illustration)**. The eighth character in the sequence is the engine designation:

R = 2.5L four-cylinder, overhead valve engine

All information concerning engine removal and installation and engine block and cylinder head overhaul can be found in Part G of this Chapter.

The following repair procedures are based on the assumption the engine is in the vehicle. If the engine has been removed from the vehicle and mounted on a stand, many of the steps outlined in this Part of Chapter 2 will not apply.

The Specifications included in this Part of Chapter 2 apply only to the procedures contained in this Part. Part G of this Chapter contains the specifications necessary for cylinder head and engine block rebuilding.

2   Repair operations possible with the engine in the vehicle

Many major repair operations can be accomplished without removing the engine from the vehicle.

1.1 The vehicle identification number is located on the left front corner of the instrument panel

Clean the engine compartment and the exterior of the engine with some type of degreaser before any work is done. It'll make the job easier and help keep dirt out of the internal areas of the engine.

Depending on the components involved, it may be helpful to remove the hood to improve access to the engine as repairs are performed (refer to Chapter 11 if necessary). Cover the fenders to prevent damage to the paint. Special pads are available, but an old bedspread or blanket will also work.

If vacuum, exhaust, oil or coolant leaks develop, indicating a need for gasket or seal replacement, the repairs can generally be made with the engine in the vehicle. The intake and exhaust manifold gaskets, timing cover gasket, oil pan gasket, crankshaft oil seals and cylinder head gasket are all accessible with the engine in place.

Exterior engine components, such as the intake and exhaust manifolds, the oil pan (and the oil pump), the water pump, the starter motor, the alternator and the fuel system components can be removed for repair with the engine in place.

Since the cylinder head can be removed without pulling the engine, valve component servicing can also be accomplished with the engine in the vehicle.

In extreme cases caused by a lack of necessary equipment, repair or replacement of piston rings, pistons, connecting rods and rod bearings is possible with the engine in the vehicle. However, this practice is not recommended because of the cleaning and prepa-ration work that must be done to the components involved.

3   Valve cover - removal and installation

Removal

1   Remove the air cleaner assembly, tagging each hose to be disconnected with a piece of numbered tape to simplify installation.
2   Disconnect and remove the PCV valve (see Chapter 1).
3   Remove the EGR valve (see Chapter 6).
4   Remove the spark plug wires and the clips from the valve cover.
5   Remove the valve cover bolts.
6   Detach the valve cover from the cylinder head using special tools. The use of these tools will prevent distorting the valve cover but if they are not available, carefully tap on the cover with a wood block and a hammer. Don't pry between the cover and the cylinder head, as this would bend the sealing flange causing oil leaks.

Installation

7   The mating surfaces of the cylinder head and valve cover must be perfectly clean when the cover is installed. Use a gasket scraper to remove all traces of sealant or old gasket, then clean the mating surfaces with lacquer thinner or acetone (if there's sealant or oil on the mating surfaces when the cover

4.2a Removing the rocker arm pivot bolt

4.2b Loosen the rocker arm bolt, rotate the rocker arm to one side and lift out the pushrod

s installed, oil leaks may develop).

Clean the mounting bolt threads with a die if necessary to remove any corrosion and restore damaged threads. Make sure the threaded holes in the head are clean - run a tap into them if necessary to remove corrosion and restore damaged threads.

Apply a 3/16 inch bead of RTV-type sealant all the way around the sealing flange on the cover. Allow the sealant to dry or "set up" for approximately 1 minute.

10 Place the valve cover on the cylinder head and install the mounting bolts. Tighten the bolts a little at a time to the torque listed in this Chapter's Specifications.

11 Complete the installation by reversing the removal procedure.

4   Rocker arms and pushrods - removal, inspection and installation

Refer to illustrations 4.2a, 4.2b and 4.4

Removal

1 Refer to Section 3 and detach the valve cover from the cylinder head.

2 Beginning at the front of the cylinder head, loosen the rocker arm bolts **(see illustration)**. **Note:** *If the pushrods are the only items being removed, rotate the rocker arms to one side so the pushrods can be lifted out* **(see illustration)**.

3 Remove the bolts, the rocker arms and the pivot balls and store them in marked containers (they must be reinstalled in their original locations).

4 Remove the pushrods and store them separately to make sure they don't get mixed up during installation **(see illustration)**.

5 If the pushrod guides must be removed for any reason, make sure they're marked so they can be reinstalled in their original locations.

Inspection

6 Check each rocker arm for wear, cracks and other damage, especially where the pushrods and valve stems contact the rocker arm faces.

7 Make sure the hole at the pushrod end of each rocker arm is open.

8 Check each rocker arm pivot area for wear, cracks and galling.

If the rocker arms are worn or damaged, replace them with new ones and use new pivot balls as well.

9 Inspect the pushrods for cracks and excessive wear at the ends. Roll each pushrod across a piece of plate glass to see if it's bent (if it wobbles, it's bent).

Installation

10 Lubricate the lower ends of the pushrods with clean engine oil or moly-base grease and install them in their original locations. Make sure each pushrod seats completely in the lifter socket.

4.4 A perforated cardboard box can be used to store the pushrods to ensure they're reinstalled in their original locations - note the label indicating the front (drivebelt end) of the engine

11 Apply moly-base grease to the ends of the valve stems and the upper ends of the pushrods before positioning the rocker arms and installing the nuts.

12 Install the pivot balls onto the rocker arm bolts. Apply moly-base grease to the pivot balls to prevent damage to the mating surfaces before engine oil pressure builds up. Insert the bolt and pivot ball assemblies into their respective rocker arms, then install them to the cylinder head. Make sure the pushrods seat properly in the rocker arms. Tighten the bolts to the torque listed in this Chapter's Specifications.

13 Install the valve cover (see Section 3).

5   Valve springs, retainers and seals - replacement

Refer to illustrations 5.4, 5.9a, 5.9b and 5.17
Note: *Broken valve springs and defective valve stem seals can be replaced without removing the cylinder head. Two special tools and a compressed air source are normally required to perform this operation, so read through this Section carefully and rent or buy the tools before beginning the job. If compressed air isn't available, a length of nylon rope can be used to keep the valves from falling into the cylinder during this procedure.*

1 Refer to Section 3 and remove the valve cover.

2 Remove the spark plug from the cylinder which has the defective component. If all of the valve stem seals are being replaced, all of the spark plugs should be removed.

3 Turn the crankshaft until the piston in the affected cylinder is at top dead center on the compression stroke (refer to Chapter 2, Part G, for instructions). If you're replacing all of the valve stem seals, begin with cylinder number one and work on the valves for one cylinder at a time. Move from cylinder-to-cylinder following the firing order sequence (see this Chapter's Specifications).

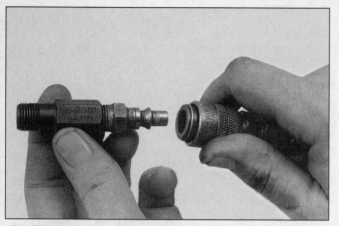

5.4 This is what the air hose adapter that threads into the spark plug hole looks like - they're commonly available at auto parts stores

5.9a A lever-type valve spring compressor is used to compress the spring and remove the keepers to replace valve seals or springs with the head installed – note that a rocker arm bolt has been reinstalled to be used for leverage

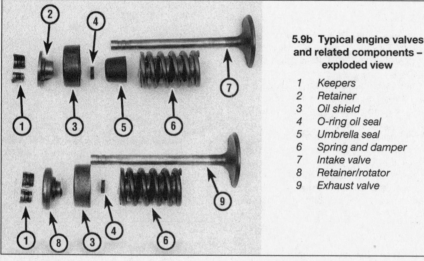

5.9b Typical engine valves and related components – exploded view

1   Keepers
2   Retainer
3   Oil shield
4   O-ring oil seal
5   Umbrella seal
6   Spring and damper
7   Intake valve
8   Retainer/rotator
9   Exhaust valve

5.17 Keepers don't always stay in place, so apply a small dab of grease to each one as shown here before installation - it'll hold them in place on the valve stem as the spring is released

4   Thread an adapter into the spark plug hole **(see illustration)** and connect an air hose from a compressed air source to it. Most auto parts stores can supply the air hose adapter. **Note:** *Many cylinder compression gauges utilize a screw-in fitting that may work with your air hose quick-disconnect fitting.*

5   Remove the bolt, pivot ball and rocker arm for the valve with the defective part and pull out the pushrod. If all of the valve stem seals are being replaced, all of the rocker arms and pushrods should be removed (refer to Section 4).

6   Apply compressed air to the cylinder. **Warning:** *The piston may be forced down by compressed air, causing the crankshaft to turn suddenly. If the wrench used when positioning the number one piston at TDC is still attached to the bolt in the crankshaft nose, it could cause damage or injury when the crankshaft moves.*

7   The valves should be held in place by the air pressure. If the valve faces or seats are in poor condition, leaks may prevent air pressure from retaining the valves - a "valve job" is necessary to correct this problem.

8   If you don't have access to compressed air, an alternative method can be used. Position the piston at a point just before TDC on the compression stroke, then feed a long piece of nylon rope through the spark plug hole until it fills the combustion chamber. Be sure to leave the end of the rope hanging out of the engine so it can be removed easily. Use a large ratchet and socket to rotate the crankshaft in the normal direction of rotation (clockwise) until slight resistance is felt.

9   Stuff shop rags into the cylinder head holes above and below the valves to prevent parts and tools from falling into the engine, then use a valve spring compressor to compress the spring **(see illustration)**. Remove the keepers with small needle-nose pliers or a magnet **(see illustration)**. **Note:** *A couple of different types of tools are available for compressing the valve springs with the head in place. One type, shown here, utilizes the rocker arm bolt for leverage, while the other type grips the lower spring coils and presses on the retainer as the knob is turned. Both types work very well, although the lever type is usually less expensive.*

10   Remove the spring retainer and valve spring, then remove the valve guide seal.

Note: *If air pressure fails to hold the valve in the closed position during this operation, the valve face or seat is probably damaged. If so, the cylinder head will have to be removed for additional repair operations.*

11   Wrap a rubber band or tape around the top of the valve stem so the valve won't fall into the combustion chamber, then release the air pressure. **Note:** *If a rope was used instead of air pressure, turn the crankshaft slightly in the direction opposite normal rotation.*

12   Inspect the valve stem for damage. Rotate the valve in the guide and check the end for eccentric movement, which would indicate the valve stem is bent.

13   Move the valve up-and-down in the guide and make sure it doesn't bind. If the valve stem binds, either the valve is bent or the guide is damaged. In either case, the head will have to be removed for repair.

14   Reapply air pressure to the cylinder to retain the valve in the closed position, then remove the tape or rubber band from the valve stem. If a rope was used instead of air pressure, rotate the crankshaft in the normal direction of rotation until slight resistance is felt.

15   Lubricate the valve stem with engine oil

6.9 Pry the engine forward then insert the bolt through the slave hole on the torque rod and into the bracket

material off the intake manifold and head gasket mating surfaces.

Installation

12 Installation is the reverse of removal. Be sure to use a new gasket. Tighten the nuts/bolts to the torque listed in this Chapter's Specifications, following the correct sequence **(see illustration 6.10a and 6.10b)**.
13 Add coolant, run the engine and check for leaks and proper operation.

7 Exhaust manifold - removal and installation

Refer to illustration 7.11

Removal

1 Disconnect the negative cable from the battery. **Caution:** *On models equipped with the Theftlock audio system, be sure the lockout feature is turned off before performing any procedure which requires disconnecting the battery.*
2 Unplug the electrical connector from the oxygen sensor.
3 Remove the drivebelt and the alternator (see Chapters 1 and 5).
4 Remove the torque rod.
5 Raise the front of the vehicle, support it securely on jackstands and apply the parking brake. Block the rear wheels to keep the vehicle from rolling off the jackstands. Unbolt the exhaust pipe from the manifold. Lower the vehicle.
6 Remove the oil level indicator tube.
7 Remove the exhaust manifold-to-cylinder head nuts/bolts, pull the manifold off the engine and lift it out of the exhaust pipe flange. Remove and discard the gasket.
8 Scrape all traces of gasket material off the exhaust manifold and cylinder head mating surfaces.

Installation

9 Clean all bolt and stud threads before installation. A wire brush can be used on the manifold mounting studs, while a tap works well when cleaning the cylinder head bolt holes.

d install a new valve guide seal. **Note:** *ake and exhaust valve seals are different.*

Install the spring in position over the lve.

Install the valve spring retainer. Com- ess the valve spring and carefully install the epers in the groove. Apply a small dab of ease to the inside of each keeper to hold it place if necessary **(see illustration)**. move the pressure from the spring tool d make sure the keepers are seated.

Disconnect the air hose and remove the apter from the spark plug hole. If a rope as used in place of air pressure, pull it out the cylinder.

Refer to Section 4 and install the rocker m(s) and pushrod(s).

Install the spark plug(s) and hook up the re(s).

Refer to Section 3 and install the valve ver.

Start and run the engine, then check for leaks and unusual sounds coming from e valve cover area.

Intake manifold - removal and installation

fer to illustrations 6.9, 6.10a and 6.10b

emoval

Relieve the fuel pressure (see Chapter 4),

then disconnect the negative battery cable from the battery. **Caution:** *On models equipped with the Theftlock audio system, be sure the lockout feature is turned off before performing any procedure which requires disconnecting the battery.*
2 Remove the TBI cover/air cleaner assembly (see Chapter 4).
3 Raise the front of the vehicle and support it securely on jackstands. Drain the coolant (refer to Chapter 1).
4 Label and disconnect any wires and vacuum hoses which will interfere with manifold removal.
5 Remove the throttle and TV cables and brackets (see Chapter 4).
6 Remove the TBI assembly and fuel lines (see Chapter 4).
7 Remove the power steering pump (if equipped) and tie it aside in an upright position (see Chapter 10).
8 Remove the heater hose.
9 Pry the engine forward to make extra room for intake manifold clearance. First remove the torque strut-to-engine bracket nut and bolt. Then use a prybar at the bracket and move the engine forward. Next, support the torque strut by installing a bolt through the slave hole and into the bracket **(see illustration)**.
10 Remove the mounting fasteners from the intake manifold **(see illustrations)**.
11 Separate the intake manifold and gasket from the engine. Scrape all traces of gasket

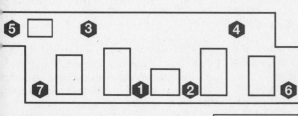

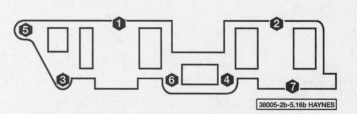

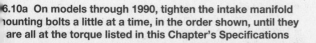

6.10a On models through 1990, tighten the intake manifold mounting bolts a little at a time, in the order shown, until they are all at the torque listed in this Chapter's Specifications

6.10b Use this intake manifold bolt tightening sequence on 1991 and later models

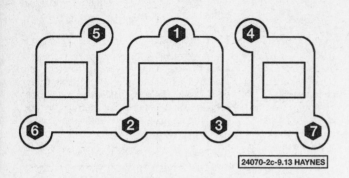

7.11 Tightening sequence for the exhaust manifold mounting bolts

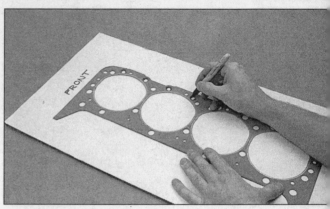

8.11 To avoid mixing up the head bolts, use the new gasket to transfer the bolt hole pattern to a piece of cardboard, then punch holes in the cardboard to accept the bolts

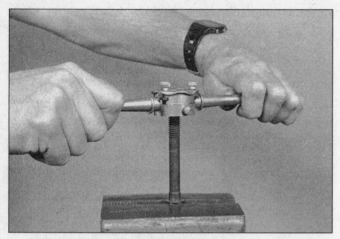

8.17 A die should be used to clean the head bolt threads prior to installation

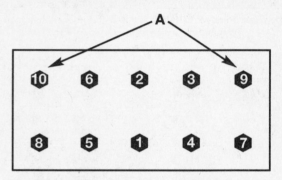

A - Apply sealing compound to threads on bolts 9 and 10

8.21 Cylinder head bolt tightening sequence

10   If a new manifold is being installed, transfer the oxygen sensor from the old manifold to the new one.

11   Installation is the reverse of removal. Be sure to use a new gasket and tighten the bolts, using the correct sequence, to the torque listed in this Chapter's Specifications **(see illustration)**.

8   Cylinder head - removal and installation

Caution: *Allow the engine to cool completely before loosening the cylinder head bolts.*
Note: *On vehicles with high mileage and during an engine overhaul, camshaft lobe height should be checked prior to cylinder head removal (see Chapter 2, Part G, for instructions).*

Removal
Refer to illustration 8.11

1   Relieve the fuel pressure and remove the TBI cover (see Chapter 4). Disconnect the cable from the negative battery terminal. **Caution:** *On models equipped with the Theft-lock audio system, be sure the lockout feature*

is turned off before performing any procedure which requires disconnecting the battery.

2   Remove the alternator and brackets as described in Chapter 5.

3   Remove the intake manifold as described in Section 6.

4   Remove the exhaust manifold as described in Section 7.

5   Detach the dipstick tube from the exhaust manifold.

6   Unbolt the power steering pump (if equipped) and set it aside without disconnecting the hoses (see Chapter 10).

7   Disconnect any remaining wires, hoses, fuel and vacuum lines from the cylinder head. Be sure to label them to simplify reinstallation.

8   Disconnect the spark plug wires and remove the spark plugs. Be sure the plug wires are labeled to simplify reinstallation.

9   Remove the valve cover (see Section 3).

10   Remove the rocker arms and pushrods (see Section 4).

11   Using the new head gasket, outline the cylinders and bolt pattern on a piece of cardboard **(see illustration)**. Be sure to indicate the front (radiator side) of the engine for reference. Punch holes at the bolt locations. Loosen each of the cylinder head mounting

bolts 1/4-turn at a time until they can be removed by hand. Store the bolts in the cardboard holder as they're removed - this will ensure they are reinstalled in their original locations, which is absolutely essential.

12   Lift the head off the engine. If it's stuck don't attempt to pry it off - you could damage the sealing surfaces. Instead, use a hammer and block of wood to tap the head and break the gasket seal. Place the head on a block of wood to prevent damage to the gasket surface.

13   Remove the cylinder head gasket.

14   Refer to Chapter 2, Part G, for cylinder head disassembly and valve service procedures.

Installation
Refer to illustrations 8.17 and 8.21

15   If a new cylinder head is being installed transfer all external parts from the old cylinder head to the new one.

16   If not already done, thoroughly clean the gasket surfaces on the cylinder head and engine block. Do not gouge or otherwise damage the gasket surfaces.

17   To get the proper torque readings, the threads of the head bolts must be clean **(see illustration)**. This also applies to the

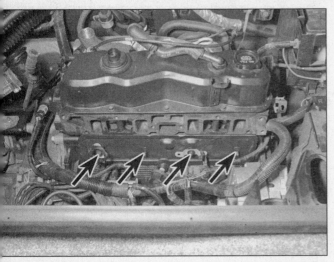

9.6 The pushrod cover is held in place with four nuts (arrows) (the exhaust manifold is removed for clarity)

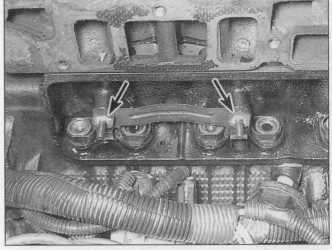

9.7a Location of the pushrod cover stud locknuts (arrows) - manifold removed for clarity

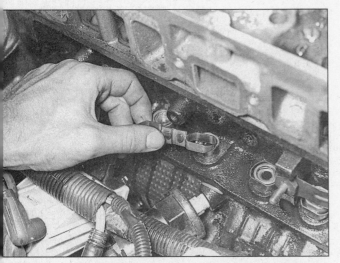

9.7b Remove the lifter guide - if you're removing more than one guide, keep them in order to prevent mix-ups during installation

9.8 On engines that haven't become sticky with sludge and varnish, the lifters can usually be removed by hand

threaded holes in the engine block. Run a tap through the holes to ensure they are clean.

18 Place the gasket in position over the engine block dowel pins. Note any marks like "THIS SIDE UP" and install the gasket accordingly.

19 Carefully lower the cylinder head onto the engine, over the dowel pins and the gasket.

20 Install the bolts finger tight. Don't tighten any of the bolts at this time.

21 Tighten each of the bolts in 1/4-turn increments in the recommended sequence (see illustration), until the stage 1 torque listed in this Chapter's Specifications is reached. Mark each bolt with a felt-tip marker each time you tighten it to make sure none of the bolts have been left out of the sequence. Repeat the sequence, tightening the head bolts to the stage 2 torque listed in this Chapter's Specifications. Repeat the sequence again, tightening all the bolts the amount of angle listed in this Chapter's Specifications.

22 The remaining installation steps are the reverse of removal.

23 Be sure to refill the cooling system and change the oil and filter (see Chapter 1).

9 Hydraulic lifters - removal, inspection and installation

Removal

Refer to illustrations 9.6, 9.7a, 9.7b, 9.8 and 9.9

1 A noisy valve lifter can be isolated when the engine is idling. Place a length of hose or tubing near the position of each valve while listening at the other end. Or, remove the valve cover and, with the engine idling, place a finger on each of the valve spring retainers, one at a time. If a valve lifter is defective, it'll be evident from the shock felt at the retainer as the valve seats.

2 The most likely cause of a noisy valve

lifter is a piece of dirt trapped between the plunger and the lifter body.

3 Remove the valve cover (see Section 3).

4 Loosen both rocker arm bolts at the cylinder with the noisy lifter and rotate the rocker arms away from the pushrods.

5 Remove the pushrod guides and pushrods (see Section 4).

6 Remove the pushrod cover (see illustration).

7 Remove the lifter guide retainer by unscrewing the locknuts on the pushrod cover studs. Remove the lifter guide (see illustrations).

8 There are several ways to extract a lifter from its bore. A special removal tool is available, but isn't always necessary. On newer engines without a lot of varnish buildup, lifters can often be removed with a small magnet or even with your fingers (see illustration). A scribe can also be used to pull the lifter out of the bore. Caution: *Don't use pliers of any type to remove a lifter unless you*

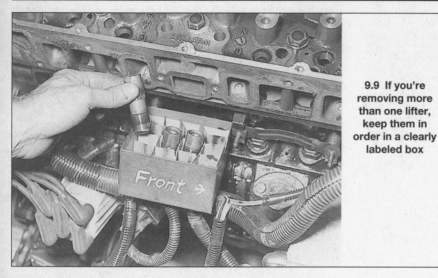

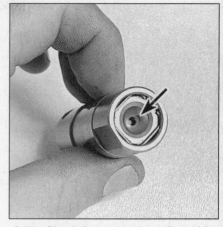

9.9 If you're removing more than one lifter, keep them in order in a clearly labeled box

9.11a Check the pushrod seat (arrow) in the top of each lifter for wear

intend to replace it with a new one - they will damage the precision machined and hardened surface of the lifter, rendering it useless.
9   Store the lifters in a clearly labeled box to insure their reinstallation in the same lifter bores **(see illustration)**.

Inspection

Refer to illustrations 9.11a and 9.11b
10   Clean the lifters with solvent and dry them thoroughly without mixing them up.
11   Check each lifter wall and pushrod seat for scuffing, score marks and uneven wear **(see illustration)**. Check the rollers carefully for wear and damage and make sure they turn freely without excessive play **(see illustration)**. If the lifter walls are damaged or worn (which isn't very likely), inspect the lifter bores in the engine block as well. If the pushrod seats are worn, check the pushrod ends.
12   Used roller lifters can be reinstalled with a new camshaft and the original camshaft can be used if new roller lifters are installed.

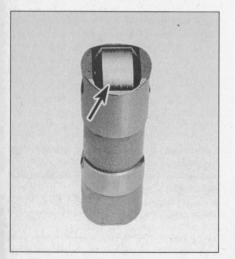

9.11b The roller on roller lifters must turn freely - check for wear and excessive play as well

Installation

Refer to illustration 9.19
13   If the same lifters are being used, they must be installed in their original bores. Coat them with moly-base grease or engine assembly lube.
14   Lubricate the bearing surfaces of the lifter bores with engine oil.
15   Install the lifter(s) in the lifter bore(s). **Note:** *Make sure the oil orifice is facing toward the front of the engine.*
16   Install the lifter guide(s) and retainer(s) **(see illustrations 9.7a and 9.7b)**.
17   Install the pushrods, pushrod guides, rocker arms and rocker arm retaining bolts (see Section 4).
18   Tighten the rocker arm bolts to the torque listed in this Chapter's Specifications.
19   Clean the pushrod cover and its mating surface on the cylinder block. Apply a continuous 3/16 inch (5 mm) diameter bead of RTV sealant to the pushrod cover **(see illustration)**, then install the pushrod cover. Tighten the nuts to the torque listed in this Chapter's Specifications.
20   Install the valve cover (see Section 3).
21   The remainder of installation is the reverse of removal.

10 Crankshaft pulley - removal and installation

Refer to illustrations 10.9, 10.10 and 10.11

Removal

1   Remove the cable from the negative battery terminal. **Caution:** *On models equipped with the Theftlock audio system, be sure the lockout feature is turned off before performing any procedure which requires disconnecting the battery.*
2   Remove the torque strut bolt from the cylinder head bracket.
3   Remove the drivebelt (see Chapter 1).
4   With the parking brake applied and the shifter in Park (automatic), raise the front of the vehicle and support it securely on jackstands. Remove the right front wheel and the fender inner splash shield.
5   Disconnect the right side control arm from the steering knuckle (see Chapter 10).
6   If necessary, support the engine with a floor jack positioned under the oil pan. Place a wood block on the jack head to serve as a cushion.
7   If necessary, remove the two right side

9.19 The valve cover is sealed with RTV – make sure the sealant is still tacky and the semi-circular cutout (arrow) is facing down

10.9 Remove the crankshaft hub bolt using a breaker bar and socket

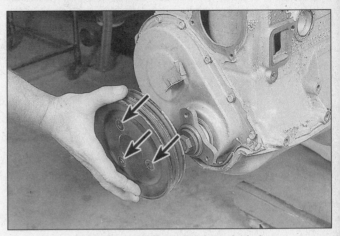

10.10 Separate the pulley from the hub

frame attaching bolts and loosen, but do not remove, the bolts and nuts from the left side frame member.

8   If necessary, lower the engine slightly with the floorjack.

9   Remove the crankshaft hub bolt **(see illustration)**. A breaker bar will probably be needed, since the bolt is very tight. If necessary, remove the bellhousing cover and insert a large screwdriver into the teeth of the starter ring gear to prevent the crankshaft from turning.

10   Remove the pulley bolts and remove the crankshaft pulley from the crankshaft hub **(see illustration)**.

11   Using a puller, remove the crankshaft hub from the crankshaft **(see illustration)**.

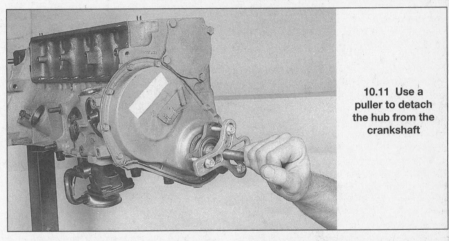

10.11 Use a puller to detach the hub from the crankshaft

Installation

12   Refer to Section 11 for the oil seal replacement procedure.

13   Apply a thin layer of multi-purpose grease to the seal contact surface of the hub.

14   Position the hub on the crankshaft and slide it through the seal until it bottoms against the crankshaft gear. Note that the slot (keyway) in the hub must be aligned with the Woodruff key in the end of the crankshaft. Install the pulley-to-crankshaft bolt.

15   Tighten the pulley-to-crankshaft bolt to the torque listed in this Chapter's Specifications.

16   The remaining installation steps are the reverse of removal.

11   Crankshaft front oil seal - replacement

1   Remove the crankshaft pulley (see Section 10).

2   Pry the old oil seal out with a seal removal tool or a screwdriver. Be very careful not to nick or otherwise damage the crankshaft in the process. Wrap the screwdriver tip with vinyl tape to protect the crankshaft.

3   Apply a thin coat of RTV-type sealant to the outer edge of the new seal. Lubricate the seal lip with moly-based grease or clean engine oil.

4   Place the seal squarely in position in the bore and drive it into place with special tool.

5   If you don't have the special tool, carefully tap the seal into place with a large socket (or piece of pipe) and a hammer. The outer diameter of the socket or pipe should be the same size as the seal outer diameter. Make sure the seal is seated completely in the bore.

6   Install the crankshaft pulley (see Section 10).

7   Reinstall the remaining parts in the reverse order of removal.

8   Start the engine and check for oil leaks at the seal.

12   Timing cover, timing chain, sprockets and tensioner - removal, inspection and installation

Note: *On models through 1990, timing gears are used. On 1991 and later models, a timing chain and sprockets are used. The timing gears cannot be removed in the vehicle on* 1990 *and earlier models. On* 1991 *and later models, the timing chain and sprockets can be removed in the vehicle.*

Timing cover (all models)

Removal

1   Remove the crankshaft pulley as described in Section 10.

2   Remove the timing gear cover-to-block bolts, then detach the cover.

3   Using a scraper and degreaser, remove all old sealant from the sealing surfaces of the timing gear cover, engine block and oil pan.

Installation

4   If necessary, replace the front oil seal by carefully prying it out of the cover with a large screwdriver. Don't distort the cover.

5   Install the new seal with the spring side toward the inside of the cover. Drive the seal into place using a seal installation tool or a large socket and hammer. A block of wood will also work.

6   Apply a 1/4 inch wide by 1/8 inch thick bead of RTV sealant to the front cover-to-block mating surface.

7   Apply a 3/8 inch wide by 3/16 inch thick bead of RTV sealant to the oil pan-to-front cover mating surface.

8   Place the cover in position over the dowel pins on the block. Slide the crankshaft pulley onto the end of the crankshaft to align the cover.

9   Install the bolts that secure the cover to the block, tightening them a little at a time until the cover is seated, then tighten all of them to the torque listed in this Chapter's Specifications. Follow a criss-cross pattern to avoid distorting the cover.

10   Install the crankshaft pulley.

Timing chain and sprockets (1991 and 1992 only)

Inspection

11   The timing chain should be replaced with a new one if the engine has high mileage, the chain has visible damage, or total freeplay midway between the sprockets exceeds one-inch. **Caution:** *Failure to replace a worn timing chain may result in erratic engine performance, loss of power and decreased fuel mileage. Loose chains can "jump" timing. In the worst case, chain "jumping" or breakage will result in severe engine damage.*

12   Inspect the sprockets for wear and damage. Look for teeth that are deformed, chipped, pitted and cracked.

13   Inspect the timing chain tensioner for cracks or wear and replace it if necessary.

Removal

14   Loosen the camshaft bolt.

15   Turn the engine to Top Dead Center (TDC) and check that the camshaft and crankshaft timing marks are aligned (see illustration 12.14).

16   Remove the camshaft sprocket bolt. Do not turn the camshaft in the process (if you do, realign the timing marks before the bolt is removed).

17   Slip the timing chain and sprocket assembly off the engine.

18   Timing chains and sprockets should be replaced in sets. Be sure to align the key in the crankshaft with the keyway in the sprocket during installation.

19   Clean the timing chain and sprockets with solvent and dry them with compressed air (if available). **Warning:** *Wear eye protection when using compressed air.*

Installation

20   Installation is the reverse of removal. Be sure to tighten fasteners to the torques listed in this Chapter's Specifications and lubricate the chain and sprockets with clean engine oil before installing the timing chain cover.

---

13 Camshaft and bearings - removal, inspection and installation

Due to the fact the engine is mounted transversely in the vehicle, there isn't enough

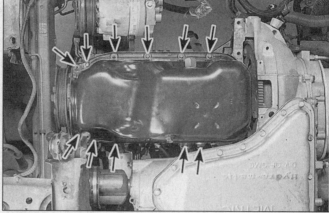

14.18 Oil pan mounting bolt locations (some bolts are not visible)

room to remove the camshaft with the engine in place. Therefore, the procedure is covered in Chapter 2, Part G.

---

14 Oil pan - removal and installation

Refer to illustration 14.18

Removal

1   Warm up the engine, then drain the oil and remove the oil filter (see Chapter 1).

2   Detach the cable from the negative battery terminal. **Caution:** *On models equipped with the Theftlock audio system, be sure the lockout feature is turned off before performing any procedure which requires disconnecting the battery.*

3   Remove the coolant reservoir (see Chapter 3).

4   Remove the engine torque strut.

5   Remove the air cleaner assembly and the air inlet tube (see Chapter 4).

6   Remove the drivebelt (see Chapter 1).

7   Remove the dipstick tube.

8   Raise the vehicle and support it securely on jackstands. Support the transaxle and engine assembly with a floor jack and a piece of wood placed under the transaxle.

9   On air conditioned models, remove the air conditioner from the brackets and set it aside, without disconnecting the hoses (see Chapter 3).

10   Remove the starter (see Chapter 5).

11   Remove the lower bellhousing cover.

12   Turn the front wheels all the way to the right and remove the engine wiring harness retaining screws under the oil pan on the right and left sides.

13   Remove the splash pan from the right side of the engine.

14   Remove the front engine mount bracket bolts and nuts.

15   Remove the transaxle mount nuts and raise the engine and transaxle assembly approximately 2 inches with the floor jack.

16   Remove the front engine mount and bracket.

17   Loosen the frame bolts.

18   Remove the bolts and nuts securing the oil pan to the engine block **(see illustration)**.

19   Tap on the pan with a soft-face hammer to break the gasket seal, then detach the oil pan from the engine. Don't pry between the oil pan and the block, as the pan could become bent or the block could become gouged.

20   Using a gasket scraper, remove all traces of old sealant from the engine block and oil pan. Make sure the threaded bolt holes in the block are clean. Wash the oil pan with solvent and dry it thoroughly.

Installation

21   Check the sealing flanges for distortion, particularly around the bolt holes. If necessary, place the pan on a block of wood and use a hammer to flatten and restore the gasket surfaces. Clean the mating surfaces with lacquer thinner or acetone.

22   Place a continuous bead of RTV sealant on the oil pan-to-block sealing flanges and the oil pan-to-front cover surface. Do not apply excessive amounts of the sealer and make sure the sealer is applied to the inside of the bolt holes.

23   Apply a thin coat of RTV sealant to the ends of the rear oil pan seal down to the ears.

24   Carefully place the oil pan against the block.

25   Install the bolts and tighten them in 1/4-turn increments to the torque listed in this Chapter's Specifications. Start with the bolts closest to the center of the pan and work out in a spiral pattern. Don't overtighten them or leakage may occur.

26   Reinstall components removed for access to the oil pan.

27   Add oil and install a new filter, run the engine and check for oil leaks.

---

15 Oil pump - removal and installation

Removal

1   Remove the oil pan (see Section 14).

2   Place a large drain pan under the engine.

3   Unbolt the oil pump cover from the force balancer assembly. **Warning:** *Use caution if you remove the pressure regulator valve - it is under considerable spring pressure.*

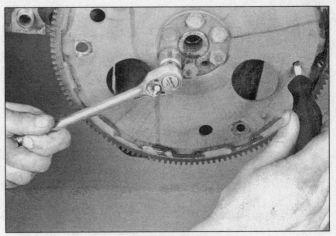

17.2 Most driveplates have locating dowels (arrow) - if the one you're working on doesn't have one, make some marks to ensure correct installation

17.3 A large screwdriver wedged in one of the holes in the driveplate can be used to keep the driveplate from turning as the mounting bolts are removed and installed

4   Lower the pump cover from the force balancer assembly, then remove the gears from the balancer housing.

Installation

5   Pack the cavities between the gears with petroleum jelly to prime it.
6   Attach the oil pump to the force balancer assembly. Align the top end of the oil pump drive with the lower end of the oil pump. When aligned properly, it should slip into place easily.
7   Install the pump mounting bolts and tighten them to the torque listed in this Chapter's Specifications.
8   Install the oil pan and add oil (see Section 14).

16   Crankshaft force balancer - removal and installation

Removal

1   Remove the oil pan (see Section 14).
2   Remove the bolts that retain the force balancer assembly to the engine block and carefully lower the assembly.

Installation

3   Rotate the engine until number 1 or 4 cylinder is at Top Dead Center (see Chapter 2G).
4   Install the balancer assembly with the balance weights at BDC (plus or minus one-half gear tooth). **Note:** *When installing the force balancer assembly, the end of the housing without the dowel pins must contact the engine block surface evenly. If it loses contact, the gears will not engage properly and the unit will become defective.*
5   Install the retaining bolts and tighten them to the torque listed in this Chapter's Specifications, using an "X" crossover pattern.
6   Install the oil pan (see Section 14).

17   Driveplate - removal and installation

Refer to illustrations 17.2 and 17.3

1   Raise the vehicle and support it securely on jackstands, then refer to Chapter 7 and remove the transaxle. If it's leaking, now would be a very good time to replace the front pump seal and O-ring (automatic transaxle only).
2   If there is no dowel pin, make some marks on the driveplate and crankshaft to ensure correct alignment during reinstallation **(see illustration)**.
3   Remove the bolts that secure the driveplate to the crankshaft **(see illustration)**. If the crankshaft turns, wedge a screwdriver through the openings in the driveplate. Since the driveplate is fairly heavy, be sure to support it while removing the last bolt.
4   Remove the driveplate from the crankshaft.
5   Clean the driveplate to remove grease and oil. Inspect it for cracks and cracked and broken ring gear teeth. Lay the driveplate on a flat surface and use a straightedge to check for warpage.
6   Clean and inspect the mating surfaces of the driveplate and the crankshaft. If the crankshaft rear seal is leaking, replace it before reinstalling the driveplate.
7   Position the driveplate against the crankshaft. Be sure to align the dowel or marks made during removal. Before installing the bolts, apply thread locking compound to the threads.
8   Keep the driveplate from turning as described above while you tighten the bolts to the torque listed in this Chapter's Specifications.
9   The remainder of installation is the reverse of the removal procedure.

18   Rear main oil seal - replacement

Refer to illustration 18.2

1   Remove the driveplate (see Section 17).
2   Using a thin screwdriver or seal removal tool, carefully remove the oil seal from the engine block **(see illustration)**. Be very careful not to damage the crankshaft surface while prying the seal out.
3   Clean the bore in the block and the seal

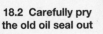

18.2 Carefully pry the old oil seal out

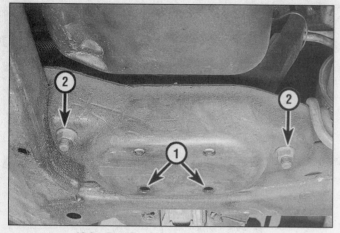

19.9a Details of the front engine mount

1	Four-cylinder engine mounting nuts	2	V6 engine mounting nuts

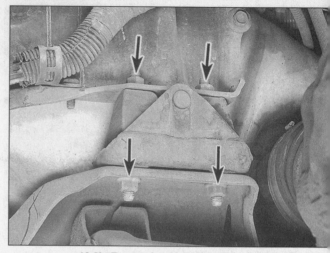

19.9b Transaxle mount bolts (arrows)

contact surface on the crankshaft. Check the seal contact surface on the crankshaft for scratches and nicks that could damage the new seal lip and cause oil leaks - if the crankshaft is damaged, the only alternative is a new or different crankshaft. Inspect the seal bore for nicks and scratches. Carefully smooth it with a fine file if necessary, but don't nick the crankshaft in the process.

4   A special tool is recommended to install the new oil seal. Lubricate the oil seal lips. Slide the seal onto the mandril until the dust lip bottoms squarely against the collar of the tool. **Note:** *If the special tool isn't available, carefully work the seal lip over the crankshaft and tap it into place with a hammer and punch.*

5   Align the dowel pin on the tool with the dowel pin hole in the crankshaft and attach the tool to the crankshaft by hand-tightening the bolts.

6   Turn the tool handle until the collar bottoms against the case, seating the seal.

7   Loosen the tool handle and remove the bolts. Remove the tool.

8   Check the seal and make sure it's seated squarely in the bore.

9   Install the driveplate (see Section 17).

10   Install the transaxle.

19   Engine mounts - check and replacement

Refer to illustrations 19.9a and 19.9b

Warning: *A special tool is available to support the engine during repair operations. Similar fixtures are available from rental yards. Improper lifting methods or devices are hazardous and could result in severe injury or death. DO NOT place any part of your body under the engine/transaxle when it's supported only by a jack. Failure of the lifting device could result in severe injury or death.*

1   Engine mounts seldom require attention, but broken or deteriorated mounts should be replaced immediately or the added strain placed on the driveline components may cause damage or wear.

Check

2   During the check, the engine must be raised slightly to remove the weight from the mounts.

3   Raise the vehicle and support it securely on jackstands, then position a jack under the engine oil pan. Place a large block of wood between the jack head and the oil pan, then carefully raise the engine just enough to take the weight off the mounts.

4   Check the mounts to see if the rubber cracked, hardened or separated from th metal plates. Sometimes the rubber will sp right down the center.

5   Check for relative movement between th mount plates and the engine or frame (use large screwdriver or pry bar to attempt to mov the mounts). If movement is noted, lower th engine and tighten the mount fasteners.

6   Rubber preservative should be applie to the mounts to slow deterioration.

Replacement

7   Disconnect the negative battery cabl from the battery, then raise the vehicle an support it securely on jackstands (if n already done). **Caution:** *On models equippe with the Theftlock audio system, be sure th lockout feature is turned off before performing any procedure which requires disconnecting the battery.*

8   Remove the engine torque strut (se **illustration 6.9)**.

9   Raise the engine slightly with a jack hoist. Remove the fasteners and detach th mount from the frame **(see illustrations)**.

10   Remove the mount-to-block bracke bolts/nuts and detach the mount.

11   Installation is the reverse of remova Use thread locking compound on the thread and be sure to tighten everything securely.

Chapter 2 Part C
2.3 liter four-cylinder (Quad-4) engine

Contents

Specifications

General

Cylinder numbers (timing chain end-to-transaxle end)	1-2-3-4
Firing order	1-3-4-2

Front

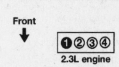

2.3L engine

Cylinder location

Camshafts and housings

Lobe lift (intake and exhaust)	
VIN D	0.3750 in (9.525 mm)
VIN A	0.4100 in (10.414 mm)
Lobe taper limit	0.0018 to 0.0033 inch per 0.5512 in (0.046 to 0.083 mm per 14.0 mm)
Endplay	0.0009 to 0.0088 in (0.025 to 0.225 mm)
Journal diameter	
No. 1	1.5728 to 1.5720 in (39.95 to 39.93 mm)
All others	1.3751 to 1.3760 in (34.93 to 34.95 mm)
Bearing oil clearance	0.0019 to 0.0043 in (0.050 to 0.110 mm)

Lifters

Bore diameter	1.3775 to 1.3787 in (34.989 to 34.019 mm)
Outside diameter	1.3763 to 1.3770 in (34.959 to 34.975 mm)
Lifter-to-bore clearance	0.0006 to 0.0024 in (0.014 to 0.060 mm)
Camshaft housing warpage limit	0.001 inch per 3.937 in (0.025 mm per 100 mm)

Oil pump gear backlash

Oil pump gear backlash	0.0091 to 0.0201 in (0.23 to 0.51 mm)

Torque specifications

	Ft-lbs (unless otherwise indicated)
Camshaft housing-to-cylinder head bolts	
Step 1	11
Step 2	Turn bolts an additional 75-degrees
Camshaft sprocket-to-camshaft bolt	40
Crankshaft pulley-to-crankshaft bolt	
Step 1	74
Step 2	Turn bolt an additional 90-degrees
Cylinder head bolts	
Short bolts	
Step 1	26
Step 2	Turn bolts an additional 100-degrees
Long bolts	
Step 1	26
Step 2	Turn bolts an additional 110-degrees
Exhaust manifold-to-cylinder head nuts	27
Flywheel-to-crankshaft bolts	
Step 1	22
Step 2	Turn bolts an additional 45-degrees
Intake manifold-to-cylinder head nuts/bolts	18
Oil pan baffle nuts/bolts	30
Oil pan bolts	
6 mm	106 in-lbs
8 mm	17
Oil pump-to-block bolts	33
Oil pump cover-to-oil pump body bolts	106 in-lbs
Oil pump screen assembly-to-pump bolts	22
Oil pump screen assembly-to-brace bolts	106 in-lbs
Crankshaft rear main oil seal housing bolts	106 in-lbs
Timing chain cover-to-housing bolts	106 in-lbs
Timing chain housing bolts	19
Timing chain housing-to-block stud	19
Timing chain tensioner bolts	115 in-lbs

1   General information

This Part of Chapter 2 is devoted to in-vehicle repair procedures for the 2.3 liter four-cylinder (Quad-4) engine. The 2.3L Quad-4 engine became available in 1990 and 1991, in the Pontiac Grand Prix and Oldsmobile Cutlass Supreme models only. All information concerning engine removal and installation and engine block and cylinder head overhaul can be found in Part G of this Chapter.

The following repair procedures are based on the assumption the engine is installed in the vehicle. If the engine has been removed from the vehicle and mounted on a stand, many of the steps outlined in this Part of Chapter 2 will not apply.

The Specifications included in this Part of Chapter 2 apply only to the procedures contained in this Part. Part G of Chapter 2 contains the Specifications necessary for cylinder head and engine block rebuilding.

The Quad-4 engine utilizes a number of advanced design features to increase power output and improve durability. The aluminum cylinder head contains four valves per cylinder. A double-row timing chain drives two overhead camshafts - one for intake and one for exhaust. Lightweight bucket-type hydraulic lifters actuate the valves. Rotators are used on all valves for extended service life.

The Quad-4 engine is easily identified by

the designation molded onto the ignition module cover. It can also be identified by the vehicle identification number on the left front corner of the instrument panel. This plate is visible from outside the vehicle (**see illustration 1.1** in Chapter 2, Part B). The eighth character in the sequence is the engine designation:

A or D = 2.3L four-cylinder, Quad-4 engine

2   Repair operations possible with the engine in the vehicle

Many major repair operations can be accomplished without removing the engine from the vehicle.

Clean the engine compartment and the exterior of the engine with some type of degreaser before any work is done. It'll make the job easier and help keep dirt out of the internal areas of the engine.

Depending on the components involved, it may be helpful to remove the hood to improve access to the engine as repairs are performed (refer to Chapter 11 if necessary). Cover the fenders to prevent damage to the paint. Special pads are available, but an old bedspread or blanket will also work.

If vacuum, exhaust, oil or coolant leaks develop, indicating a need for gasket or seal replacement, the repairs can generally be made with the engine in the vehicle. The

intake and exhaust manifold gaskets, timing chain housing gasket, oil pan gasket, crankshaft oil seals and cylinder head gasket are all accessible with the engine in place.

Exterior engine components, such as the intake and exhaust manifolds, the oil pan (and the oil pump), the water pump, the starter motor, the alternator and the fuel system components can be removed for repair with the engine in place.

Since the cylinder head can be removed without pulling the engine, camshaft and valve component servicing can also be accomplished with the engine in the vehicle. Replacement of the timing chain and sprockets is also possible with the engine in the vehicle.

In extreme cases caused by a lack of necessary equipment, repair or replacement of piston rings, pistons, connecting rods and rod bearings is possible with the engine in the vehicle. However, this practice is not recommended because of the cleaning and preparation work that must be done to the components involved.

3   Intake manifold - removal and installation

Removal

Refer to illustration 3.6

1   Relieve the fuel system pressure as described in Chapter 4.

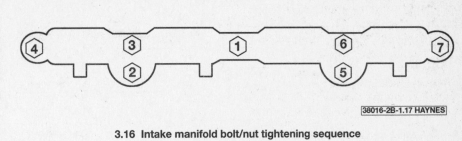

3.16 Intake manifold bolt/nut tightening sequence

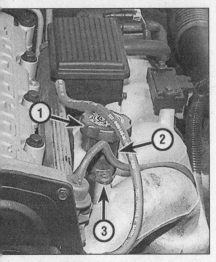

3.6 Oil fill tube components

1   Oil fill cap and dipstick
2   Oil fill tube
3   Bolt

2   Disconnect the negative battery cable from the battery, then refer to Chapter 1 and drain the cooling system. **Caution:** *On models equipped with the Theftlock audio system, be sure the lockout feature is turned off before performing any procedure which requires disconnecting the battery.*
3   Remove the throttle body (see Chapter 4).
4   Label and disconnect the vacuum and breather hoses and electrical wires.
5   Remove the oil/air separator (see Chapter 6).
6   Remove the oil fill cap and dipstick assembly **(see illustration)**.
7   Unbolt the oil fill tube and detach it from the engine block, rotating it as necessary to gain clearance between the intake tubes.
8   Remove the intake manifold brace **(see illustration)**.
9   Loosen the manifold mounting nuts/bolts in 1/4-turn increments until they can be removed by hand.
10   The manifold will probably be stuck to the cylinder head and force may be required to break the gasket seal. If necessary, dislodge the manifold with a soft-face hammer. **Caution:** *Don't pry between the head and manifold or damage to the gasket sealing surfaces will result and vacuum leaks could develop.*

Installation

Refer to illustration 3.16
Note: *The mating surfaces of the cylinder head and manifold must be perfectly clean when the manifold is installed. Gasket removal solvents in aerosol cans are available at most auto parts stores and may be helpful when removing old gasket material stuck to the head and manifold (since the components are made of aluminum, aggressive scraping can cause damage). Be sure to follow the directions printed on the container.*

11   Use a gasket scraper to remove all traces of sealant and old gasket material, then clean the mating surfaces with lacquer thinner or acetone. If there's old sealant or oil on the mating surfaces when the manifold is reinstalled, vacuum leaks may develop.
12   Use a tap of the correct size to chase the threads in the bolt holes, then use compressed air (if available) to remove the debris from the holes. **Warning:** *Wear safety glasses or a face shield to protect your eyes when using compressed air. Use a die to clean and restore the stud threads.*
13   Position the gasket on the cylinder head. Make sure all intake port openings, coolant passage holes and bolt holes are aligned correctly. **Note:** *Install the new manifold gasket with the numbers that are stamped on the gasket faced toward the manifold surface.*
14   Install the manifold, taking care to avoid damaging the gasket.
15   Thread the nuts/bolts into place by hand.
16   Tighten the nuts/bolts to the torque listed in this Chapter's Specifications following the recommended sequence **(see illustration)**. Work up to the final torque in three steps.
17   The remaining installation steps are the reverse of removal. Start the engine and check carefully for leaks at the intake manifold joints.

---

4   Exhaust manifold - removal and installation

Refer to illustrations 4.4 and 4.10

Removal

Warning: *Allow the engine to cool completely before performing this procedure.*
1   Disconnect the negative battery cable from the battery. **Caution:** *On models equipped with the Theftlock audio system, be sure the lockout feature is turned off before performing any procedure which requires disconnecting the battery.*
2   Unplug the oxygen sensor (see Chapter 6).
3   Remove the manifold heat shields.
4   Raise the vehicle and support it securely

on jackstands, then remove the exhaust pipe-to-manifold nuts. The nuts are usually rusted in place, so penetrating oil should be applied to the stud threads before attempting to remove them. Loosen them a little at a time, working from side-to-side to prevent the flange from jamming **(see illustration)**.
5   Remove the exhaust manifold brace.
6   Separate the exhaust pipe flange nuts (spring loaded) from the manifold studs, then pull the pipe down slightly to break the seal at the manifold joint.
7   Loosen the exhaust manifold mounting nuts 1/4-turn at a time each, working from the inside out, until they can be removed by hand.
8   Separate the manifold from the head and remove it.

Installation

9   The manifold and cylinder head mating surfaces must be clean when the manifold is reinstalled. Use a gasket scraper to remove all traces of old gasket material and carbon deposits.
10   Using a new gasket, install the manifold and hand tighten the fasteners. Working from the center out following the recommended sequence **(see illustration)**, tighten the bolts/nuts to the torque listed in this Chapter's Specifications.

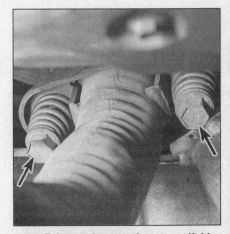

4.4 Exhaust pipe-to-exhaust manifold flange bolts – make sure you do not lose the springs when removing the bolts

11   The remaining installation steps are the reverse of removal.

5   Crankshaft pulley - removal and installation

Refer to illustrations 5.4, 5.5 and 5.8

1   Remove the cable from the negative battery terminal. **Caution:** *On models equipped with the Theftlock audio system, be sure the lockout feature is turned off before performing any procedure which requires disconnecting the battery.*
2   Remove the drivebelt (see Chapter 1).
3   With the parking brake applied and the shifter in Park (automatic) or in gear (manual), raise the front of the vehicle and support it securely on jackstands.
4   Remove the bolt from the front of the crankshaft. A breaker bar will probably be necessary, since the bolt is very tight. Use GM special tool J-38122 (or equivalent) to keep the crankshaft from turning. If the special tool isn't available, insert a bar through a hole in the pulley to prevent the crankshaft from turning **(see illustration)**.
5   Using a puller (tool J-24420-B or equivalent), remove the crankshaft pulley from the crankshaft **(see illustration)**.
6   Check the oil seal and replace it if necessary. Refer to Section 8.
7   Apply a thin layer of clean multi-purpose grease to the seal contact surface of the crankshaft pulley hub.
8   Position the pulley on the crankshaft and slide it through the seal until it bottoms against the crankshaft sprocket. Note that the slot (keyway) in the hub must be aligned with the Woodruff key in the end of the crankshaft **(see illustration)**. The crankshaft bolt can also be used to press the pulley into position.
9   Tighten the crankshaft bolt to the torque and angle of rotation listed in this Chapter's Specifications.
10   The remaining installation steps are the reverse of removal.

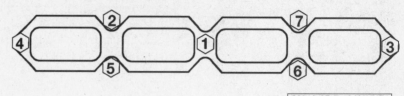

4.10 Exhaust manifold bolt tightening sequence

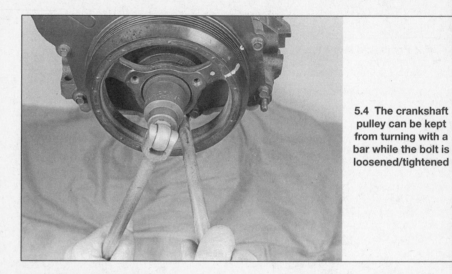

5.4 The crankshaft pulley can be kept from turning with a bar while the bolt is loosened/tightened

6   Crankshaft front oil seal - replacement

Refer to illustrations 6.2 and 6.5

1   Remove the crankshaft pulley (see Section 5).
2   Pry the old oil seal out with a seal removal tool **(see illustration)** or a screwdriver. Be very careful not to nick or otherwise damage the crankshaft in the process

and don't distort the timing chain cover.
3   Apply a thin coat of RTV-type sealant to the outer edge of the new seal. Lubricate the seal lip with moly-base grease or clean engine oil.
4   Place the seal squarely in position in the bore with the spring side facing in.
5   Carefully tap the seal into place with a large socket or section of pipe and a hammer **(see illustration)**. The outer diameter of the socket or pipe should be the same size as the seal outer diameter.

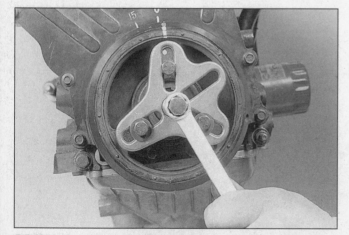

5.5 Use a bolt-type puller hub as shown here - if a jaw-type puller that applies force to the outer edge is used, the pulley will be damaged

5.8 Align the keyway in the crankshaft pulley hub with the Woodruff key in the crankshaft (arrow)

6.2 Pry the old seal out with a seal removal tool (shown here)
or a screwdriver

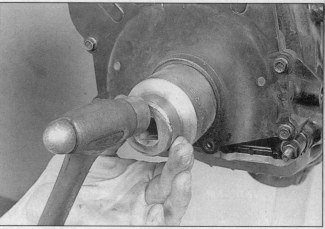

6.5 Install the new seal with a large socket or piece of pipe
and a hammer

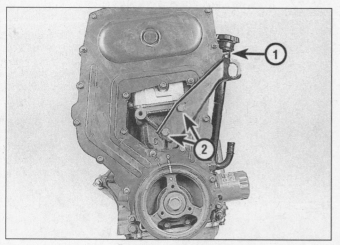

7.3 Remove the vent hose and engine lifting bracket bolts

1   Vent hose fitting   2   Engine lifting bracket bolts

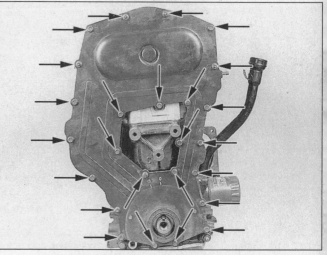

7.5 Timing chain cover fastener locations (arrows)

6   Install the crankshaft pulley (see Section 5).
7   Start the engine and check for oil leaks at the seal.

---

7   Timing chain and sprockets - removal, inspection and installation

Note: *Special tools are required for this procedure, so read through it before beginning work.*

Removal

Refer to illustrations 7.3, 7.5, 7.8, 7.10, 7.11, 7.12, 7.14, 7.17 and 7.18

1   Disconnect the negative battery cable from the battery. **Caution:** *On models equipped with the Theftlock audio system, be sure the lockout feature is turned off before performing any procedure which requires disconnecting the battery.*
2   Remove the coolant reservoir (see Chapter 3).

7.8 Note how it's installed,
then remove the oil slinger
(arrow) from the crankshaft

3   Detach the timing chain cover vent hose and unbolt the engine lifting bracket at the drivebelt end of the engine **(see illustration)**.
4   Remove the crankshaft pulley (see Section 5).
5   Working from above, remove the upper timing chain cover fasteners **(see illustration)**.

6   Working from below, remove the lower timing chain cover fasteners.
7   Detach the cover and gaskets from the housing.
8   Slide the oil slinger off the crankshaft **(see illustration)**.
9   Temporarily reinstall the crankshaft pul-

7.10 Insert 8 mm (5/16-inch) bolts (arrows) or pins to hold the
camshaft sprockets

7.11 The mark on the crankshaft sprocket must align with the
mark on the block (arrows)

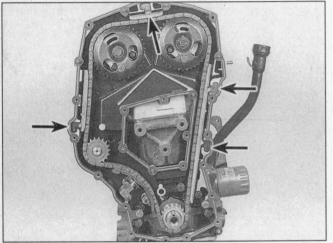

7.12 The timing chain guides are wedged into the housing at four
points (arrows) - just pull them out

7.14 The timing chain tensioner shoe is held in place by an E-clip

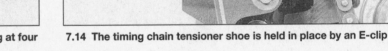

1   Tensioner mounting bolts   2   E-clip

ley bolt to use when turning the crankshaft.
10 Turn the crankshaft clockwise until the
camshaft sprocket's timing pin holes line up
with the holes in the timing chain housing.
Insert 8 mm pins or bolts into the holes to
maintain alignment (see illustration).
11 The mark on the crankshaft sprocket
should line up with the mark on the engine
block (see illustration). The crankshaft
sprocket keyway should point up and line up

with the centerline of the cylinder bores.
12 Remove the three timing chain guides
(see illustration).
13 Detach the timing chain tensioner sleeve
and spring.
14 Make sure all the slack in the timing
chain is above the tensioner assembly, then
remove the chain tensioner shoe (see illustration). The timing chain must be disengaged from the wear grooves in the tensioner

shoe in order to remove the shoe. Slide a
screwdriver blade under the timing chain
while pulling the shoe out. Note: If difficulty is
encountered when removing the chain tensioner shoe, proceed as follows:
a) Hold the intake camshaft sprocket with a
   special tool and remove the sprocket
   bolt and washer.
b) Remove the washer from the bolt and
   thread the bolt back into the camshaft
   by hand.
c) Remove the intake camshaft sprocket,
   using a three-jaw puller in the three relief
   holes in the sprocket, if necessary. Caution: Don't try to pry the sprocket off the
   camshaft or damage to the sprocket
   could occur.
15 Remove the tensioner assembly retaining bolts and tensioner. The spring can be
caged with a special anti-release tool (see
illustration 7.31). Warning: The tensioner
plunger is spring loaded and could come out
with great force, causing personal injury.
16 Remove the chain housing-to-block
stud (timing chain tensioner shoe pivot).
17 Slip the timing chain off the sprockets

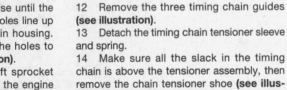

7.17 Begin removing
the chain at the
exhaust camshaft
sprocket

7.18 Mark the sprockets, then remove the bolts and pull the sprockets off

7.22 The camshaft sprocket locating dowels should be near the top (arrows) prior to sprocket installation

(see illustration).

18   To remove the camshaft sprockets (if not already done), loosen the bolts while the pins are still in place. Mark the sprockets for identification **(see illustration)**, remove the bolts and pins, then pull on the sprockets by hand until they slip off the dowels.

19   The crankshaft sprocket will slip off the crankshaft by hand.

20   The idler sprocket and bearing are pressed into place. If replacement is necessary, remove the timing chain housing (see Section 8) and take it to a dealer service department or automotive machine shop. Special tools are required and the bearing must be replaced each time it's pressed out.

Inspection

21   Visually inspect all parts for wear and damage. Look for loose pins, cracks, worn rollers and side plates. Check the sprockets for hook-shaped, chipped and broken teeth. **Note:** *Some scoring of the timing chain shoe and guides is normal.* Replace the timing chain, sprockets, chain shoe and guides as a set if the engine has high mileage or fails the visual parts inspection.

Installation

Refer to illustrations 7.22, 7.23 and 7.31

22   Turn the camshafts until the dowel pins are at the top **(see illustration)**. Install both camshaft sprockets (if removed). Apply sealant to the camshaft sprocket bolt threads and make sure the washers are in place. Keep the camshaft from turning with 8 mm bolts). Tighten the bolts to the torque listed in this Chapter's Specifications.

23   Recheck the positions of the camshaft and crankshaft sprockets for correct valve timing **(see illustration)**. **Note:** *If the camshafts are out of position and must be rotated more than 1/8-turn in order to install the alignment dowel pins:*

a) *The crankshaft must be rotated 90-degrees clockwise past Top Dead Center to give the valves adequate clearance to open.*

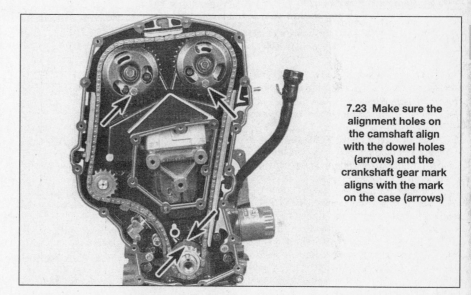

7.23 Make sure the alignment holes on the camshaft align with the dowel holes (arrows) and the crankshaft gear mark aligns with the mark on the case (arrows)

b) *Once the camshafts are in position and the dowels installed, rotate the crankshaft counterclockwise back to Top Dead Center.* **Caution:** *Do not rotate the crankshaft clockwise to TDC (valve or piston damage could occur).*

24   Slip the timing chain over the exhaust camshaft sprocket, then around the idler and crankshaft sprocket.

25   Remove the alignment dowel pin from the intake camshaft. Using a special sprocket rotating tool, rotate the intake camshaft sprocket counterclockwise enough to mesh the timing chain with it. Release the special tool. The chain run between the two camshaft sprockets will tighten. If the valve timing is correct, the intake cam alignment dowel pin should slide in easily. If it doesn't index, the camshafts aren't timed correctly; repeat the procedure.

26   Leave the dowel pins installed for now. Working under the vehicle, check the timing marks. With slack removed from the timing chain between the intake cam sprocket and the crankshaft sprocket, the timing marks on the crankshaft sprocket and the engine block should be aligned. If the marks aren't aligned,

move the chain one tooth forward or backward, remove the slack and recheck the marks.

27   Install the chain housing-to-block stud (timing chain tensioner shoe pivot), and tighten it to the torque listed in this Chapter's Specifications.

28   Reload the timing chain tensioner assembly to its "zero" position as follows:

a) *Assemble the restraint cylinder, spring and nylon plug in the plunger. While rotating the restraint cylinder clockwise, push it into the plunger until it bottoms. Keep rotating the restraint cylinder clockwise, but allow the spring to push it out of the plunger. The pin in the plunger will lock the restraint cylinder in the loaded position.*

b) *Position a special tensioner anti-release tool onto the plunger.*

c) *Install the plunger assembly in the tensioner body with the long end toward the crankshaft when installed.*

29   Install the tensioner assembly in the chain housing. Tighten the bolts to the torque listed in this Chapter's Specifications. **Note:**

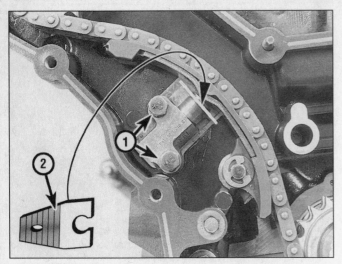

7.31 Remove the anti-release device from the tensioner and depress the shoe once to release the tensioner

1   *Tensioner mounting bolts*
2   *Tensioner anti-release tool*

8.14 Position a new gasket over the dowel pins (arrows)

Recheck the plunger assembly installation - it's correctly installed when the long end is toward the crankshaft.

30   Install the tensioner shoe, spring and sleeve.

31   Remove tensioner anti-release tool and squeeze the plunger assembly into the tensioner body to unload the plunger assembly **(see illustration)**.

32   Remove the alignment pins.

33   Slowly rotate the crankshaft clockwise two full turns (720-degrees). Do not force it; if resistance is felt, back up and recheck the installation procedure. Align the crankshaft timing mark with the mark on the engine block and temporarily reinstall the 8 mm alignment pins. The pins should slide in easily if the valve timing is correct. **Caution:** *If the valve timing is incorrect, severe engine damage could occur.*

34   Install the remaining components in the reverse order of removal. Check fluid levels, start the engine and check for proper operation and coolant/oil leaks.

8   Timing chain housing - removal and installation

Refer to illustration 8.14

1   Remove the timing chain and sprockets (see Section 7).

2   Remove the exhaust manifold (see Section 4).

3   If you're installing a replacement timing chain housing, remove the water pump (see Chapter 3).

4   Remove the timing chain housing-to-belt tensioner bracket brace.

5   Remove the four oil pan-to-timing chain housing bolts.

6   Remove the timing chain housing-to-block lower fasteners.

7   Remove the lowest cover retaining stud from the timing chain housing.

8   Remove the rear engine mount nut (see Section 16).

9   Loosen the front engine mount nut, leaving about three threads remaining in contact.

10   Remove the eight chain housing-to-

camshaft housing bolts.

11   Position a floor jack under the oil pan, using an 18-inch long piece of wood (2x4) on the jack pad to distribute the weight.

12   Raise the engine off the front and rear mounts until the front mount bracket contacts the nut.

13   Remove the timing chain housing and gaskets. Thoroughly clean the mating surfaces to remove any traces of old sealant or gasket material.

14   Install the timing chain housing with new gaskets **(see illustration)**. Tighten the bolts to the torque listed in this Chapter's Specifications.

15   The remaining steps are the reverse of removal.

9   Camshaft, lifters and housing - removal, inspection and installation

Note: *Special tools are required for this procedure, so read through it before beginning work.*

Removal

Refer to illustrations 9.7, 9.10, 9.11 and 9.14

1   Disconnect the cable from the negative battery terminal. **Caution:** *On models equipped with the Theftlock audio system, be sure the lockout feature is turned off before performing any procedure which requires disconnecting the battery.*

2   Remove the timing chain and sprockets (see Section 7).

3   Remove the timing chain housing-to-camshaft housing bolts (see Section 8).

4   Remove the ignition coil and module assembly (see Chapter 5).

Intake (front)

5   Disconnect the idle speed power steer

9.7 The power steering pump drive pulley must be removed with a puller

9.10 Oil pressure sending unit mounting hole (1) and engine lifting bracket bolts (2)

9.11 Gently lift the camshaft housing off the cylinder head and turn it over so the lifters won't fall out

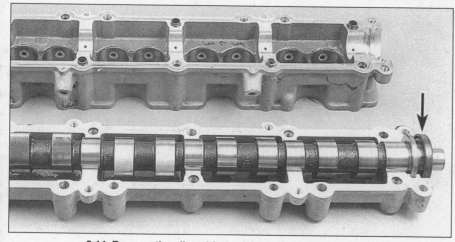

9.14 Remove the oil seal (arrow) from the intake camshaft

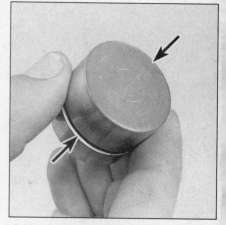

9.17a Check the camshaft lobe surfaces and the bore surfaces (arrows) of the lifters for wear

ing pressure switch.

6 Remove the power steering pump and brackets (see Chapter 10).

7 Remove the power steering pump drive pulley **(see illustration)**. **Caution:** *The power steering pump drive pulley must be removed following this procedure or damage to the pulley will result. If any other removal procedure is used, a new pulley must be installed.*

8 Remove the oil/air separator (PCV System) as an assembly, with the hoses attached (see Chapter 6).

9 Remove the fuel rail from the cylinder head and set it aside (see Chapter 4).

Exhaust (rear)

10 Unplug the oil pressure sending unit wire and unbolt the engine lifting bracket **(see illustration)**.

Intake and exhaust

11 Loosen the camshaft housing-to-cylinder head bolts in 1/4-turn increments, following the tightening sequence in reverse **(see illustration 9.28)**. Leave the two cover-to-housing bolts in place temporarily. Lift the housing off

the cylinder head **(see illustration)**.

12 Remove the two camshaft cover-to-housing bolts. Push the cover off the housing by threading four of the housing-to-head bolts into the tapped holes in the cover. Carefully lift the camshaft out of the housing.

13 Remove all traces of old gasket material from the mating surfaces and clean them with lacquer thinner or acetone to remove any traces of oil.

14 Remove the oil seal from the intake camshaft **(see illustration)** and discard it.

15 Remove the lifters and store them in order so they can be reinstalled in their original locations. To minimize lifter bleed-down, store the lifters valve-side up, submerged in clean engine oil.

Inspection

Refer to illustrations 9.17a, 9.17b, 9.18 and 9.19

16 Refer to Chapter 2, Part G, for camshaft inspection procedures, but use this Chapter's Specifications. Do not attempt to salvage camshafts. Whenever a camshaft is replaced,

replace all the lifters actuated by the camshaft as well.

17 Visually inspect the lifters for wear, galling, score marks and discoloration from overheating **(see illustrations)**.

9.17b Check the valve-side of the lifters too

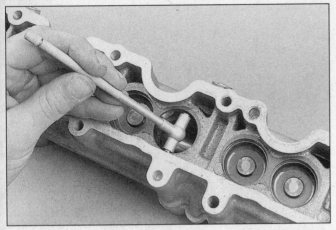

9.18 Use a telescoping gauge and micrometer to measure the lifter bores . . .

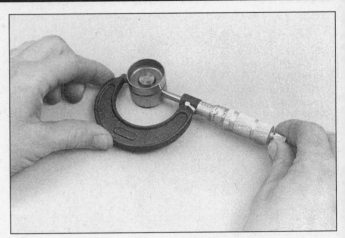

9.19 . . . then measure the lifters with a micrometer - subtract each lifter diameter from the corresponding bore diameter to obtain the lifter-to-bore clearances

9.24 The dowel pins (arrows) should be at the top (12 o'clock position)

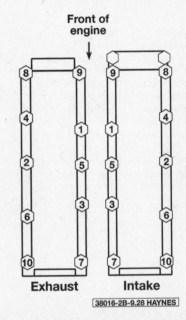

9.28 Camshaft housing-to-cylinder head bolt tightening sequence

18   Measure each lifter bore inside diameter and record the results **(see illustration)**.

19   Measure each lifter outside diameter and record the results **(see illustration)**.

20   Subtract the lifter outside diameter from the corresponding bore inside diameter to determine the clearance. Compare the results to this Chapter's Specifications and replace parts as necessary.

Installation

Refer to illustrations 9.24, 9.28 and 9.29

21   Using a new gasket, position the camshaft housing on the cylinder head and temporarily hold it in place with one bolt.

22   Coat the camshaft journals and lobes and the lifters with Camshaft and Lifter Prelube or equivalent and install them in their original locations.

23   On the intake camshaft only, lubricate the lip of the oil seal, then position the seal on the camshaft journal with the spring side facing in.

24   Install the camshaft in the housing with the sprocket dowel pin UP (12 o'clock position) **(see illustration)**. Position the cover on the housing, holding it in place with the two bolts, as described above.

25   Apply Pipe Sealant or equivalent to the threads of the camshaft housing and cover bolts.

26   Install new housing seals. Each seal is pre-fit.

27   Install the camshaft cover and bolts while positioning the oil seal (intake side only). Be sure the seal is installed to the depth of 0.020 inch (0.50 mm) into the cover and cylinder head.

28   Tighten the bolts in the sequence shown **(see illustration)** to the torque and angle of rotation listed in this Chapter's Specifications.

29   Install the power steering pump pulley with a special tool **(see illustration)**.

30   Install the remaining parts in the reverse order of removal.

31   Change the oil and filter (see Chapter 1).

Note: *If new lifters have been installed or the lifters bled down while the engine was disassembled, excessive lifter noise may be experienced after startup - this is normal. Use the following procedure to purge the lifters of air:*

a) *Start the engine and allow it to warm up for five minutes.*

b) *Increase engine speed to 2000 rpm until the lifter noise is gone.*

32   Road test the vehicle and check for oil and coolant leaks.

9.29 The power steering pump drive pulley must be pressed on with a special tool

10.5 This is what the air hose adapter that threads into the spark plug hole looks like - they're commonly available from auto parts stores

10.18 Apply a small dab of grease to each keeper before installation to hold it in place on the valve stem until the spring is released

10 Valve springs, retainers and seals - replacement

Refer to illustrations 10.5 and 10.18
Note: *Broken valve springs and defective valve stem seals can be replaced without removing the cylinder head. Two special tools and a compressed air source are normally required to perform this operation, so read through this Section carefully and rent or buy the tools before beginning the job. If compressed air isn't available, a length of nylon rope can be used to keep the valves from falling into the cylinder during this procedure.*

1   Remove the spark plug from the cylinder which has the defective part. Due to the design of this engine, the intake and exhaust camshaft housings can be removed separately to service their respective components. If all of the valve stem seals are being replaced, all of the spark plugs and both camshaft housings should be removed.

2   Refer to Chapter 5 and remove the ignition coil assembly.

3   Remove the camshaft(s), lifters and housing(s) as described in Section 9.

4   Turn the crankshaft until the piston in the affected cylinder is at top dead center on the compression stroke (refer to Chapter 2, Part G, for instructions). If you're replacing all of the valve stem seals, begin with cylinder number one and work on the valves for one cylinder at a time. Move from cylinder-to-cylinder following the firing order sequence (see this Chapter's Specifications).

5   Thread an adapter into the spark plug hole **(see illustration)** and connect an air hose from a compressed air source to it. Most auto parts stores can supply the air hose adapter. **Note:** *Many cylinder compression gauges utilize a screw-in fitting that may work with your air hose quick-disconnect fitting.*

6   Apply compressed air to the cylinder. **Warning:** *The piston may be forced down by compressed air, causing the crankshaft to turn suddenly. If the wrench used when positioning the number one piston at TDC is still attached to the bolt in the crankshaft nose, it could cause damage or injury when the crankshaft moves.*

7   The valves should be held in place by the air pressure. If the valve faces or seats are in poor condition, leaks may prevent air pressure from retaining the valves - in this case a "valve job" is needed.

8   If you don't have access to compressed air, an alternative method can be used. Position the piston at a point just before TDC on the compression stroke, then feed a long piece of nylon rope through the spark plug hole until it fills the combustion chamber. Be sure to leave the end of the rope hanging out of the engine so it can be removed easily. Use a large ratchet and socket to rotate the crankshaft in the normal direction of rotation (clockwise) until slight resistance is felt.

9   Stuff shop rags into the cylinder head holes adjacent to the valves to prevent parts and tools from falling into the engine, then use a valve spring compressor to compress the spring. Remove the keepers with small needle-nose pliers or a magnet.

10   Remove the retainer and valve spring, then remove the valve guide seal and rotator. **Note:** *If air pressure fails to hold the valve in the closed position during this operation, the valve face or seat is probably damaged. If so, the cylinder head will have to be removed for additional repair operations.*

11   Wrap a rubber band or tape around the top of the valve stem so the valve won't fall into the combustion chamber, then release the air pressure. **Note:** *If a rope was used instead of air pressure, turn the crankshaft slightly in a counterclockwise direction (opposite normal rotation).*

12   Inspect the valve stem for damage. Rotate the valve in the guide and check the end for eccentric movement, which would indicate the valve stem is bent.

13   Move the valve up-and-down in the guide and make sure it doesn't bind. If the valve stem binds, either the valve is bent or the guide is damaged. In either case, the head will have to be removed for repair.

14   Reapply air pressure to the cylinder to retain the valve in the closed position, then remove the tape or rubber band from the valve stem. If a rope was used instead of air pressure, rotate the crankshaft in the normal direction of rotation until slight resistance is felt.

15   Reinstall the valve rotator.

16   Lubricate the valve stem with engine oil and install a new guide seal.

17   Install the spring in position over the valve.

18   Install the valve spring retainer. Compress the valve spring and carefully install the keepers in the groove. Apply a small dab of grease to the inside of each keeper to hold it in place if necessary **(see illustration)**. Remove the pressure from the spring tool and make sure the keepers are seated.

19   Disconnect the air hose and remove the adapter from the spark plug hole. If a rope was used in place of air pressure, pull it out of the cylinder.

20   Refer to Section 9 and install the camshaft(s), lifters and housing(s).

21   Install the spark plug(s) and coil assembly.

22   Start and run the engine, then check for oil leaks and unusual sounds coming from the camshaft housings.

11 Cylinder head - removal and installation

Removal

1   Disconnect the negative cable from the battery. **Caution:** *On models equipped with the Theftlock audio system, be sure the lock-out feature is turned off before performing any procedure which requires disconnecting the battery.*

2   Refer to Section 3 and remove the intake manifold. The cooling system must be drained to prevent coolant from getting into internal areas of the engine when the head is removed.

3   Refer to Section 4 and detach the exhaust manifold.

4   Remove the camshafts and housings as described in Section 9.

5   Make a holder for the head bolts. Using the new head gasket, outline the cylinders and bolt pattern on a piece of cardboard. Be sure to indicate the front of the engine for reference. Punch holes at the bolt locations **(see illustration 8.11 in Part B).**

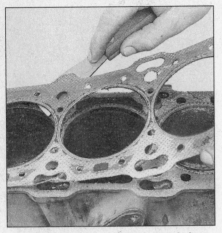

11.10 Remove the old gasket and clean the head thoroughly

11.15 Cylinder head bolt TIGHTENING sequence

6   Loosen the head bolts in 1/4-turn increments until they can be removed by hand. Work from bolt-to-bolt in a pattern that's the reverse of the tightening sequence **(see illustration 11.15)**. Store the bolts in the cardboard holder as they're removed - this will ensure they are reinstalled in their original locations.

7   Lift the head off the engine. If resistance is felt, don't pry between the head and block as damage to the mating surfaces will result. To dislodge the head, place a block of wood against the end of it and strike the wood block with a hammer. Store the head on blocks of wood to prevent damage to the gasket sealing surfaces.

8   Cylinder head disassembly and inspection procedures are covered in detail in Chapter 2, Part G.

Installation

Refer to illustrations 11.10 and 11.15

9   The mating surfaces of the cylinder head and block must be perfectly clean when the head is installed.

10   Use a gasket scraper to remove all traces of carbon and old gasket material **(see illustration)**, then clean the mating surfaces with lacquer thinner or acetone. If there's oil on the mating surfaces when the head is installed, the gasket may not seal correctly and leaks could develop. **Note:** *Since the head is made of aluminum, aggressive scraping can cause damage. Be extra careful not to nick or gouge the mating surface with the scraper. Use a vacuum cleaner to remove debris that falls into the cylinders.*

11   Check the block and head mating surfaces for nicks, deep scratches and other damage. If damage is slight, it can be removed with a flat mill file; if it's excessive, machining may be the only alternative.

12   Use a tap of the correct size to chase the threads in the head bolt holes. Mount each bolt in a vise and run a die down the threads to remove corrosion and restore the threads. Dirt, corrosion, sealant and damaged threads will affect torque readings.

12.11 The oil pan baffle is held in place by four bolts (arrows)

13   Position the new gasket over the dowel pins in the block.

14   Carefully position the head on the block without disturbing the gasket.

15   Install the bolts in their original locations and tighten them finger tight. Following the recommended sequence **(see illustration)**, tighten the bolts in several steps to the torque and angle of rotation listed in this Chapter's Specifications.

16   The remaining installation steps are the reverse of removal.

17   Refill the cooling system and change the oil and filter (see Chapter 1, if necessary).

18   Run the engine and check for leaks and proper operation.

12 Oil pan - removal and installation

Refer to illustrations 12.11 and 12.15
Note: *The following procedure is based on the assumption the engine is in place in the vehicle. If its been removed, simply unbolt the oil pan and detach it from the block.*

Removal

1   Disconnect the negative battery cable from the battery, then refer to Chapter 1 and drain the oil. **Caution:** *On models equipped*

with the Theftlock audio system, be sure th lockout feature is turned off before perform ing any procedure which requires discon necting the battery.

2   Remove the lower splash shield.

3   Detach the lower bellhousing cover.

4   Unbolt the exhaust manifold brac (manual transaxle models only - see Sec tion 4).

5   Remove the radiator outlet pipe-to-o pan bolt.

6   On manual transaxle equipped model remove the transaxle-to-oil pan nut and stu with a 7 mm socket.

7   Gently pry the spacer out from betwee the oil pan and transaxle.

8   Remove the oil pan-to-transaxle bolt.

9   Remove the oil pan mounting bolts.

10   Carefully separate the pan from th block. Don't pry between the block and pa or damage to the sealing surfaces may resu and oil leaks may develop. **Note:** *Th crankshaft may have to be rotated to ga clearance for oil pan removal.*

11   If you need to get at the crankshaft other lower end components, remove the pan baffle **(see illustration)**.

Installation

12   Clean the sealing surfaces with lacqu thinner or acetone. Make sure the bolt hol

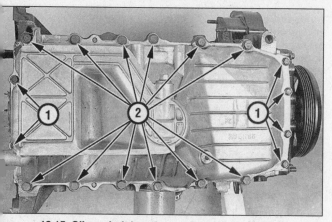

12.15 Oil pan bolt locations - viewed from below

13.5 Check the oil pump drive gear (arrow) and driven gear for wear and damage

the block are clean.

3   The gasket should be checked carefully and replaced with a new one if damage is noted. Minor imperfections can be repaired with silicone sealant. **Caution:** *Use only enough sealant to restore the gasket to its original size and shape. Excess sealant may cause part misalignment and oil leaks.*

4   Reinstall the oil pan baffle, if removed. With the gasket in position, carefully hold the pan against the block and install the bolts finger tight.

5   Tighten the bolts in three steps to the torque listed in this Chapter's Specifications **(see illustration)**. Start at the center of the pan and work out toward the ends in a spiral pattern. Note that the bolts are not all tightened to the same torque value.

6   The remaining steps are the reverse of removal. **Caution:** *Don't forget to refill the engine with oil before starting it (see Chapter 1).*

7   Start the engine and check carefully for oil leaks at the oil pan.

13 Oil pump - removal, inspection and installation

Refer to illustration 13.5

Removal

1   Remove the oil pan as described in Section 12.

2   While supporting the oil pump, remove the mounting bolts.

3   Lower the pump from the engine.

Inspection

4   Clean all parts thoroughly and remove the cover.

5   Visually inspect all parts for wear, cracks and other damage **(see illustration)**. Replace the pump if it's defective, if the engine has high mileage or if the engine is being rebuilt.

Installation

6   Position the pump on the engine and

install the mounting bolts. Tighten them to the torque listed in this Chapter's Specifications.

7   Oil pump drive gear backlash must be checked whenever the oil pump, crankshaft or engine block is replaced. Mount a dial indicator on the engine with the indicator stem touching the oil pump driven gear. Check the backlash and compare it to the Specifications in this Chapter. If the clearance is not correct, replace the oil pump.

8   Install the oil pan (and baffle, if removed).

9   Add oil and run the engine. Check for oil pressure and leaks.

14 Flywheel - removal and installation

This procedure is essentially the same for all engines. Refer to Part B and follow the driveplate removal and installation procedure outlined there. Check the surface of the flywheel for cracks, rivet grooves and other signs of damage. If necessary, have the flywheel machined by an automotive machine shop. Be sure to use the bolt torque listed in this Chapter's Specifications.

15 Rear main oil seal - replacement

Refer to illustrations 15.5, 15.6, 15.7 and 15.8
Warning: *A special tool is available to support the engine during repair operations. Similar fixtures are available from rental yards. Improper lifting methods or devices are hazardous and could result in severe injury or death. DO NOT place any part of your body under the engine/transaxle when it's supported only by a jack. Failure of the lifting device could result in serious injury or death.*
Note: *The rear main (crankshaft) oil seal is a one-piece unit that can be replaced without removing the engine. However, the transaxle must be removed and the engine must be supported as this procedure is done. A special tool is available for seal installation, but the procedure outlined here was devised to avoid having to use the tool.*

1   Remove the transaxle (see Chapter 7).
2   Remove the pressure plate and clutch disc (see Chapter 8).
3   Remove the flywheel (see Section 14).
4   Remove the oil pan (see Section 12).
5   After the oil pan has been removed, remove the bolts **(see illustration)**, detach the seal housing and peel off all the old gas-

15.5 Remove the seal housing bolts (arrows)

15.6 After removing the housing from the engine, support it on wood blocks and drive out the old seal with a punch and hammer

15.7 Drive the new seal into the housing with a block of wood or a section of pipe, if you have one large enough - don't cock the seal in the housing bore

15.8 Lubricate the seal journal and lip, then position a new gasket over the dowel pins (arrows)

16.7 Upper torque strut mounting bolt locations (arrows)

ket material.

6   Position the seal housing on a couple of wood blocks on a workbench and drive the old seal out from the back side with a punch and hammer **(see illustration)**.

7   Drive the new seal into the housing with a wood block **(see illustration)**.

8   Lubricate the crankshaft seal journal and the lip of the new seal with moly-base grease. Position a new gasket on the engine block **(see illustration)**.

9   Slowly and carefully push the new seal onto the crankshaft. The seal lip is stiff, so work it onto the crankshaft with a smooth object such as the end of an extension as you push the housing against the block.

10   Install and tighten the housing bolts to the torque listed in this Chapter's Specifications.

11   Install the flywheel and clutch components.

12   Reinstall the transaxle.

16   Engine mounts - check and replacement

Refer to illustrations 16.7, 16.8 and 16.13

Warning: *A special tool is available to support the engine during repair operations. Similar fixtures are available from rental yards. Improper lifting methods or devices are hazardous and could result in severe injury or death. DO NOT place any part of your body under the engine/transaxle when it's supported only by a jack. Failure of the lifting device could result in serious injury or death.*

1   Engine mounts seldom require attention, but broken or deteriorated mounts should be replaced immediately or the added strain placed on the driveline components may cause damage or wear.

Check

2   During the check, the engine must be

raised slightly to remove the weight from the mounts.

3   Raise the vehicle and support it securely on jackstands. Support the engine as described above. If the special support fixture is unavailable, position a jack under the engine oil pan. Place a large block of wood between the jack head and the oil pan, then carefully raise the engine just enough to take the weight off the mounts. **Warning:** *DO NOT place any part of your body under the engine when it's supported only by a jack!*

4   Check the mounts to see if the rubber is cracked, hardened or separated from the metal plates. Sometimes the rubber will split right down the center.

5   Check for movement between the mount plates and the engine or frame (use a large screwdriver or pry bar to attempt to move the mounts). If movement is noted, lower the engine and tighten the mount fasteners.

16.8 Torque strut to body bracket bolts

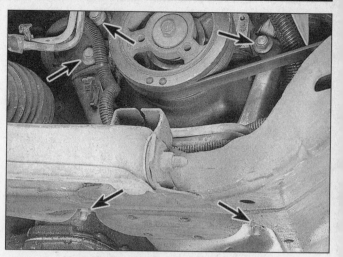

16.13 Engine mount bolt locations (arrows)

6   Rubber preservative should be applied to the mounts to slow deterioration.

Replacement

Upper mount (torque strut)

7   Detach the negative battery cable from the battery. Remove the upper torque strut bracket bolt **(see illustration)**. **Caution:** *On models equipped with the Theftlock audio system, be sure the lockout feature is turned off before performing any procedure which requires disconnecting the battery.*

8   Remove the torque strut-to-body mount bolts **(see illustration)**.
9   Remove the torque strut.
10   Place the new torque strut in position and install the bolts. Tighten the nuts securely.
11   Reconnect the negative battery cable.

Lower mount

12   Detach the negative battery cable from the battery. **Caution:** *On models equipped with the Theftlock audio system, be sure the lockout feature is turned off before perform-*

ing any procedure which requires disconnecting the battery.
13   Working under the frame mounting bracket, remove the nuts from the engine mount **(see illustration)**.
14   Raise the engine off the mount.
15   Remove the mount-to-bracket nuts.
16   Remove the mount from the vehicle.
17   Installation is the reverse of removal.
18   Gently lower the engine.
19   Tighten the nuts securely.
20   Reconnect the negative battery cable.

Notes

Chapter 2 Part D
2.8L, 3.1L and 3100 V6 engines

Contents

Specifications

General

Cylinder numbers (drivebelt end-to-transaxle end)
Front bank (radiator side) 2-4-6
Rear bank 1-3-5
Firing order 1-2-3-4-5-6

Torque specifications

Ft-lbs (unless otherwise indicated)

Camshaft sprocket bolts
2.8L and 3.1L
  1989 and earlier 18
  1990 on 21
3100
  1995 and earlier 74
  1996 and 1997 81
  1998 and later 103
Cylinder head bolts
1997 and earlier
  First step 33
  Second step .... Rotate an additional 90 degrees (1/4-turn)
1998 and later
  First step 37
  Second step .... Rotate an additional 90 degrees (1/4-turn)
Exhaust manifold-to-cylinder head bolts
2.8L and 3.1L 18
3100 12
Flywheel/driveplate-to-crankshaft bolts
2.8L and 3.1L 52
3100
  1995 and earlier 59
  1996 and 1997 61
  1998 and later 52
Lower intake manifold-to-cylinder head bolts/nuts
2.8L and 3.1L
  Step one 15
  Step two 24
3100 .... 115 in-lbs

Front ↓

2.8L and 3.1L V6 engines

1 4 6 3 2 5

1671H

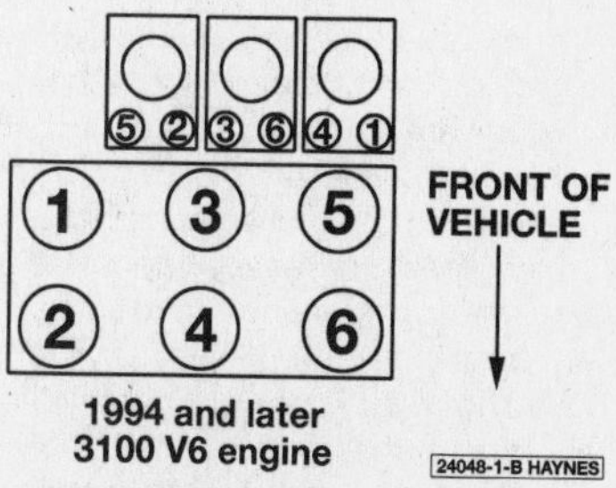

FRONT OF VEHICLE ↓

1994 and later 3100 V6 engine

24048-1-B HAYNES

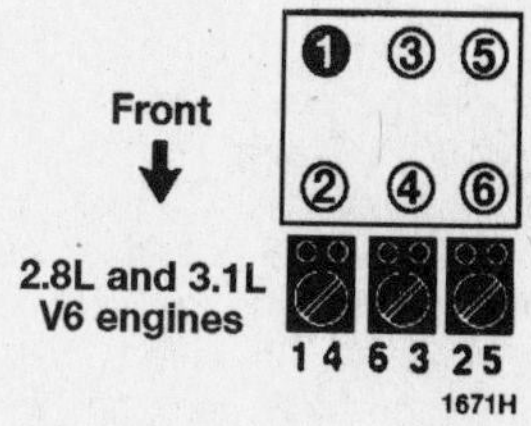

Cylinder and coil terminal locations

Torque specifications (continued)   Ft-lbs (unless otherwise indicated)

Oil pan bolts/nuts	
2.8L and 3.1L (see illustration 14.15a)	
Number 1	18
Number 2	13
All others	89 in-lbs
3100	
Retaining	18
Side	37
Oil pump mounting bolts	30
Oil pump drive bolt (3100)	27
Rocker arm nuts/bolts	
1995 and earlier	
Nuts	18
1996 and 1997	
Bolts	89 in-lbs plus an additional 30 degrees
1998 and later	
Bolts	168 in-lbs plus an additional 30 degrees
Timing chain cover bolts (see illustration 12.16)	
2.8L and 3.1L	
6 mm (number 3)	20
8 mm (number 4)	28
3100	
1997 and earlier	
Small	15
Large	35
1998 and later	
Small	15
Medium	35
Large	41
Timing chain damper bolt	
2.8L and 3.1L	15
3100	
1995 and earlier	18
1996 and later	15
Valve cover-to-cylinder head bolts	89 in-lbs
Vibration damper bolt	76

1   General information

Refer to illustration 1.2

This Part of Chapter 2 is devoted to in-vehicle repair procedures for the 2.8L, 3.1L and 3100 V6 engines. The 2.8L engine became available in 1988 and 1989, in the Pontiac Grand Prix, Oldsmobile Cutlass Supreme and the Buick Regal models only. The 3.1L engine became available in 1989 and was produced through 1994, in all four models covered by this manual. The 3100 engine became available in 1994 through 1999, it is utilized in the Pontiac Grand Prix, Oldsmobile Cutlass Supreme and the Buick Regal models only. These engines utilize cast-iron blocks with six cylinders arranged in a "V" shape at a 60-degree angle between the two banks. The overhead valve aluminum cylinder heads are equipped with replaceable valve guides and seats. Hydraulic lifters actuate the valves through tubular pushrods.

The engines are easily identified by looking for the designations printed on the aluminum intake plenum, directly on the top (see illustration). To positively identify these engines, locate the vehicle identification number on the left front corner of the instrument panel. This plate is visible from outside the vehicle (see illustration 1.1 in Chapter 2,

Part B). The eighth character in the sequence is the engine designation:

W = 2.8L V6 engine
T = 3.1L V6 engine
V = 3.1L Turbocharged V6 engine
M = 3100 V6 engine

All information concerning engine removal and installation and engine block and cylinder head overhaul can be found in Part G of this Chapter. The following repair procedures are based on the assumption the engine is installed in the vehicle. If the engine has been removed from the vehicle and mounted on a stand, many of the steps outlined in this Part of Chapter 2 will not apply.

The Specifications included in this Part of Chapter 2 apply only to the procedures contained in this Part. Part G of Chapter 2 contains the Specifications necessary for cylinder head and engine block rebuilding.

2   Repair operations possible with the engine in the vehicle

Many major repair operations can be accomplished without removing the engine from the vehicle.

Clean the engine compartment and the exterior of the engine with some type of degreaser before any work is done. It'll make the job easier and help keep dirt out of the internal areas of the engine.

Depending on the components involved, it may be helpful to remove the hood to improve access to the engine as repairs are performed (refer to Chapter 11 if necessary). Cover the fenders to prevent damage to the paint. Special pads are available, but an old bedspread or blanket will also work.

If vacuum, exhaust, oil or coolant leaks develop, indicating a need for gasket or seal replacement, the repairs can generally be done with the engine in the vehicle. The intake and exhaust manifold gaskets, timing chain cover gasket, oil pan gasket, crankshaft oil seals and cylinder head gaskets are all accessible with the engine in place.

Exterior engine components, such as the intake and exhaust manifolds, the oil pan (and the oil pump), the water pump, the starter motor, the alternator and the fuel system components can be removed for repair with the engine in place.

Since the cylinder heads can be removed without pulling the engine, valve component servicing can also be accomplished with the engine in the vehicle. Replacement of the timing chain and sprock-

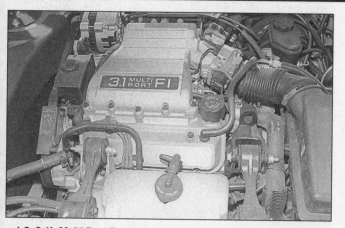

1.2 3.1L Multi Port Fuel Injected engine (2.8L and 3100 similar)

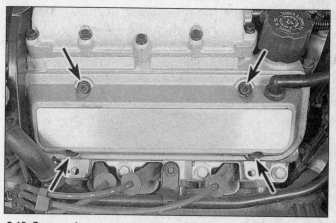

3.10 Some valve cover mounting bolts (arrows) require a Torx bit for removal and installation

ets is also possible with the engine in the vehicle.

In extreme cases caused by a lack of necessary equipment, repair or replacement of piston rings, pistons, connecting rods and rod bearings is possible with the engine in the vehicle. However, this practice is not recommended because of the cleaning and preparation work that must be done to the components involved.

3 Valve covers - removal and installation

Removal

Front cover

Refer to illustration 3.10

1   Disconnect the negative battery cable from the battery. **Caution:** *On models equipped with the Theftlock audio system, be sure the lockout feature is turned off before performing any procedure which requires disconnecting the battery.*

2   Remove the air cleaner assembly (see Chapter 4).

3   On the 3100, remove the thermostat bypass pipe clip nut.

4   On the 3100, remove the automatic transaxle vacuum modulator pipe assembly.

5   On the 3100, disconnect the right engine mount strut at the engine.

6   If necessary, remove the spark plug wires from the spark plugs (see Chapter 1). Be sure each wire is labeled before removal to ensure correct reinstallation.

7   Detach the spark plug wire harness clamps from the coolant tube.

8   Remove the PCV tube from the valve cover.

9   Drain the coolant (see Chapter 1).

10   Remove the valve cover mounting bolts. **Note:** *Some models are equipped with special bolts. Remove them with a T-30 Torx driver* **(see illustration).**

11   Detach the valve cover. **Note:** *If the*

cover sticks to the cylinder head, use a block of wood and a hammer to dislodge it. If the cover still won't come loose, pry on it carefully, but don't distort the sealing flange.

Rear cover

12   Disconnect the negative battery cable from the battery. **Caution:** *On models equipped with the Theftlock audio system, be sure the lockout feature is turned off before performing any procedure which requires disconnecting the battery.*

13   Remove the air cleaner assembly (see Chapter 4).

14   On the 2.8L and 3.1L, drain the coolant (see Chapter 1). Remove the coolant hoses at the base of the throttle valve.

15   Rotate the engine (see Chapter 1).

16   Detach the vacuum hoses at the plenum. Also, disconnect the brake booster vacuum supply at the plenum.

17   Remove the EGR tube at the crossover pipe. On the 3100, disconnect the EGR valve and move it aside.

18   Disconnect the ignition wire guide and harness at the plenum. On the 3100, remove the ignition coil.

19   If necessary, remove the bracket at the right side of the plenum.

20   Remove the serpentine drivebelt (see Chapter 1).

21   On the 3100, remove the black plastic cover from the shock tower.

22   Remove the alternator and loosen the alternator bracket (see Chapter 5).

23   Detach the breather hose from the PCV valve.

24   Remove the spark plug wires from the spark plugs (see Chapter 1). Be sure each wire is labeled before removal to ensure correct reinstallation.

25   Remove the accelerator cable, T.V. cable and also the cruise control cable, as necessary.

26   Detach the valve cover. **Note:** *If the cover sticks to the cylinder head, use a block of wood and a hammer to dislodge it. If the cover still won't come loose, pry on it carefully, but don't distort the sealing flange.*

Note: *Some models are equipped with special bolts. Remove them with a T-30 Torx driver* **(see illustration 3.10).**

Installation

27   The mating surfaces of each cylinder head and valve cover must be perfectly clean when the covers are installed. Use a gasket scraper to remove all traces of sealant or old gasket material, then clean the mating surfaces with lacquer thinner or acetone (if there's sealant or oil on the mating surfaces when the cover is installed, oil leaks may develop). The valve covers are made of aluminum, so be extra careful not to nick or gouge the mating surfaces with the scraper.

28   Clean the mounting bolt threads with a die if necessary to remove any corrosion and restore damaged threads. Use a tap to clean the threaded holes in the heads.

29   Place the valve cover and new gasket in position, then install the bolts. Tighten the bolts in several steps to the torque listed in this Chapter's Specifications.

30   Complete the installation by reversing the removal procedure. Start the engine and check carefully for oil leaks at the valve cover-to-head joints.

4 Rocker arms and pushrods - removal, inspection and installation

Refer to illustrations 4.2a, 4.2b and 4.3

Removal

1   Remove the valve cover(s) (see Section 3).

2   On 1995 and earlier models, the rocker arms are retained by nuts with a pivot ball between the nut and the rocker. The 1996 and later engines have the rocker arm on a pedestal, with a bolt through the rocker arm and pedestal. Beginning at the drivebelt end of one cylinder head, remove the rocker arm mounting nuts/bolts one at a time and detach the rocker arms, nuts, and pivot

4.2a On 1995 and earlier models remove the rocker arm nuts (arrows)

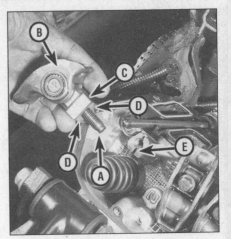

4.2b 1996 and later engines use rocker arm bolts, stamped roller rocker arms, and rocker arm pedestals - they are kept as an assembly by a small sleeve between the bolt and the pedestal - note the projections on the pedestal; they fit into grooves in the head

A   Rocker arm bolt
B   Rocker arm
C   Rocker arm pedestal
D   Pedestal projections
E   Grooves in the head

balls/pedestals (see illustrations). Store each set of rocker arm components separately in a marked plastic bag to ensure they're reinstalled in their original locations. The 1996 and later rocker arms have the pedestal mount "captured" on the rocker arm bolt by a metal sleeve inside. The components can be separated if necessary by tapping the bolt out of the pedestal, but normally all components for a particular valve will stay as an assembly. **Note:** *If you only need to remove the pushrods, loosen the rocker arm nuts/bolts and turn the rocker arms to allow room for pushrod removal.*

3   Remove the pushrods and store them separately to make sure they don't get mixed up during installation (see illustration). **Note:** *Intake and exhaust pushrods are different lengths. The intake pushrods are shorter and the exhaust pushrods are longer in length. On 1995 and earlier models the pushrods may be color coded with several paint stripes to help distinguish the intake (red or orange striped) pushrods from the exhaust (blue striped) pushrods. Never mix-up the pushrods and always install them in the original position from which they were removed or engine damage may occur.*

4.3 A perforated cardboard box can be used to store the pushrods to ensure they are reinstalled in their original locations - note the label indicating the transaxle end of the engine

Inspection

4   Inspect each rocker arm for wear, cracks and other damage, especially where the pushrods and valve stems make contact.
5   Check the pivot seat in each rocker arm and the pivot ball faces. Look for galling, stress cracks and unusual wear patterns. If the rocker arms are worn or damaged, replace them with new ones and install new pivot balls as well.
6   Make sure the hole at the pushrod end of each rocker arm is open.
7   Inspect the pushrods for cracks and excessive wear at the ends. Roll each pushrod across a piece of plate glass to see if it's bent (if it wobbles, it's bent).

Installation

8   Lubricate the lower end of each pushrod with clean engine oil or moly-base grease and install them in their original locations. Make sure each pushrod seats completely in the lifter socket.
9   Apply moly-base grease to the ends of the valve stems and the upper ends of the pushrods.
10   Apply moly-base grease to the pivot balls to prevent damage to the mating surfaces before engine oil pressure builds up. Install the rocker arms, pivot balls and nuts and tighten the nuts to the torque listed in this Chapter's Specifications. As the nuts are tightened, make sure the pushrods engage properly in the rocker arms.
11   Install the valve covers and the remaining components detached during valve cover removal (see Section 3). Start the engine and check for any unusual noises coming from the valve cover area.

5   Valve springs, retainers and seals - replacement

This procedure is essentially the same as the procedure for the 2.5 liter four-cylinder engine. Refer to Part B and follow the procedure outlined there.

6   Intake manifold - removal and installation

Removal

Refer to illustrations 6.9, 6.11a and 6.11b
1   Relieve the fuel system pressure (see Chapter 4).
2   Disconnect the negative battery cable from the battery. **Caution:** *On models equipped with the Theftlock audio system, be sure the lockout feature is turned off before performing any procedure which requires disconnecting the battery.*
3   Remove the EGR valve (see Chapter 4) at the plenum.
4   Remove the plenum, fuel rail and injectors (see Chapter 4). When disconnecting fuel line fittings, be prepared to catch some fuel with a rag, then cap the fittings to prevent contamination.
5   Remove the serpentine drivebelt and drain the cooling system (see Chapter 1).
6   Remove the alternator and loosen the bracket (see Chapter 5).
7   Unbolt the power steering pump (if equipped) and set it aside without disconnecting the hoses.
8   Disconnect the coolant tubes and hoses as necessary.
9   Label and disconnect any remaining fuel and vacuum lines (see illustration) and wires from the manifold.
10   Remove the valve covers (see Section 3).
11   Remove the manifold mounting bolts and nuts, then separate the manifold from the

6.9 Label (arrow) and disconnect all remaining hoses and lines

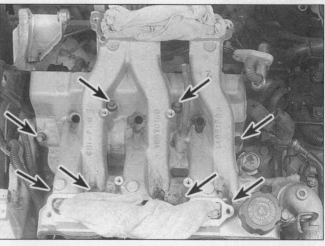

6.11a Intake manifold mounting bolt/nut locations (arrows)

engine (see illustrations). Don't pry between the manifold and heads, as damage to the soft aluminum gasket sealing surfaces may result. If you're installing a new manifold, transfer all fittings and sensors to the new manifold.

12   Loosen the rocker arm nuts/bolts, rotate the rocker arms out of the way and remove the pushrods that go through the manifold gasket (see Section 4).

Installation

Refer to illustration 6.16

Note: The mating surfaces of the cylinder heads, block and manifold must be perfectly clean when the manifold is installed. Gasket removal solvents are available at most auto parts stores and may be helpful when removing old gasket material that's stuck to the heads and manifold (since the manifold is made of aluminum, aggressive scraping can cause damage). Be sure to follow the directions printed on the container.

13   Lift the old gasket off. Use a gasket scraper to remove all traces of sealant and old gasket material, then clean the mating surfaces with lacquer thinner or acetone. If there's old sealant or oil on the mating surfaces when the manifold is installed, oil or vacuum leaks may develop. Use a vacuum cleaner to remove any gasket material that falls into the intake ports or the lifter valley.

14   Use a tap of the correct size to chase the threads in the bolt holes, if necessary, then use compressed air (if available) to remove the debris from the holes. **Warning:** Wear safety glasses or a face shield to protect your eyes when using compressed air!

15   Apply a 3/16-inch (5 mm) bead of RTV sealant or equivalent to the front and rear ridges of the engine block between the heads. **Note:** Some gasket sets include end seals.

16   Install the intake manifold gasket **(see illustration).**

17   Install the pushrods and rocker arms (see Section 4).

18   Carefully lower the manifold into place and install the mounting bolts/nuts finger tight. First tighten the four vertical bolts (in the middle of the manifold) then the four angled bolts (at the ends) **(see illustration 6.11a)**. **Note:** Coat the bolt threads with pipe thread sealant.

19   Tighten the mounting bolts until they're all at the torque listed in this Chapter's Specifications. **Caution:** To prevent oil leaks, tighten the vertical bolts first to ensure that the lower manifold stays centered on the gaskets, then tighten the angled bolts.

20   Install the remaining components in the reverse order of removal.

21   Change the oil and filter and refill the cooling system (see Chapter 1). Start the engine and check for leaks.

7   Exhaust manifolds - removal and installation

Front manifold

Refer to illustrations 7.12, 7.13 and 7.14

1   Disconnect the negative battery cable from the battery. **Caution:** On models equipped with the Theftlock audio system, be sure the lockout feature is turned off before performing any procedure which requires disconnecting the battery.

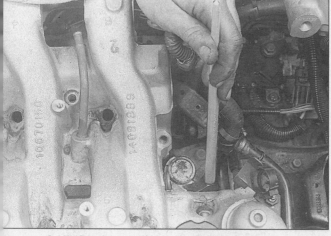

6.11b Pry the manifold loose at a casting boss - don't pry between the gasket surfaces!

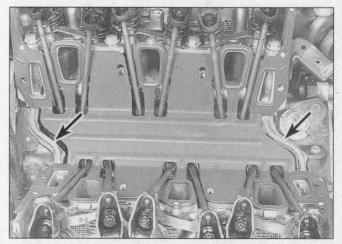

6.16 Apply a bead of sealant to, or position an end seal on, the ridges between the heads (arrows)

7.12 On early models the exhaust crossover pipe is secured by four nuts; the upper nuts (arrows) are visible here

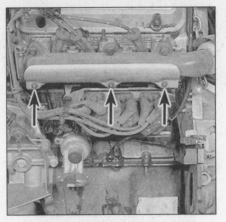

7.13 Exhaust manifold heat shield mounting bolts (arrows) (3.1L engine shown, 2.8L and 3100 engine similar)

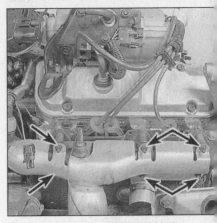

7.14 Exhaust manifold bolt locations (arrows)

2   Remove the air cleaner assembly, inlet hose and mass air flow sensor (see Chapter 4).

3   Allow the engine to cool completely, then drain the coolant (see Chapter 1) and disconnect the coolant bypass tube, if necessary.

4   On the 3100, remove the upper and lower radiator hoses.

5   On the 3100, remove the tie straps from the heater outlet pipe and ignition wire harness, and remove the heater outlet pipe.

6   On the 3100, remove the automatic transaxle vacuum modulator pipe if equipped.

7   Remove the coolant reservoir (see Chapter 3).

8   Remove the serpentine drivebelt (see Chapter 1).

9   Remove the air conditioning compressor, leaving the hoses connected to it (see Chapter 3).

10   Remove the torque strut from the right side, if necessary (see Section 18).

11   Remove the air conditioning bracket and the torque strut bracket.

12   Unbolt the crossover pipe where it joins the front manifold **(see illustration)**. **Note:** *On later models it will be necessary to remove the crossover pipe heat shield to allow access to the crossover pipe mounting nuts. The crossover pipe heat shield is mounted to the left and right exhaust manifolds by two bolts on each side (four bolts total).*

13   Remove the manifold heat shield **(see illustration)**.

14   Remove the mounting bolts and detach the manifold from the cylinder head **(see illustration)**.

15   Clean the mating surfaces to remove all traces of old gasket material, then inspect the manifold for distortion and cracks. Warpage can be checked with a precision straightedge held against the mating flange. If a feeler gauge thicker than 0.030-inch can be inserted between the straight edge and flange surface, take the manifold to an automotive machine shop for resurfacing.

16   Place the manifold in position with a new gasket and install the mounting bolts finger tight.

17   Starting in the middle and working out toward the ends, tighten the mounting bolts a little at a time until all of them are at the torque listed in this Chapter's Specifications.

18   Install the remaining components in the reverse order of removal.

19   Start the engine and check for exhaust leaks between the manifold and cylinder head and between the manifold and exhaust pipe.

Rear manifold

20   Disconnect the negative battery cable from the battery. **Caution:** *On models equipped with the Theftlock audio system, be sure the lockout feature is turned off before performing any procedure which requires disconnecting the battery.*

21   Allow the engine to cool completely, then drain the coolant (see Chapter 1).

22   Remove the coolant reservoir, if necessary (see Chapter 3).

23   Rotate the engine to gain working clearance at the rear of the engine (see Chapter 1).

24   Remove the air cleaner assembly (see Chapter 4).

25   On some models it may be necessary to remove the accelerator cable and T.V. cable and brackets for clearance (see Chapters 4 and 7).

26   Unbolt the crossover pipe (or front manifold extension pipe) where it joins the rear manifold **(see illustration 7.12)**.

27   Remove the exhaust manifold heat shield **(see illustration 7.13)**. Note that the rear manifold heat shield typically has two bolts retaining it to the manifold, not three bolts like the front manifold heat shield.

28   Set the parking brake, block the rear wheels and raise the front of the vehicle, supporting it securely on jackstands.

29   Working under the vehicle, remove the exhaust pipe-to-manifold bolts. You may have to apply penetrating oil to the fastener threads - they're usually corroded.

30   On the 3100, it may be necessary to lower the drivetrain/front suspension frame assembly. Place a floor jack under the frame front center crossmember. Loosen the rear

frame bolts - DO NOT REMOVE THEM. Remove the front frame bolts and lower the front of the frame.

31   On the 3100, disconnect the EGR tube assembly from the exhaust manifold.

32   On the 3100, remove the automatic transaxle vacuum modulator pipe (if equipped) and the filler tube.

33   Remove the oxygen sensor from the manifold if equipped.

34   Unbolt and remove the exhaust manifold **(see illustration 7.14)**.

35   Clean the mating surfaces to remove all traces of old sealant, then check for warpage and cracks. Warpage can be checked with a precision straightedge held against the mating flange. If a feeler gauge thicker than 0.030-inch can be inserted between the straightedge and flange surface, take the manifold to an automotive machine shop for resurfacing.

36   Place the manifold in position with a new gasket and install the bolts finger tight.

37   Starting in the middle and working out toward the ends, tighten the mounting bolts a little at a time until all of them are at the torque listed in this Chapter's Specifications.

38   Install the remaining components in the reverse order of removal.

39   Start the engine and check for exhaust leaks between the manifold and cylinder head and between the manifold and exhaust pipe.

8   Cylinder heads - removal and installation

Refer to illustrations 8.3, 8.13, 8.16a, 8.16b, 8.18 and 8.19

Caution: *Allow the engine to cool completely before loosening the cylinder head bolts.*

Note: *On engines with high mileage and during an overhaul, camshaft lobe height should be checked prior to cylinder head removal (see Chapter 2, Part G for instructions).*

Removal

1   Disconnect the negative battery cable from the battery. **Caution:** *On models equipped with the Theftlock audio system, be*

ure the lockout feature is turned off before
erforming any procedure which requires dis-
onnecting the battery.

Remove the air cleaner assembly (see
Chapter 4) and then remove the intake mani-
old as described in Section 6.

If you're removing the front cylinder
ead, remove the oil dipstick tube mounting
olt **(see illustration)**.

Disconnect all wires and vacuum hoses
rom the cylinder head(s). Be sure to label
hem to simplify reinstallation.

Disconnect the spark plug wires and
emove the spark plugs (see Chapter 1). Be
ure the plug wires are labeled to simplify
einstallation.

Detach the exhaust manifold from the
ylinder head being removed (see Section 7).

Remove the valve cover(s) (see Sec-
ion 3).

Remove the rocker arms and pushrods
see Section 4).

Using the new head gasket, outline the
ylinders and bolt pattern on a piece of card-
oard **(see illustration 8.11 in Part B)**. Be
ure to indicate the front (drivebelt end) of the
ngine for reference. Punch holes at the bolt
ocations. Loosen each of the cylinder head
nounting bolts 1/4-turn at a time until they
an be removed by hand - work from bolt-to-
olt in a pattern that's the *reverse* of the tight-
ning sequence **(see illustration 8.19)**. Store
ne bolts in the cardboard holder as they're
emoved - this will ensure they are reinstalled
n their original locations, which is absolutely
ssential.

0   Lift the head(s) off the engine. If resis-
ance is felt, don't pry between the head and
lock as damage to the mating surfaces will
esult. Recheck for head bolts that may have
een overlooked, then use a hammer and
lock of wood to tap up on the head and
reak the gasket seal. Be careful because
here are locating dowels in the block which
osition each head. As a last resort, pry each
ead up at the rear corner only and be careful
ot to damage anything. After removal, place
ne head on blocks of wood to prevent dam-
ge to the gasket surfaces.

1   Refer to Chapter 2, Part G, for cylinder
ead disassembly, inspection and valve ser-
ce procedures.

nstallation

2   The mating surfaces of each cylinder
ead and block must be perfectly clean when
ne head is installed.

3   Use a gasket scraper to remove all
aces of carbon and old gasket material **(see
lustration)**, then clean the mating surfaces
ith lacquer thinner or acetone. If there's oil
n the mating surfaces when the head is
stalled, the gasket may not seal correctly
nd leaks may develop. When working on the
lock, it's a good idea to cover the lifter valley
ith shop rags to keep debris out of the
ngine. Use a shop rag or vacuum cleaner to
move any debris that falls into the cylinders.

4   Check the block and head mating sur-
ces for nicks, deep scratches and other

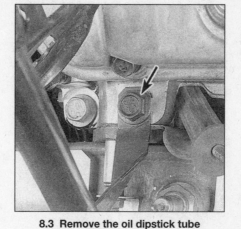

8.3   Remove the oil dipstick tube mounting bolt (arrow)

8.13   Remove the old gasket and carefully scrape off all old gasket material and sealant

8.16a   Position the new gasket over the dowel pins (arrows) . . .

8.16b   . . . with the correct side facing up

damage. If damage is slight, it can be
removed with a file; if it's excessive, machining may be the only alternative.

15   Use a tap of the correct size to chase
the threads in the head bolt holes. Dirt, corrosion, sealant and damaged threads will affect
torque readings.

16   Position the new gasket over the dowel
pins in the block. Some gaskets are marked
TOP or THIS SIDE UP to ensure correct
installation **(see illustrations)**.

17   Carefully position the head on the block
without disturbing the gasket.

18   Apply RTV sealant or equivalent to the
threads and the undersides of the bolt heads.
Install the bolts in the correct locations - two
different lengths are used **(see illustration)**.
Here's where the cardboard holder comes in
handy.

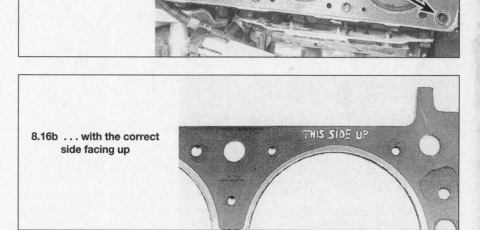

8.18   Two different length cylinder head bolts are used - apply sealant to the threads and the undersides of the bolt heads (arrows)

8.19 Cylinder head bolt TIGHTENING SEQUENCE (reverse the sequence when loosening the bolts)

9.4 Remove the bolts (A) and pull up the roller lifter guides (B)

19   Tighten the bolts, using the recommended sequence **(see illustration)**, to the torque listed in this Chapter's Specifications. Then, using the same sequence, turn each bolt the amount of angle listed in this Chapter's Specifications.

20   The remaining installation steps are the reverse of removal.

21   Change the engine oil and filter (see Chapter 1).

9   Hydraulic lifters - removal, inspection and installation

Refer to illustrations 9.4, 9.5 and 9.6

1   A noisy valve lifter can be isolated when the engine is idling. Hold a mechanic's stethoscope or a length of hose near the location of each valve while listening at the other end. Another method is to remove the valve cover and, with the engine idling, touch each of the valve spring retainers, one at a time. If a valve lifter is defective, it'll be evident from the shock felt at the retainer each time the valve seats.

2   The most likely causes of noisy valve lifters are dirt trapped inside the lifter and lack of oil flow, viscosity or pressure. Before condemning the lifters, check the oil for fuel contamination, correct level, cleanliness and correct viscosity.

Removal

3   Remove the valve cover(s) and intake manifold as described in Sections 3 and 6.

4   Remove the rocker arms and pushrods (see Section 4). On 3100 engines remove the bolts holding the roller lifter guide to the block, and remove the two roller lifter guides **(see illustration)**. Mark the guides as to which side they came from.

5   There are several ways to extract the lifters from the bores. A special tool designed to grip and remove lifters is manufactured by many tool companies and is widely available, but it may not be required in every case. On

newer engines without a lot of varnish buildup, the lifters can often be removed with a small magnet or even with your fingers. A machinist's scribe with a bent end can be used to pull the lifters out by positioning the point under the retainer ring in the top of each lifter **(see illustration)**. **Caution:** *Don't use pliers to remove the lifters unless you intend to replace them with new ones (along with the camshaft). The pliers may damage the precision machined and hardened lifters, rendering them useless.*

6   Before removing the lifters, arrange to store them in a clearly labeled box to ensure they're reinstalled in their original locations. Remove the lifters and store them where they won't get dirty **(see illustration)**. **Note:** *Some engines may have both standard and 0.010-inch oversize lifters installed at the factory. They are marked on the block at the lifter boss and will have 0.25 mm OS stamped on it.*

Inspection and installation

7   Parts for valve lifters are not available separately. The work required to remove them from the engine again if cleaning is

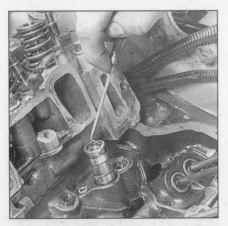

9.5 A magnetic pick-up tool or a scribe can be used to remove the lifters

unsuccessful outweighs any potential savings from repairing them. Refer to Chapter 2 Part B, for lifter inspection procedures and Part G for camshaft inspection procedures. If the lifters are worn, they must be replaced with new ones. **Note:** *If the lifters are non roller type, the camshaft must be replaced as well - never install used non-roller lifters with a new camshaft or new non-roller lifters with a used camshaft.*

8   When reinstalling used lifters, make sure they're replaced in their original bores. Soak new lifters in oil to remove trapped air. Coat all lifters with moly-base grease or engine assembly lube prior to installation.

9   The remaining installation steps are the reverse of removal.

10   Run the engine and check for oil leaks.

10   Vibration damper - removal and installation

Refer to illustrations 10.7, 10.8 and 10.9
Caution: *On some engines a rubber sleeve connects the inertia weight to the vibration damper. Take care when working on the*

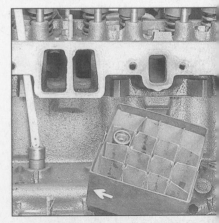

9.6 Store the lifters in a box like this to ensure installation in their original locations

10.7 Remove the vibration damper-to-crankshaft bolt (arrow) - it's very tight, so use a six-point socket and a breaker bar

10.8 Use a puller that bolts to the vibration damper hub; jaw-type pullers will damage the vibration damper

10.9 The damper keyway must be aligned with the woodruff key (arrow) in the crankshaft nose

vibration damper that you do not accidentally shift the inertia weight's position relative to the sleeve or damper, as this will upset the tuning of the vibration damper.

Removal

1 Disconnect the negative battery cable from the battery. **Caution:** *On models equipped with the Theftlock audio system, be sure the lockout feature is turned off before performing any procedure which requires disconnecting the battery.*
2 Loosen the lug nuts on the right front wheel.
3 Raise the vehicle and support it securely on jackstands.
4 Remove the right front wheel.
5 Remove the right front inner fender splash shield.
6 Remove the serpentine drivebelt (see Chapter 1).
7 On automatic transaxle equipped models, remove the driveplate cover and position a large screwdriver in the ring gear teeth to keep the crankshaft from turning while an assistant removes the vibration damper-to-crankshaft bolt **(see illustration)**. On manual transaxle equipped models, engage high gear and apply the brakes while an assistant removes the vibration damper bolt. The bolt is normally very tight, so use a large breaker bar and a six-point socket. **Note:** *On later models it may be necessary to lower the right side of the drivetrain/front suspension frame assembly to allow access to the vibration damper bolt. Place a floor jack under the frame front center crossmember. Loosen the right side frame bolts - DO NOT REMOVE THEM! and lower the right side of the frame to access the vibration damper.*
8 Pull the damper off the crankshaft with a bolt-type puller **(see illustration)**. Leave the Woodruff key in place in the end of the crankshaft.

Installation

9 Installation is the reverse of removal. Be sure to apply moly-base grease to the seal

contact surface of the damper hub (if it isn't lubricated, the seal lip could be damaged and oil leakage would result). Align the damper hub keyway with the Woodruff key **(see illustration)**.
10 Tighten the vibration damper-to-crankshaft bolt to the torque listed in this Chapter's Specifications.
11 Reinstall the remaining parts in reverse order of removal.

11 Crankshaft front oil seal - replacement

Refer to illustrations 11.2 and 11.3

1 Remove the vibration damper (see Section 10).
2 Note how the seal is installed - the new one must be installed to the same depth and facing the same way. Carefully pry the oil seal out of the cover with a seal puller or a large screwdriver **(see illustration)**. Be very careful not to distort the cover or scratch the crankshaft! Wrap electrician's tape around the tip of the screwdriver to avoid damage to the crankshaft.

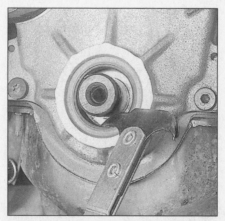

11.2 Carefully pry the old seal out of the timing chain cover - don't damage the crankshaft in the process

3 Apply clean engine oil or multi-purpose grease to the outer edge of the new seal, then install it in the cover with the lip (spring side) facing IN. Drive the seal into place **(see illustration)** with a large socket and a hammer (if a large socket isn't available, a piece of pipe will also work). Make sure the seal enters the bore squarely and stop when the front face is at the proper depth.
4 Reinstall the vibration damper.

12 Timing chain cover - removal and installation

Refer to illustrations 12.5 and 12.16

1 Disconnect the negative battery cable from the battery. **Caution:** *On models equipped with the Theftlock audio system, be sure the lockout feature is turned off before performing any procedure which requires disconnecting the battery.*
2 Loosen, but do not remove, the water pump pulley bolts, then remove the serpentine drivebelt (see Chapter 1).
3 Remove the water pump pulley (see Chapter 3).

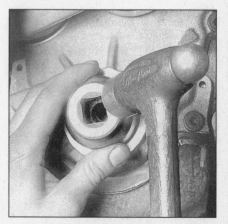

11.3 Drive the new seal into place with a large socket and hammer

12.5 The drivebelt tensioner is secured to the timing chain cover by a bolt (arrow)

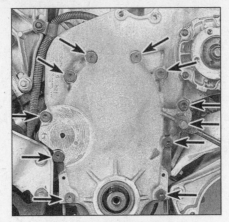

12.16 Timing chain cover bolt locations (arrows)

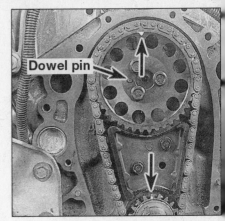

13.3a On early models the timing marks (arrows) on the sprockets should be at the top (12 o'clock position) - a straight line should pass through the camshaft sprocket timing mark, the center of the camshaft, the crankshaft sprocket timing mark and the center of the crankshaft

4   Remove the vibration damper (see Section 10).
5   Unbolt the drivebelt tensioner **(see illustration)** and idler, if equipped.
6   Drain the coolant and engine oil (see Chapter 1).
7   Remove the alternator and loosen the mounting bracket (see Chapter 5).
8   Unbolt the power steering pump (if equipped) and tie it aside (see Chapter 10). Leave the hoses connected.
9   Unbolt the flywheel/driveplate cover below the transaxle.
10   Remove the starter (see Chapter 5).
11   Remove the oil pan (see Section 14).
12   Disconnect the coolant hoses from the fill pipe and water pump.
13   Unbolt the spark plug wire shield at the water pump.
14   Disconnect the canister purge hose and tie it aside.
15   If necessary, remove the crankshaft position sensor assembly from the timing chain cover (see Chapter 6).
16   Remove the timing chain cover-to-engine block bolts **(see illustration)**. Note that some of the bolts require T-40 and T-50 Torx bits for removal.
17   Separate the cover from the engine. If it's stuck, tap it with a soft-face hammer, but don't try to pry it off.
18   Use a gasket scraper to remove all traces of old gasket material and sealant from the cover and engine block. The cover is made of aluminum, so be careful not to nick or gouge it. Clean the gasket sealing surfaces with lacquer thinner or acetone.
19   Apply a thin layer of RTV sealant or equivalent to both sides of the new gasket, then position the gasket on the engine block (the dowel pins should keep it in place). Attach the cover to the engine and install the bolts.
20   Apply sealant to the bottom of the gasket. Follow a criss-cross pattern when tightening the fasteners and work up to the torque

listed in this Chapter's Specifications in three steps.
21   The remainder of installation is the reverse of removal.
22   Add oil and coolant, start the engine and check for leaks.

13   Timing chain and sprockets - inspection, removal and installation

Refer to illustrations 13.3a and 13.3b

Inspection

1   The timing chain should be replaced with a new one if the engine has high mileage, the chain has visible damage, or total freeplay midway between the sprockets exceeds one-inch. Failure to replace a worn timing chain may result in erratic engine performance, loss of power and decreased fuel mileage. Loose chains can "jump" timing. In the worst case, chain "jumping" or breakage will result in severe engine damage.

Removal

2   Remove the timing chain cover (see Section 12) and loosen the camshaft sprocket bolt(s) two turns but do not remove them. **Note:** *Earlier models use three bolts to retain the camshaft sprocket while later models use a single bolt to retain the camshaft sprocket.*
3   Temporarily install the vibration damper bolt and turn the crankshaft with the bolt to align the timing marks on the crankshaft and camshaft sprockets **(see illustrations)**. On early models the timing marks should be at the top (12 o'clock position). On later models when number 1 piston is at TDC, the crankshaft sprocket timing mark should align at the 12 o'clock position with the mark on

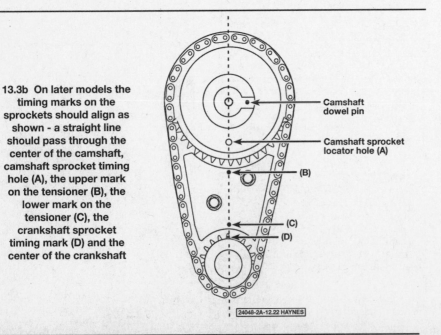

13.3b On later models the timing marks on the sprockets should align as shown - a straight line should pass through the center of the camshaft, camshaft sprocket timing hole (A), the upper mark on the tensioner (B), the lower mark on the tensioner (C), the crankshaft sprocket timing mark (D) and the center of the crankshaft

Camshaft dowel pin
Camshaft sprocket locator hole (A)
(B)
(C)
(D)

24048-2A-12.22 HAYNES

14.14 The front engine mount (A) is between the front of the oil pan and the chassis - remove the bolts (arrows) holding the bracket to the oil pan

14.15a On 2.8L and 3.1L engines simply remove the bolts around the perimeter of the oil pan

the bottom of the chain tensioner plate, and the small hole in the camshaft sprocket should be at the 6 o'clock position, aligned with the timing mark in the top of the chain tensioner plate.

4 Remove the camshaft sprocket bolts. Do not turn the camshaft in the process (if you do, realign the timing marks before the bolts are removed).

5 Use two large screwdrivers to carefully pry the camshaft sprocket off the camshaft dowel pin. Slip the timing chain and camshaft sprocket off the engine.

6 Timing chains and sprockets should be replaced in sets. If you intend to install a new timing chain, remove the crankshaft sprocket with a puller and install a new one. Be sure to align the key in the crankshaft with the keyway in the sprocket during installation.

7 Inspect the timing chain damper (guide) for cracks and wear and replace it if necessary. The damper is mounted in between the camshaft and crankshaft sprockets and is held to the engine block by two bolts **(see illustrations 13.3a or 13.3b).**

8 Clean the timing chain and sprockets with solvent and dry them with compressed air (if available). **Warning:** *Wear eye protection when using compressed air.*

9 Inspect the components for wear and damage. Look for teeth that are deformed, chipped, pitted and cracked.

Installation

10 If the camshaft was moved during this procedure, rotate the camshaft to position the dowel pin at 9 o'clock on early models or at 3 o'clock on late models engines. Mesh the timing chain with the camshaft sprocket, then engage it with the crankshaft sprocket. The timing marks should be aligned as shown in **illustration 13.3a and 13.3b. Note:** *If the crankshaft has been disturbed, turn it until the "O" stamped on the crankshaft sprocket is exactly at the top.*

11 Install the camshaft sprocket bolt(s) and tighten them to the torque listed in this Chapter's Specifications.

12 Lubricate the chain and sprocket with clean engine oil. Install the timing chain cover

(see Section 12).

13 The remaining installation steps are the reverse of removal.

14 Oil pan - removal and installation

Refer to illustrations 14.14, 14.15a and 14.15b.

Note: *On vehicles equipped with an automatic transaxle, remove the transaxle (see Chapter 7) to make clearance for the oil pan.*

Removal

1 Disconnect the cable from the negative battery terminal. **Caution:** *On models equipped with the Theftlock audio system, be sure the lockout feature is turned off before performing any procedure which requires disconnecting the battery.*

2 Remove the serpentine drivebelt (see Chapter 1).

3 Remove the air conditioning compressor, and the compressor bracket if equipped.

4 On the 3100, remove the electric cooling fan assemblies.

5 Raise the front of the vehicle and place it securely on jackstands. Apply the parking brake and block the rear wheels to keep it from rolling off the stands. Remove the lower splash pan and drain the engine oil (refer to Chapter 1 if necessary).

6 On the 3100, remove the front exhaust pipe and disconnect the connector from the oil level sensor at the bottom of the oil pan. Then remove the transaxle to oil pan braces, if equipped.

7 Remove the steering gear pinch bolt (see Chapter 10). **Caution:** *Be sure to separate the steering gear from the rack and pinion stub shaft to avoid damage to the steering gear and intermediate shaft.*

8 Remove the transaxle mount to frame retaining nuts (see Section 18), then remove the engine-to-frame mount retaining nuts.

9 On 2.8L and 3.1L engines remove the engine mount bracket (horse shoe type bracket) from the front of the engine block (see Section 18).

10 Remove the flywheel/driveplate lower cover.

11 Remove the starter (see Chapter 5).

12 Place a floor jack under the frame front center crossmember.

13 Loosen (but do not remove) the rear drivetrain/front suspension frame bolts, then remove the front frame bolts and lower the front of the frame.

14 On 3100 engines remove the front engine mount from the oil pan **(see illustration).**

15 Remove the oil pan bolts and nuts, then carefully separate the oil pan from the block **(see illustrations). Note:** *The oil pan rear side bolts on 3100 engines, are difficult to*

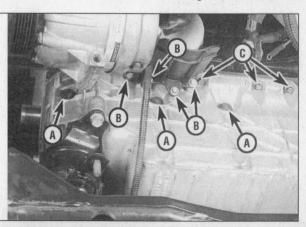

14.15b On 3100 engines remove six oil pan side bolts (A indicates three bolts on the radiator side) - also remove the oil filter shield bolts (B) and the nuts and bolts around the perimeter of the oil pan (C)

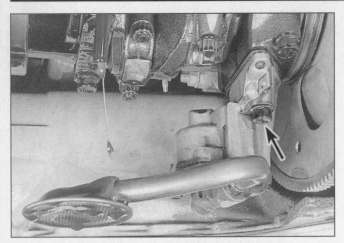

15.2 Oil pump mounting bolt location (arrow)

18.1 V6 engine torque strut (front)

remove, *but a box wrench with the correct offset can remove them.* Don't pry between the block and the pan or damage to the sealing surfaces could occur and oil leaks may develop. Instead, tap the pan with a soft-face hammer to break the gasket seal. Don't pry between the block and the pan or damage to the sealing surfaces could occur and oil leaks may develop. Instead, tap the pan with a soft-face hammer to break the gasket seal.

Installation

16   Clean the pan with solvent and remove all old sealant and gasket material from the block and pan mating surfaces. Clean the mating surfaces with lacquer thinner or acetone and make sure the bolt holes in the block are clear. Check the oil pan flange for distortion, particularly around the bolt holes. If necessary, place the pan on a block of wood and use a hammer to flatten and restore the gasket surface.

17   Always use a new gasket whenever the oil pan is installed. Apply a bead of RTV sealant to the front of the gasket, where it contacts the front cover.

18   Place the oil pan in position on the block and install the nuts/bolts finger tight.

19   After the fasteners are installed, tighten them to the torque listed in this Chapter's Specifications. Starting at the center, follow a criss-cross pattern and work up to the final torque in three steps. **Note:** *On 3100 engines it will be necessary to tighten all the bolts around the perimeter (vertical mounted bolts) of the oil pan first and tighten the side bolts (horizontal mounted bolts) last.*

20   The remaining steps are the reverse of the removal procedure.

21   Refill the engine with oil, run it until normal operating temperature is reached and check for leaks.

15   Oil pump - removal and installation

Refer to illustration 15.2

1   Remove the oil pan (see Section 14).

2   Unbolt the oil pump and lower it from the engine **(see illustration)**.

3   If the pump is defective, replace it with a new one - don't reuse the original or attempt to rebuild it.

4   Prime the pump by pouring motor oil into the pick-up screen while turning the pump driveshaft.

5   To install the pump, turn the hexagonal driveshaft so it mates with the oil pump drive.

6   Install the pump mounting bolt and tighten it to the torque listed in this Chapter's Specifications.

16   Flywheel/driveplate - removal and installation

Refer to Part B and follow the procedure outlined there. However, use the bolt torque listed in this Chapter's Specifications.

17   Rear main oil seal - replacement

Refer to Part B and follow the procedure outlined there.

18   Engine mounts - check and replacement

Refer to illustration 18.1

Refer to Part B and follow the procedure outlined there.

Chapter 2 Part E
3.4 liter V6 engine

Contents

Specifications

General

Cylinder numbers (drivebelt end-to-transaxle end)
Front bank (radiator side)	2-4-6
Rear bank	1-3-5
Firing order	1-2-3-4-5-6

Camshafts

Lobe lift
Intake	0.370 in
Exhaust	0.370 in

Journal diameter
1994 and earlier	2.165 to 2.166 in
1995 and later	2.1643 to 2.154 in

Journal clearance
1994 and earlier	0.0015 to 0.0035 in
1995 and later	0.0019 to 0.0040 in

Oil pump

Gear backlash	0.0037 to 0.0077 in
Gear length	1.199 to 1.200 in
Gear diameter	1.498 to 1.500 in
Gear housing depth	1.202 to 1.205 in
Gear housing inner diameter	1.504 to 1.506 in
Gear side clearance	0.003 to 0.004 in
Gear end clearance	0.002 to 0.006 in
Pressure valve-to-bore clearance	0.0015 to 0.0035 in

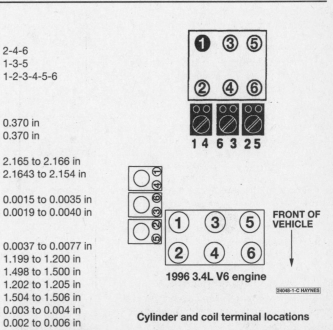

FRONT OF VEHICLE

1996 3.4L V6 engine

24048-1-C HAYNES

Cylinder and coil terminal locations

Torque specifications

	Ft-lbs (unless otherwise indicated)
Cam carrier cover bolts	97 in-lb
Cam carrier-to-cylinder head bolts	20
Camshaft sprocket bolts	96
Cylinder head bolts	
First step	44
Second step	Rotate an additional 90-degrees (1/4-turn)
Exhaust manifold-to-cylinder head bolts	
Nut	
1995 and earlier	18
1996 and later	116 in-lbs
Stud	13
Flywheel/driveplate-to-crankshaft bolts	60
Intake manifold-to-cylinder head bolts/nuts	
1994 and earlier	30
1995	22
1996 and later	116 in-lbs
Oil pan bolts/nuts	
Pan bolt (rear)	18 to 20
Others	
1994 and earlier	89 in-lbs
1995 and later	97 in-lbs
Oil pump	
Mounting bolt	40 to 43
Drive bolt	27
Cover bolt	89 in-lbs
Timing belt cover bolts	89 in-lbs
Timing chain tensioner bolt	20
Vibration damper bolt	79

1   General information

Refer to illustration 1.4

This Part of Chapter 2 is devoted to in-vehicle repair procedures for the 3.4 liter V6 engine. The 3.4L engine became available in 1990 and was produced through 1996; it is utilized in the Pontiac Grand Prix, Oldsmobile Cutlass Supreme and the Chevrolet Lumina models only. This engine utilizes a cast-iron block with six cylinders arranged in a "V" shape at a 60-degree angle between the two banks.

The overhead cam aluminum cylinder heads have two exhaust valves and two intake valves for each cylinder, and have pressed-in valve guides and valve seats.

The camshafts are located inside aluminum cam carrier housings that are located on top of each cylinder head. Each cam carrier contains two camshafts, one for the exhaust valves and one for the intake valves. The camshaft thrust plates are at rear of the carriers. The aluminum of the carriers serves as the bearing surface for the camshafts.

The engine is easily identified by looking at the designation printed on the fuel rail cover **(see illustration)**. To positively identify this engine, locate the vehicle identification number on the left front corner of the instrument panel. This plate is visible from outside the vehicle **(see illustration 1.1 in Chapter 2, Part B)**. The eighth character in the sequence is the engine designation:

X = 3.4L DOHC V6 engine

All information concerning engine removal and installation and engine block and cylinder head overhaul can be found in Part G of this Chapter. The following repair

1.4 3.4 liter double overhead cam (DOHC) V6 engine

procedures are based on the assumption the engine is installed in the vehicle. If the engine has been removed from the vehicle and mounted on a stand, many of the steps outlined in this Part of Chapter 2 will not apply.

The Specifications included in this Part of Chapter 2 apply only to the procedures contained in this Part. Part G of Chapter 2 contains the Specifications necessary for cylinder head and engine block rebuilding.

2   Repair operations possible with the engine in the vehicle

Many major repair operations can be accomplished without removing the engine from the vehicle.

Clean the engine compartment and the exterior of the engine with some type of degreaser before any work is done. It'll make the job easier and help keep dirt out of the internal areas of the engine.

Depending on the components involved, it may be helpful to remove the hood to improve access to the engine as repairs are performed (refer to Chapter 11 if necessary). Cover the fenders to prevent damage to the paint. Special pads are available, but an old bedspread or blanket will also work.

If vacuum, exhaust, oil or coolant leaks develop, indicating a need for gasket or seal replacement, the repairs can generally be done with the engine in the vehicle. The intake and exhaust manifold gaskets, timing chain cover gasket, oil pan gasket, crankshaft oil seals and cylinder head gaskets are all accessible with the engine in place.

Exterior engine components, such as the intake and exhaust manifolds, the oil pan (and the oil pump), the water pump, the starter motor, the alternator and the fuel system components can be removed for repair with the engine in place.

Since the cylinder heads can be removed without pulling the engine, valve component servicing can also be accomplished with the engine in the vehicle. Replacement of the timing chain and sprockets is also possible with the engine in the vehicle.

In extreme cases caused by a lack of necessary equipment, repair or replacement of piston rings, pistons, connecting rods and rod bearings is possible with the engine in the vehicle. However, this practice is not recommended because of the cleaning and preparation work that must be done to the components involved.

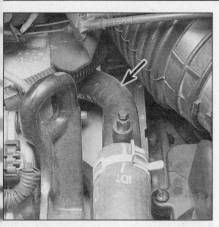

3.4 Thermostat housing outlet (arrow)

4.3 Remove the exhaust crossover pipe
from the manifold

1   Exhaust manifold
2   Exhaust crossover pipe
3   Engine torque strut

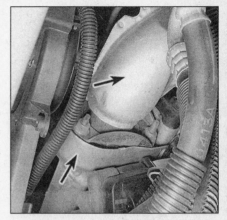

4.15 The exhaust crossover pipe (upper
arrow) attaches to each exhaust manifold
with nuts (lower arrow)

3   Intake manifold - removal and installation

Removal

Refer to illustration 3.4

1   Disconnect the negative battery cable from the battery. **Caution:** *On models equipped with the Theftlock audio system, be sure the lockout feature is turned off before performing any procedure which requires disconnecting the battery.*

2   Relieve the fuel system pressure, then remove the fuel rail (see Chapter 4). **Note:** *When disconnecting fuel line fittings, be prepared to catch some fuel with a rag, then cap the fittings to prevent contamination.*

3   Remove the plenum (see Chapter 4).

4   Remove the radiator hose from the thermostat housing **(see illustration)**.

5   Remove the connector from the coolant temperature sensor.

6   Remove the heater pipe nut at the throttle body.

7   Remove the manifold mounting bolts and nuts, then separate the manifold from the engine. Don't pry between the manifold and heads, as damage to the soft aluminum gasket sealing surfaces may result. If you're installing a new manifold, transfer all fittings and sensors to the new manifold.

Installation

Note: *The mating surfaces of the cylinder heads, block and manifold must be perfectly clean when the manifold is installed. Gasket removal solvents are available at most auto parts stores and may be helpful when removing old gasket material that's stuck to the heads and manifold (since the manifold is made of aluminum, aggressive scraping can cause damage). Be sure to follow the directions printed on the container.*

8   Lift the old gasket off. Use a gasket scraper to remove all traces of sealant and old gasket material, then clean the mating surfaces with lacquer thinner or acetone. If there's old sealant or oil on the mating surfaces when the manifold is installed, oil or vacuum leaks may develop. Use a vacuum

cleaner to remove any gasket material that falls into the intake ports or the lifter valley.

9   Use a tap of the correct size to chase the threads in the bolt holes, if necessary, then use compressed air (if available) to remove the debris from the holes. **Warning:** *Wear safety glasses or a face shield to protect your eyes when using compressed air!*

10   Install the intake manifold gasket **(see illustration 3.8)**.

11   Carefully lower the manifold into place and install the mounting bolts/nuts finger tight. **Note:** *To ease reassembly, you can temporarily install two M8 x 1.25 x 50 mm bolts, with washers, in the vertical holes in the intake manifold. This will help align the intake bolt grommet bores with the threaded holes in the cylinder heads.*

12   Tighten the mounting bolts/nuts in two steps, working from the center out, in a circular pattern, until they're all at the torque listed in this Chapter's Specifications. If necessary, remove the two bolts from the vertical holes in the intake manifold.

13   Install the remaining components in the reverse order of removal.

14   Change the oil and filter and refill the cooling system (see Chapter 1). Start the engine and check for leaks.

4   Exhaust manifolds - removal and installation

Front Manifold

Refer to illustration 4.3

1   Disconnect the negative battery cable from the battery. **Caution:** *On models equipped with the Theftlock audio system, be sure the lockout feature is turned off before performing any procedure which requires disconnecting the battery.*

2   Remove the air cleaner assembly (see Chapter 4).

3   Remove the exhaust crossover pipe from the manifold **(see illustration)**.

4   Remove the cooling fans (see Chapter 3).

5   On manual transmission cars, remove the front hose from the air pipe.

6   Remove the mounting nuts, then remove the front exhaust manifold from the cylinder head.

7   Clean the mating surfaces to remove all traces of old gasket material, then inspect the manifold for distortion and cracks. Warpage can be checked with a precision straightedge held against the mating flange. If a feeler gauge thicker than 0.030-inch can be inserted between the straightedge and flange surface, take the manifold to an automotive machine shop for resurfacing.

8   Place the manifold in position with a new gasket and install the mounting nuts finger tight.

9   Starting in the middle and working out toward the ends, tighten the mounting nuts a little at a time until all of them are at the torque listed in this Chapter's Specifications.

10   Install the remaining components in the reverse order of removal.

11   Start the engine and check for exhaust leaks between the manifold and cylinder head and between the manifold and exhaust pipe.

Rear manifold

Refer to illustration 4.15

12   Disconnect the negative battery cable from the battery. **Caution:** *On models equipped with the Theftlock audio system, be sure the lockout feature is turned off before performing any procedure which requires disconnecting the battery.*

13   Remove the air cleaner and duct assembly.

14   If the car is equipped with an automatic transaxle, remove the rear cam carrier (see Section 12).

15   Remove the exhaust crossover pipe from the manifold **(see illustration)**.

16   Remove the EGR tube from the exhaust manifold (see Chapter 4).

17   Set the parking brake, block the rear wheels and raise the front of the car supporting it securely on jackstands.

18   Remove the front exhaust pipe and catalytic converter assembly.

19   Remove the oxygen sensor.

20   On manual transaxle cars, remove the exhaust pipe front heat shield.

21   If necessary, remove the rear alternator bracket.

22   On automatic transaxle cars, remove the dipstick tube.

23   If necessary, remove the steering gear intermediate shaft.

24   Remove the exhaust manifold nuts.

25   If necessary, support the engine from underneath, remove the rear cradle bolts, then lower the engine and remove the steering gear heat shield.

26   Remove the exhaust manifold.

27   Clean the mating surfaces to remove all traces of old gasket material, then inspect the manifold for distortion and cracks. Warpage can be checked with a precision straightedge held against the mating flange. If a feeler gauge thicker than 0.030-inch can be inserted between the straightedge and flange surface, take the manifold to an automotive machine shop for resurfacing.

28   Place the manifold in position with a new gasket and install the mounting nuts finger tight.

29   Starting in the middle and working out toward the ends, tighten the mounting nuts a little at a time until all of them are at the torque listed in this Chapter's Specifications.

30   Install the remaining components in the reverse order of removal.

31   Start the engine and check for exhaust leaks between the manifold and cylinder head and between the manifold and exhaust pipe.

---

5   Crankshaft pulley and vibration damper - removal and installation

Removal

Refer to illustrations 5.7 and 5.9

1   Disconnect the negative battery cable from the battery. **Caution:** *On models equipped with the Theftlock audio system, be sure the lockout feature is turned off before performing any procedure which requires disconnecting the battery.*

2   Loosen the lug nuts on the right front wheel.

3   Raise the vehicle and support it securely on jackstands.

4   Remove the right front wheel.

5   Remove the right front inner fender splash shield.

6   Remove the serpentine drivebelt (see Chapter 1).

7   On automatic transaxle equipped models, remove the driveplate cover and position a large screwdriver in the ring gear teeth to keep the crankshaft from turning while an

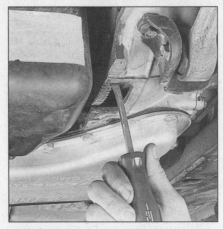

5.7 Have an assistant hold the ring gear with a large screwdriver as the pulley-to-crankshaft bolt is loosened/tightened

assistant removes the crankshaft bolt **(see illustration)**. On manual transaxle equipped models, engage high gear and apply the brakes while an assistant removes the damper bolt. **Note:** *The bolt is normally very tight, so use a large breaker bar and a six-point socket.*

8   Remove the pulley-to-damper bolts, then remove the pulley from the damper.

9   Pull the damper off the crankshaft with a bolt-type puller **(see illustration)**. Leave the Woodruff key in place in the end of the crankshaft.

Installation

10   Installation is the reverse of removal. Be sure to apply moly-base grease to the seal contact surface of the damper hub (if it isn't lubricated, the seal lip could be damaged and oil leakage would result). Align the hub keyway with the Woodruff key.

11   Tighten the damper-to-crankshaft bolt to the torque listed in this Chapter's Specifications.

12   Install the pulley on the damper and torque its bolts to 37 ft-lb.

6.2 Carefully pry the old seal out with a seal removal tool (shown) or a large screwdriver

5.9 Use a puller that bolts to the vibration damper hub; jaw type pullers will damage the vibration damper

13   Reinstall the remaining parts in the reverse order of removal.

---

6   Crankshaft front oil seal - replacement

Refer to illustrations 6.2 and 6.3

1   Remove the crankshaft pulley and vibration damper (see Section 5).

2   Note how the seal is installed - the new one must be installed to the same depth and facing the same way. Carefully pry the oil seal out of the cover with a seal puller **(see illustration)** or a large screwdriver. Be very careful not to distort the cover or scratch the crankshaft! Wrap electrician's tape around the tip of the screwdriver to avoid damage to the crankshaft.

3   Apply clean engine oil or multi-purpose grease to the outer edge of the new seal, then install it in the cover with the lip (spring side) facing IN. Drive the seal into place **(see illustration)** with a large socket and a hammer (if a large socket isn't available, a piece of pipe will also work). Make sure the seal enters the bore squarely; stop when the front

6.3 Drive the new seal into place with a large socket and hammer

9.3 To check the camshaft timing belt length, insert a small ruler between the tensioner pulley and the front cover flange (arrow), and measure the depth

1   *Tensioner pulley*
2   *Intermediate shaft sprocket*

8.8 Timing belt tensioner side plate (arrow)

face is at the proper depth.
4   Reinstall the vibration damper and the crankshaft pulley.

7   Timing belt covers - removal and installation

Front and rear covers

1   Remove the spark plug wire cover (front cover only). On 1996 and later models remove the ignition coils and the coil mounting bracket to access the rear cover.
2   Remove the retaining bolts.
3   Remove the covers.

Center Cover

4   Remove the ECM harness cover.
5   Remove the serpentine belt (see Chapter 1) and the serpentine belt tensioner.
6   Remove the clip holding the power steering line at the alternator.
7   Remove the cover bolts, then the cover.
8   Installation is the reverse of removal.

8   Timing belt tensioner plate and actuator - removal and installation

Removal

Refer to illustration 8.8

1   Remove the plenum (see Chapter 4).
2   Remove the serpentine belt (see Chapter 1).
3   If necessary, move the ECM aside, then remove its mounting bracket.
4   Remove the timing belt covers (see Section 7).
5   Align the timing marks (see Section 14).
6   Loosely clamp the two camshaft sprockets on each side of the engine together using clamping pliers or equivalent.
7   If the timing belt is not going to be removed, use a C-clamp and a protective cloth to hold the belt in position on the rear exhaust cam sprocket (see illustration 7.2).

Note: *Make sure there is no deflection in the sprocket. If there is, set up your clamp again.*
8   Remove the retaining bolts from the tensioner side plate, then remove the side plate **(see illustration)**.
9   Remove the tensioner actuator from its base with a rotating motion. **Note:** *When you do this the tensioner will extend to its maximum travel.* **Caution:** *A tapered bushing is positioned between the actuator and the mounting base; be careful not to lose it when removing the actuator.*

Installation

10   Lightly clamp the body of the tensioner actuator in a vise with its rod tip facing down. **Caution:** *Do not position the actuator in a manner that will cause damage to the rod tip or the rubber boot.*
11   Leave the actuator in this position for at least five minutes to allow the oil to drain into the boot end.
12   Find a standard paper clip - one with no serrations - and straighten it out so that you have a length of about one inch. Form the remaining portion of the paper clip into a loop.
13   Remove the rubber end plug from the rear of the actuator. Do not remove the vent plug.
14   Push your paper clip through the center hole in the vent plug, and on into the pilot hole.
15   Retract the tensioner plunger by pushing the rod tip against a table top while turning the screw at the rear of the actuator clockwise.
16   When the tensioner plunger is fully retracted, align the screw slot with the vent hole and push the paper clip into the slot to retain the plunger.
17   If oil was lost from the tensioner, fill it to the bottom of the plug hole with a synthetic 5W-30 engine oil. **Note:** *Fill the tensioner only when it is fully retracted and the paper clip is installed.*
18   Install the rubber end plug; be sure it is snapped fully into place and fits flush against the body of the tensioner.
19   Check to be sure that all bushings and the holes they fit into are in good condition

and properly installed, then fit the actuator onto its base. Be sure the actuator is free and can rotate under its own weight. **Note:** *Do not oil the bushings.* **Caution:** *Make sure the tensioner's tapered fulcrum is properly seated in its bushing.*
20   Install the tensioner plate assembly and tighten the bolts to the torque listed in this Part's Specifications.
21   Make sure the actuator rod tip is seated in the tensioner pulley socket, then remove the paper clip; this will allow the actuator to extend to its normal position.
22   Gently rotate the pulley counterclockwise into the belt.
23   Remove the clamps from the sprockets and belt.
24   Seat the belt by rotating the engine three full turns clockwise (the direction of crankshaft rotation). **Caution:** *Do not rotate the engine counterclockwise.*
25   Align the timing marks on the crankshaft pulley and timing chain cover, then inspect the camshaft timing marks to be sure the cam timing is correct.
26   Reassembly is reverse of disassembly.

9   Timing belt - inspection, removal and installation

Refer to illustration 9.3

Inspection

1   Remove the timing belt covers (see Section 7).
2   Inspect the timing belt for signs of wear, such as cracks or tears, and for oil contamination. Make sure the belt's teeth are in good condition. Check also for fraying or for wear around one edge of the belt, which would indicate misalignment of cam/belt drive components. Replace the belt as necessary.
3   Check the camshaft belt length. Insert a very thin, narrow ruler under the tensioner pulley until it contacts the tensioner base **(see illustration)**. If the belt is too long, the ruler will drop into the groove in the tensioner bracket. Check this by recording the length indicated on the ruler. If the depth from the edge of the pulley to the base of the ten-

10.4 After you get the pivot bolt started, rotate the arm counterclockwise until the square lug is at the 6 o'clock position

11.3 Remove the camshaft carrier cover bolts with a socket, ratchet and extension

sioner measures 42.6 mm or less, the length of the belt is OK. If the ruler shows 45.1 mm or more, it means it dropped into the groove in the tensioner bracket, indicating that the belt is too long and must be replaced.

Removal

4   Remove the timing belt tensioner and pulley (see Section 8).
5   If you're going to reuse the belt, mark the direction of rotation on the belt.
6   Remove the belt by sliding it carefully off the sprockets and pulleys.
Caution: *Do not kink, fold, twist or pry on the belt or you will cause damage.*

Installation

7   Start at the intermediate shaft sprocket **(see illustration 9.3)** and route the belt around the other sprockets or pulleys in a counterclockwise direction. **Note:** *Make sure you accumulate slack in the belt as you go.*
8   Install the tensioner assembly and complete the procedure by following steps 20 through 26 of Section 8.

10 Timing belt tensioner pulley and arm assembly - removal and installation

Removal

1   Remove the timing belt tensioner plate and actuator (see Section 8).
2   Remove the tensioner pulley **(see illustration 9.3)**. **Note:** *The pivot bushing inside the pulley can fall out. Hold it in place with tape or a magnet.*

Installation

Refer to illustration 10.4
Note: *The timing belt should be in place before you install the pulley assembly.*
3   Use a magnet or tape to hold the tensioner pulley's pivot tube in place while you fit the pulley into position. Remove the magnet or tape and thread the pivot bolt into it's hole.
4   After you get the pivot bolt started,

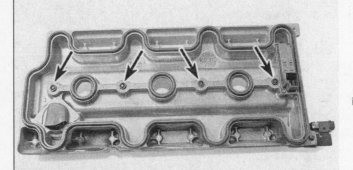

11.11 Make sure bolt isolators (arrows) are fully seated before installing the cover

rotate the arm counterclockwise until the square lug **(see illustration)** is at the 6 O'clock position. **Note:** *The pivot bushing and tube should be clean, but should not be lubricated.*
5   Tighten the tensioner pulley pivot bolt to 37 ft-lb (50 Nm).
6   Install the timing belt tensioner plate and actuator (see Section 8).

11 Camshaft carrier cover - removal and installation

Removal

Front cover
Refer to illustration 11.3
1   Remove the breather hose from the cover.
2   Remove the spark plug wires from the spark plugs. **Note:** *Be sure each wire is labeled before removal to ensure correct reinstallation.*
3   Remove the carrier cover bolts **(see illustration)**, then remove the cover. **Note:** *If the cover sticks to the cylinder head, use a block of wood and a hammer to dislodge it. If the cover still won't come loose, pry on it carefully, but don't distort the sealing flange.*
4   Remove the gasket and O-rings from the cover.

Rear Cover
5   Remove the plenum (see Chapter 4).

6   Remove the rear timing belt cover (see Section 7).
7   Remove the spark plug wires from the spark plugs (see Chapter 1). **Note:** *Be sure each wire is labeled before removal to ensure correct reinstallation.*
8   Remove the breather hose from the cover.
9   Remove the carrier cover bolts, then remove the cover. **Note:** *If the cover sticks to the cylinder head, use a block of wood and a hammer to dislodge it. If the cover still won't come loose, pry on it carefully, but don't distort the sealing flange.*
10   Remove the gasket and O-rings from the cover.

Installation

Refer to illustration 11.11
11   Installation is the reverse of removal. When installing the cover retaining bolts, be sure the bolt isolators **(see illustration)** are fully seated in the cover, then tighten the bolts to 89 lb-in.

12 Camshaft carrier - removal and installation

Removal

Front camshaft carrier
Refer to illustration 12.10
1   Remove the front camshaft carrier cover

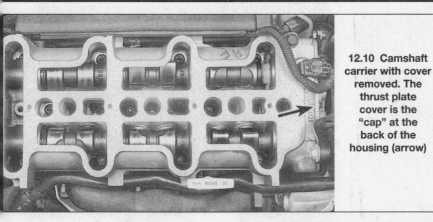

12.10 Camshaft carrier with cover removed. The thrust plate cover is the "cap" at the back of the housing (arrow)

(see Section 11).

2 Remove the timing belt (see Section 9).

3 Remove the exhaust crossover pipe.

4 Drain the coolant and remove the upper radiator hose.

5 Remove the heater pipe hose at the plenum.

6 Remove the front exhaust manifold (see Section 4).

7 Remove the engine torque strut **(see illustration 4.3)**.

8 Remove the front engine coolant pipe.

9 Install fuel line hoses under the camshafts and between the lifters to hold the lifters in place when the carrier is removed. **Note:** *You will need 6 lengths of hose. On the exhaust side the hose dimensions should be 6 x 3/16 inch. On the intake side the hose dimensions should be 3 x 5/32 inch.*

10 Remove the cam carrier mounting bolts, then remove the cam carrier **(see illustration)**.

Rear camshaft carrier

11 Remove the rear camshaft carrier cover (see Section 11).

12 Remove the timing belt (see Section 9).

13 Install fuel line hoses under the camshafts and between the lifters to hold the lifters in place when the carrier is removed (see Step 9). **Note:** *You will need 6 lengths of hose. On the exhaust side, the hose dimensions should be 6 x 3/16 inch. On the intake side, the hose dimensions should be 3 x 5/32 inch.*

14 Remove the camshaft carrier mounting bolts, then remove the cam carrier **(see illustration 12.10)**.

Installation

15 Installation is the reverse of removal. **Caution:** *Before installing the cam carrier, be sure to remove any oil that seeped into the cam carrier-to-cylinder head bolt holes.*

13 Camshaft, lifters, sprockets and oil seal - removal, inspection and installation

Note: *This procedure will require special tools available at automotive tool companies and some auto part stores. If the specialty*

tools, camshaft timing clamps, camshaft sprocket holding tool and camshaft and intermediate shaft sprocket remover, are not available, obtain similar tools and adapt them for this application.

Removal

1 Remove the cam carrier (see Section 12).

2 Remove the lifters from the cam carrier housing. **Note:** *Keep the lifters in order so that they can be replaced in their original positions.*

3 Turn the camshafts so the "flats" are facing up, then install the special camshaft timing clamp tool (see Step 9 in Section 12).

4 Remove the camshaft sprocket bolts. Use special sprocket holding tool to hold the sprockets while you loosen the bolts.

5 Remove the sprockets with a puller, then remove the flat ring from the sprocket bore.

6 Remove the camshaft thrust plate cover, then the thrust plate **(see illustration 12.10)**.

7 Remove the timing clamp, then remove the camshafts by carefully withdrawing them through the back of the cam carrier. **Caution:** *Be sure you don't damage the camshaft journals inside the carrier when you remove the camshafts.*

8 Remove the seals from the carrier by carefully prying them it out with a screwdriver. **Caution:** *The aluminum seal seating surface in the carrier is easily damaged.*

Inspection

9 Check for damage or pitting on the camshaft lobes. Check the nose of the camshaft for brinelling. Check the camshaft journals inside the carrier for wear or damage. For camshaft lobe lift inspection and journal diameter inspection, refer to Part G, Section 13. For lifter inspection, refer to Part C, Section 9.

Installation

Caution: *When installing the camshaft, exercise care to avoid damaging the seal.*
Note: *Coat the camshaft lobes and journals with moly-base grease or engine assembly lube prior to installation. When installing the thrust plate, be sure the arrow points up.*

10 Install the camshaft seal using a large socket and a hammer (if a large socket isn't available, a piece of pipe will also work). Make sure the seal enters the bore squarely; stop when the front face is at the same depth as the original seal was. **Note:** *Coat the seal with clean engine oil or multi-purpose grease prior to installation.*

11 The remainder of installation is the reverse of removal. Be sure to check cam timing after reassembly and before starting the engine (see Section 14).

14 Camshaft timing procedure

Refer to illustrations 14.1a and 14.1b
Note: *This procedure will require special tools available at automotive tool companies and some auto part stores. If the specialty tools, camshaft timing clamps, camshaft sprocket holding tool and camshaft and intermediate shaft sprocket remover, are not available, obtain similar tools and adapt them for this application.*

1 Some special tools and procedures are required to adjust the camshaft timing. If you suspect that the camshaft timing is out of adjustment, check to see if the timing marks on the camshaft sprockets are properly aligned when the engine is at top dead center (TDC) on the number 1 cylinder's exhaust stroke. When at #1 TDC exhaust, the timing marks should be aligned **(see illustrations)**. If the timing marks are not aligned as shown,

14.1a Timing marks at #1 TDC on exhaust stroke

1 Camshaft sprocket timing marks (front bank shown). Make sure the camshafts are aligned this way.

2 Intermediate shaft sprocket timing mark

3 Engine front cover mark

first try to determine if the cause is a slipped timing belt. If the belt has slipped, find the cause and replace the belt as required (see Section 9 for belt inspection and replacement). If the belt is good and you suspect the camshafts are out of time, they will have to be checked and/or adjusted as follows. **Note:** *The following procedure is based on the timing belt and tensioner already being in place on the engine.*

2   Turn the camshafts so the "flats" are facing up, then install the timing clamps (see Section 13).

3   Loosen the camshaft sprockets bolts so the sprockets can "freewheel." Use special camshaft timing clamp tools to hold the sprockets while you loosen the bolts.

4   Position the crankshaft timing mark at the TDC position **(see illustration 14.1b)**.

5   If necessary, scribe new timing marks on the camshaft sprockets using a straight-edge and a scribe, and remove the old timing marks.

6   Tighten the sprocket bolts at the rear bank camshafts and remove the timing clamp. **Note:** *The "running torque" (the torque required to turn the bolt before it is seated) should be 44-66 lb-ft. If the running torque is either more or less than this specification, replace the shim ring and lock ring and inspect the camshaft for brinelling or damaged threads. The sprocket bolt is fully seated when the edge of the lock ring is flush with the sprocket.*

7   Turn the crankshaft 360 degrees and realign the crankshaft timing mark. **Note:** *The camshafts on the rear bank should now turn with the crankshaft, while the camshafts sprockets on the front bank should still free-wheel.*

8   Repeat steps 5 and 6 on the front bank. **Note:** *When the cam timing procedure is complete, the position of the timing flats on the front bank of camshafts should differ from those of the rear bank camshafts by 180 degrees.*

---

15   Valve springs, retainers and seals - replacement

Refer to illustrations 15.4, 15.8 and 15.16
Note: *Broken valve springs and defective valve stem seals can be replaced without removing the cylinder head. Two special tools and a compressed air source are normally required to perform this operation, so read through this Section carefully and rent or buy the tools before beginning the job. If compressed air isn't available, a length of nylon rope can be used to keep the valves from falling into the cylinder during this procedure.*

1   Remove the cam carrier (see Section 12).

2   Remove the spark plugs from the cylinders which have the defective components. If all of the valve stem seals are being replaced, all of the spark plugs should be removed.

3   Turn the crankshaft until the piston in

14.1b When the camshaft and intermediate sprocket timing marks are aligned as shown in figure 14.1a, the crank timing mark should be aligned with the pointer on the front cover (arrow)

the affected cylinder is at top dead center. If you're replacing all of the valve stem seals, begin with cylinder number one and work on the valves for one cylinder at a time. Move from cylinder-to-cylinder following the firing order sequence (see this Chapter's Specifications).

4   Thread an adapter into the spark plug hole **(see illustration)** and connect an air hose from a compressed air source to it. Most auto parts stores can supply the air hose adapter. **Note:** *Many cylinder compression gauges utilize a screw-in fitting that may work with your air hose quick-disconnect fitting.*

5   Apply compressed air to the cylinder. **Warning:** *The piston may be forced down by compressed air, causing the crankshaft to turn suddenly. If the wrench used when positioning the number one piston at TDC is still attached to the bolt in the crankshaft nose, it could cause damage or injury when the crankshaft moves.*

6   The valves should be held in place by the air pressure. If the valve faces or seats are in poor condition, leaks may prevent air

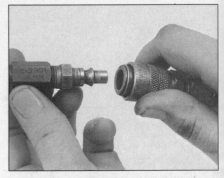

15.4 This is what the typical air hose adapter that threads into the spark plug hole looks like - they're commonly available at auto parts stores

pressure from retaining the valves - a "valve job" is necessary to correct this problem.

7   If you don't have access to compressed air, an alternative method can be used. Position the piston at a point just before TDC on the compression stroke, then feed a long piece of nylon rope through the spark plug hole until it fills the combustion chamber. Be sure to leave the end of the rope hanging out of the engine so it can be removed easily. Rotate the crankshaft in the normal direction of rotation (clockwise) until slight resistance is felt.

8   Stuff shop rags into the cylinder head holes above and below the valves to prevent parts and tools from falling into the engine, then use a valve spring compressor to compress the spring. Remove the keepers **(see illustration)** with small needle-nose pliers or a magnet.

9   Remove the spring retainer and valve spring, then remove the valve guide seal.

10   Wrap a rubber band or tape around the top of the valve stem so the valve won't fall into the combustion chamber, then release the air pressure. **Note:** *If a rope was used instead of air pressure, turn the crankshaft slightly in the direction opposite normal rotation.*

11   Inspect the valve stem for damage.

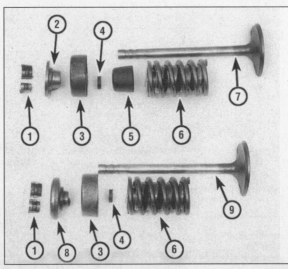

15.8 Typical engine valves and related components

1   *Keepers*
2   *Retainer*
3   *Oil shield*
4   *O-ring stem seal*
5   *Umbrella or positive type seal*
6   *Spring and damper*
7   *Intake valve*
8   *Retainer/rotator*
9   *Exhaust valve*

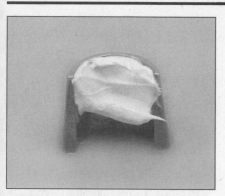

15.16 Keepers don't always stay in place, so apply a small dab of grease to each one as shown here before installation - the grease will hold the keepers in place on the valve stem

Rotate the valve in the guide and check the end for eccentric movement, which would indicate the valve stem is bent.

12   Move the valve up-and-down in the guide and make sure it doesn't bind. If the valve stem binds, either the valve is bent or the guide is damaged. In either case, the head will have to be removed for repair.

13   Reapply air pressure to the cylinder to retain the valve in the closed position, then remove the tape or rubber band from the valve stem. If a rope was used instead of air pressure, rotate the crankshaft in the normal direction of rotation until slight resistance is felt.

14   Lubricate the valve stem with engine oil and install a new valve guide seal.

15   Install the spring in position over the valve.

16   Install the valve spring retainer. Compress the valve spring and carefully install the keepers in the groove. Apply a small dab of grease to the inside of each keeper to hold it in place if necessary **(see illustration)**. Remove the pressure from the spring tool and make sure the keepers are seated.

17   Disconnect the air hose and remove the adapter from the spark plug hole. If a rope was used in place of air pressure, pull it out of the cylinder.

18   Install the spark plug(s) and hook up the wire(s).

19   Install the cam carrier.

20   Start and run the engine, then check for oil leaks and unusual sounds coming from the valve cover area.

16   Cylinder head - removal and installation

Refer to illustrations 16.12 and 16.14

Caution: *Allow the engine to cool completely before loosening the cylinder head bolts.*

Note: *On engines with high mileage and during an overhaul, camshaft lobe height should be checked prior to cylinder head removal (see Part G for instructions).*

16.12 Position the new gasket over the dowel pins (arrows)

Removal

1   Remove the cam carrier (see Section 12).

2   Remove the intake manifold (see Section 3).

3   If you're removing the front cylinder head, remove the oil dipstick tube mounting bolt and the electrical connector for the temperature sender. If your removing the rear cylinder head, remove the electrical connector for the oxygen sensor.

4   If you're removing the front cylinder head, remove the exhaust manifold (see Section 4). If you're removing the rear cylinder head, remove the exhaust crossover pipe, separate the exhaust pipe at the exhaust manifold and, on manual transmission cars, remove the rear air hose from the air pipe.

5   If you're removing the rear cylinder head, remove the timing belt tensioner bracket (see Section 8).

6   Lift the head off the engine. If resistance is felt, don't pry between the head and block as damage to the mating surfaces will result. Recheck for head bolts that may have been overlooked, then use a hammer and block of wood to tap up on the head and break the gasket seal. Be careful because there are locating dowels in the block which position each head. As a last resort, pry each head up at the rear corner only and be careful not to damage anything. After removal, place the head on blocks of wood to prevent damage to the gasket surfaces.

7   Refer to Part G for cylinder head disassembly, inspection and valve service procedures.

Installation

8   The mating surfaces of each cylinder head and block must be perfectly clean when the head is installed.

9   Use a gasket scraper to remove all traces of carbon and old gasket material, then clean the mating surfaces with lacquer thinner or acetone. If there's oil on the mating surfaces when the head is installed, the gasket may not seal correctly and leaks may develop. When working on the block, it's a good idea to cover any holes with shop rags to keep debris out of the engine. Use a shop rag or vacuum cleaner to remove any debris

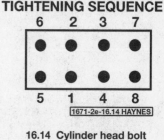

TIGHTENING SEQUENCE

6	2	3	7
●	●	●	●
●	●	●	●
5	1	4	8

1671-2e-16.14 HAYNES

16.14 Cylinder head bolt tightening sequence

that falls into the cylinders.

10   Check the block and head mating surfaces for nicks, deep scratches and other damage. If damage is slight, it can be removed with a file; if it's excessive, machining may be the only alternative.

11   Use a tap of the correct size to chase the threads in the head bolt holes. Dirt, corrosion, sealant and damaged threads will affect torque readings.

12   Position the new gasket over the dowel pins in the block **(see illustration)**. Be sure the metal tabs between the cylinders are facing up.

13   Carefully position the head on the block without disturbing the gasket.

14   Install the bolts and tighten them in the sequence shown **(see illustration)** to the torque listed in this Chapter's Specifications. Then, using the same sequence, turn each bolt the amount of angle listed in this Chapter's Specifications.

15   The remaining installation steps are the reverse of removal.

16   Change the engine oil and filter (see Chapter 1).

17   Intermediate shaft timing belt sprocket and oil seal - removal and installation

Removal

1   Align the cam timing marks by turning the engine to cylinder #1 Top Dead Center on the exhaust stroke, then remove the timing belt (see Section 9).

18.5 Engine lift bracket (arrow)

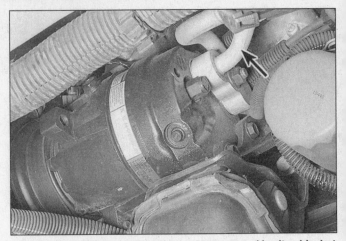

18.16 Unbolt the air conditioning compressor and lay it aside, but DO NOT disconnect the refrigerant lines (arrow)

2   Raise the car and support it on jack-stands.

3   Remove the flywheel inspection cover.

4   On cars with automatic transaxle, position a large screwdriver in the ring gear teeth to keep the crankshaft from turning while an assistant removes the intermediate shaft timing belt sprocket bolt **(see illustration 9.3)**. On cars with manual transaxle, engage high gear and apply the brakes while an assistant removes the sprocket bolt. **Note:** *The intermediate shaft timing belt sprocket bolt is normally very tight, so use a large breaker bar and a six-point socket.*

5   Note the relationship of the sprocket's timing mark to the timing chain cover, then remove the sprocket with a puller. **Caution:** *Avoid pounding on the intermediate shaft or prying on the sprocket as these actions can damage the thrust bearing.*

6   Note how the oil seal is installed - the new one must be installed to the same depth and facing the same way. Carefully pry the oil seal out of the cover with a seal puller or a large screwdriver **(see illustration 6.2)**. Be very careful not to distort the cover or scratch the intermediate shaft! Wrap electrician's tape around the tip of the screwdriver to avoid damage to the intermediate shaft.

Installation

7   Apply clean engine oil or multi-purpose grease to the outer edge of the new seal, then install it in the cover with the lip (spring side) facing IN. Drive the seal into place with a large socket and a hammer **(see illustration 6.3)**. If a large socket isn't available, a piece of pipe will also work. Make sure the seal enters the bore squarely and stop when the front face is at the proper depth.

8   Lubricate the seal "running surface" of the sprocket, then carefully fit the sprocket onto the intermediate shaft, through the seal and into the timing chain cover. **Note:** *Be sure the locating tangs of the intermediate shaft timing belt sprocket align with those on the timing chain sprocket (hidden behind the timing chain cover). To be sure the tangs are*

aligned, measure the distance from the front of the intermediate shaft sprocket to the timing chain cover. If the distance is more than 42 mm, the tangs are not aligned.

9   Check that the timing mark on the sprocket is aligned with the reference mark on the timing chain cover **(see illustration 14.1a)**. If necessary, reposition the sprocket.

10   Lubricate the intermediate shaft O-ring with clean engine oil, then carefully place it into position on the end of the shaft.

11   Put a little clean engine oil on the threads of the sprocket bolt, then install the sprocket bolt and washer. Torque the bolt to 96 ft-lb while an assistant prevents the crankshaft from turning.

12   The remainder of installation is the reverse of removal.

18 Timing chain cover - removal and installation

Removal

Refer to illustrations 18.5 and 18.16

1   Disconnect the negative battery cable. **Caution:** *On models equipped with the Theft-lock audio system, be sure the lockout feature is turned off before performing any procedure which requires disconnecting the battery.*

2   Remove the serpentine drivebelt (see Chapter 1).

3   Remove the camshaft timing belt tensioner and it's bracket (see Section 8).

4   Remove the timing belt (see Section 9) and the timing belt idler pulleys. **Note:** *Mark the direction of rotation on the timing belt before removing it.*

5   Remove the front engine lift bracket **(see illustration)**.

6   Remove the cooling fans.

7   Disconnect the coolant hoses from the water pump.

8   Remove the heater pipe retaining screws from the frame.

9   Remove the starter (see Chapter 5).

10   Remove the vibration damper (see Section 5).

11   Remove the alternator (see Chapter 5).

12   On manual transmission cars, remove the front-to-rear engine mount bracket brace.

13   Remove the oil filter.

14   Remove the oil cooler assembly.

15   Remove the oil pan front nuts and bolts, and loosen the remaining oil pan fasteners.

16   Remove the air conditioning compressor and lay it aside without disconnecting the refrigerant lines **(see illustration)**.

17   Remove the lower timing chain cover bolts.

18   Remove the intermediate shaft timing belt sprocket (see Section 17).

19   Remove the water pump pulley.

20   On manual transmission cars, remove the front exhaust air pipe check valve.

21   Remove any wiring or relays that will interfere with removal of the cover.

22   Remove the timing chain cover-to-engine block bolts.

23   Separate the cover from the engine. If it's stuck, tap it with a soft-face hammer, but don't try to pry it off.

24   Use a gasket scraper to remove all traces of old gasket material and sealant from the cover and engine block. The cover is made of aluminum, so be careful not to nick or gouge it. Clean the gasket sealing surfaces with lacquer thinner or acetone.

Installation

25   Install the new gasket.

26   Apply a thin layer of RTV sealant or equivalent to the lower edges of the timing chain cover, then install the cover.

27   Apply thread sealant to the upper timing chain cover bolts then install them. Draw the timing chain cover against the block by tightening the bolts in a criss-cross pattern. Tighten the bolts to the torque listed in this Chapter's Specifications.

28   The remainder of installation is the reverse of removal.

29   Add oil and coolant, start the engine and check for leaks.

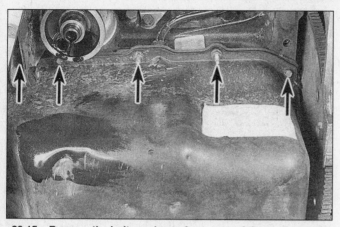

20.15a Remove the bolts and nuts from around the perimeter of the oil pan . . .

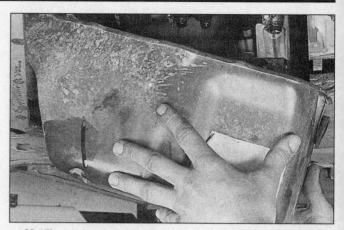

20.15b . . . then carefully separate the oil pan from the block

19 Timing chain, sprockets and tensioner - inspection, removal and installation

Note: *This procedure will require special tools available at automotive tool companies and some auto part stores. If the specialty tools, camshaft timing clamps, camshaft sprocket holding tool, camshaft and intermediate shaft sprocket remover and tensioner retractor tool, are not available, obtain similar tools and adapt them for this application.*

Inspection

1   The timing chain should be replaced with a new one if the engine has high mileage, the chain has visible damage, or has too much freeplay. Failure to replace a worn timing chain may result in erratic engine performance, loss of power and decreased fuel mileage. A worn, damaged or loose chain can break or "jump" time. In the worst case, chain breakage or "jumping" will result in severe engine damage.
2   The timing chain tensioner should be replaced if it is worn, cracked, or displays other damage.

Removal

3   Remove the timing chain cover (see Section 18).
4   Mark the positions of the sprockets to the chain.
5   Remove the timing chain tensioner bolts.
6   Remove the chain and sprockets as an assembly using special camshaft sprocket tools to pull the crankshaft sprocket off the crankshaft.
7   Remove the timing chain tensioner.

Installation

8   Install the tensioner on the block. **Note:** *Use the upper attaching bolt as the primary locator.*
9   Thread the remaining tensioner bolts into their holes finger tight.

10   Torque the bolt in the slotted hole in the tensioner to 20 ft-lb, then torque the remaining bolts to 18 ft-lb.
11   Retract the tensioner shoe using special timing chain tensioner retractor tool, and insert an appropriately-sized rivet or nail into the spring pin hole in the tensioner to hold it in position. **Caution:** *Avoid using any type of tool that will damage the tensioner or mar the surface of the tensioner shoe.* **Note:** *Be sure the rivet or nail is stout enough to maintain the shoe in it's retracted position.*
12   Apply a light coating of clean engine oil to the tensioner's chain contact surfaces
13   Carefully install the timing chain and sprockets assembly. **Note:** *Try to keep the two sprockets parallel as you install them on their respective shafts. The crank sprocket should be installed with it's large chamfer and counterbore toward the crankshaft. The intermediate shaft sprocket should be installed with it's splines facing away from the block. The crankshaft sprocket will need to be pressed on for the last 8 mm of it's travel; use a special tool. Finally, be sure the timing marks are aligned after installation.*
Caution: *Be sure the tensioner shoe and guide do not become dislodged or damaged during installation of the sprocket and chain assembly.*
14   When the sprockets and chain assembly are in place and properly timed, remove the retaining pin from the tensioner.
15   Install the timing chain cover in the reverse order of removal.

20 Oil pan - removal and installation

Note: *On vehicles equipped with an automatic transaxle, it may be necessary to remove the transaxle (see Chapter 7) to make clearance for the oil pan.*

Removal

Refer to illustrations 20.15a and 20.15b
1   Disconnect the cable from the negative battery terminal. **Caution:** *On models*

equipped with the Theftlock audio system, be sure the lockout feature is turned off before performing any procedure which requires disconnecting the battery.
2   Raise the front of the vehicle and place it securely on jackstands. Apply the parking brake and block the rear wheels to keep it from rolling off the stands.
3   Remove the front wheels and remove the lower splash pan.
4   Drain the engine oil (see Chapter 1) and remove the oil filter, the oil cooler assembly, and the oil level sensor.
5   Drain the coolant (see Chapter 1) and remove the coolant recovery tank.
6   Remove the steering gear bolts and use some wire to hang the steering gear from the body.
7   Remove the right and left lower ball joint nuts, then separate the ball joints from the control arms.
8   Disconnect the power steering cooler line clamps at the frame.
9   Remove the engine mount nuts at the frame.
10   If necessary, remove the flywheel/driveplate lower cover.
11   Remove the starter (see Chapter 5).
12   Place a floor jack under the frame front center crossmember.
13   Loosen the rear frame bolts - DO NOT REMOVE THEM!
14   Remove the front frame bolts and lower the front of the frame.
15   Remove the bolts and nuts, then carefully separate the oil pan from the block **(see illustrations)**. Don't pry between the block and the pan or damage to the sealing surfaces could occur and oil leaks may develop. Instead, tap the pan with a soft-face hammer to break the gasket seal.

Installation

16   Clean the pan with solvent and remove all old sealant and gasket material from the block and pan mating surfaces. Clean the mating surfaces with lacquer thinner or acetone and make sure the bolt holes in the block are clear. Check the oil pan flange for

distortion, particularly around the bolt holes. If necessary, place the pan on a block of wood and use a hammer to flatten and restore the gasket surface.

17   Always use a new gasket whenever the oil pan is installed. Apply a bead of RTV sealant to the front of the gasket, where it contacts the timing chain cover.

18   Place the oil pan in position on the block and install the nuts/bolts.

19   After the fasteners are installed, tighten them to the torque listed in this Chapter's Specifications. Starting at the center, follow a criss-cross pattern and work up to the final torque in three steps.

20   The remaining steps are the reverse of the removal procedure.

21   Refill the engine with oil, run it until normal operating temperature is reached and check for leaks.

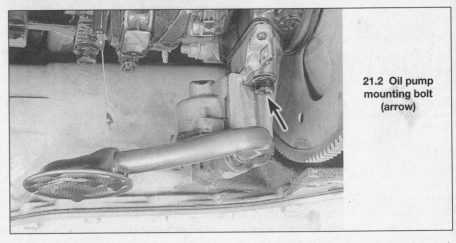

21.2 Oil pump mounting bolt (arrow)

21   Oil pump - removal, inspection and installation

Refer to illustration 21.2

1   Remove the oil pan (see Section 20) and the oil pan baffle.

2   Unbolt the oil pump and lower it from the engine **(see illustration)** along with the drive extension.

3   Inspect the pump body and cover for cracks, scoring or other damage.

4   Remove the pump cover and the gears. Check the gears for wear, galling or chips. Check the length and diameter of the gears. Check the lash between the gears with a feeler gauge. Compare your measurements with the dimensions given in this Chapter's Specifications.

5   Check the idler gear shaft and the drive shaft extension for looseness.

6   Check the pressure regulator for scoring or sticking, and check it's spring for tension.

7   Check that the suction pipe is not bent or otherwise damaged.

8   If the pump is found defective in any of the checks above, replace it with a new one - don't reuse the original or attempt to rebuild it.

9   Prime the pump by pouring motor oil into the pick-up screen while turning the pump driveshaft.

10   To install the pump, turn the hexagonal driveshaft so it mates with the oil pump drive.

11   Install the pump mounting bolt and tighten it to the torque listed in this Chapter's Specifications.

22   Flywheel/driveplate - removal and installation

This procedure is essentially the same as for the 2.5L four-cylinder engine. Refer to Part B and follow the procedure outlined

there. However, use the bolt torque listed in this Chapter's Specifications.

23   Rear main oil seal - replacement

This procedure is essentially the same as that for the 2.5L four-cylinder engine. Refer to Part B and follow the procedure outlined there.

24   Engine mounts - check and replacement

1   There are four mounts that connect the drivetrain to the chassis, two at the timing chain end (a front and a rear), one at the transaxle end, and one upper mount connecting the engine torque strut to the body near the top of the radiator.

2   Engine mounts seldom require attention, but broken or deteriorated mounts should be replaced immediately or the added strain placed on the driveline components may cause damage or wear.

Check

3   During the check, the engine must be raised slightly to remove the weight from the mounts.

4   Raise the vehicle and support it securely on jackstands, then position a jack under the engine oil pan. Place a large block of wood between the jack head and the oil pan, then carefully raise the engine just enough to take the weight off the mounts. **Warning:** *DO NOT place any part of your body under the engine when it's supported only by a jack!*

5   Check the mount insulators (the rubber part between the engine and the chassis brackets) to see if the rubber is cracked, hardened or separated from the metal plates. Sometimes the rubber will split right down the center.

6   Check for relative movement between the mount plates and the engine or frame (use a large screwdriver or prybar to attempt

to move the mounts). If movement is noted, lower the engine and tighten the mount fasteners.

7   Rubber preservative should be applied to the insulators to slow deterioration.

Replacement

8   Disconnect the negative battery cable from the battery, then raise the vehicle and support it securely on jackstands (if not already done). **Caution:** *On models equipped with a Delco Theftlock audio system, be sure the lockout feature is turned off before performing any procedure which requires disconnecting the battery.*

All except the right (rear) engine mount

9   Raise the engine slightly with a jack or hoist. Remove the bolts/nuts holding the mount to the engine and to the chassis brackets, then raise the engine more until you can remove the mount nuts, remove the old mount and insert the new one, then lower the engine and tighten the nuts to Specifications.

Right (rear) engine mount

10   The procedure for replacing the rear engine mount is more involved. The hood must be removed and a three-bar engine support must be used to support the engine from above. Raise and support the front of the vehicle on jackstands.

11   Refer to Chapter 3 and remove the engine cooling fan.

12   Disconnect the engine torque strut from its mount.

13   Refer to Chapter 8 and remove the right driveaxle, then refer to Chapter 10 and disconnect the right balljoint at the control arm.

14   Remove the nuts from the right engine mount (it attaches the transaxle-to-engine brace to the front subframe), raise the engine and replace the mount.

15   Installation is the reverse of removal. Use thread-locking compound on the mount bolts and be sure to tighten them securely to this Chapter's Specifications.

Chapter 2 Part F
3800 V6 Engine

Contents

Specifications

General

Cylinder numbers (drivebelt end-to-transaxle end)	
Front bank (radiator side)	1-3-5
Rear bank	2-4-6
Firing order	1-6-5-4-3-2

Valve lifters

Diameter	0.8420 to 0.8427 in
Lifter-to-bore clearance	0.0008 to 0.0025 in

Oil pump

1986 and later models	
Outer gear-to-housing clearance	0.008 to 0.015 in
Inner gear-to-outer gear clearance	0.006 in
Gear end clearance	0.001 to 0.0035 in
Pump cover warpage limit	0.002 in

Torque specifications

Ft-lbs (unless otherwise indicated)

Camshaft sprocket bolts	
1991	
Step one	52
Step two	Turn an additional 110-degrees
1992	
Step one	74
Step two	Turn an additional 105-degrees
1993 on	
Step one	74
Step two	Turn an additional 90-degrees

Cylinder head bolts	
1994 and earlier	
Step one	35
Step two	Tighten an additional 130-degrees
Step three	Turn an additional 30 degrees
1995 and later	
Step one	37
Step two	Tighten an additional 120-degrees
Step three (only the center four bolts!)	Turn an additional 30 degrees

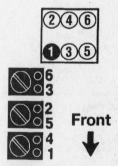

Cylinder and coil
terminal locations

Front

Torque specifications (continued)

Ft-lbs (unless otherwise indicated)

Driveplate-to-crankshaft bolts	
1991	
Step one	10 Nm
Step two	Turn an additional 90-degrees
1992 on	
Step one	11
Step two	Turn an additional 50-degrees
Exhaust manifold-to-cylinder head bolts	
1995 and earlier	38
1996 and later	22
Intake manifold-to-cylinder head bolts	
1997 and earlier	89 in-lbs
1998 and later	132 in-lbs
Oil pan bolts	120 to 125 in-lbs
Oil filter adapter-to-timing chain cover bolts	24
Oil pump	
Cover-to-timing chain cover bolts	96 to 98 in-lbs
Pickup tube and screen assembly bolts	
1995 and earlier	96 in-lbs
1996 and later	132 in-lbs
Rocker arm cover nuts/bolts	89 in-lbs
Rocker arm pivot bolts	
1991 and 1992	28
1993 through 1995	
Step one	18
Step two	Turn an additional 70-degrees
1996 and later	
Step one	11
Step two	Turn an additional 90-degrees
Timing chain cover bolts	
1995 and earlier	22
1996 through 1997	
Step one	11
Step two	Turn an additional 40-degrees
1998 and later	
Step one	15
Step two	Turn an additional 40-degrees
Vibration damper-to-crankshaft bolt	
1990	219
1991	
Step one	105
Step two	Turn an additional 56-degrees
1992 and later	
Step one	111
Step two	Turn an additional 76-degrees

1   General information

Note: *On models equipped with the Delco Loc II audio system, be sure the lockout feature is turned off before performing any procedure which requires disconnecting the battery.*

This Part of Chapter 2 is devoted to in-vehicle repair procedures for the "3800" V6 engines. The 3800 engine became available in 1990 through 1999, it is utilized in the Pontiac Grand Prix and the Buick Regal models only. This engine utilizes a cast-iron block with six cylinders arranged in a "V" shape at a 60-degree angle between the two banks. The overhead valve cast iron cylinder heads are equipped with integral valve guides and seats. Hydraulic lifters actuate the valves through tubular pushrods.

The engines are easily identified by looking for the designations printed on the upper intake plenum or cover. To positively identify this engine, locate the vehicle identification number on the left front corner of the instrument panel. This plate is visible from outside the vehicle (**see illustration 1.1** in Chapter 2, Part B). The eighth character in the sequence is the engine designation:

L or K = 3800 V6 engine
1 = 3800 Supercharged V6 engine

Information concerning camshaft, balance shaft and engine removal and installation, as well as engine block and cylinder head overhaul is in Part G of this Chapter.

The following repair procedures are based on the assumption the engine is installed in the vehicle. If the engine has been removed from the vehicle and mounted on a stand, many of the steps included in this Part of Chapter 2 will not apply.

The Specifications included in this Part of Chapter 2 apply only to the procedures in this Part. The Specifications necessary for rebuilding the block and cylinder heads are found in Part G.

2   Repair operations possible with the engine in the vehicle

Many major repair operations can be accomplished without removing the engine from the vehicle.

Clean the engine compartment and the exterior of the engine with some type of pressure washer before any work is done. A clean engine will make the job easier and will help keep dirt out of the internal areas of the engine.

Depending on the components involved, it may be a good idea to remove the hood to improve access to the engine as repairs are performed (refer to Chapter 11 if necessary).

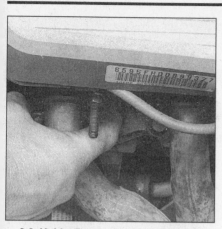

3.6 Hold a finger over the spark plug opening until you feel air escaping . . .

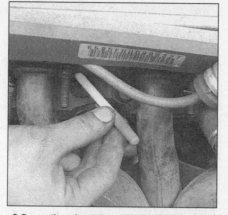

3.8 . . . then insert a soft plastic pen into the hole to detect piston movement

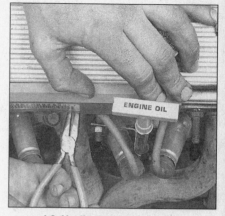

4.2 Unclip the harness cover

If vacuum, exhaust, oil or coolant leaks develop, indicating a need for gasket or seal replacement, the repairs can generally be made with the engine in the vehicle. The intake and exhaust manifold gaskets, oil pan gasket and cylinder head gaskets are all accessible with the engine in place.

Exterior engine components such as the intake and exhaust manifolds, the oil pan, the oil pump, the water pump, the starter motor, the alternator and the fuel injection system can be removed for repair with the engine in place. The timing chain and sprockets can also be replaced with the engine in the vehicle, but the camshaft and balance shaft cannot be removed with the engine in place.

Since the cylinder heads can be removed without pulling the engine, valve component servicing can also be accomplished with the engine in the vehicle.

In extreme cases caused by a lack of necessary equipment, repair or replacement of piston rings, pistons, connecting rods and rod bearings is possible with the engine in the vehicle. However, this practice is not recommended because of the cleaning and preparation work that must be done to the components involved.

3 Top Dead Center (TDC) for number one piston - locating

Refer to illustrations 3.6 and 3.8

1 Top Dead Center (TDC) is the highest point in the cylinder each piston reaches as it travels up-and-down when the crankshaft turns. Each piston reaches TDC on the compression stroke and again on the exhaust stroke, but TDC generally refers to piston position on the compression stroke.

2 Positioning the piston(s) at TDC is an essential part of certain procedures such as timing chain/sprocket removal and camshaft removal.

3 Before beginning this procedure, be sure to place the transaxle in Park, apply the parking brake and block the rear wheels. Raise the front of the vehicle and support it securely on jackstands.

4.4 Remove the acoustic engine cover from the top of the plenum - On 1996 and later models twist out the oil filler neck, then pull up and forward on the cover to remove it

4 Remove the spark plugs (see Chapter 1).

5 When looking at the drivebelt end of the engine, normal crankshaft rotation is clockwise. In order to bring any piston to TDC, the crankshaft must be turned with a socket and ratchet attached to the bolt threaded into the center of the vibration damper on the crankshaft.

6 Have an assistant turn the crankshaft with a socket and ratchet as described above while you hold a finger over the number one spark plug hole **(see illustration)**. **Note:** *See the cylinder numbering diagram in the specifications for this Chapter.*

7 When the piston approaches TDC, air pressure will be felt at the spark plug hole. Instruct your assistant to turn the crankshaft slowly.

8 Insert a plastic pen into the spark plug hole **(see illustration)**. As the piston rises the pen will be pushed out. Note the point where the pen stops moving out - this is TDC.

9 After the number one piston has been positioned at TDC on the compression stroke, TDC for any of the remaining pistons can be located by repeating the procedure described above and following the firing order.

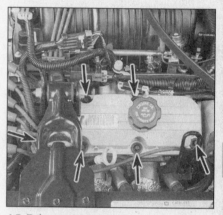

4.5 Remove the valve cover bolts (arrows)

4 Rocker arm covers - removal and installation

Refer to illustrations 4.2, 4.4 and 4.5

Front cover removal

1 Disconnect the negative battery cable from the battery. **Caution:** *On models equipped with the Theftlock audio system, be sure the lockout feature is turned off before performing any procedure which requires disconnecting the battery.*

2 Remove the spark plug wires from the spark plugs and remove the harness cover **(see illustration)**. Number each wire before removal to ensure correct reinstallation.

3 Remove the drivebelt (see Chapter 1) and the alternator brace.

4 Remove the acoustic cover over the top of the intake plenum. On 1996 and later models it will be necessary to twist out the oil filler tube from the front valve cover, pull up the front of the acoustic cover and pull it forward (towards the radiator) to release it from the bracket at the rear of the plenum **(see illustration)**.

5 Remove the engine lift bracket from the front exhaust manifold. Remove the rocker arm cover mounting bolts/nuts **(see illustration)**.

5.2a First loosen the rocker arm bolts

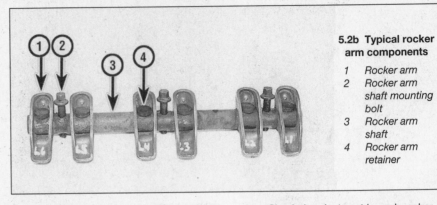

5.2b Typical rocker arm components

1 *Rocker arm*
2 *Rocker arm shaft mounting bolt*
3 *Rocker arm shaft*
4 *Rocker arm retainer*

6   Detach the rocker arm cover. **Note:** *If the cover sticks to the cylinder head, use a soft-face hammer to dislodge it.*

Rear cover removal

7   Disconnect the negative battery cable from the battery. **Caution:** *On models equipped with the Theftlock audio system, be sure the lockout feature is turned off before performing any procedure which requires disconnecting the battery.*

8   Remove the spark plug wires from the spark plugs and remove the wire holder. Be sure each wire is labeled before removal to ensure correct reinstallation.

9   Remove the serpentine drivebelt (see Chapter 1) and the power steering pump, if necessary (see Chapter 10).

10   Remove the acoustic engine cover if equipped (see Step 4) and the EGR pipe, valve and adapter (see Chapter 6).

11   Remove the engine lift bracket from the rear exhaust manifold. Remove the rocker arm cover mounting bolts/nuts.

12   Detach the rocker arm cover. **Note:** *If the cover sticks to the cylinder head, use a soft-face hammer to dislodge it.*

Installation

13   The mating surfaces of the cylinder head and rocker arm cover must be perfectly clean when the covers are installed. Use a gasket scraper to remove all traces of sealant or old gasket, then clean the mating surfaces with lacquer thinner or acetone (if there's sealant or oil on the mating surfaces when the cover is installed, oil leaks may develop). The rocker arm covers are made of aluminum, so be extra careful not to nick or gouge the mating surfaces with the scraper.

14   Apply a thread sealer to the mounting bolt threads. Place the rocker arm cover and new gasket in position, then install the bolts.

15   Tighten the bolts/nuts in several steps to the torque listed in this Chapter's specifications.

16   Complete the installation by reversing the removal procedure. Be sure to add coolant if it was drained.

17   Start the engine and check for oil leaks at the rocker arm cover-to-head joints.

5   Rocker arms and pushrods - removal, inspection and installation

Refer to illustrations 5.2a, 5.2b and 5.3

Removal

1   Refer to Section 4 and detach the rocker arm covers from the cylinder heads.

2   On models with pedestal-mounted rocker arms, loosen the rocker arm pivot bolts one at a time and detach the rocker arms, bolts, pivots and pivot retainers **(see illustrations)**. On models with rocker arm shafts, loosen the three bolts on each shaft a little at a time, working from the center out. Keep track of the rocker arm positions, since they must be returned to the same locations. Store each set of rocker components separately in a marked plastic bag to ensure that they're reinstalled in their original locations.

3   Remove the pushrods and store them separately to make sure they don't get mixed up during installation **(see illustration)**.

Inspection

4   Check each rocker arm for wear, cracks and other damage, especially where the pushrods and valve stems contact the rocker arm.

5.3 If more than one pushrod is being removed, store them in a perforated cardboard box to prevent mix-ups during installation - note the label indicating the front of the engine

5   Check the pivot seat in each rocker arm and the pivot faces. Look for galling, stress cracks and unusual wear patterns. If the rocker arms are worn or damaged, replace them with new ones and install new pivots or shafts as well.

6   Make sure the hole at the pushrod end of each rocker arm is open.

7   Inspect the pushrods for cracks and excessive wear at the ends. Roll each pushrod across a piece of plate glass to see if it's bent (if it wobbles, it's bent).

Installation

8   Lubricate the lower end of each pushrod with clean engine oil or moly-base grease and install them in their original locations. Make sure each pushrod seats completely in the lifter socket.

9   Apply moly-base grease to the ends of the valve stems, the upper ends of the pushrods and to the pivot faces to prevent damage to the mating surfaces before engine oil pressure builds up.

10   Coat the rocker arm pivot bolts with a light oil. Install the rocker arms, pivots and pivot retainers or shafts and bolts. Tighten the bolts to the torque listed in this Chapter's specifications (on models with rocker arm shafts, tighten the bolts a little at a time until the specified torque is reached). As the bolts

6.4 This is what the air hose adapter that threads into the spark plug hole looks like - they're commonly available from auto parts stores

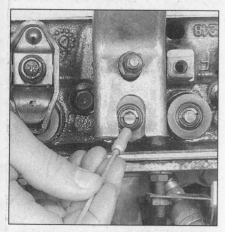

6.8 After compressing the valve spring, remove the keepers with a magnet (as shown here) or small needle-nose pliers

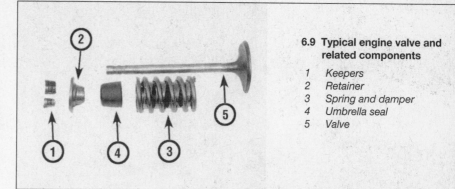

6.9 Typical engine valve and related components

1   *Keepers*
2   *Retainer*
3   *Spring and damper*
4   *Umbrella seal*
5   *Valve*

are tightened, make sure the pushrods seat properly in the rocker arms.
11   Install the rocker arm covers (see Section 4).

6   Valve springs, retainers and seals - replacement

Refer to illustrations 6.4, 6.8, 6.9 and 6.17
Note: *Broken valve springs and defective valve stem seals can be replaced without removing the cylinder heads. Two special tools and a compressed air source are normally required to perform this operation, so read through this Section carefully and rent or buy the tools before beginning the job. If compressed air is not available, a length of nylon rope can be used to keep the valves from falling into the cylinder during this procedure.*
1   Refer to Section 4 and remove the rocker arm cover from the affected cylinder head. If all of the valve stem seals are being replaced, remove both rocker arm covers.
2   Remove the spark plug from the cylinder which has the defective component. If all of the valve stem seals are being replaced, all of the spark plugs should be removed.
3   Turn the crankshaft until the piston in the affected cylinder is at Top Dead Center on the compression stroke (refer to Section 3 for instructions). If you're replacing all of the valve stem seals, begin with cylinder number one and work on the valves for one cylinder at a time. Move from cylinder-to-cylinder following the firing order sequence (1-6-5-4-3-2).
4   Thread an adapter into the spark plug hole **(see illustration)** and connect an air hose from a compressed air source to it. Most auto parts stores can supply the air hose adapter. **Note:** *Many cylinder compression gauges utilize a screw-in fitting that may work with your air hose quick-disconnect fitting.*
5   Remove the bolt, pivot and rocker arm for the valve with the defective part and pull out the pushrod (see Section 5). If all of the valve stem seals are being replaced, all of the

rocker arms and pushrods should be removed.
6   Apply compressed air to the cylinder. The valves should be held in place by the air pressure. If the valve faces or seats are in poor condition, leaks may prevent the air pressure from retaining the valves - refer to the alternative procedure below.
7   If you don't have access to compressed air, an alternative method can be used. Position the piston at a point just before TDC on the compression stroke, then feed a long piece of nylon rope through the spark plug hole until it fills the combustion chamber. Be sure to leave the end of the rope hanging out of the engine so it can be removed easily. Use a large breaker bar and socket to rotate the crankshaft in the normal direction of rotation until slight resistance is felt.
8   Stuff shop rags into the cylinder head holes above and below the valves to prevent parts and tools from falling into the engine, then use a valve spring compressor to compress the spring. Remove the keepers with small needle-nose pliers or a magnet **(see illustration)**. **Note:** *A couple of different types of tools are available for compressing the valve springs with the head in place. One type grips the lower spring coils and presses on the retainer as the knob is turned, while the other type, shown here, utilizes the rocker arm bolt for leverage. Both types work very well, although the lever type is usually less expensive.*
9   Remove the spring retainer and valve spring, then remove the guide seal **(see illustration)**. **Note:** *If air pressure fails to hold the valve in the closed position during this operation, the valve face or seat is probably damaged. If so, the cylinder head will have to be removed for additional repair operations.*
10   Wrap a rubber band or tape around the top of the valve stem so the valve won't fall into the combustion chamber, then release the air pressure. **Note:** *If a rope was used instead of air pressure, turn the crankshaft slightly in the direction opposite normal rotation.*
11   Inspect the valve stem for damage. Rotate the valve in the guide and check the end for eccentric movement, which would indicate that the valve is bent.
12   Move the valve up-and-down in the guide and make sure it doesn't bind. If the

6.17 Apply a small dab of grease to each keeper as shown here before installation - it will hold them in place on the valve stem as the spring is released

valve stem binds, either the valve is bent or the guide is damaged. In either case, the head will have to be removed for repair.
13   Reapply air pressure to the cylinder to retain the valve in the closed position, then remove the tape or rubber band from the valve stem. If a rope was used instead of air pressure, rotate the crankshaft in the normal direction of rotation until slight resistance is felt.
14   Lubricate the valve stem with clean engine oil and install a new valve guide seal. Using a hammer and a deep socket or seal installation tool, gently tap the seal into place until it's completely seated on the guide.
15   Install the spring in position over the valve.
16   Install the valve spring retainer and compress the valve spring.
17   Position the keepers in the upper groove. Apply a small dab of grease to the inside of each keeper to hold it in place if necessary **(see illustration)**. Remove the pressure from the spring tool and make sure the keepers are seated.
18   Disconnect the air hose and remove the adapter from the spark plug hole. If a rope was used in place of air pressure, pull it out of the cylinder.
19   Install the rocker arm(s) and pushrod(s).
20   Install the spark plug(s) and connect the wire(s).
21   Refer to Section 4 and install the rocker arm cover(s).

7.6 Pinch the tabs (arrows) together with pliers to detach the cables from the bracket

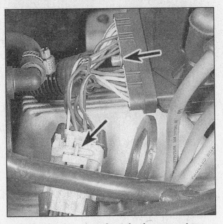

7.12 Unplug the electrical connectors (arrows) and move the harness aside

7.14 Carefully pry up on a casting boss - don't pry between gasket surfaces

22   Start and run the engine, then check for oil leaks and unusual sounds coming from the rocker arm cover area.

7   Intake manifold - removal and installation

Refer to illustrations 7.6, 7.12, 7.14, 7.15, 7.19a and 7.19b

Removal

1   Relieve the fuel system pressure (see Chapter 4).
2   Disconnect the negative battery cable from the battery. **Caution:** *On models equipped with the Theftlock audio system, be sure the lockout feature is turned off before performing any procedure which requires disconnecting the battery.*
3   Remove the air intake duct, the mass airflow sensor and the throttle body (see Chapter 4).
4   Remove the power steering pump-to-manifold brace (Chapter 10). On 1995 and later models remove the acoustic cover **(see illustration 4.4)**.
5   Referring to Chapter 4, remove the fuel rail and injectors. On 1995 and later models

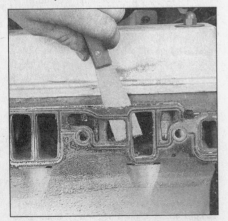

7.15 Remove all traces of gasket material, but don't gouge the soft aluminum

remove the upper plenum.
6   Remove the control cables from their brackets **(see illustration)**.
7   Remove the drivebelt (see Chapter 1).
8   Disconnect and remove the alternator and brackets (see Chapter 5).
9   Drain the cooling system (see Chapter 1).
10   Disconnect the coolant tubes and hoses at the manifold.
11   Remove the ignition module/coil assembly (See Chapter 5). On 1995 and later engines remove the EGR tube from the lower manifold if equipped.
12   Label and disconnect the fuel and vacuum lines and electrical wires at the manifold **(see illustration)**. When disconnecting fuel line fittings, be prepared to catch some fuel, then cap the fittings to prevent contamination.
13   Remove the manifold mounting bolts.
14   Separate the manifold from the engine **(see illustration)**. Do not pry between the manifold and heads, as damage to the gasket sealing surfaces may result. If you're installing a new manifold, transfer all fittings and sensors to the new manifold.

Installation

Note: *The mating surfaces of the cylinder heads, block and manifold must be perfectly clean when the manifold is installed. Gasket removal solvents in aerosol cans are available at most auto parts stores and may be helpful when removing old gasket material that's stuck to the heads and manifold (since the manifold is made of aluminum, aggressive scraping can cause damage). Be sure to follow the directions printed on the container.*

15   Use a gasket scraper to remove all traces of sealant and old gasket material **(see illustration)**, then clean the mating surfaces with lacquer thinner or acetone. If there's old sealant or oil on the mating surfaces when the manifold is installed, oil or vacuum leaks may develop. Use a vacuum cleaner to remove any gasket material that falls into the intake ports or the lifter valley.
16   Use a tap of the correct size to chase the threads in the bolt holes, then use com-

pressed air (if available) to remove the debris from the holes. **Warning:** *Wear safety glasses or a face shield to protect your eyes when using compressed air.*
17   If steel manifold gaskets are used, apply contact cement or a spray adhesive to both sides of the gaskets before positioning them on the heads. Apply RTV sealant to the ends of the new manifold-to-block seals, then install them. Make sure the pointed end of the seal fits snugly against both the head and block.
18   Carefully lower the manifold into place. Apply thread lock compound to the mounting bolt threads and install the bolts finger tight.
19   Tighten the mounting bolts in two stages, following the recommended sequence **(see illustrations)**, until they're all at the torque listed in this Chapter's specifications.
20   Install the remaining components in the reverse order of removal.
21   Change the oil and filter and fill the cooling system (see Chapter 1). Start the engine and check for oil and vacuum leaks.

8   Valve lifters - removal, inspection and installation

Refer to illustrations 8.6a, 8.6b, 8.6c, 8.7, 8.8, 8.11 and 8.12

1   A noisy valve lifter can be isolated when the engine is idling. Hold a mechanic's stethoscope or a length of hose near the position of each valve while listening at the other end.
2   The most likely causes of noisy valve lifters are dirt trapped between the plunger and the lifter body or lack of oil flow, viscosity or pressure. Before condemning the lifters, we recommend checking the oil for fuel contamination, proper level, cleanliness and correct viscosity.

Removal

3   Remove the rocker arm cover(s) as described in Section 4.
4   Remove the rocker arms and pushrods (see Section 5).

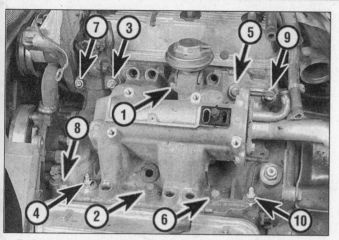

7.19a Intake manifold tightening sequence (1994 and earlier)

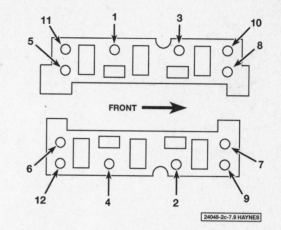

7.19b Intake manifold tightening sequence (1995 and later)

8.6a The guide plate is held in place by two bolts (arrows)

8.6b The lifter guides slip over the lifters

8.6c On 1995 and later engines the roller lifters are retained by guides bolted to the sides of the lifter valley – remove the four bolts to take off the guides and remove the lifters

5   Remove the intake manifold as described in Section 7.

6   Earlier models have conventional lifters. Later production engines have roller lifters. Roller lifters are kept from turning by a guide plate and lifter guide (see illustrations), which must be removed to access the lifters.

7   There are several ways to extract the lifters from the bores. A special tool designed to grip and remove lifters is manufactured by many tool companies and is widely available, but it may not be required in every case (see illustration). On newer engines without a lot of varnish buildup, the lifters can often be removed with a small magnet or even with your fingers. A machinist's scribe with a bent end can be used to pull the lifters out by positioning the point under the retainer ring in the top of each lifter. Caution: *Don't use pliers to remove the lifters unless you intend to replace them with new ones. The pliers will damage the precision machined and hardened lifters, rendering them useless.*

8   Before removing the lifters, arrange to store them in a clearly labeled box to ensure that they're reinstalled in their original locations. Remove the lifters and store them where they won't get dirty (see illustration).

8.7 Many times you can remove the lifters by hand

Inspection

9   Parts for valve lifters are not available separately. The work required to remove them from the engine again if repair is unsuccessful outweighs any potential savings from repairing them.

10   Clean the lifters with solvent and dry them thoroughly without mixing them up.

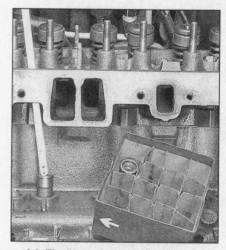

8.8 The lifters on an engine that has accumulated many miles may have to be removed with a special tool - be sure to store the lifters in an organized manner to make sure they're reinstalled in their original locations

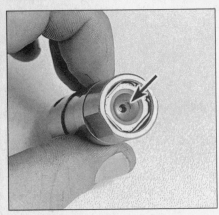

8.11 Check the pushrod seat (arrow) in the top of each lifter for wear

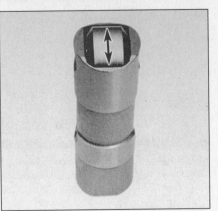

8.12 The roller must turn freely - check for wear and excessive play as well

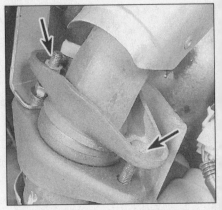

9.4 The exhaust manifolds are bolted to the crossover pipe (arrows)

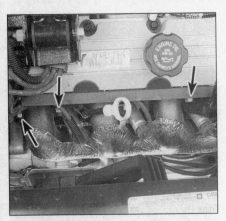

9.8 Remove the front exhaust manifold bolts/studs (arrows) - 3 bolts (out of 6 total) visible in this photo

10.3a The alternator bracket is attached to the cylinder head with two nuts (arrows)

10.3b On some models, the air conditioner compressor bracket is attached to the cylinder head with one Torx T-50 bolt (arrow)

Conventional lifters

11   Check each lifter wall, pushrod seat and foot for scuffing, score marks and uneven wear **(see illustration)**. Each lifter foot (the surface that rides on the cam lobe) must be slightly convex, although this can be difficult to determine by eye. If the base of the lifter is concave, the lifters and camshaft must be replaced. If the lifter walls are damaged or worn (which isn't very likely), inspect the lifter bores in the engine block as well. If the pushrod seats **(see illustration)** are worn, check the pushrod ends.

Roller lifters

12   Check the rollers carefully for wear and damage and make sure they turn freely without excessive play **(see illustration)**. The inspection procedure for conventional lifters also applies to roller lifters.

Installation

13   Used roller lifters can be reinstalled with a new camshaft and the original camshaft can be used if new lifters are installed.
14   If the conventional lifters are worn, they must be replaced with new ones and the camshaft must be replaced as well - never install used conventional lifters with a new camshaft or new lifters with a used camshaft.

15   When reinstalling used lifters, make sure they're replaced in their original bores. Soak new lifters in oil to remove trapped air. Coat all lifters with moly-base grease or engine assembly lube prior to installation.
16   The remaining installation steps are the reverse of removal.
17   Run the engine and check for oil leaks.

9   Exhaust manifolds - removal and installation

Warning: *Allow the engine to cool completely before beginning this procedure.*
Note: *Exhaust system fasteners are frequently difficult to remove - they get frozen in place because of the heating/cooling cycle to which they're constantly exposed. To ease removal, apply penetrating oil to the threads of all exhaust manifold and exhaust pipe fasteners and allow it to soak in.*

Removal - front manifold

Refer to illustrations 9.4 and 9.8
1   Disconnect the negative battery cable.
Caution: *On models equipped with the Theft-lock audio system, be sure the lockout feature is turned off before performing any procedure which requires disconnecting the battery.*

2   Remove the cooling fan for access (see Chapter 3).
3   Remove the mass airflow sensor, air duct and crankcase ventilation pipe, if necessary (see Chapter 4). Remove the engine mount struts. On 1996 and later models remove the acoustic cover **(see illustration 4.4)**.
4   Unbolt the manifold from the crossover pipe **(see illustration)**.
5   Remove the dipstick tube hold-down nut and wiggle the dipstick tube out of the block.
6   Detach the spark plug wires from the front spark plugs (see Chapter 1).
7   Remove the heat shield, if equipped.
8   Unbolt and remove the exhaust manifold **(see illustration)**.

Removal - rear manifold

9   Disconnect the negative battery cable
Caution: *On models equipped with the Theft lock audio system, be sure the lockout feature is turned off before performing any procedure which requires disconnecting the battery.*
10   Drain the coolant from the radiator (see Chapter 1).
11   Remove the mass airflow sensor, air duct and crankcase ventilation pipe, if necessary (see Chapter 4).

10.3c Drivebelt tensioner bolt locations (arrows)

10.9 Cylinder head bolt LOOSENING sequence

12   Remove the two nuts attaching the crossover pipe to the rear exhaust manifold.

13   Disconnect the spark plug wires from the rear spark plugs (see Chapter 1).

14   Remove the EGR pipe and transaxle dipstick tube if it's in the way (see Chapters 6 and 7).

15   Unbolt the power steering pump (some models) without removing the hoses and hold it to one side with wire (see Chapter 10).

16   Detach the alternator support bracket. Refer to Chapter 5 as necessary.

17   Remove the heater hoses and coolant tube brackets and tubing above the exhaust manifold. On 1996 and later models remove the acoustic cover (see illustration 4.4) and the acoustic cover mounting bracket.

18   Remove the manifold heat shield, if equipped.

19   Detach the IAC wiring from the throttle body.

20   Set the parking brake, block the rear wheels and raise the front of the vehicle, supporting it securely on jackstands.

21   Working under the vehicle, remove the two exhaust pipe-to-manifold bolts.

22   Disconnect the oxygen sensor wire (the sensor is threaded into the manifold), then lower the vehicle.

23   Remove the six bolts and detach the manifold from the head.

Installation

24   Clean the mating surfaces of the manifold and cylinder head to remove all traces of old gasket material, then check the manifold for warpage and cracks. If the manifold gasket was blown, take the manifold to an automotive machine shop for resurfacing.

25   Place the manifold in position with a new gasket and install the bolts finger tight.

26   Starting in the middle and working out toward the ends, tighten the mounting bolts a little at a time until all of them are at the specified torque.

27   Install the remaining components in the reverse order of removal.

28   Start the engine and check for exhaust leaks between the manifold and cylinder head and between the manifold and exhaust pipe.

10 Cylinder heads - removal and installation

Refer to illustrations 10.3a, 10.3b, 10.3c, 10.9, 10.10, 10.13, 10.16 and 10.19

Removal

1   Disconnect the negative battery cable at the battery. Caution: On models equipped with the Theftlock audio system, be sure the lockout feature is turned off before performing any procedure which requires disconnecting the battery.

2   Disconnect the spark plug wires and remove the spark plugs (see Chapter 1). Be sure to label the plug wires to simplify reinstallation.

3   Remove the nuts/bolts holding the brackets to the head (see illustrations).

4   Disconnect all wires and hoses from the cylinder head(s). Be sure to label them to simplify reinstallation.

5   Remove the intake manifold as described in Section 7.

6   Detach the exhaust manifold(s) from the cylinder head(s) being removed (see Section 9).

7   Remove the rocker arm cover(s) (see Section 4).

8   Remove the rocker arms and pushrods

(see Section 5).

9   Loosen the head bolts in 1/4-turn increments until they can be removed by hand. Work from bolt-to-bolt as shown (see illustration).

10   Lift the head off the engine. If resistance is felt, don't pry between the head and block as damage to the mating surfaces will result. Recheck for head bolts that may have been overlooked, then use a hammer and block of wood to tap the head and break the gasket seal. Be careful because there are locating dowels in the block which position each head. As a last resort, pry each head up at the rear corner only and be careful not to damage anything (see illustration). After removal, place the head on blocks of wood to prevent damage to the gasket surfaces.

11   Refer to Chapter 2, Part G, for cylinder head disassembly, inspection and valve service procedures.

Installation

12   The mating surfaces of the cylinder heads and block must be perfectly clean when the heads are installed.

13   Use a gasket scraper to remove all traces of carbon and old gasket material (see illustration), then clean the mating surfaces with lacquer thinner or acetone. If there's oil on the mating surfaces when the heads are

10.10 Pry carefully - don't force a tool between the gasket surfaces

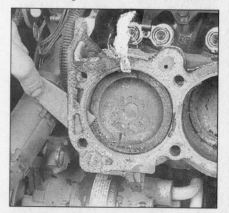

10.13 Carefully remove all traces of old gasket material

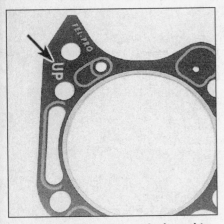

10.16 Look for gasket marks (arrow) to ensure correct installation

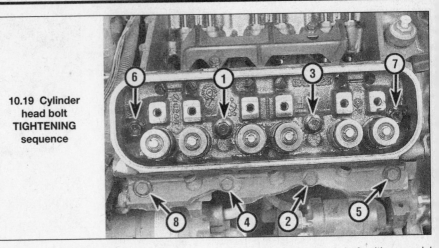

10.19 Cylinder head bolt TIGHTENING sequence

installed, the gaskets may not seal correctly and leaks may develop. When working on the block, it's a good idea to cover the lifter valley with shop rags to keep debris out of the engine. Use a shop rag or vacuum cleaner to remove any debris that falls into the cylinders.

14   Check the block and head mating surfaces for nicks, deep scratches and other damage. If damage is slight, it can be removed with a file; if it's excessive, machining may be the only alternative.

15   Use a tap of the correct size to chase the threads in the head bolt holes. Dirt, corrosion, sealant and damaged threads will affect torque readings.

16   Position the new gaskets over the dowel pins in the block. If steel gaskets are used, apply contact cement or spray adhesive. Install "non-retorquing" type gaskets dry (no sealant), unless the manufacturer states otherwise. Most gaskets are marked UP or TOP **(see illustration)** because they must be installed a certain way.

17   Carefully position the heads on the block without disturbing the gaskets.

18   Use NEW head bolts - don't reinstall the old ones - and apply thread sealer to the threads and the undersides of the bolt heads.

19   Tighten the bolts as directed in this Chapter's specifications in the sequence shown **(see illustration)**. This must be done in three steps, following the sequence each time.

20   The remaining installation steps are the reverse of removal.

21   Change the oil and filter (see Chapter 1).

11 Vibration damper - removal and installation

Refer to illustrations 11.7a, 11.7b and 11.9

Caution: The vibration damper is serviced as an assembly. Do not attempt to separate the pulley from the balancer hub.

1   Disconnect the negative cable from the battery. **Caution:** On models equipped with the Theftlock audio system, be sure the lockout feature is turned off before performing any procedure which requires disconnecting the battery.

2   Loosen the lug nuts on the right front wheel.

3   Raise the vehicle and support it securely on jackstands.

4   Remove the right front wheel.

5   Remove the right front fender inner splash shield.

6   Remove the drivebelt (see Chapter 1).

7   Remove the lower bellhousing cover

plate and hold the crankshaft with a special flywheel locking tool. If this tool is unavailable, position a large screwdriver in the ring gear teeth **(see illustrations)** to keep the crankshaft from turning while an assistant removes the crankshaft balancer bolt **(see illustration)**. The bolt is normally quite tight, so use a large breaker bar and a six-point socket. **Note:** On later models it may be necessary to lower the right side of the drivetrain/front suspension frame assembly to allow access to the vibration damper bolt. Place a floor jack under the frame front center crossmember. Loosen the right side frame bolts - DO NOT REMOVE THEM! - and lower the right side of the frame to access the vibration damper.

8   The damper should pull off the crankshaft by hand. Leave the Woodruff key in place in the end of the crankshaft.

9   Installation is the reverse of removal. Align the keyway with the key **(see illustration)** and avoid bending the metal tabs. Be sure to apply moly-base grease to the seal contact surface on the back side of the damper (if it isn't lubricated, the seal lip could be damaged and oil leakage would result).

10   Apply sealant to the threads and tighten the crankshaft bolt to the torque listed in this Chapter's Specifications.

11   Reinstall the remaining components in the reverse order of removal.

11.7a Wedge a large screwdriver in the teeth to hold the flywheel/driveplate in place

11.7b Remove the bolt in the center of the hub (arrow)

11.9 Be sure to align the keyway (arrow) with the key in the crankshaft

12.2 Pry out the old seal with a seal removal tool (shown here) or a screwdriver

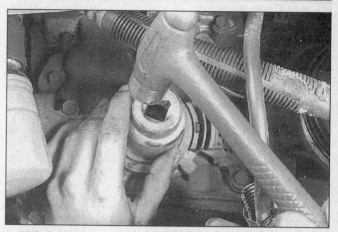

12.5 Gently drive the new seal into place with a hammer and large socket

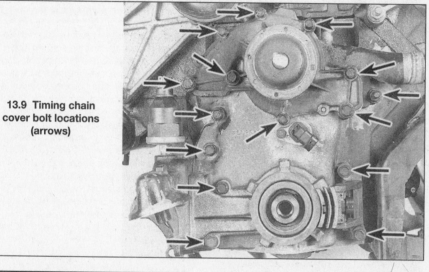

13.9 Timing chain cover bolt locations (arrows)

13.10 Once the timing cover has been removed, remove the camshaft button from the end of the camshaft (1994 and earlier models)

12 Crankshaft front oil seal - replacement

Refer to illustrations 12.2 and 12.5

1   Remove the vibration damper (see Section 11) and the crankshaft position sensor shield if equipped. If there's a groove worn into the seal contact surface on the vibration damper, sleeves are available that fit over the groove, restoring the contact surface to like-new condition. These sleeves are sometimes included with the seal kit. Check with your parts supplier for details.

2   Pry the old oil seal out with a seal removal tool or a screwdriver **(see illustration)**. Be very careful not to nick or otherwise damage the crankshaft in the process.

3   Apply a thin coat of RTV-type sealant to the outer edge of the new seal. Lubricate the seal lip with moly-base grease or clean engine oil.

4   Place the seal squarely in position in the bore and press it into place with special tool. Make sure the seal enters the bore squarely and seats completely.

5   If the special tool is unavailable, carefully guide the seal into place with a large

socket or piece of pipe and a hammer **(see illustration)**. The outer diameter of the socket or pipe should be the same size as the seal outer diameter.

6   Install the vibration damper (see Section 11).

7   Reinstall the remaining parts in the reverse order of removal.

8   Start the engine and check for oil leaks at the seal.

13 Timing chain cover - removal and installation

Refer to illustrations 13.9, 13.10, 13.11 and 13.12

1   Disconnect the negative battery cable from the battery. **Caution:** *On models equipped with the Theftlock audio system, be sure the lockout feature is turned off before performing any procedure which requires disconnecting the battery.*

2   Set the parking brake and put the transmission in Park. Raise the front of the vehicle and support it securely on jackstands. Remove the splash shield from the right inner fender.

3   Drain the oil and coolant (see Chapter 1).

4   Remove the coolant hoses from the timing chain cover.

5   Remove the large (8 mm) water pump-to-block bolts, leaving the smaller (6 mm) diameter bolts in place.

6   Remove the vibration damper (see Section 11) and the crankshaft position sensor shield if equipped.

7   Unplug the connectors from the oil pressure, camshaft and crankshaft sensors on distributorless ignition models or remove the distributor, if equipped (see Chapter 5).

8   Remove the oil pan (see Section 17).

9   Remove the timing chain cover-to-engine block bolts **(see illustration)**. Note that two of the bolts also secure the crankshaft sensor. Lift the sensor off when removing these bolts.

10   Separate the cover from the front of the engine. On earlier models there is a spring and button that controls camshaft end play inside the cam sprocket. If they are missing, look for them in the oil pan **(see illustration)**.

11   Use a gasket scraper to remove all traces of old gasket material and sealant from

13.11 Remove all traces of old gasket material

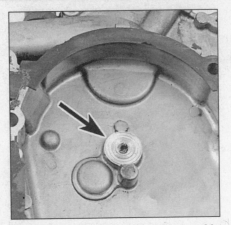

13.12 The camshaft thrust surface on this cover is worn away (arrow) which means that a new cover must be installed (1994 and earlier models)

14.3 Carefully pull the oil filter adapter away from the timing cover - the pressure regulator (arrow) is spring loaded and may spring out when the adapter is removed

the cover and engine block **(see illustration)**. The cover is made of aluminum, so be careful not to nick or gouge it. Clean the gasket sealing surfaces with lacquer thinner or acetone.

12   On 1994 and earlier models check the camshaft thrust surface in the cover for excessive wear **(see illustration)**. If it's worn, a new cover will be required (see the note at the beginning of this Section).

13   The oil pump cover must be removed and the cavity packed with petroleum jelly as described in Section 15 before the cover is installed.

14   Apply a thin layer of RTV sealant to both sides of the new gasket, then position the gasket on the engine block (the dowel pins should keep it in place). Make sure the spring and button are in place in the end of the camshaft (hold them in place with grease), then attach the cover to the engine. The oil pump drive must engage with the crankshaft or distributor gear.

15   Apply thread sealant to the bolt threads, then install them finger tight. Install the crankshaft sensor, but leave the bolts finger tight until Step 16. Follow a criss-cross pattern when tightening the other bolts and work up to the torque listed in this Chapter's specifications in three steps to avoid warping the cover.

16   Use a special crankshaft sensor adjuster tool to position the crankshaft sensor on the timing cover. Tighten the bolts to the torque listed in this Chapter's specifications for the timing chain cover.

17   The remainder of installation is the reverse of removal.

18   Add oil and coolant, start the engine and check for leaks.

14   Oil filter adapter and pressure regulator valve - removal and installation

Refer to illustrations 14.3 and 14.4

1   Remove the oil filter (see Chapter 1).

2   Remove the timing chain cover (see Section 13).

3   Remove the four bolts holding the oil filter adapter to the timing chain cover **(see illustration)**. The cover is spring loaded, so remove the bolts while keeping pressure on the cover, then release the spring pressure carefully.

4   Remove the pressure regulator valve and

spring (see illustration). Use a gasket scraper to remove all traces of the old gasket

5   Clean all parts with solvent and dry them with compressed air (if available). **Warning** *Wear eye protection. Check for wear, score marks and valve binding.*

6   Installation is the reverse of removal. Be sure to use a new gasket. **Caution:** *If a new timing chain cover is being installed on the engine, make sure the oil pressure relief valve supplied with the new cover is used. If the old style relief valve is installed in a new cover, oil pressure problems will result.*

7   Tighten the bolts to the torque listed in this Chapter's Specifications.

8   Run the engine and check for oil leaks.

15   Oil pump - removal, inspection and installation

Refer to illustrations 15.4, 15.10 and 15.11

Removal

1   Remove the oil filter (see Chapter 1).

2   Remove the oil filter adapter, pressure regulator valve and spring (see Section 14).

3   Remove the timing chain cover (see Section 13).

4   Remove the oil pump cover-to-timing chain cover bolts **(see illustration)**.

5   Lift out the cover and oil pump gears as an assembly.

Inspection

6   Clean the parts with solvent and dry them with compressed air (if available). **Warning:** *Wear eye protection!*

7   Inspect all components for wear and score marks. Replace any worn out or damaged parts.

8   Refer to Section 14 for bypass valve information.

9   Reinstall the gears in the timing chain cover.

10   Measure the outer gear-to-housing

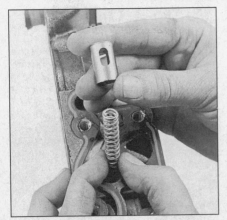

14.4 Remove the pressure regulator valve and spring, then check the valve for wear and damage

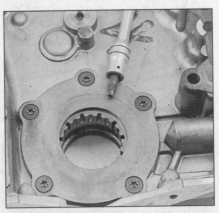

15.4 The oil pump cover is attached to the inside of the timing chain cover - a T-30 Torx driver is required for removal of the screws

15.10 Measuring the outer gear-to-housing clearance with a feeler gauge

15.11 Measuring the inner gear-to-outer gear clearance with a feeler gauge

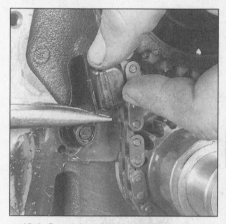

16.1 The shim has a notch in it to fit over the key in the crankshaft (1994 and earlier models)

16.3 The marks on the crankshaft and camshaft sprockets (arrows) must be aligned adjacent to each other as shown here (the mark on the camshaft sprocket is a dimple - the one on the crankshaft sprocket is a raised dot)

16.4 Carefully release the spring with needle-nose pliers

clearance with a feeler gauge (see illustration).

1   Measure the inner gear-to-outer gear clearance at several points (see illustration).

2   Use a dial indicator or straightedge and feeler gauges to measure the gear end clearance (distance from the gear to the gasket surface of the cover).

3   Check for pump cover warpage by laying a precision straightedge across the cover and trying to slip a feeler gauge between the cover and straightedge.

4   Compare the measurements to this Chapter's Specifications. Replace all worn or damaged components with new ones.

Installation

5   Remove the gears and pack the pump cavity with petroleum jelly.

6   Install the gears - make sure petroleum jelly is forced into every cavity. Failure to do so could cause the pump to lose its prime when the engine is started, causing damage from lack of oil pressure.

7   Install the pump cover, using a new gasket only - its thickness is critical for maintaining the correct clearances.

8   Install the pressure regulator spring and valve.

9   Install the timing chain cover.

10   Install the oil filter and check the oil level. Start and run the engine and check for correct oil pressure, then look carefully for oil leaks at the timing chain cover.

16   Timing chain and sprockets - removal and installation

Refer to illustrations 16.1, 16.3, 16.4, 16.5 and 16.6

Removal

Remove the timing chain cover (see Section 13), then slide the shim off the nose of the crankshaft (see illustration).

The timing chain should be replaced with a new one if the total free play midway between the sprockets exceeds one inch. Failure to replace the timing chain may result in erratic engine performance, loss of power and lowered fuel mileage.

3   Temporarily install the vibration damper bolt and turn the crankshaft clockwise to align the timing marks on the crankshaft and camshaft sprockets directly opposite each other (see illustration).

4   Remove the timing chain damper. **Note:** *On some earlier models it will be necessary to detach the spring and rotate the damper out of the way* (see illustration). *On all other models simply remove the damper bolt to detach it from the block.*

5   Remove the camshaft sprocket bolts (see illustration). Try not to turn the camshaft in the process (if you do, realign the timing marks after the bolts are loosened).

6   Alternately pull the camshaft sprocket and then the crankshaft sprocket forward

16.5 Earlier models use two bolts (arrow) to retain the camshaft sprocket while later model engines use a single bolt to retain the sprocket

16.6 Guide the chain and sprockets off as an assembly

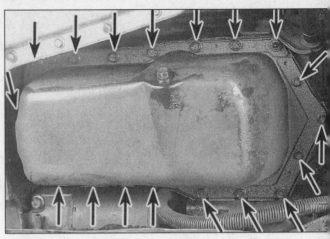

17.4 The oil pan bolts (arrows) are located around the perimeter of the oil pan (viewed from below)

and remove the sprockets and timing chain as an assembly **(see illustration)**.

7   Remove the camshaft gear.

8   Clean the timing chain components with solvent and dry them with compressed air (if available). **Warning:** *Wear eye protection.*

9   Inspect the components for wear and damage. Look for teeth that are deformed, chipped, pitted, polished or discolored.

Installation

Note: *If the crankshaft has been disturbed, install the sprocket temporarily and turn the crankshaft until the mark on the crankshaft sprocket is exactly at the top. If the camshaft was disturbed, install the sprocket temporarily and turn the camshaft until the timing mark is at the bottom, opposite the mark on the crankshaft sprocket* **(see illustration 16.3)**.

10   Assemble the timing chain on the sprockets, then slide the sprocket and chain assembly onto the shafts with the timing marks aligned as shown in **illustration 16.3**. **Note:** *Be sure the timing marks on the balance shaft gears are aligned before installing the camshaft timing chain sprocket. Alignment of the balance shaft gears is covered in Chapter 2G.*

11   Install the camshaft sprocket bolt(s) and tighten them to the torque listed in this Chapter's specifications.

12   Attach the timing chain damper assembly to the block and install the spring.

13   On 1994 and earlier models, install the camshaft thrust button and spring. Hold it in place with grease.

14   Lubricate the chain and sprocket with clean engine oil and install the timing chain cover (see Section 13).

17 Oil pan - removal and installation

Refer to illustration 17.4

1   Disconnect the cable from the negative battery terminal. **Caution:** *On models equipped with the Theftlock audio system, be*

sure the lockout feature is turned off before performing any procedure which requires disconnecting the battery.

2   Raise the vehicle and place it securely on jackstands. Drain the engine oil and replace the oil filter (refer to Chapter 1 if necessary). On 1995 and later models, remove the front exhaust pipe.

3   Remove the driveplate inspection cover and starter, if necessary for access (refer to Chapter 5). On 1992 and later models, disconnect the electrical connector from the oil lever sensor. On 1995 models, remove the front and rear engine mounts and the torque arm strut. On 1996 and later models, remove the passenger side engine mount and bracket from the engine (see Section 21).

4   Remove the oil pan mounting bolts **(see illustration)** and carefully separate the oil pan from the block. Don't pry between the block and the pan or damage to the sealing surfaces may result and oil leaks may develop. Instead, tap the pan with a soft-face hammer to break the gasket seal.

5   Clean the pan with solvent and remove all old sealant and gasket material from the block and pan mating surfaces. Clean the mating surfaces with lacquer thinner or acetone and make sure the bolt holes in the block are clear. Check the oil pan flange for distortion, particularly around the bolt holes. If necessary, place the pan on a block of wood and use a hammer to flatten and restore the gasket surface.

6   Always use a new gasket whenever the oil pan is installed.

7   Place the oil pan in position on the block and install the bolts.

8   After the bolts are installed, tighten them to the torque listed in this Chapter's specifications. Starting at the center, follow a crisscross pattern and work up to the final torque in three steps.

9   The remaining steps are the reverse of the removal procedure.

10   Refill the engine with oil. Run the engine until normal operating temperature is reached and check for leaks.

18 Oil pump pickup tube and screen assembly - removal and installation

Refer to illustration 18.2

1   Remove the oil pan (see Section 17).

2   Unbolt the oil pump pickup tube and screen assembly and detach it from the engine **(see illustration)**.

3   Clean the screen and housing assembly with solvent and dry it with compressed air, available. **Warning:** *Wear eye protection.*

4   If the oil screen is damaged or has metal chips in it, replace it. An abundance of metal chips indicates a major engine problem which must be corrected.

5   Make sure the mating surfaces of the pipe flange and the engine block are clean and free of nicks and install the screen assembly with a new gasket.

6   Install the oil pan (see Section 17).

7   Be sure to refill the engine with oil before starting it.

19 Driveplate - removal and installation

Refer to illustration 19.2

1   Refer to Chapter 7 and remove the transaxle.

2   Place a prybar or large screwdriver through a hole in the driveplate to keep the crankshaft from turning, then remove the mounting bolts **(see illustration)**.

3   Pull straight back on the driveplate to detach it from the crankshaft. Inspect the driveplate for cracks and chipped teeth. Replace it if it's damaged.

4   Reinstall the driveplate. The holes in the driveplate are staggered to ensure correct positioning on the crankshaft. Use Locktite on the bolt threads and tighten them to the specified torque in a crisscross pattern.

5   Reinstall the transaxle (see Chapter 7).

18.2 Remove the bolts and lower the pickup tube and screen assembly

19.2 Place a prybar or large screwdriver through one of the holes in the driveplate to keep the crankshaft from turning as the bolts are loosened/tightened

0 Crankshaft rear oil seal - replacement

Refer to illustration 20.3

Remove the transaxle (see Chapter 7). Remove the driveplate (see Section 19).

Using a thin screwdriver or seal removal tool, carefully remove the oil seal from the engine block **(see illustration)**. Be very careful not to damage the crankshaft surface while prying the seal out.

Clean the bore in the block and the seal contact surface on the crankshaft. Check the seal contact surface on the crankshaft for scratches and nicks that could damage the new seal lip and cause oil leaks - if the crankshaft is damaged, the only alternative is a new or different crankshaft. Inspect the seal bore for nicks and scratches. Carefully smooth if with a fine file if necessary, but don't nick the crankshaft in the process.

A special tool is recommended to install the new oil seal. Lubricate the lips of the seal with clean engine oil. Slide the seal onto the mandril until the dust lip bottoms squarely against the collar of the tool. **Note:** *If the spe-cial tool isn't available, carefully work the seal lip over the crankshaft and tap it into place with a hammer and punch.*

6   Align the dowel pin on the tool with the dowel pin hole in the crankshaft and attach the tool to the crankshaft by hand-tightening the bolts.

7   Turn the tool handle until the collar bottoms against the case, seating the seal.

8   Loosen the tool handle and remove the bolts. Remove the tool.

9   Check the seal and make sure it's seated squarely in the bore.

10   Install the driveplate (see Section 19).

11   Install the transaxle (see Chapter 7).

21 Engine mount - check and replacement

Refer to illustrations 21.4, 21.10 and 21.11
Note: *See Chapter 7 for transaxle mount information.*
Warning: *A special engine support fixture should be used to support the engine during repair operations. Similar fixtures are available from rental yards. Improper lifting methods or devices are hazardous and could result in severe injury or death. DO NOT place any part of your body under the engine/transaxle when it's supported only by a jack. Failure of the lifting device could result in serious injury or death.*

1   Engine mounts seldom require attention, but broken or deteriorated mounts should be replaced immediately or the added strain placed on the driveline components may cause damage or wear. The 3800 engine has a transaxle mount (see Chapter 7), two engine torque struts and a mount at the front (timing belt end) of the engine that attaches to a "cradle" type bracket under the front of the oil pan.

Check

2   During the check, the engine must be raised slightly to remove the weight from the mounts.

3   Raise the vehicle and support it securely on jackstands, then position a jack under the engine oil pan. Place a large block of wood between the jack head and the oil pan, then carefully raise the engine just enough to take the weight off the mounts. **Warning:** *DO NOT place any part of your body under the engine when it's supported only by a jack!*

4   Check the mounts to see if the rubber is cracked, hardened or separated from the metal plates. Sometimes the rubber will split right down the center **(see illustration)**.

5   Check for relative movement between the mount plates and the engine or frame (use a large screwdriver or prybar to attempt to move the mounts). If movement is noted, lower the engine and tighten the mount fasteners.

6   Rubber preservative may be applied to the mounts to slow deterioration.

Replacement

7   Disconnect the negative battery cable from the battery, then raise the vehicle and

20.3 Carefully pry the old oil seal out

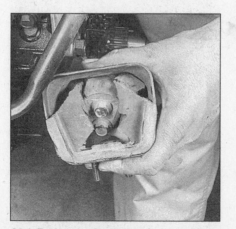

21.4 Broken engine mount (removed from vehicle for clarity)

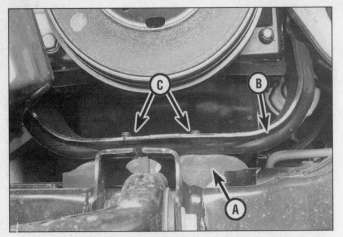

21.10 The front engine mount (A) is located between the front subframe and the engine front mount cradle (B) - remove the two nuts (C) at the cradle and two nuts below the subframe to remove the mount

21.11 Remove the throughbolts (A) and the engine torque struts (B) for replacement

support it securely on jackstands (if not already done). **Caution:** *On models equipped with the Theftlock audio system, be sure the lockout feature is turned off before performing any procedure which requires disconnecting the battery.*

8   On 1996 and later models remove the right front wheel and loosen the inner fender splash panel.

9   Raise the engine slightly with a jack or hoist. Install an engine support as described

in the **Warning** above. Remove the fasteners and detach the mount from the frame bracket

10   Remove the mount-to-block bracket bolts/nuts and detach the mount. **Note:** *1995 and earlier models are equipped with a front and rear engine mounts while 1996 and later models are equipped with a single mount and bracket located directly below the crankshaft balancer* **(see illustration)**.

11   The engine torque struts mount between the engine and the upper radiator

support on the body **(see illustration)**. The can be replaced without jacking up the engine, by removing the throughbolts and the bolts attaching the struts to the mounts or the radiator support. **Note:** *When reinstalling the mounts, you may have to use a prybar to rotate the engine toward or away from the radiator to align the throughbolts.*

12   Installation is the reverse of removal Use thread locking compound on the thread and be sure to tighten everything securely.

Chapter 2 Part G
General engine overhaul procedures

Contents

Specifications

2.2 liter four-cylinder engine

General

Displacement	134 cubic inches
Cylinder compression pressure	
Minimum	100 psi
Maximum variation between cylinders	30-percent
Oil pressure (minimum)	15 psi at 1200 rpm

Cylinder head

Warpage limit	0.005 inch

Valves and related components

Valve face angle	45-degrees
Valve seat	
Angle	46-degrees
Width	
Intake	0.049 to 0.059 inch
Exhaust	0.063 to 0.075 inch
Runout	0.002 inch
Margin width	1/32 inch minimum
Valve stem-to-guide clearance	
Intake	0.0011 to 0.0026 inch
Exhaust	0.0014 to 0.003 inch
Valve spring free length	1.89 inch
Valve spring pressure and length (intake and exhaust)	
Valve closed	79 to 85 lbs at 1.637 inch
Valve open	225 to 233 lbs at 1.247 inch
Valve spring installed height	1.637 inch

Crankshaft and connecting rods

Crankshaft endplay	0.002 to 0.008 inch
Connecting rod side clearance (endplay)	0.004 to 0.015 inch
Main bearing journal	
Diameter	2.4945 to 2.4954 inch
Taper/out-of-round limits	0.0002 inch
Main bearing oil clearance	0.0006 to 0.0019 inch
Connecting rod journal	
Diameter	1.9983 to 1.9994 inch
Taper/out-of-round limits	0.0002 inch
Connecting rod bearing oil clearance	0.001 to 0.0031 inch

2.2 liter four-cylinder engine (continued)

Engine block
Cylinder bore
    Diameter .... 3.5036 to 3.5043 inch
    Out-of-round limit .... 0.0005 inch
    Taper limit (maximum) .... 0.0005 inch
Block deck warpage limit .... 0.005 inch (**Note:** *If more than 0.010 inch must be removed, replace the block*)

Pistons and rings
Piston-to-bore clearance .... 0.0007 to 0.0017 inch
Piston ring side clearance
    Compression rings .... 0.0019 to 0.0027 inch
    Oil control ring .... 0.0019 to 0.0082 inch
Piston ring end gap
    Compression rings .... 0.010 to 0.020 inch
    Oil control ring .... 0.010 to 0.050 inch

Camshaft
Lobe lift
    Intake .... 0.259 inch
    Exhaust .... 0.250 inch
Bearing journal diameter .... 1.867 to 1.869 inch
Bearing oil clearance .... 0.001 to 0.0039 inch

Torque specifications   **Ft-lbs**
Main bearing cap bolts 66
Connecting rod cap nuts .... 20 plus an additional 50-degrees rotation
Camshaft thrust plate-to-block bolts 9
Note: *Refer to Part A for additional torque specifications.*

2.5 liter four-cylinder engine

General
RPO sales code LR8
VIN code .... R
Displacement .... 151 cubic inches
Compression pressure
    Minimum .... 100 psi
    Maximum variation between cylinders 30-percent
Firing order 1-3-4-2
Oil pressure (minimum) .... 26 psi at 800 rpm

Cylinder head
Warpage limit .... 0.005 in (0.127 mm)

Valves and related components
Valve face angle 45-degrees
Valve seat
    Angle 46-degrees
    Width
        Intake .... 0.035 to 0.075 in (0.889 to 1.905 mm)
        Exhaust .... 0.058 to 0.105 in (1.473 to 2.667 mm)
    Runout .... 0.002 in (0.05 mm)
Margin width .... 1/32 in minimum
Valve stem-to-guide clearance
    Intake .... 0.0011 to 0.0026 in (0.028 to 0.071 mm)
    Exhaust .... 0.0013 to 0.0041 in (0.033 to 0.104 mm)
Valve spring free length .... 2.01 in (51.0 mm)
Valve spring pressure (intake and exhaust)
    Valve closed .... 75 lb @ 1.68 in (332 N @ 42.6 mm)
    Valve open .... 173 lb @ 1.24 in (770 N @ 31.5 mm)
Installed height .... 1.68 in (42.64 mm)

Crankshaft and connecting rods
Crankshaft endplay .... 0.0005 to 0.018 in (0.013 to 0.26 mm)
Connecting rod side clearance (endplay) .... 0.006 to 0.024 in (0.15 to 0.6 mm)
Main bearing journal
    Diameter .... 2.3000 in (58.399 to 58.400 mm)
    Taper/out-of-round limits .... 0.005 in (0.013 mm)
Main bearing oil clearance .... 0.0005 to 0.0220 in (0.013 to 0.56 mm)

Connecting rod journal
Diameter .... 2.0000 in (50.708 to 50.805 mm)
Taper/out-of-round limits .... 0.0005 in (0.013 mm)
Connecting rod bearing oil clearance .... 0.0005 to 0.0030 in (0.013 to 0.070 mm)

Engine block
Cylinder bore
Diameter .... 4.0 in (101.6 mm)
Out-of-round limit .... 0.001 in (0.02 mm)
Taper limit .... 0.005 in (0.13 mm)
Block deck warpage limit .... If more than 0.010 in (0.25 mm) must be removed, replace the block

Pistons and rings
Piston-to-bore clearance .... 0.0014 to 0.0022 in (0.036 to 0.056 mm)
Piston ring side clearance
Top compression ring .... 0.002 to 0.003 in (0.05 to 0.08 mm)
Second compression ring .... 0.001 to 0.003 in (0.03 to 0.08 mm)
Oil control ring .... 0.015 to 0.055 in (0.38 to 1.40 mm)
Piston ring end gap
Compression rings .... 0.010 to 0.020 in (0.30 to 0.50 mm)
Oil control ring .... 0.020 to 0.060 in (0.50 to 1.50 mm)

Camshaft
Lobe lift (intake and exhaust) .... 0.248 in (6.302 mm)
Bearing journal diameter .... 1.869 in (47.4726 mm)
Bearing oil clearance .... 0.0007 to 0.0027 in (0.01778 to 0.0685 mm)
Thrust plate end clearance .... 0.0015 to 0.0050 in (0.0381 to 0.1270 mm)

Torque specifications* — **Ft-lbs** (unless otherwise indicated)
Main bearing cap bolts 65
Connecting rod cap nuts 29
Camshaft thrust plate-to-block bolts .... 89 in-lbs

Note: *Refer to Part B for additional torque specifications.*

2.3 liter four-cylinder (Quad-4) engine

General
RPO sales code .... LGO and LD2
VIN engine code .... A (LGO) or D (LD2)
Displacement .... 138 cubic inches
Firing order 1-3-4-2
Cylinder compression pressure
Minimum .... 100 psi
Maximum variation between cylinders 30-percent
Oil pressure
At 900 rpm .... 15 psi minimum
At 2000 rpm .... 30 psi minimum

Cylinder head
Warpage limit .... 0.008 in (0.203 mm)

Valves and related components
Valve face
Angle
Intake 44-degrees
Exhaust 44.5-degrees
Runout limit .... 0.0015 in (0.038 mm)
Valve seats
Angle (intake and exhaust) 45-degrees
Width
Intake .... 0.0370 to 0.0748 in (0.94 to 1.90 mm)
Exhaust .... 0.0037 to 0.0748 in (0.094 to 1.90 mm)
Valve margin width .... 1/32 in minimum
Valve stem diameter
Intake .... 0.27512 to 0.27445 in (6.990 to 6.972 mm)
Exhaust .... 0.2740 to 0.2747 in (6.959 to 6.977 mm)
Valve stem-to-guide clearance
Intake .... 0.0010 to 0.0027 in (0.025 to 0.069 mm)
Exhaust .... 0.0015 to 0.0032 in (0.038 to 0.081 mm)

2.3 liter four-cylinder (Quad-4) engine (continued)

Valves
  Length
    Intake .... 4.3300 in (109.984 mm)
    Exhaust .... 4.3103 in (109.482 mm)
  Installed height* .... 0.9840 to 1.0040 in (25.00 to 25.50 mm)
  Stem length exposed beyond retainer .... 0.1190 to 0.1367 in (3.023 to 3.473 mm)
Valve spring free length .... Not available
Valve spring pressure (intake and exhaust)
  Valve closed .... 71 to 79 lbs @ 1.4370 in (314 to 353 N @ 36.5 mm)
  Valve open .... 193 to 207 lbs @ 1.0433 in (857 to 922 N @ 26.08 mm)

Measured from tip of stem to top of camshaft housing mounting surface

Crankshaft and connecting rods

Crankshaft
  Endplay .... 0.0034 to 0.0095 in (0.087 to 0.243 mm)
  Runout
    At center main journal .... 0.00098 in (0.025 mm)
    At flywheel flange .... 0.00098 in (0.025 mm)
  Main bearing journal
    Diameter .... 2.0470 to 2.0480 in (51.996 to 52.020 mm)
    Out-of-round/taper limits .... 0.0005 in (0.0127 mm)
  Main bearing oil clearance .... 0.0005 to 0.0023 in (0.013 to 0.058 mm)
  Connecting rod bearing journal
    Diameter .... 1.8887 to 1.8897 in (47.975 to 48.00 mm)
    Out-of-round/taper limits .... 0.0005 in (0.127 mm)
  Connecting rod bearing oil clearance .... 0.0005 to 0.0020 in (0.013 to 0.053 mm)
  Seal journal
    Diameter .... 3.2210 to 3.2299 in (81.96 to 82.04 mm)
    Runout limit .... 0.0012 in (0.03 mm)
  Connecting rod side clearance (endplay) .... 0.0059 to 0.0177 in (0.150 to 0.450 mm)

Engine block

Cylinder bore
  Diameter .... 3.6217 to 3.6223 in (91.992 to 92.008 mm)
  Out-of-round limit .... 0.0004 in (0.010 mm)
  Taper limit (thrust side) .... 0.0003 in (0.008 mm) measured 4.173 in (106 mm) down the bore
Block deck warpage limit .... If more than 0.010 in (0.25 mm) must be removed, replace the block
Runout (rear face of block-to-crankshaft centerline) .... 0.002 in (0.050 mm) maximum

Pistons and rings

Piston diameter .... 3.6203 to 3.6210 in (91.957 to 91.973 mm) at 70-degrees F (21-degrees C)
Piston-to-bore clearance .... 0.0007 to 0.0020 in (0.019 to 0.051 mm)
Piston ring end gap
  Top compression ring .... 0.0138 to 0.0236 in (0.35 to 0.60 mm)
  Second compression ring .... 0.0157 to 0.0256 in (0.40 to 0.65 mm)
  Oil control ring .... 0.0157 to 0.0551 in (0.40 to 1.40 mm)
Piston ring side clearance
  Top compression ring
    VIN D .... 0.00197 to 0.00394 in (0.050 to 0.100 mm)
    VIN A .... 0.0027 to 0.0047 in (0.070 to 0.120 mm)
  Second compression ring .... 0.00157 to 0.00315 in (0.040 to 0.080 mm)

Camshaft

Lobe lift (intake and exhaust)
  VIN D .... 0.375 in (9.525 mm)
  VIN A .... 0.410 in (10.414 mm)
Journal diameter
  Number 1 .... 1.5728 to 1.5720 in (39.95 to 39.93 mm)
  Numbers 2 through 5 .... 1.3751 to 1.3760 in (34.93 to 34.95 mm)
Endplay .... 0.0009 to 0.0088 in (0.025 to 0.225 mm)

Torque specifications**   **Ft-lbs**

Main bearing cap bolts
  Step 1 15
  Step 2 .... Rotate an additional 90-degrees
Connecting rod cap nuts
  Step 1 18
  Step 2 .... Rotate an additional 80-degrees

** **Note:** *Refer to Part C for additional torque specifications.*

V6 engines

General

RPO sales code

2.8 liter	LB6
3.1 liter	LHO
3100	L82
3.4 liter	LQ1
3800	L27 and L36

VIN code

2.8 liter	W
3.1 liter	T (LHO) or V (turbo)
3100	M
3.4 liter	X
3800	L (L27) or K (L36)

Displacement

2.8 liter	173 cubic inches
3.1 liter	192 cubic inches
3100	192 cubic inches
3.4 liter	204 cubic inches
3800	231 cubic inches
Cylinder compression pressure	100 psi minimum
Maximum variation between cylinders	30-percent

Firing order

V6 engines (except 3800)	1-2-3-4-5-6
3800 engines	1-6-5-4-3-2
Oil pressure	15 psi at 1100 rpm

Cylinder head

Warpage limit

3.4 liter engine	0.004 in (0.1 mm)*
Others	0.005 in (0.127 mm)*

*If more than 0.010 in (0.25 mm) must be removed, replace the head

Valves and related components

Valve margin width

3.4 liter	0.029 in (0.75 mm)
3100	
Intake	0.083 in (2.10 mm)
Exhaust	0.106 in (2.70 mm)
3800	0.025 in (0.635 mm)
Others	0.031 in (0.793 mm) minimum
Valve face angle	45-degrees

Valve seat angle

3.4 liter	46-degrees
Others	45-degrees

Valve stem-to-guide clearance

3.4 liter	
Intake	0.0011 - 0.0026 in (0.028 - 0.066 mm)
Exhaust	0.0018 - 0.0033 in (0.046 - 0.084 mm)
3800	
Intake	0.0015 - 0.0035 in (0.038 - 0.089 mm)
Exhaust	0.0015 - 0.0032 in (0.038 - 0.081 mm)
Others	0.001 to 0.0027 in (0.026 to 0.068 mm)

Valve spring free length (intake and exhaust)

3100	1.89 in (48.5 mm)
3.4 liter	1.6551 in (42.04 mm)
3800	
1995 through 1997	1.981 in (50.32 mm)
1998 and later	1.960 in (49.78 mm)
Others	1.91 in (48.5 mm)

Valve spring pressure

Closed

3.4 liter	65 lbs @ 1.400 in (289 N @ 35.56 mm)
3100	80 lbs @ 1.710 in (356 N @ 43 mm)
3800	80 lbs @ 1.75 in (356 N @ 43.68 mm)
Others	90 lbs @ 1.701 in (400 N @ 43 mm)

V6 engines (continued)

Open
   3.4 liter .... 160 lbs @ 1.030 in (711 N @ 26.16 mm)
   3100 .... 250 lbs @ 1.239 in (1111 N @ 31.5 mm)
   3800 .... 210 lbs @ 1.315 in (935 N @ 33.4 mm)
   Others .... 215 lbs @ 1.291 in (956 N @ 33 mm)
Installed height
  2.8 and 3.1 liter
   1988 .... 1.701 in (43.20 mm)
   1989 on .... 1.5748 in (40 mm)
  3100 .... 1.710 in (43 mm)
  3.4 liter
   Intake .... 1.654 +/- 0.019 in (42.006 +/- 0.475 mm)
   Exhaust .... 1.653 +/- 0.019 in (41.991 +/- 0.475 mm)
  3800 .... 1.690 - 1.720 in (42.93 - 44.45 mm)

Crankshaft and connecting rods

Connecting rod journal
  Diameter
   3.4 liter and 3100 .... 1.9987 to 1.9994 in (50.768 to 50.784 mm)
   3800 .... 2.2487 to 2.2499 in (57.117 to 57.147 mm)
   Others .... 1.9994 to 1.9983 in (50.784 to 50.758 mm)
  Bearing oil clearance
   3100
    1995 and earlier... 0.0011 to 0.0030 in (0.028 to 0.076 mm)
    1996 and later .... 0.0007 to 0.0024 in (0.018 to 0.062 mm)
   3.4 liter .... 0.0011 to 0.0032 in (0.028 to 0.082 mm)
   3800
    1995 and earlier... 0.0008 to 0.0022 in (0.020 to 0.055 mm)
    1996 and later .... 0.0005 to 0.0026 in (0.0127 to 0.0660 mm)
   Others
    1988... 0.0013 to 0.0026 in (0.033 to 0.066 mm)
    1989... 0.0014 to 0.0036 in (0.038 to 0.093 mm)
    1990 on .... 0.0011 to 0.0034 in (0.028 to 0.086 mm)
Connecting rod side clearance (endplay)
  3.4 liter and 3100.. 0.007 to 0.017 in (0.18 to 0.44 mm)
  3800
   1997 and earlier .... 0.003 to 0.015 in (0.076 to 0.381 mm)
   1998 and later .... 0.004 to 0.020 in (0.102 to 0.508 mm)
  Others
   1988... 0.006 to 0.017 in (0.152 0.432 mm)
   1989 on .... 0.14 to 0.027 in (0.36 to 0.68 mm)
Main bearing journal
  Diameter
   3.4 liter .... 2.6472 to 2.6479 in (67.239 to 67.257 mm)
   3800 .... 2.4988 to 2.4998 in (63.470 to 63.495 mm)
   Others .... 2.6473 to 2.6483 in (67.241 to 67.265 mm)
  Bearing oil clearance
   3100 and 3.4 liter .... 0.0008 to 0.0025 in (0.019 to 0.064 mm)
   3800 .... 0.0008 to 0.0022 in (0.020 to 0.055 mm)
   Others
    1988... 0.0016 to 0.0032 in (0.041 to 0.081 mm)
    1989... 0.0012 to 0.0027 in (0.032 to 0.069 mm)
    1990 on .... 0.0012 to 0.0030 in (0.032 to 0.077 mm)
  Taper/out-of-round limit
   3800 .... 0.0003 in (0.0008 mm)
   Others .... 0.0002 in (0.005 mm)
Crankshaft endplay (at thrust bearing)
  3800 .... 0.003 to 0.011 in (0.076 to 0.279 mm)
  Others .... 0.0024 to 0.0083 in (0.06 to 0.21 mm)

Engine block

Cylinder bore
  Diameter
   2.8 liter .... 3.503 to 3.506 in (88.992 to 89.070 mm)
   3.1 liter and 3100 .... 3.5046 to 3.5033 in (89.016 to 89.034 mm)
   3.4 liter .... 3.6228 to 3.6235 in (92.020 to 92.038 mm)
   3800 .... 3.8 in (96.5 mm)

Out-of-round limit
    3.4 liter and 3800 .... 0.0004 in (0.010 mm)
    Others .... 0.0005 in (0.013 mm)
Taper limit (thrust side)
    3100 .... 0.0008 in (0.020 mm)
    Others .... 0.0005 in (0.013 mm)
Block deck warpage limit .... If more than 0.010 in (0.25 mm) must be removed, replace the block

Pistons and rings

Piston-to-bore clearance
    3.4 liter and 3100 .... 0.0013 to 0.0027 in (0.032 to 0.068 mm)
    3800 .... 0.0004 to 0.0022 in (0.010 to 0.056 mm)
    Others
        1988 .... 0.0020 to 0.0028 in (0.051 to 0.073 mm)
        1989 on .... 0.00093 to 0.00222 in (0.0235 to 0.0565 mm)
Piston ring end gap
    Top compression ring
        3100
            1995 and earlier .... 0.007 to 0.016 in (0.18 to 0.41 mm)
            1996 and later .... 0.006 to 0.014 in (0.15 to 0.36 mm)
        3.4 liter
            1995 and earlier .... 0.0098 to 0.0197 in (0.25 to 0.50 mm)
            1996 and later .... 0.008 to 0.018 in (0.20 to 0.45 mm)
        3800 .... 0.012 to 0.022 in (0.305 to 0.559 mm)
        Others .... 0.010 to 0.020 in (0.25 to 0.50 mm)
    Second compression ring
        3100 .... 0.0197 to 0.0280 in (0.5 to 0.71 mm)
        3.4 liter .... 0.022 to 0.032 in (0.56 to 0.81 mm)
        3800
            1998 and earlier .... 0.030 to 0.040 in (0.762 to 1.016 mm)
            1999 .... 0.023 to 0.033 in (0.58 to 0.84 mm)
        Others
            1988 and 1989 .... 0.010 to 0.020 in (0.025 to 0.50 mm)
            1990 on .... 0.020 to 0.028 in (0.050 to 0.71 mm)
    Oil control ring
        3.4 liter and 3100 .... 0.0098 to 0.0299 in (0.25 to 0.76 mm)
        Others
            1988 and 1989 (except 3.1L) .... 0.020 to 0.055 in (0.51 to 1.40 mm)
            1989 (3.1L only) .... 0.010 to 0.050 in (0.25 to 1.27 mm)
            1990 on .... 0.010 to 0.030 in (0.25 to 0.75 mm)
Piston ring side clearance
    Compression ring
        3.4 liter
            1995 and earlier .... 0.0016 to 0.0035 in (0.04 to 0.09 mm)
            1996 and later .... 0.0013 to 0.0031 in (0.033 to 0.079 mm)
        3800 .... 0.0013 to 0.0031in (0.033 to 0.079 mm)
        Others
            1988 and 1989 (except 3.1L) .... 0.0010 to 0.0030 in (0.03 to 0.08 mm)
            1989 (3.1L only) and 1990 on .... 0.0020 to 0.0035 in (0.05 to 0.09 mm)
    Oil control ring
        3.4 liter
            1995 and earlier .... 0.0019 to 0.0080 in (0.048 to 0.20 mm)
            1996 and later .... 0.0011 to 0.0081 in (0.028 to 0.206 mm)
        3800 .... 0.0011 to 0.0081 in (0.028 to 0.206 mm)
        Others .... 0.008 in (0.20 mm)

Camshaft

Bearing journal diameter
    3100 .... 1.868 to 1.869 in (47.45 to 47.48 mm)
    3.4 liter... 2.1643 to 2.1654 in (54.973 to 55.001 mm)
    3800
        1997 and earlier .... 1.785 to 1.786 in (45.339 to 45.364 mm)
        1998 and later .... 1.8462 to 1.8448 in (47.655 to 46.858 mm)
    Others... 1.8678 to 1.8815 in (47.44 to 47.79 mm)
Bearing oil clearance
    3.4 liter... 0.0019 to 0.0040 in (0.049 to 0.102 mm)
    3800
        1997 and earlier .... 0.0005 to 0.0035 in (0.013 to 0.089 mm)
        1998 and later .... 0.0016 to 0.0047 in (0.041 to 0.119 mm)
    Others... 0.001 to 0.004 in (0.026 to 0.101 mm)

V6 engines (continued)

Lobe lift

3100 .... 0.2727 in (6.9263 mm)

3.4 liter

    Intake .... 0.370 in (9.398 mm)

    Exhaust .... 0.370 in (9.398 mm)

3800

    Intake

        1997 and earlier .... 0.250 in (6.43 mm)

        1998 and later .... 0.258 in (6.55 mm)

    Exhaust .... 0.255 in (6.48 mm)

Others

    Intake .... 0.2626 in (6.67 mm)

    Exhaust .... 0.2732 in (6.92 mm)

Balance shaft (3800 only)

Endplay .... 0.0 to 0.0067 in (0.0 to 0.171 mm)

Drive gear backlash .... 0.002 to 0.005 in (0.050 to 0.127 mm)

Rear journal diameter .... 1.4994 to 1.5002 in (38.085 to 38.105 mm)

Rear bearing oil clearance .... 0.0005 to 0.0043 in (0.0127 to 0.109 mm)

Torque specifications***   Ft-lbs

Main bearing caps

3100

    Step 1 37

    Step 2 .... Tighten an additional 120 degrees

3.4 liter

    Bolts

        Step 1 37

        Step 2 .... Rotate an additional 75 degrees

    Studs

        Step 1 37

        Step 2 .... Rotate an additional 77 degrees

3800

    1995 and earlier

        Step 1 26

        Step 2 .... Tighten an additional 45 degrees

    1996 and later

        Step 1 30

        Step 2 .... Tighten an additional 110 degrees

Others

    1988 and 1989 70

    1990 on 73

Connecting rod caps

3.4 liter and 3100

    Step 1 15

    Step 2 .... Rotate an additional 75 degrees

3800

    Step 1 20

    Step 2 .... Rotate an additional 50 degrees

Others

    1988 and 1989 37

    1990 on 39

Camshaft/intermediate shaft retainer bolts .... 89 in-lbs

Balance shaft retainer bolts 22

Balance shaft driven gear bolt

    Step 1 16

    Step 2 .... Tighten an additional 70 degrees

*** **Note:** *Refer to Parts D, E or F for additional torque specifications.*

1   General information

Included in this portion of Chapter 2 are the general overhaul procedures for the cylinder head(s) and internal engine components.

The information ranges from advice concerning preparation for an overhaul and the purchase of replacement parts to detailed, step-by-step procedures covering removal and installation of internal engine components and the inspection of parts.

The following Sections have been written based on the assumption the engine has been removed from the vehicle. For information concerning in-vehicle engine repair, as well as removal and installation of the exter-nal components necessary for the overhaul, see Parts A through F (depending on engine type and size) of this Chapter and Section 8 of this Part.

The Specifications included in this Part are only those necessary for the inspection and overhaul procedures which follow. Refer to Parts A through F (depending on engine type and size) for additional Specifications.

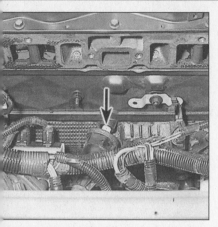

2.4a On 2.2 and 2.5 liter four-cylinder engines, the oil pressure sending unit is located at the lower center edge of the hydraulic lifter and valve cover (arrow) (view with intake manifold removed)

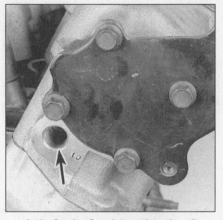

2.4b On the Quad-4 engine, the oil pressure sending unit is located in the transaxle end of the cylinder head

2.4c On V6 engines, the oil pressure sending unit (arrow) is located adjacent to the oil filter housing (typical V6 shown) - remove the sending unit . . .

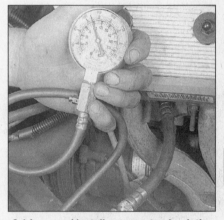

2.4d . . . and install a gauge to check the engine oil pressure

2 Engine overhaul - general information

Refer to illustrations 2.4a, 2.4b, 2.4c and 2.4d

It's not always easy to determine when, or if, an engine should be completely overhauled, as a number of factors must be considered.

High mileage isn't necessarily an indication an overhaul is needed, while low mileage doesn't preclude the need for an overhaul. Frequency of servicing is probably the most important consideration. An engine that's had regular and frequent oil and filter changes, as well as other required maintenance, will most likely give many thousands of miles of reliable service. Conversely, a neglected engine may require an overhaul very early in its life.

Excessive oil consumption is an indication that piston rings, valve seals and/or valve guides are in need of attention. Make sure oil leaks aren't responsible before deciding the rings and/or guides are bad. Perform a cylinder compression check to determine the extent of the work required (see Section 3).

Remove the oil pressure sending unit and check the oil pressure with a gauge installed in its place **(see illustrations)**. Compare the results to this Chapter's Specifications. As a general rule, engines should have ten psi oil pressure for every 1,000 rpm's. If the pressure is extremely low, the bearings and/or oil pump are probably worn out.

Loss of power, rough running, knocking or metallic engine noises, excessive valve train noise and high fuel consumption rates may also point to the need for an overhaul, especially if they're all present at the same time. If a complete tune-up doesn't remedy the situation, major mechanical work is the only solution.

An engine overhaul involves restoring the internal parts to the specifications of a new engine. During an overhaul, the piston rings are replaced and the cylinder walls are reconditioned (rebored and/or honed). If a rebore is done by an automotive machine shop, new oversize pistons will also be installed. The main bearings, connecting rod bearings and camshaft bearings are generally replaced with new ones and, if necessary, the crankshaft may be reground to restore the journals. Generally, the valves are serviced as well, since they're usually in less-than-perfect condition at this point. While the engine is being overhauled, other components, such as the starter and alternator, can be rebuilt as well. The end result should be a like new engine that will give many trouble free miles. **Note:** *Critical cooling system components such as the hoses, drivebelts, thermostat and water pump MUST be replaced with new parts when an engine is overhauled. The radiator should be checked carefully to ensure it isn't clogged or leaking (see Chapter 3). Also, we don't recommend overhauling the oil pump - always install a new one when an engine is rebuilt.*

Before beginning the engine overhaul, read through the entire procedure to familiarize yourself with the scope and requirements of the job. Overhauling an engine isn't particularly difficult, if you follow all of the instructions carefully, have the necessary tools and equipment and pay close attention to all specifications; however, it can be time consuming. Plan on the vehicle being tied up for a minimum of two weeks, especially if parts must be taken to an automotive machine shop for repair or reconditioning. Check on availability of parts and make sure any necessary special tools and equipment are obtained in advance. Most work can be done with typical hand tools, although a number of precision measuring tools are required for inspecting parts to determine if they must be replaced. Often an automotive machine shop will handle the inspection of parts and offer advice concerning reconditioning and replacement. **Note:** *Always wait until the engine has been completely disassembled and all components, especially the engine*

block, have been inspected before deciding what service and repair operations must be performed by an automotive machine shop. Since the block's condition will be the major factor to consider when determining whether to overhaul the original engine or buy a rebuilt one, never purchase parts or have machine work done on other components until the block has been thoroughly inspected. As a general rule, time is the primary cost of an overhaul, so it doesn't pay to install worn or substandard parts.

As a final note, to ensure maximum life and minimum trouble from a rebuilt engine, everything must be assembled with care in a spotlessly clean environment.

3 Cylinder compression check

Refer to illustration 3.6

1 A compression check will tell you what mechanical condition the upper end (pistons, rings, valves, head gaskets) of the engine is in. Specifically, it can tell you if the compression is down due to leakage caused by worn piston rings, defective valves and seats or a blown head gasket. **Note:** *The engine must*

3.6 A compression gauge with a threaded fitting for the spark plug hole is preferred over the type that requires hand pressure to maintain the seal

4.7a The timing marks should be aligned as shown here

4.7b Timing marks (arrows) on the Quad-4 engine

be at normal operating temperature and the battery must be fully charged for this check.

2   Begin by cleaning the area around the spark plugs before you remove them. Compressed air should be used, if available, otherwise a small brush or even a bicycle tire pump will work. The idea is to prevent dirt from getting into the cylinders as the compression check is being done.

3   Remove all of the spark plugs from the engine (see Chapter 1).

4   Block the throttle wide open.

5   Disable the fuel system by removing the ECM/fuel pump fuse from the fuse block located under the left side of the dash. Disable the ignition system by removing the DIS (Direct Ignition System) fuse, located in the RS electrical center on the right front strut tower.

6   Install the compression gauge in the number one spark plug hole **(see illustration)**.

7   Crank the engine over at least seven compression strokes and watch the gauge. The compression should build up quickly in a healthy engine. Low compression on the first stroke, followed by gradually increasing pressure on successive strokes, indicates worn piston rings. A low compression reading on the first stroke, which doesn't build up during successive strokes, indicates leaking valves or a blown head gasket (a cracked head could also be the cause). Deposits on the undersides of the valve heads can also cause low compression. Record the highest gauge reading obtained.

8   Repeat the procedure for the remaining cylinders and compare the results to this Chapter's Specifications.

9   If the readings are below normal, add some engine oil (about three squirts from a plunger-type oil can) to each cylinder, through the spark plug hole, and repeat the test.

10   If the compression increases after the oil is added, the piston rings are definitely worn. If the compression doesn't increase significantly, the leakage is occurring at the valves or head gasket. Leakage past the valves may be caused by burned valve seats and/or faces or warped, cracked or bent valves.

11   If two adjacent cylinders have equally low compression, there's a strong possibility the head gasket between them is blown. The appearance of coolant in the combustion chambers or the crankcase would verify this condition.

12   If one cylinder is about 20 percent lower than the others, and the engine has a slightly rough idle, a worn exhaust lobe on the camshaft could be the cause.

13   If the compression is unusually high, the combustion chambers are probably coated with carbon deposits. If that's the case, the cylinder head(s) should be removed and decarbonized.

14   If compression is way down or varies greatly between cylinders, it would be a good idea to have a leak-down test performed by an automotive repair shop. This test will pinpoint exactly where the leakage is occurring and how severe it is.

15   Install the fuses and drive the vehicle to restore the block learn memory.

---

4   Top Dead Center (TDC) for number one piston - locating

Refer to illustrations 4.7a and 4.7b

1   Top Dead Center (TDC) is the highest point in the cylinder each piston reaches as it travels up-and-down when the crankshaft turns. Each piston reaches TDC on the compression stroke and again on the exhaust stroke, but TDC generally refers to piston position on the compression stroke.

2   Positioning the piston(s) at TDC is an essential part of certain procedures such as camshaft removal and timing chain/sprocket removal.

3   Before beginning this procedure, be sure to place the transaxle in Neutral (or Park on automatic transaxle models), apply the parking brake and block the rear wheels.

4   Remove the spark plugs (see Chapter 1).

5   When looking at the drivebelt end of the engine, normal crankshaft rotation is clockwise. In order to bring any piston to TDC, the crankshaft must be turned with a socket and

ratchet attached to the bolt threaded into the center of the lower drivebelt pulley (vibration damper) on the crankshaft.

6   Have an assistant turn the crankshaft with a socket and ratchet as described above while you hold a finger over the number one spark plug hole. **Note:** *See the Specifications section in Parts A through F (depending on engine type and size) for the number one cylinder location.*

7   When the piston approaches TDC, pressure will be felt at the spark plug hole. Have your assistant stop turning the crankshaft when the timing marks are aligned **(see illustrations)**.

8   If the timing marks are bypassed, turn the crankshaft two complete revolutions clockwise until the timing marks are properly aligned.

9   After the number one piston has been positioned at TDC on the compression stroke, TDC for any of the remaining pistons can be located by turning the crankshaft one-half turn (180-degrees) on four-cylinder engines or one-third turn (120-degrees) on V6 engines to get to TDC for the next cylinder in the firing order.

---

5   Engine removal - methods and precautions

If you've decided the engine must be removed for overhaul or major repair work, several preliminary steps should be taken.

Locating a suitable place to work is extremely important. Adequate work space along with storage space for the vehicle, will be needed. If a shop or garage isn't available, at the very least a flat, level, clean work surface made of concrete or asphalt is required.

Cleaning the engine compartment and engine before beginning the removal procedure will help keep tools clean and organized.

An engine hoist or A-frame will also be necessary. Make sure the equipment is rated in excess of the combined weight of the engine and transaxle. Safety is of primary importance, considering the potential hazards involved in lifting the engine out of the vehicle.

If the engine is being removed by

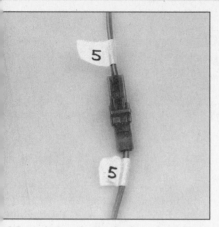

6.5a Label each wire before unplugging the connector

6.5b The ECM is located on the driver's side of the engine compartment (Quad-4 engine shown). Remove the plastic cover and unbolt the ECM from the mounting frame

6.13 Unbolt the power steering pump and move it aside, then use wire or rope to hold it in place (typical V6 engine shown, others similar)

novice, a helper should be available. Advice and aid from someone more experienced would also be helpful. There are many instances when one person cannot simultaneously perform all of the operations required when lifting the engine out of the vehicle.

Plan the operation ahead of time. Arrange for or obtain all of the tools and equipment you'll need prior to beginning the job. Some of the equipment necessary to perform engine removal and installation safely and with relative ease are (in addition to an engine hoist) a heavy duty floor jack, complete sets of wrenches and sockets as described in the front of this manual, wood blocks and plenty of rags and cleaning solvent for mopping up spilled oil, coolant and gasoline. If the hoist must be rented, be sure to arrange for it in advance and perform all of the operations possible without it beforehand. This will save you money and time.

Plan for the vehicle to be out of use for quite a while. A machine shop will be required to perform some of the work which the do-it-yourselfer can't accomplish without special equipment. These shops often have a busy schedule, so it would be a good idea to consult them before removing the engine in order to accurately estimate the amount of time required to rebuild or repair components that may need work.

Always be extremely careful when removing and installing the engine. Serious injury can result from careless actions. Plan ahead, take your time and a job of this nature, although major, can be accomplished successfully.

6   Engine - removal and installation

Refer to illustrations 6.5a, 6.5b, 6.13, 6.23 and 6.30

Warning: *Gasoline is extremely flammable, so take extra precautions when disconnecting any part of the fuel system. Don't smoke or allow open flames or bare light bulbs in or near the work area and don't work in a garage where a natural gas appliance (such as a clothes dryer or water heater) is installed. If*

you spill gasoline on your skin, rinse it off immediately. Have a fire extinguisher rated for gasoline fires handy and know how to use it! Also, the air conditioning system is under high pressure - have a dealer service department or service station discharge the system before disconnecting any of the hoses or fittings.
Note: *Read through the following steps carefully and familiarize yourself with the procedure before beginning work.*

Removal

1   Carefully inspect the routing of the air conditioning system refrigerant lines before beginning the engine removal to see if line disconnection, and therefore professional discharging is necessary. Have the air conditioning system discharged by a dealer service department or service station.

All models

2   Refer to Chapter 4 and relieve the fuel system pressure, then disconnect the negative cable from the battery. **Caution:** *On models equipped with the Theftlock audio system, be sure the lockout feature is turned off before performing any procedure which requires disconnecting the battery.*

3   Cover the fenders and cowl and remove the hood (see Chapter 11). Special pads are available to protect the fenders, but an old bedspread or blanket will also work.

4   Remove the air cleaner assembly (and MAF sensor on V6 models so equipped) (see Chapter 4).

5   Remove the plastic firewall covers. Label the vacuum lines, emissions system hoses, wiring connectors, ground straps and fuel lines to ensure correct reinstallation, then detach them. The relay panel and bracket can be detached as an assembly. Pieces of masking tape with numbers or letters written on them work well **(see illustration)**. If there's any possibility of confusion, make a sketch of the engine compartment and clearly label the lines, hoses and wires. **Note:** *The ECM wiring can be unplugged and the*

harness pulled through the firewall. The engine harness plugs into the firewall connector. **Note:** *On some models, the plastic wire harness must be detached from the body and the ECM and the fuse block placed on top of the engine* **(see illustration).**

6   Raise the vehicle and support it securely on jackstands. Drain the cooling system (see Chapter 1).

7   Label and detach all coolant hoses from the engine.

8   Remove the coolant reservoir, cooling fan, shroud and radiator (see Chapter 3).

9   Remove the drivebelt and idler, if equipped (see Chapter 1). On 2.5L four-cylinder and V6 models, remove the crankshaft pulley/vibration damper (see Parts A through F, depending on engine type/size).

10   Disconnect the fuel lines running from the engine to the chassis (see Chapter 4). Plug or cap all open fittings/lines.

11   Remove the torque struts on the engine (see Engine mounts - removal and installation in Parts A through F, depending on engine type/size).

12   Disconnect the accelerator cable (and TV cable/cruise control cable, if equipped) from the engine (see Chapters 4 and 7).

13   Unbolt the power steering pump and set it aside (see Chapter 10). Leave the lines/hoses attached and make sure the pump is kept in an upright position in the engine compartment **(see illustration).**

14   Remove the steering gear pinch bolt and separate the intermediate shaft from the rack and pinion stub shaft (steering gear). **Caution:** *Be sure to separate the steering gear from the rack and pinion stub shaft to avoid damage to the steering gear and intermediate shaft. This damage could cause loss of steering control which could cause injury.*

15   Unbolt the air conditioning compressor (see Chapter 3) and set it aside.

16   Drain the engine oil and remove the filter (see Chapter 1).

17   Remove the starter and the alternator (see Chapter 5).

18   On Quad-4 models, remove the oil/air

6.23 Attach the chain or hoist cable to the engine brackets (arrows)

6.30 Lift the engine off the mounts and guide it carefully around any obstacles as an assistant raises the hoist until it clears the front of the vehicle

separator (see Chapter 6).

19   Check for clearance and remove the brake master cylinder, if necessary, to allow clearance for the transaxle (see Chapter 9).

20   Disconnect the exhaust system from the engine (see Chapter 4). On Quad-4 models, remove the exhaust manifold (see Chapter 2C). **Note:** *On V6 and Quad-4 models, the engine and transaxle are removed from the vehicle as a unit. 2.5 liter four-cylinder engines should be removed from the vehicle after the transaxle has been removed.*

V6 and Quad-4 models only

21   Disconnect all the components attaching the transaxle to the vehicle (including driveaxles, intermediate shaft, cables, wiring, linkage, etc. - see Chapter 7).

22   Support the transaxle with a jack. Position a block of wood on the jack head to prevent damage to the transaxle.

All models

23   Attach an engine sling or a length of chain to the lifting brackets on the engine **(see illustration).**

24   Roll the hoist into position and connect the sling to it. Take up the slack in the sling or chain, but don't lift the engine. **Warning:** *DO NOT place any part of your body under the engine when it's supported only by a hoist or other lifting device.*

25   If you're working on a vehicle with an automatic transaxle, refer to Chapter 7 and remove the torque converter-to-driveplate fasteners.

26   Remove the transaxle-to-body mount nuts and pry the mount out of the frame bracket.

27   On 2.5 liter four-cylinder models, remove the transaxle (see Chapter 7).

28   Remove the engine mount-to-chassis bolts/nuts.

29   Recheck to be sure nothing is still connecting the engine to the vehicle. Disconnect anything still remaining.

30   Raise the engine (or engine/transaxle assembly) slightly to disengage the mounts. Slowly raise the engine out of the vehicle **(see illustration).** Check carefully to make

sure nothing is hanging up as the hoist is raised.

31   On V6 and Quad-4 models, once the engine/transaxle assembly is out of the vehicle, remove the transaxle-to-engine block bolts. Carefully separate the engine from the transaxle. If you're working on a vehicle with an automatic transaxle, be sure the torque converter stays in place (clamp a pair of locking pliers to the housing to keep the converter from sliding out). If you're working on a vehicle with a manual transaxle, the input shaft must be completely disengaged from the clutch.

32   Remove the clutch and flywheel or driveplate and mount the engine on an engine stand.

Installation

33   Check the engine and transaxle mounts. If they're worn or damaged, replace them.

34   If you're working on a manual transaxle equipped vehicle, install the clutch and pressure plate (see Chapter 7). Now is a good time to install new clutch components. Apply a dab of high-temperature grease to the input shaft.

35   **Caution:** *DO NOT use the transaxle-to-engine bolts to force the transaxle and engine together. If you're working on an automatic transaxle equipped vehicle, take great care when installing the torque converter, following the procedure outlined in Chapter 7.*

36   Carefully lower the engine into the engine compartment - make sure the mounts line up. Reinstall the remaining components in the reverse order of removal. Double-check to make sure everything is hooked up right.

37   Add coolant, oil, power steering and transmission fluid as needed. If the brake master cylinder was removed, bleed the brakes (see Chapter 9). Recheck the fluid level and test the brakes.

38   Run the engine and check for leaks and proper operation of all accessories, then install the hood and test drive the vehicle.

39   If the air conditioning system was discharged, have it evacuated, recharged and leak tested by the shop that discharged it.

7   Engine rebuilding alternatives

The home mechanic is faced with a number of options when performing an engine overhaul. The decision to replace the engine block, piston/connecting rod assemblies and crankshaft depends on a number of factors, with the number one consideration being the condition of the block. Other considerations are cost, access to machine shop facilities, parts availability, time required to complete the project and the extent of prior mechanical experience.

Some of the rebuilding alternatives include:

Individual parts - If the inspection procedures reveal the engine block and most engine components are in reusable condition, purchasing individual parts may be the most economical alternative. The block, crankshaft and piston/connecting rod assemblies should all be inspected carefully. Even if the block shows little wear, the cylinder bores should be surface honed.

Short block - A short block consists of an engine block with a crankshaft and piston/connecting rod assemblies already installed. All new bearings are incorporated and all clearances will be correct. The existing camshaft, valve train components, cylinder head(s) and external parts can be bolted to the short block with little or no machine shop work necessary.

Long block - A long block consists of a short block plus an oil pump, oil pan, cylinder head(s), valve cover(s), camshaft and valve train components, timing sprockets and chain and timing chain cover. All components are installed with new bearings, seals and gaskets incorporated throughout. The installation of manifolds and external parts is all that's necessary.

Give careful thought to which alternative is best for you and discuss the situation with local automotive machine shops, auto parts dealers and experienced rebuilders before ordering or purchasing replacement parts.

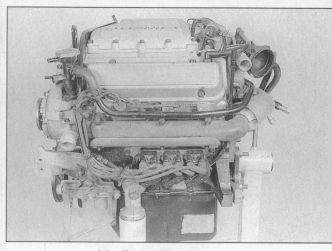

8.3a Typical V6 engine - front

8.3b Typical V6 engine - drivebelt end

8 Engine overhaul - disassembly sequence

Refer to illustrations 8.3a, 8.3b and 8.3c

1   It's much easier to disassemble and work on the engine if it's mounted on a portable engine stand. A stand can often be rented quite cheaply from an equipment rental yard. Before it's mounted on a stand, the flywheel/driveplate should be removed from the engine.

2   If a stand isn't available, it's possible to disassemble the engine with it blocked up on the floor. Be extra careful not to tip or drop the engine when working without a stand.

3   If you're going to obtain a rebuilt engine, all external components **(see illustrations)** must come off first, to be transferred to the replacement engine, just as they will if you're doing a complete engine overhaul yourself. These include:

Alternator and brackets
Emissions control components
Ignition coil/module assembly, spark plug
    wires and spark plugs
Thermostat and housing cover
Water pump
EFI components

Intake/exhaust manifolds
Oil filter
Engine mounts
Clutch and flywheel/driveplate
Engine rear plate (if equipped)

Note: *When removing the external components from the engine, pay close attention to details that may be helpful or important during installation. Note the installed position of gaskets, seals, spacers, pins, brackets, washers, bolts and other small items.*

4   If you're obtaining a short block, which consists of the engine block, crankshaft, pistons and connecting rods all assembled, then the cylinder head(s), oil pan and oil pump will have to be removed as well. See Engine rebuilding alternatives for additional information regarding the different possibilities to be considered.

5   If you're planning a complete overhaul, the engine must be disassembled and the internal components removed in the following general order:

2.2 and 2.5 liter four-cylinder engine

Valve cover
Intake and exhaust manifolds
Rocker arms and pushrods

Valve lifters
Cylinder head
Timing gear cover
Timing gears, /chain and sprockets
Camshaft
Oil pan
Oil pump
Piston/connecting rod assemblies
Crankshaft and main bearings

2.3 liter four-cylinder (Quad-4) engine

Timing chain and sprockets
Timing chain housing
Cylinder head and camshafts
Oil pan
Oil pump
Piston/connecting rod assemblies
Rear main oil seal housing
Crankshaft and main bearings

V6 engines

Valve covers
Intake and exhaust manifolds
Timing belt (3.4L only)
Cam carriers and camshafts (3.4L only)
Rocker arms and pushrods (if equipped)
Valve lifters (except 3.4L)
Cylinder heads
Timing chain cover
Timing chain and sprockets
Camshaft (except 3.4L)
Oil pan
Oil pump
Piston/connecting rod assemblies
Rear main oil seal housing
Intermediate shaft (3.4L only)
Crankshaft and main bearings

6   Before beginning the disassembly and overhaul procedures, make sure the following items are available. Also, refer to *Engine overhaul - reassembly sequence* for a list of tools and materials needed for engine reassembly.

Common hand tools
Small cardboard boxes or plastic bags for
    storing parts

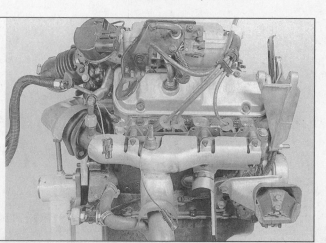

8.3c Typical V6 engine - rear

Gasket scraper
Ridge reamer
Vibration damper puller
Micrometers
Telescoping gauges
Dial indicator set
Valve spring compressor
Cylinder surfacing hone
Piston ring groove cleaning tool
Electric drill motor
Tap and die set
Wire brushes
Oil gallery brushes
Cleaning solvent

9   Cylinder head - disassembly

Refer to illustrations 9.2, 9.3 and 9.4

Note: *New and rebuilt cylinder heads are commonly available for most engines at dealerships and auto parts stores. Due to the fact that some specialized tools are necessary for the disassembly and inspection procedures, and replacement parts aren't always readily available, it may be more practical and economical for the home mechanic to purchase replacement head(s) rather than taking the time to disassemble, inspect and recondition the original(s).*

1   Cylinder head disassembly involves removal of the intake and exhaust valves and related components. If you're working on a Quad-4 engine or a 3.4L V6, the camshafts and housings must be removed before beginning the cylinder head disassembly procedure (see Parts C or E of this Chapter). If you're working on a 2.2 or 2.5 liter four-cylinder engine, or a V6 engine other than the 3.4L, remove the rocker arm nuts, pivot balls and rocker arms from the cylinder head studs. Label the parts or store them separately so they can be reinstalled in their original locations.

2   Before the valves are removed, arrange to label and store them, along with their related components, so they can be kept separate and reinstalled in their original locations **(see illustration)**.

3   Compress the springs on the first valve with a spring compressor and remove the keepers **(see illustration)**. Carefully release the valve spring compressor and remove the retainer, the spring and the spring seat (if used). Note that valve rotators are installed under the springs on the Quad-4 engine.

4   Pull the valve out of the head, then remove the oil seal from the guide. If the valve binds in the guide (won't pull through), push it back into the head and deburr the area around the keeper groove with a fine file or whetstone **(see illustration)**.

5   Repeat the procedure for the remaining valves. Remember to keep all the parts for each valve together so they can be reinstalled in the same locations.

6   Once the valves and related components have been removed and stored in an organized manner, the head should be thoroughly cleaned and inspected. If a complete

9.2 A small plastic bag, with an appropriate label, can be used to store the valve train components so they can be kept together and reinstalled in the original position

engine overhaul is being done, finish the engine disassembly procedures before beginning the cylinder head cleaning and inspection process.

10   Cylinder head - cleaning and inspection

1   Thorough cleaning of the cylinder head(s) and related valve train components, followed by a detailed inspection, will enable you to decide how much valve service work must be done during the engine overhaul. **Note:** *If the engine was severely overheated, the cylinder head is probably warped (see Step 12).*

Cleaning

2   Scrape all traces of old gasket material and sealant off the head gasket, intake manifold and exhaust manifold mating surfaces. Be very careful not to gouge the cylinder head. Special gasket removal solvents that soften gaskets and make removal much easier are available at auto parts stores.

3   Remove all built up scale from the coolant passages.

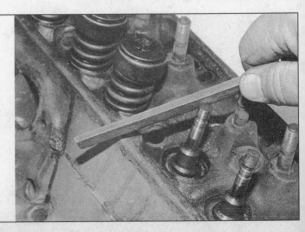

9.4 If the valve won't pull through the guide, deburr the edge of the stem end and the area around the top of the keeper groove with a file or whetstone

9.3 Use a valve spring compressor to compress the spring, then remove the keepers from the valve stem

4   Run a stiff wire brush through the various holes to remove deposits that may have formed in them.

5   Run an appropriate size tap into each of the threaded holes to remove corrosion and thread sealant that may be present. If compressed air is available, use it to clear the holes of debris produced by this operation. **Warning:** *Wear eye protection when using compressed air!*

6   Clean the rocker arm pivot stud threads with a wire brush (except Quad-4 engine or 3.4 liter V6).

7   Clean the cylinder head with solvent and dry it thoroughly. Compressed air will speed the drying process and ensure that all holes and recessed areas are clean. **Note:** *Decarbonizing chemicals are available and may prove very useful when cleaning cylinder heads and valve train components. They're very caustic and should be used with caution. Be sure to follow the instructions on the container.*

8   Clean the rocker arms, pivot balls, nuts and pushrods (except Quad-4 engine or 3.4 liter V6) with solvent and dry them thoroughly (don't mix them up during the cleaning process). Compressed air will speed the drying process and can be used to clean out the oil passages.

9   Clean all the valve springs, spring seats, rotators (where applicable), keepers and retainers with solvent and dry them thor-

10.12 Check the cylinder head gasket surface for warpage by trying to slip a feeler gauge under the straightedge (see this Chapter's Specifications for the maximum warpage allowed and use a feeler gauge of that thickness)

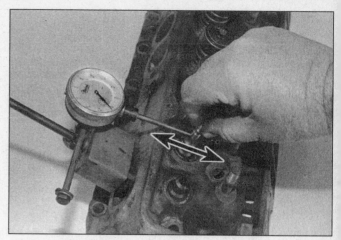

10.14 A dial indicator can be used to determine the valve stem-to-guide clearance (move the valve stem as indicated by the arrows)

oughly. Do the components from one valve at a time to avoid mixing up the parts.

10   Scrape off any heavy deposits that may have formed on the valves, then use a motorized wire brush to remove deposits from the valve heads and stems. Again, make sure the valves don't get mixed up.

Inspection

Refer to illustrations 10.12, 10.14, 10.15, 10.16, 10.17, and 10.18
Note: *Be sure to perform all of the following inspection procedures before concluding machine shop work is required. Make a list of the items that need attention.*

Cylinder head

11   Inspect the head very carefully for cracks, evidence of coolant leakage and other damage. If cracks are found, check with an automotive machine shop concerning repair. If repair isn't possible, a new cylinder head should be obtained.

12   Using a straightedge and feeler gauge, check the head gasket mating surface for warpage **(see illustration)**. If the warpage exceeds the limit listed in this Chapter's Specifications, it can be resurfaced at an automotive machine shop. **Note:** *If the V6 engine heads are resurfaced, the intake manifold flanges will also require machining.*

13   Examine the valve seats in each of the combustion chambers. If they're pitted, cracked or burned, the head will require valve service that's beyond the scope of the home mechanic.

14   Check the valve stem-to-guide clearance by measuring the lateral movement of the valve stem with a dial indicator attached securely to the head **(see illustration)**. The valve must be in the guide and approximately 1/16-inch off the seat. The total valve stem movement indicated by the gauge needle must be divided by two to obtain the actual clearance. After this is done, if there's still some doubt regarding the condition of the valve guides, they should be

checked by an automotive machine shop (the cost should be minimal).

Valves

15   Carefully inspect each valve face for uneven wear, deformation, cracks, pits and burned areas. Check the valve stem for scuffing and galling and the neck for cracks. Rotate the valve and check for any obvious indication that it's bent. Look for pits and excessive wear on the end of the stem. The presence of any of these conditions **(see illustration)** indicates the need for valve service by an automotive machine shop.

16   Measure the margin width on each valve **(see illustration)**. Any valve with a margin narrower than specified in this Chapter will have to be replaced with a new one.

Valve components

17   Check each valve spring for wear (on the ends) and pits. Measure the free length and compare it to this Chapter's Specifications **(see illustration)**. Any springs that are shorter than specified have sagged and

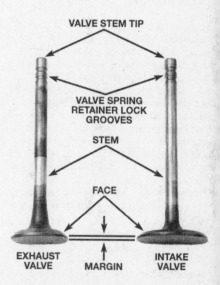

10.15 Check for valve wear patterns at the various points

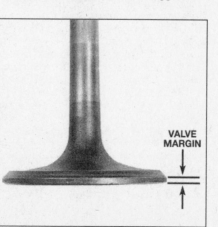

10.16 The margin width on each valve must be as specified (if no margin exists, the valve cannot be reused)

10.17 Measure the free length of each valve spring with a dial or vernier caliper

10.18 Check each valve spring for squareness

12.4 Gently tap the new seal onto the guide using a deep socket

12.6 Apply a small dab of grease to each keeper as shown here before installation - it'll hold them in place on the valve stem as the spring is released

shouldn't be reused. The tension of all springs should be checked with a special fixture before deciding they're suitable for use in a rebuilt engine (take the springs to an automotive machine shop for this check).

18   Stand each spring on a flat surface and check it for squareness (see illustration). If any of the springs are distorted or sagged, replace all of them with new parts.

19   Check the spring retainers and keepers for obvious wear and cracks. Any questionable parts should be replaced with new ones, as extensive damage will occur if they fail during engine operation. Make sure the rotators (Quad-4 engine and 3.4 liter V6 only) operate smoothly with no binding or excessive play.

Rocker arm components (all except Quad-4 engine and 3.4 liter V6)

20   Check the rocker arm faces (the areas that contact the pushrod ends and valve stems) for pits, wear, galling, score marks and rough spots. Check the rocker arm pivot contact areas and pivot balls as well. Look for cracks in each rocker arm and nut.

21   Inspect the pushrod ends for scuffing and excessive wear. Roll each pushrod on a flat surface, like a piece of plate glass, to determine if it's bent.

22   Check the rocker arm studs in the cylinder heads for damaged threads and secure installation.

23   Any damaged or excessively worn parts must be replaced with new ones.

Camshafts, lifters and housings (Quad-4 engine and 3.4 liter V6)

24   Refer to Part C or E for the inspection procedures for these components.

All components

25   If the inspection process indicates the valve components are in generally poor condition and worn beyond the limits specified, which is usually the case in an engine that's being overhauled, reassemble the valves in the cylinder head and refer to Section 11 for valve servicing recommendations.

11   Valves - servicing

1   Because of the complex nature of the job and the special tools and equipment needed, servicing of the valves, the valve seats and the valve guides, commonly known as a valve job, should be done by a professional.

2   The home mechanic can remove and disassemble the head, do the initial cleaning and inspection, then reassemble and deliver it to a dealer service department or an automotive machine shop for the actual service work. Doing the inspection will enable you to see what condition the head and valvetrain components are in and will ensure that you know what work and new parts are required when dealing with an automotive machine shop.

3   The dealer service department, or automotive machine shop, will remove the valves and springs, recondition or replace the valves and valve seats, recondition the valve guides, check and replace the valve springs, rotators, spring retainers and keepers (as necessary), replace the valve seals with new ones, reassemble the valve components and make sure the installed spring height is correct. The cylinder head gasket surface will also be resurfaced if it's warped.

4   After the valve job has been performed by a professional, the head will be in like new condition. When the head is returned, be sure to clean it again before installation on the engine to remove any metal particles and abrasive grit that may still be present from the valve service or head resurfacing operations. Use compressed air, if available, to blow out all the oil holes and passages.

12   Cylinder head - reassembly

Refer to illustrations 12.4 and 12.6

1   Regardless of whether or not the head was sent to an automotive repair shop for valve servicing, make sure it's clean before beginning reassembly.

2   If the head was sent out for valve servic-

ing, the valves and related components will already be in place. Begin the reassembly procedure with Step 8.

3   Install the spring seats or valve rotators (if equipped) before the valve seals.

4   Install new seals on each of the valve guides. Using a hammer and a deep socket or seal installation tool, gently tap each seal into place until it's completely seated on the guide (see illustration). Don't twist or cock the seals during installation or they won't seal properly on the valve stems.

5   Beginning at one end of the head, lubricate and install the first valve. Apply moly-base grease or clean engine oil to the valve stem.

6   Position the valve springs (and shims, if used) over the valves. Compress the springs with a valve spring compressor and carefully install the keepers in the groove, then slowly release the compressor and make sure the keepers seat properly. Apply a small dab of grease to each keeper to hold it in place if necessary (see illustration).

7   Repeat the procedure for the remaining valves. Be sure to return the components to their original locations - don't mix them up!

8   Check the installed valve spring height with a ruler graduated in 1/32-inch increments or a dial caliper. If the head was sent out for service work, the installed height should be correct (but don't automatically assume it is). The measurement is taken from the top of each spring seat, rotator or top shim to the bottom of the retainer. If the height is greater than specified in this Chapter, shims can be added under the springs to correct it. **Caution:** *Do not, under any circumstances, shim the springs to the point where the installed height is less than specified.*

9   Apply moly-base grease to the rocker arm faces and the pivot balls, then install the rocker arms and pivots on the cylinder head studs (except Quad-4 engine and 3.4L V6 - see Parts A, B, D or F, depending on engine type and size).

10   If you're working on a Quad-4 engine or 3.4L V6, refer to Parts C or E and install the camshafts, lifters and housings on the head.

13.3 When checking the camshaft lobe lift, the dial indicator plunger must be positioned directly above and in line with the pushrod

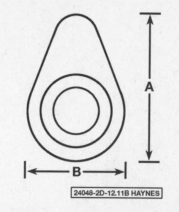

24048-2D-12.11B HAYNES

13.9 To verify camshaft lobe lift, measure the major (A) and minor (B) diameters of each lobe with a micrometer or vernier caliper - subtract each minor diameter from the major diameter to arrive at the lobe lift

13.12a The V6 engine oil pump drive (arrow) is located at the transaxle-end of the block

13 Camshaft, intermediate shaft (3.4L) and balance shaft (3800) - removal and inspection

Camshaft lobe lift check
With cylinder head installed

Refer to illustration 13.3

1   In order to determine the extent of cam lobe wear, the lobe lift should be checked prior to camshaft removal. Refer to Part A, B, D or F and remove the valve cover(s).

2   Position the number one piston at TDC on the compression stroke (see Section 4).

3   Beginning with the number one cylinder valves, mount a dial indicator on the engine and position the plunger against the top surface of the first rocker arm. The plunger should be directly above and in line with the pushrod **(see illustration)**.

4   Zero the dial indicator, then very slowly turn the crankshaft in the normal direction of rotation (clockwise) until the indicator needle stops and begins to move in the opposite direction. The point at which it stops indicates maximum cam lobe lift.

5   Record this figure for future reference,

then reposition the piston at TDC on the compression stroke.

6   Move the dial indicator to the remaining number one cylinder rocker arm and repeat the check. Be sure to record the results for each valve.

7   Repeat the check for the remaining valves. Since each piston must be at TDC on the compression stroke for this procedure, work from cylinder-to-cylinder following the firing order sequence.

8   After the check is complete, compare the results to this Chapter's Specifications. If camshaft lobe lift is less than specified, cam lobe wear has occurred and a new camshaft should be installed.

With cylinder head removed

Refer to illustration 13.9

9   If the cylinder head(s) have already been removed, an alternate method of lobe measurement can be used. Remove the camshaft as described below. Using a micrometer, measure the lobe at its highest point. Then measure the base circle perpendicular (90-degrees) to the lobe **(see illustration)**. Do this for each lobe and record the results.

10   Subtract the base circle measurement from the lobe height. The difference is the lobe lift. See Step 8 above.

Removal

Camshaft (all engines except Quad-4 and 3.4L) and intermediate shaft (3.4L)

Refer to illustrations 13.12a, 13.12b, 13.12c, 13.13, 13.14 and 13.15

Note 1: *The intermediate shaft on 3.4L engines is removed and installed much like a conventional camshaft, although it does not operate the lifters or valves.*

Note 2: *For camshaft removal procedures on Quad-4 and 3.4L engines, refer to Chapter 2C or 2E.*

11   Remove the lifters and pushrods. On models so equipped, remove the timing chain and sprockets. Refer to the appropriate Section in Parts A, B, D or F, depending on the engine.

12   On the 2.2L four-cylinder and on V6 engines (except 3800), remove the oil pump drive **(see illustration)**. If it's stuck, remove the oil pump and push the drive out with a long socket extension **(see illustration)**. Be sure to replace the O-ring before reassembly **(see illustration)**.

13.12b If the oil pump drive is stuck, remove the pump and push the drive out with a long socket extension

13.12c Be sure to replace the O-ring (arrow)

13.13 Thread long bolts into the sprocket bolt holes to use as a handle when removing and installing the camshaft (V6 engine)

13.14 On 2.5L engines the camshaft thrust plate bolts can be removed through the access holes in the camshaft timing gear

13.15 Support the camshaft near the block to avoid damaging the bearings

13.19 Remove the retaining bolts and the balance shaft thrust plate

13   On the 2.2L four-cylinder and on V6 engines, thread long bolts into the camshaft sprocket bolt holes to use as a handle when removing the camshaft from the block **(see illustration)**. On 3800 engines check the balance shaft drive gear back lash, then remove the drive gear before removing the camshaft (see Steps 16 through 18).

14   Remove the camshaft thrust plate. On 2.5L engines through 1990, rotate the camshaft so that you can get access to the thrust plate bolts through the holes in the timing gear **(see illustration)**.

15   Slide the camshaft or intermediate shaft straight out of the engine. Support the shaft near the block and be careful not to scrape or nick the bearings **(see illustration)**.

Balance shaft (3800 only)

Refer to illustrations 13.19 and 13.20

16   Refer to Chapter 2 Part F and remove the timing chain and sprockets (lifters should already be removed and stored in marked plastic bags).

17   Check the gear backlash between the balance shaft drive gear (behind the camshaft sprocket) and the balance shaft driven gear. Set up a dial indicator against the top teeth of the balance shaft driven gear (the one on the front of the balance shaft) and rock the gear by hand. Compare the backlash to this Chapter's Specifications. The gears will have to be replaced if the backlash is greater than specified.

18   Remove the balance shaft drive gear (behind the camshaft sprocket). Set up a dial indicator on the nose of the balance shaft and zero it. Reaching inside the lifter valley, grab the balance shaft and move it forward and back in the block while watching the dial indicator. Compare this endplay measurement with this Chapter's Specifications. If it is incorrect, a new balance shaft thrust plate will have to be installed.

19   Remove the two balance shaft thrust plate bolts and take off the thrust plate **(see illustration)**.

20   The balance shaft's front bearing fits tightly into the front of the block, and a slide hammer must be used to remove the shaft

13.20 A slide hammer must be threaded into the front of the balance shaft to pull the balance shaft front bearing out of the block

and its bearing **(see illustration)**. Once the bearing is free of the block, withdraw the balance shaft straight out by hand, being careful to not cock it against its rear bearing.

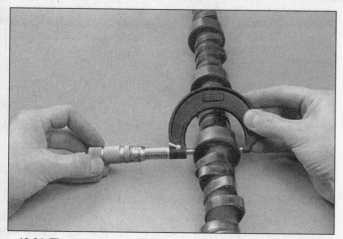

13.24 The camshaft bearing journal diameters are checked to pinpoint excessive wear and out-of-round conditions

14.1 A ridge reamer is required to remove the ridge from the top of each cylinder - do this before removing the pistons!

Inspection

Refer to illustration 13.24

21   After the camshaft, intermediate shaft or balance shaft has been removed, clean it with solvent and dry it, then inspect the bearing journals for uneven wear, pitting and evidence of seizure. If the journals are damaged, the bearing inserts in the block are probably damaged as well. Both the shaft and bearings will have to be replaced.

22   On 2.5L engines through 1990, if the timing gear must be removed from the camshaft, it must be pressed off. If you don't have access to a press, take it to an automotive machine shop. The thrust plate must be positioned so the Woodruff key in the shaft does not damage it when the shaft is pressed off.

23   If the gear has been removed, it must be pressed back on prior to installation of the camshaft.

a) *Support the camshaft in a press by placing press plate adapters behind the front journal.*

b) *Place the gear spacer ring and the thrust plate over the end of the shaft.*

c) *Install the Woodruff key in the shaft keyway.*

d) *Position the camshaft gear and press it onto the shaft until it bottoms against the gear spacer ring.*

e) *Use a feeler gauge to check the end clearance. It should be 0.0015 to 0.0050 inch. If the clearance is less than 0.0015 inch, the spacer ring should be replaced. If the clearance is more than 0.0050 inch, the thrust plate should be replaced.*

24   Measure the bearing journals with a micrometer **(see illustration)** to determine if they're excessively worn or out-of-round. Compare your readings with the values listed in this Chapter's Specifications.

25   Check the camshaft lobes and oil pump drive and driven gears for heat discoloration, score marks, chipped areas, pitting and uneven wear. If the lobes and gears are in

14.3 Check the connecting rod endplay with a feeler gauge as shown

good condition and if the lobe lift measurements are as specified, the components can be reused.

26   Check the bearings in the block for wear and damage. Look for galling, pitting and discolored areas.

27   The inside diameter of each bearing can be determined with a small hole gauge and outside micrometer or an inside micrometer. Subtract the camshaft bearing journal diameter(s) from the corresponding bearing inside diameter(s) to obtain the bearing oil clearance. If it's excessive, new bearings will be required regardless of the condition of the originals.

28   Camshaft and balance shaft bearing replacement requires special tools and expertise that place it outside the scope of the home mechanic. Take the block to an automotive machine shop to ensure the job is done correctly.

14   Pistons and connecting rods - removal

Refer to illustrations 14.1, 14.3 and 14.6
Note: *Prior to removing the piston/connecting rod assemblies, remove the cylinder head(s), the oil pan and the oil pump by referring to the*

appropriate Sections in Parts A through F, depending on the engine of Chapter 2.

1   Use your fingernail to feel if a ridge has formed at the upper limit of ring travel (about 1/4-inch down from the top of each cylinder). If carbon deposits or cylinder wear have produced ridges, they must be completely removed with a special tool **(see illustration)**. Follow the manufacturer's instructions provided with the tool. Failure to remove the ridges before attempting to remove the piston/connecting rod assemblies may result in piston breakage.

2   After the cylinder ridges have been removed, turn the engine upside-down so the crankshaft is facing up.

3   Before the connecting rods are removed, check the endplay with feeler gauges. Slide them between the first connecting rod and the crankshaft throw until the play is removed **(see illustration)**. The endplay is equal to the thickness of the feeler gauge(s). If the endplay exceeds the service limit, new connecting rods will be required. If new rods (or a new crankshaft) are installed, the endplay may fall under the minimum specified in this Chapter (if it does, the rods will have to be machined to restore it - consult an automotive machine shop for advice if necessary). Repeat the procedure for the remaining connecting rods.

14.6 To prevent damage to the crankshaft journals and cylinder walls, slip sections of rubber or plastic hose over the rod bolts before removing the pistons

15.3 Checking crankshaft endplay with a feeler gauge

4   Check the connecting rods and caps for identification marks. If they aren't plainly marked, use a small center-punch to make the appropriate number of indentations on each rod and cap (1, 2, 3, etc., depending on the engine type and cylinder they're associated with).

5   Loosen each of the connecting rod cap nuts 1/2-turn at a time until they can be removed by hand. Remove the number one connecting rod cap and bearing insert. Don't drop the bearing insert out of the cap.

6   Slip a short length of plastic or rubber hose over each connecting rod cap bolt to protect the crankshaft journal and cylinder wall as the piston is removed **(see illustration)**.

7   Remove the bearing insert and push the connecting rod/piston assembly out through the top of the engine. Use a wooden or plastic hammer handle to push on the upper bearing surface in the connecting rod. If resistance is felt, double-check to make sure all of the ridge was removed from the cylinder.

8   Repeat the procedure for the remaining cylinders.

9   After removal, reassemble the connecting rod caps and bearing inserts in their respective connecting rods and install the cap nuts finger tight. Leaving the old bearing inserts in place until reassembly will help prevent the connecting rod bearing surfaces from being accidentally nicked or gouged.

10   Don't separate the pistons from the connecting rods (see Section 19 for additional information).

15   Crankshaft - removal

Refer to illustrations 15.3, 15.4a, and 15.4b
Note: *The crankshaft can be removed only after the engine has been removed from the vehicle. It's assumed the flywheel or driveplate, crankshaft balancer/vibration damper, timing chain, timing gear, oil pan, oil pump and piston/connecting rod assemblies have already been removed. The rear main oil seal housing (Quad-4 only) must be unbolted and separated from the block before proceeding with crankshaft removal.*

1   Before the crankshaft is removed, check the endplay. Mount a dial indicator with the stem in line with the crankshaft and just touching one of the crank throws.

2   Push the crankshaft all the way to the rear and zero the dial indicator. Next, pry the crankshaft to the front as far as possible and check the reading on the dial indicator. The distance it moves is the endplay. If it's greater than specified in this Chapter, check the crankshaft thrust surfaces for wear. If no wear is evident, new main bearings should correct the endplay.

3   If a dial indicator isn't available, feeler gauges can be used. Gently pry or push the crankshaft all the way to the front of the engine. Slip feeler gauges between the crankshaft and the front face of the thrust main bearing to determine the clearance **(see illustration)**.

4   Check the main bearing caps to see if they're marked to indicate their locations. They should be numbered consecutively from the front of the engine to the rear. If they aren't, mark them with number stamping dies or a center punch **(see illustration)**. Main bearing caps generally have a cast-in arrow, which points to the front of the engine **(see**

15.4a Use a center-punch or number stamping dies to mark the main bearing caps to ensure installation in their original locations on the block (make the punch marks near one of the bolt heads)

15.4b The arrow on the main bearing cap indicates the front of the engine

16.3 Remove all covers and plugs from the engine block (rear of V6 engine shown)

1   *Camshaft rear cover*
2   *Threaded oil passage plug*
3   *Core plugs*

16.4 Use a hammer and large punch to knock the core plugs sideways in their bores, then pull them out with pliers

illustration). Loosen the main bearing cap bolts 1/4-turn at a time each, until they can be removed by hand. Note if any stud bolts are used and make sure they're returned to their original locations when the crankshaft is reinstalled.

5   Gently tap the caps with a soft-face hammer, then separate them from the engine block. If necessary, use the bolts as levers to remove the caps. Try not to drop the bearing inserts if they come out with the caps.

6   Carefully lift the crankshaft out of the engine. It may be a good idea to have an assistant available, since the crankshaft is quite heavy. With the bearing inserts in place in the engine block and main bearing caps, return the caps to their respective locations on the engine block and tighten the bolts finger tight.

16   Engine block - cleaning

Refer to illustrations 16.3, 16.4, 16.8 and 16.10

1   Remove the main bearing caps and separate the bearing inserts from the caps and the engine block. Tag the bearings, indicating which cylinder they were removed from and whether they were in the cap or the block, then set them aside.

2   Using a gasket scraper, remove all traces of gasket material from the engine block. Be very careful not to nick or gouge the gasket sealing surfaces.

3   Remove all of the covers and threaded oil gallery plugs from the block (**see illustration**). The plugs are usually very tight - they may have to be drilled out and the holes retapped. Use new plugs when the engine is reassembled.

4   Remove the core plugs from the engine block. To do this, knock one side of the plug into the block with a hammer and punch, then grasp them with large pliers and pull them back through the holes (**see illustration**). **Caution:** *The core plugs (also known as freeze or soft plugs) may be difficult or impossible to retrieve if they're driven into the block coolant passages.*

16.8 All bolt holes in the block - particularly the main bearing cap and head bolt holes - should be cleaned and restored with a tap (be sure to remove debris from the holes after this is done)

5   If the engine is extremely dirty, it should be taken to an automotive machine shop to be steam cleaned or hot tanked.

6   After the block is returned, clean all oil holes and oil galleries one more time. Brushes specifically designed for this purpose are available at most auto parts stores. Flush the passages with warm water until the water runs clear, dry the block thoroughly and wipe all machined surfaces with a light, rust preventive oil. If you have access to compressed air, use it to speed the drying process and blow out all the oil holes and galleries. **Warning:** *Wear eye protection when using compressed air!*

7   If the block isn't extremely dirty or sludged up, you can do an adequate cleaning job with hot soapy water and a stiff brush. Take plenty of time and do a thorough job. Regardless of the cleaning method used, be sure to clean all oil holes and galleries very thoroughly, dry the block completely and coat all machined surfaces with light oil.

8   The threaded holes in the block must be clean to ensure accurate torque readings during reassembly. Run the proper size tap

16.10 A large socket on an extension can be used to drive the new core plugs into the bores

into each of the holes to remove rust, corrosion, thread sealant or sludge and restore damaged threads (**see illustration**). If possible, use compressed air to clear the holes of debris produced by this operation. Now is a good time to clean the threads on the head bolts and the main bearing cap bolts as well.

9   Reinstall the main bearing caps and tighten the bolts finger tight.

10   After coating the sealing surfaces of the new core plugs with Permatex no. 2 sealant, install them in the engine block (**see illustration**). Make sure they're driven in straight and seated properly or leakage could result. Special tools are available for this purpose, but a large socket, with an outside diameter that will just slip into the core plug, a 1/2-inch drive extension and a hammer will work just as well.

11   Apply non-hardening sealant (such as Permatex no. 2 or Teflon pipe sealant) to the new oil gallery plugs and thread them into the holes in the block. Make sure they're tightened securely.

12   If the engine isn't going to be reassembled right away, cover it with a large plastic trash bag to keep it clean.

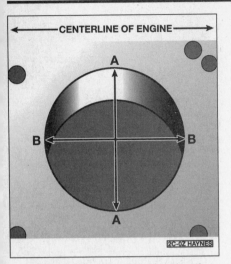

17.4a Measure the diameter of each cylinder at a right angle to the centerline (A), and parallel to the engine centerline (B) - the out-of-round specification is the difference between A and B; taper is the difference between the diameter at the top of the cylinder and the diameter at the bottom of the cylinder

17   Engine block - inspection

Refer to illustrations 17.4a, 17.4b and 17.4c
Note: *The manufacturer recommends checking the block deck and transaxle mounting bolt hole bosses for warpage and the main bearing bore concentricity and alignment. Since special measuring tools are needed, the checks should be done by an automotive machine shop. Also, if you're working on a Quad-4 engine, it may be a good idea to verify the condition of the oil flow check valve located in the oil passage in the right front corner of the block deck.*

1   Before the block is inspected, it should be cleaned as described in Section 16.

2   Visually check the block for cracks, rust and corrosion. Look for stripped threads in the threaded holes. It's also a good idea to have the block checked for hidden cracks by an automotive machine shop that has the special equipment to do this type of work. If defects are found, have the block repaired, if possible, or replaced.

3   Check the cylinder bores for scuffing and scoring.

4   Measure the diameter of each cylinder at the top (just under the ridge area), center and bottom of the cylinder bore, parallel to the crankshaft axis **(see illustrations)**. **Note:** *These measurements should not be made with the bare block mounted on an engine stand - the cylinders will be distorted and the measurements will be inaccurate.*

5   Next, measure each cylinder's diameter at the same three locations across the crankshaft axis. Compare the results to this Chapter's Specifications.

6   If the required precision measuring tools aren't available, the piston-to-cylinder clearances can be obtained, though not quite as accurately, using feeler gauge stock. Feeler gauge stock comes in 12-inch lengths and various thicknesses and is generally available at auto parts stores.

7   To check the clearance, select a feeler gauge and slip it into the cylinder along with the matching piston. The piston must be positioned exactly as it normally would be. The feeler gauge must be between the piston and cylinder on one of the thrust faces (90-degrees to the piston pin bore).

8   The piston should slip through the cylinder (with the feeler gauge in place) with moderate pressure.

9   If it falls through or slides through easily, the clearance is excessive and a new piston will be required. If the piston binds at the lower end of the cylinder and is loose toward the top, the cylinder is tapered. If tight spots are encountered as the piston/feeler gauge is

rotated in the cylinder, the cylinder is out-of-round.

10   Repeat the procedure for the remaining pistons and cylinders.

11   If the cylinder walls are badly scuffed or scored, or if they're out-of-round or tapered beyond the limits given in this Chapter's Specifications, have the engine block rebored and honed at an automotive machine shop. If a rebore is done, oversize pistons and rings will be required.

12   If the cylinders are in reasonably good condition and not worn to the outside of the limits, and if the piston-to-cylinder clearances can be maintained properly, they don't have to be rebored. Honing is all that's necessary (see Section 18).

18   Cylinder honing

Refer to illustrations 18.3a and 18.3b
1   Prior to engine reassembly, the cylinder bores must be honed so the new piston rings will seat correctly and provide the best possible combustion chamber seal. **Note:** *If you don't have the tools or don't want to tackle the honing operation, most automotive machine shops will do it for a reasonable fee.*

2   Before honing the cylinders, install the main bearing caps and tighten the bolts to the torque listed in this Chapter's Specifications.

3   Two types of cylinder hones are commonly available - the flex hone or "bottle brush" type and the more traditional surfacing hone with spring-loaded stones. Both will do the job, but for the less experienced mechanic the "bottle brush" hone will probably be easier to use. You'll also need some honing oil (kerosene will work if honing oil isn't available), rags and an electric drill motor. Proceed as follows:

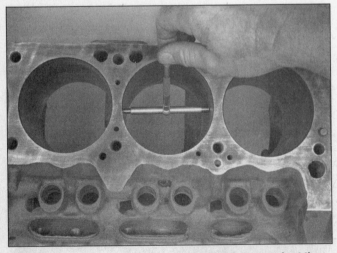

17.4b The ability to "feel" when the telescoping gauge is at the correct point will be developed over time, so work slowly and repeat the check until you're satisfied the bore measurement is accurate

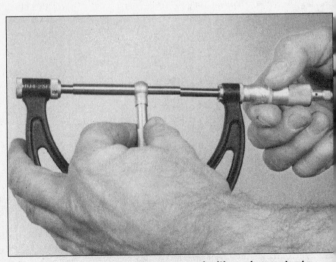

17.4c The gauge is then measured with a micrometer to determine the bore size

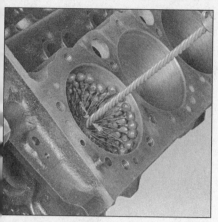

18.3a A "bottle brush" hone will produce better results if you've never honed cylinders before

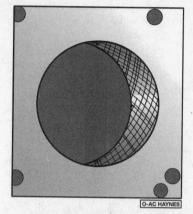

18.3b The cylinder hone should leave a smooth, crosshatch pattern with the lines intersecting at approximately a 60-degree angle

a) Mount the hone in the drill motor, compress the stones and slip it into the first cylinder (see illustration). Be sure to wear safety goggles or a face shield!

b) Lubricate the cylinder with plenty of honing oil, turn on the drill and move the hone up-and-down in the cylinder at a pace that will produce a fine crosshatch pattern on the cylinder walls. Ideally, the crosshatch lines should intersect at approximately a 60-degree angle (see illustration). Be sure to use plenty of lubricant and don't take off any more material than is absolutely necessary to produce the desired finish. Note: Piston ring manufacturers may specify a smaller crosshatch angle than the traditional 60-degrees - read and follow any instructions included with the new rings.

c) Don't withdraw the hone from the cylinder while it's running. Instead, shut off the drill and continue moving the hone up-and-down in the cylinder until it comes to a complete stop, then compress the stones and withdraw the hone. If you're using a "bottle brush" type hone, stop the drill motor, then turn the chuck in the normal direction of rotation while withdrawing the hone from the cylinder.

d) Wipe the oil out of the cylinder and repeat the procedure for the remaining cylinders.

4 After the honing job is complete, chamfer the top edges of the cylinder bores with a small file so the rings won't catch when the pistons are installed. Be very careful not to nick the cylinder walls with the end of the file.

5 The entire engine block must be washed again very thoroughly with warm, soapy water to remove all traces of the abrasive grit produced during the honing operation. Note: The bores can be considered clean when a lint-free white cloth - dampened with clean engine oil - used to wipe them out doesn't pick up any more honing residue, which will show up as gray areas on the cloth. Be sure to run a brush through all oil holes and galleries and flush them with running water.

6 After rinsing, dry the block and apply a coat of light rust preventive oil to all machined surfaces. Wrap the block in a plastic trash bag to keep it clean and set it aside until reassembly.

19 Pistons and connecting rods - inspection

Refer to illustrations 19.4a, 19,4b, 19.10 and 19.11

1 Before the inspection process can be carried out, the piston/connecting rod assemblies must be cleaned and the original piston rings removed from the pistons. Note: Always use new piston rings when the engine is reassembled.

2 Using a piston ring installation tool, carefully remove the rings from the pistons. Be careful not to nick or gouge the pistons in the process.

3 Scrape all traces of carbon from the top of the piston. A hand held wire brush or a piece of fine emery cloth can be used once the majority of the deposits have been scraped away. Do not, under any circumstances, use a wire brush mounted in a drill motor to remove deposits from the pistons. The piston material is soft and may be eroded away by the wire brush.

4 Use a piston ring groove cleaning tool to remove carbon deposits from the ring grooves. If a tool isn't available, a piece broken off the old ring will do the job. Be very careful to remove only the carbon deposits - don't remove any metal and do not nick or scratch the sides of the ring grooves (see illustrations).

5 Once the deposits have been removed, clean the piston/rod assemblies with solvent and dry them with compressed air (if available). Warning: Wear eye protection. Make sure the oil return holes in the back sides of the ring grooves are clear.

6 If the pistons and cylinder walls aren't damaged or worn excessively, and if the engine block isn't rebored, new pistons won't be necessary. Normal piston wear appears as even vertical wear on the piston thrust surfaces and slight looseness of the top ring in its groove. New piston rings, however, should always be used when an engine is rebuilt.

7 Carefully inspect each piston for cracks

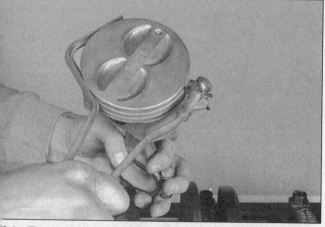

19.4a The piston ring grooves can be cleaned with a special tool, as shown here . . .

19.4b . . . or a section of a broken ring

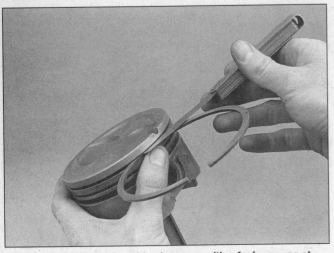

19.10 Check the ring side clearance with a feeler gauge at several points around the groove

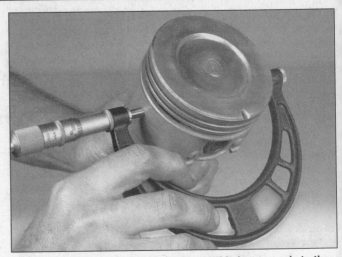

19.11 Measure the piston diameter at a 90-degree angle to the piston pin and in line with it

20.1 Use a wire or stiff plastic bristle brush to clean the oil passages in the crankshaft

around the skirt, at the pin bosses and at the ring lands.

8   Look for scoring and scuffing on the thrust faces of the skirt, holes in the piston crown and burned areas at the edge of the crown. If the skirt is scored or scuffed, the engine may have been suffering from over-heating and/or abnormal combustion, which caused excessively high operating temperatures. The cooling and lubrication systems should be checked thoroughly. A hole in the piston crown is an indication that abnormal combustion (preignition) was occurring. Burned areas at the edge of the piston crown are usually evidence of spark knock (detonation). If any of the above problems exist, the causes must be corrected or the damage will occur again. The causes may include intake air leaks, incorrect fuel/air mixture, low octane fuel, ignition timing and EGR system malfunctions.

9   Corrosion of the piston, in the form of small pits, indicates coolant is leaking into

the combustion chamber and/or the crankcase. Again, the cause must be corrected or the problem may persist in the rebuilt engine.

10   Measure the piston ring side clearance by laying a new piston ring in each ring groove and slipping a feeler gauge in beside it **(see illustration)**. Check the clearance at three or four locations around each groove. Be sure to use the correct ring for each groove - they are different. If the side clearance is greater than specified in this Chapter, new pistons will have to be used.

11   Check the piston-to-bore clearance by measuring the bore (see Section 17) and the piston diameter. Make sure the pistons and bores are correctly matched. Measure the piston across the skirt, at a 90-degree angle to the piston pin **(see illustration)**. The measurement must be taken at a specific point, depending on the engine type, to be accurate.

a) *The piston diameter is measured 3/4-inch (19 mm) below the center of the piston pin hole.*
b) *If you're working on a Quad-4 engine, measure the piston 0.4724-inch (12.0 mm) up from the lower edge of the skirt.*

12   Subtract the piston diameter from the bore diameter to obtain the clearance. If it's greater than specified, the block will have to be rebored and new pistons and rings installed.

13   Check the piston-to-rod clearance by twisting the piston and rod in opposite directions. Any noticeable play indicates excessive wear, which must be corrected. The piston/connecting rod assemblies should be taken to an automotive machine shop to have the pistons and rods resized and new pins installed.

14   If the pistons must be removed from the connecting rods for any reason, they should be taken to an automotive machine shop. While they are there have the connecting rods checked for bend and twist, since automotive machine shops have special equipment for this

purpose. **Note:** *Unless new pistons and/or connecting rods must be installed, do not disassemble the pistons and connecting rods.*

20   Crankshaft - inspection

Refer to illustrations 20.1, 20.3, 20.4, 20.6 and 20.8

1   Clean the crankshaft with solvent and dry it with compressed air (if available). **Warning:** *Wear eye protection when using compressed air. Be sure to clean the oil holes with a stiff brush **(see illustration)** and flush them with solvent.*

2   Check the main and connecting rod bearing journals for uneven wear, scoring, pits and cracks.

3   Rub a penny across each journal several times **(see illustration)**. If a journal picks up copper from the penny, it's too rough and must be reground.

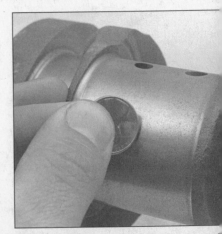

20.3 Rubbing a penny lengthwise on each journal will reveal its condition - if copper rubs off and is embedded in the crankshaft, the journals should be reground

20.4 The oil holes should be chamfered so sharp edges don't gouge or scratch the new bearings

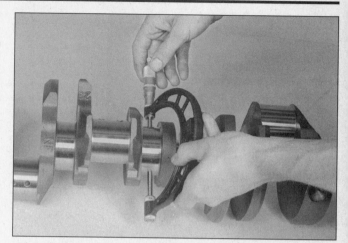

20.6 Measure the diameter of each crankshaft journal at several points to detect taper and out-of-round conditions

4   Remove all burrs from the crankshaft oil holes with a stone, file or scraper **(see illustration)**.

5   Check the rest of the crankshaft for cracks and other damage. It should be magnafluxed to reveal hidden cracks - an automotive machine shop will handle the procedure.

6   Using a micrometer, measure the diameter of the main and connecting rod journals and compare the results to this Chapter's Specifications **(see illustration)**. By measuring the diameter at a number of points around each journal's circumference, you'll be able to determine whether or not the journal is out-of-round. Take the measurement at each end of the journal, near the crank throws, to determine if the journal is tapered.

7   If the crankshaft journals are damaged, tapered, out-of-round or worn beyond the limits given in the Specifications, have the crankshaft reground by an automotive machine shop. Be sure to use the correct size bearing inserts if the crankshaft is reconditioned.

8   Check the oil seal journals at each end of the crankshaft for wear and damage. If the seal has worn a groove in the journal, or if it's nicked or scratched **(see illustration)**, the new seal may leak when the engine is reassembled. In some cases, an automotive machine shop may be able to repair the journal by pressing on a thin sleeve. If repair isn't feasible, a new or different crankshaft should be installed.

9   If you're working on a Quad-4 engine, check the oil pump drive gear for wear and damage. If replacement is required, take it to a dealer service department or an automotive machine shop. The old gear must be drilled and chiseled off and the new gear must be heated in an oven prior to installation.

10   Refer to Section 21 and examine the main and rod bearing inserts.

21   Main and connecting rod bearings - inspection

Refer to illustration 21.1

1   Even though the main and connecting rod bearings should be replaced with new ones during the engine overhaul, the old bearings should be retained for close examination, as they may reveal valuable information about the condition of the engine **(see illustration)**.

2   Bearing failure occurs because of lack

20.8 If the seals have worn grooves in the crankshaft journals, or if the seal contact surfaces are nicked or scratched, the new seals will leak

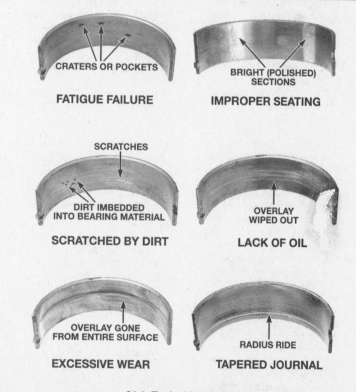

21.1 Typical bearing failures

of lubrication, the presence of dirt or other foreign particles, overloading the engine and corrosion. Regardless of the cause of bearing failure, it must be corrected before the engine is reassembled to prevent it from happening again.

3   When examining the bearings, remove them from the engine block, the main bearing caps, the connecting rods and the rod caps and lay them out on a clean surface in the same general position as their location in the engine. This will enable you to match any bearing problems with the corresponding crankshaft journal.

4   Dirt and other foreign particles get into the engine in a variety of ways. It may be left in the engine during assembly, or it may pass through filters or the PCV system. It may get into the oil, and from there into the bearings. Metal chips from machining operations and normal engine wear are often present. Abrasives are sometimes left in engine components after reconditioning, especially when parts aren't thoroughly cleaned using the proper cleaning methods. Whatever the source, these foreign objects often end up embedded in the soft bearing material and are easily recognized. Large particles won't embed in the bearing and will score or gouge the bearing and journal. The best prevention for this cause of bearing failure is to clean all parts thoroughly and keep everything spotlessly clean during engine assembly. Frequent and regular engine oil and filter changes are also recommended.

5   Lack of lubrication (or lubrication breakdown) has a number of interrelated causes. Excessive heat (which thins the oil), overloading (which squeezes the oil from the bearing face) and oil leakage or throw off (from excessive bearing clearances, worn oil pump or high engine speeds) all contribute to lubrication breakdown. Blocked oil passages, which usually are the result of misaligned oil holes in a bearing shell, will also oil starve a bearing and destroy it. When lack of lubrication is the cause of bearing failure, the bearing material is wiped or extruded from the steel backing of the bearing. Temperatures may increase to the point where the steel backing turns blue from overheating.

6   Driving habits can have a definite effect on bearing life. Full throttle, low speed operation (lugging the engine) puts very high loads on bearings, which tends to squeeze out the oil film. These loads cause the bearings to flex, which produces fine cracks in the bearing face (fatigue failure). Eventually the bearing material will loosen in pieces and tear away from the steel backing. Short trip driving leads to corrosion of bearings because insufficient engine heat is produced to drive off the condensed water and corrosive gases. These products collect in the engine oil, forming acid and sludge. As the oil is carried to the engine bearings, the acid attacks and corrodes the bearing material.

7   Incorrect bearing installation during engine assembly will lead to bearing failure as well. Tight fitting bearings leave insuffi-

23.3 When checking piston ring end gap, the ring must be square in the cylinder bore (this is done by pushing the ring down with the top of a piston as shown)

cient oil clearance and will result in oil starvation. Dirt or foreign particles trapped behind a bearing insert result in high spots on the bearing which lead to failure.

22   Engine overhaul - reassembly sequence

1   Before beginning engine reassembly, make sure you have all the necessary new parts, gaskets and seals as well as the following items on hand:

    Common hand tools
    Torque wrench (1/2-inch drive)
    Piston ring installation tool
    Piston ring compressor
    Vibration damper installation tool
    Short lengths of rubber or plastic hose to
        fit over connecting rod bolts
    Plastigage
    Feeler gauges
    Fine-tooth file
    New engine oil
    Engine assembly lube or moly-base
        grease
    Gasket sealant
    Thread locking compound

2   In order to save time and avoid problems, engine reassembly must be done in the following general order:

2.2 and 2.5 liter four-cylinder engine

    New camshaft bearings (must be done by
        automotive machine shop)
    Crankshaft and main bearings
    Piston/connecting rod assemblies
    Camshaft and lifters
    Timing gear, /chain and sprockets
    Timing gear cover
    Oil pump
    Oil pan
    Cylinder head, pushrods and rocker arms
    Intake and exhaust manifolds
    Valve cover
    Engine rear plate
    Flywheel/driveplate

2.3 liter four-cylinder (Quad-4) engine

    Crankshaft and main bearings
    Rear main oil seal housing
    Piston/connecting rod assemblies
    Oil pump
    Oil pan
    Cylinder head and camshafts
    Timing chain housing
    Timing chain and sprockets

V6 engines

    Crankshaft and main bearings
    Intermediate shaft (3.4L only)
    Rear main oil seal housing
    Piston/connecting rod assemblies
    Oil pump
    Oil pan
    Camshaft (except 3.4L)
    Timing chain and sprockets
    Timing chain cover
    Cylinder heads
    Valve lifters (except 3.4L)
    Rocker arms and pushrods (if equipped)
    Cam carriers and camshafts (3.4L only)
    Timing belt (3.4L only)
    Intake and exhaust manifolds
    Valve covers

23   Piston rings - installation

Refer to illustrations 23.3, 23.4, 23.5, 23.9a, 23.9b and 23.12

1   Before installing the new piston rings, the ring end gaps must be checked. It's assumed the piston ring side clearance has been checked and verified correct (see Section 19).

2   Lay out the piston/connecting rod assemblies and the new ring sets so the ring sets will be matched with the same piston and cylinder during the end gap measurement and engine assembly.

3   Insert the top (number one) ring into the first cylinder and square it up with the cylinder walls by pushing it in with the top of the piston **(see illustration)**. The ring should be near the bottom of the cylinder, at the lower limit of ring travel.

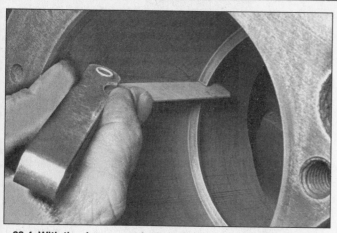

23.4 With the ring square in the cylinder, measure the end gap with a feeler gauge

23.5 If the end gap is too small, clamp a file in a vise and file the ring ends (from the outside in only) to enlarge the gap slightly

23.9a Installing the spacer/expander in the oil control ring groove

23.9b DO NOT use a piston ring installation tool when installing the oil ring side rails

4   To measure the end gap, slip feeler gauges between the ends of the ring until a gauge equal to the gap width is found **(see illustration)**. The feeler gauge should slide between the ring ends with a slight amount of drag. Compare the measurement to this Chapter's Specifications. If the gap is larger or smaller than specified, double-check to make sure you have the correct rings before proceeding.

5   If the gap is too small, it must be enlarged or the ring ends may come in contact with each other during engine operation, which can cause serious engine damage. The end gap can be increased by filing the ring ends very carefully with a fine file. Mount the file in a vise equipped with soft jaws, slip the ring over the file with the ends contacting the file teeth and slowly move the ring to remove material from the ends. When performing this operation, file only from the outside in **(see illustration)**.

6   Excess end gap isn't critical unless it's greater than 0.040-inch. Again, double-check to make sure you have the correct rings for the engine.

7   Repeat the procedure for each ring that will be installed in the first cylinder and for each ring in the remaining cylinders. Remem-

ber to keep rings, pistons and cylinders matched up.

8   Once the ring end gaps have been checked/corrected, the rings can be installed on the pistons.

9   The oil control ring (lowest one on the piston) is usually installed first. It's composed of three separate components. Slip the spacer/expander into the groove **(see illustration)**. If an anti-rotation tang is used, make sure it's inserted into the drilled hole in the ring groove. Next, install the lower side rail. Don't use a piston ring installation tool on the oil ring side rails, as they may be damaged. Instead, place one end of the side rail into the groove between the spacer/expander and the ring land, hold it firmly in place and slide a finger around the piston while pushing the rail into the groove **(see illustration)**. Next, install the upper side rail in the same manner.

10   After the three oil ring components have been installed, check to make sure both the upper and lower side rails can be turned smoothly in the ring groove.

11   The number two (middle) ring is installed next. It's usually stamped with a mark, which must face up, toward the top of the piston. **Note:** *Always follow the instructions printed*

on the ring package or box - different manufacturers may require different approaches. Don't mix up the top and middle rings, as they have different cross sections.

12   Use a piston ring installation tool and make sure the identification mark is facing the top of the piston, then slip the ring into the middle groove on the piston **(see illustration)**. Don't expand the ring any more than

23.12 Installing the compression rings with a ring expander - the mark (arrow) must face up

24.11 Lay the Plastigage strips (arrow) on the main bearing journals, parallel to the crankshaft centerline

24.15 Compare the width of the crushed Plastigage to the scale on the envelope to determine the main bearing oil clearance (always take the measurement at the widest point of the Plastigage); be sure to use the correct scale - standard and metric ones are included

necessary to slide it over the piston.

13   Install the number one (top) ring in the same manner. Make sure the mark is facing up. Be careful not to confuse the number one and number two rings.

14   Repeat the procedure for the remaining pistons and rings.

24   Crankshaft - installation and main bearing oil clearance check

Refer to illustrations 24.11 and 24.15

1   Crankshaft installation is the first step in engine reassembly. It's assumed at this point that the engine block and crankshaft have been cleaned, inspected and repaired or reconditioned.

2   Position the engine with the bottom facing up.

3   Remove the main bearing cap bolts and lift out the caps. Lay them out in the proper order to ensure correct installation.

4   If they're still in place, remove the original bearing inserts from the block and the main bearing caps. Wipe the bearing surfaces of the block and caps with a clean, lint-free cloth. They must be kept spotlessly clean.

Main bearing oil clearance check

Note: *Don't touch the faces of the new bearing inserts with your fingers. Oil and acids from your skin can etch the bearings.*

5   Clean the back sides of the new main bearing inserts and lay one in each main bearing saddle in the block. If one of the bearing inserts from each set has a large groove in it, make sure the grooved insert is installed in the block. Lay the other bearing from each set in the corresponding main bearing cap. Make sure the tab on the bearing insert fits into the recess in the block or cap. **Caution:** *The oil holes in the block must line up with the oil holes in the bearing inserts. Do not hammer the bearing into place and don't nick or gouge the bearing faces. No*

lubrication should be used at this time.

6   On the 2.2 liter four cylinder engine, the flanged thrust bearing must be installed in the number four cap and saddle (counting from the front of the engine). On 2.5 liter four-cylinder engines, the thrust bearing must be installed in the number five cap and saddle. On Quad-4 engines, the thrust bearing must be installed in the number three (center) cap and saddle. On the 3800, the thrust bearing must be installed in the number two cap and saddle. On other V6 engines, install the thrust bearing in the number three cap and saddle.

7   Clean the faces of the bearings in the block and the crankshaft main bearing journals with a clean, lint-free cloth.

8   Check or clean the oil holes in the crankshaft, as any dirt here can go only one way - straight through the new bearings.

9   Once you're certain the crankshaft is clean, carefully lay it in position in the main bearings.

10   Before the crankshaft can be permanently installed, the main bearing oil clearance must be checked.

11   Cut several pieces of the appropriate size Plastigage (they should be slightly shorter than the width of the main bearings) and place one piece on each crankshaft main bearing journal, parallel with the journal axis **(see illustration).**

12   Clean the faces of the bearings in the caps and install the caps in their original locations (don't mix them up) with the arrows pointing toward the front of the engine. Don't disturb the Plastigage.

13   Starting with the center main and working out toward the ends, tighten the main bearing cap bolts, in three steps, to the torque figure listed in this Chapter's Specifications. Don't rotate the crankshaft at any time during this operation.

14   Remove the bolts and carefully lift off the main bearing caps. Keep them in order. Don't disturb the Plastigage or rotate the crankshaft. If any of the main bearing caps are difficult to remove, tap them gently from side-to-side with a soft-face hammer to loosen them.

15   Compare the width of the crushed Plastigage on each journal to the scale printed on the Plastigage envelope to obtain the main bearing oil clearance **(see illustration)**. Check the Specifications to make sure it's correct.

16   If the clearance is not as specified, the bearing inserts may be the wrong size (which means different ones will be required). Before deciding different inserts are needed, make sure no dirt or oil was between the bearing inserts and the caps or block when the clearance was measured. If the Plastigage was wider at one end than the other, the journal may be tapered (refer to Section 20).

17   Carefully scrape all traces of the Plastigage material off the main bearing journals and/or the bearing faces. Use your fingernail or the edge of a credit card - don't nick or scratch the bearing faces.

Final crankshaft installation

18   Carefully lift the crankshaft out of the engine.

19   Clean the bearing faces in the block, then apply a thin, uniform layer of moly-based grease or engine assembly lube to each of the bearing surfaces. Be sure to coat the thrust faces as well as the journal face of the thrust bearing.

20   Make sure the crankshaft journals are clean, then lay the crankshaft back in place in the block.

21   Clean the faces of the bearings in the caps, then apply lubricant to them.

22   Install the caps in their original locations with the arrows pointing toward the front of the engine.

23   Install the bolts.

24   Tighten all except the thrust bearing cap bolts to the specified torque (work from the center out and approach the final torque in three steps).

25   Tighten the thrust bearing cap bolts to 10-to-12 ft-lbs.

26   Tap the ends of the crankshaft forward and backward with a lead or brass hammer to line up the main bearing and crankshaft thrust surfaces.

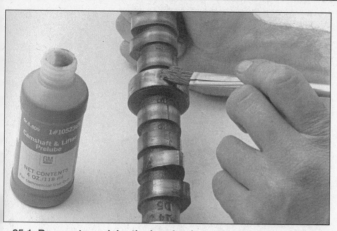

25.1 Be sure to prelube the bearing journals and lobes prior to camshaft installation

25.2 The camshaft and crankshaft gears must be positioned so the timing marks (arrows) line up

27   Retighten all main bearing cap bolts to the torque specified in this Chapter, starting with the center main and working out toward the ends.

28   Rotate the crankshaft a number of times by hand to check for any obvious binding.

29   The final step is to check the crankshaft endplay with feeler gauges or a dial indicator as described in Section 15. The endplay should be correct if the crankshaft thrust faces aren't worn or damaged and new bearings have been installed.

30   Refer to Section 26 and install the new rear main oil seal.

25   Camshaft, intermediate shaft (3.4L) and balance shaft (3800) - installation

Camshaft and intermediate shaft (3.4L)

Refer to illustrations 25.1 and 25.2

Note: *This procedure does not apply to the Quad-4 or 3.4 liter V6 engines. For camshaft installation procedures on Quad-4 and 3.4L engines, refer to Chapter 2C or 2E.*

1   Lubricate the camshaft bearing journals and cam lobes with moly-base grease or engine assembly lube **(see illustration)**.

2   Slide the camshaft into the engine, using a long bolt (the same thread as the camshaft sprocket bolt) screwed into the front of the camshaft as a "handle." Support the cam near the block and be careful not to scrape or nick the bearings. Install the camshaft retainer plate or intermediate shaft plate (3.4L) and tighten the bolts to the torque listed in this Chapter's Specifications. On 2.5L engines through 1990, align the timing marks on the camshaft gear and the crankshaft gear **(see illustration)**. Install the thrust plate bolts and tighten them to the torque listed in this Chapter's Specifications.

3   On 2.2L four cylinder and V6 engines (except 3800), dip the gear portion of the oil pump drive in engine oil and insert it into the block. It should be flush with its mounting boss before inserting the retaining bolt. **Note:** *Position a new O-ring on the oil pump driveshaft before installation.*

4   Refer to Part A, B, D or F (depending on engine type and size) to complete the installation of the lifters, timing chain and sprockets. On 3800 engines, perform the following procedure for balance shaft installation before installing the timing chain and sprockets.

Balance shaft (3800 only)

Refer to illustrations 25.5, 25.8 and 25.9

5   Lubricate the front bearing and rear journal of the balance shaft with engine oil and insert the balance shaft carefully into the block. When the front bearing approaches the insert in the front of the block, use a hammer and an appropriate-size socket to drive the front bearing into its insert. Drive it in just enough to allow installation of the balance shaft bearing retainer **(see illustration)**. Tighten the balance shaft retainer bolts to the torque listed in this Chapter's Specifications.

6   Install the balance shaft driven gear and its bolt.

7   Turn the camshaft so that with the camshaft sprocket temporarily installed, the timing mark is straight down.

8   With the camshaft sprocket and the camshaft gear removed, turn the balance shaft so the timing mark on the gear points straight down **(see illustration)**.

9   Install the camshaft gear (that drives the balance shaft) onto the cam, aligning it with

25.5 Drive the front balance shaft bearing in (it's attached to the balance shaft) just until the retainer plate can be installed

25.8 The balance shaft gear mark (arrow) should point straight down

25.9 Align the marks (arrows) on both balance shaft gears as shown

26.3 Tap around the outer edge of the new oil seal with a hammer and blunt punch to seat it squarely in the bore

the keyway and align the marks on the balance shaft gear and the camshaft gear (see illustration) by turning the balance shaft.
10   After the camshaft sprocket, crankshaft sprocket and timing chain have been installed (see Chapter 2 part F), tighten the balance shaft driven gear bolt to the torque listed in this Chapter's Specifications.

26   Rear main oil seal - installation

2.2 and 2.5 liter four-cylinder and V6 engines

Refer to illustration 26.3
1   Clean the bore in the block/cap and the seal journal on the crankshaft. Check the crankshaft journal for scratches and nicks that could damage the new seal lip and cause oil leaks. If the crankshaft is damaged, the only alternative is a new or different crankshaft.
2   Apply a light coat of engine oil or multi-purpose grease to the outer edge of the new seal. Lubricate the seal lip with moly-base grease or engine assembly lube.
3   Press the new seal into place with the special tool, if available (see Part A of this Chapter). The seal lip must face toward the front of the engine. If the special tool isn't available, carefully work the seal lip over the end of the crankshaft and tap the seal in with a hammer and blunt punch until it's seated in the bore (see illustration).

2.3 liter four-cylinder (Quad-4) engine

4   This engine is equipped with a one-piece seal that fits into a housing attached to the block. The crankshaft must be installed first and the main bearing caps bolted in place, then the new seal should be installed in the housing and the housing bolted to the block (see Part C of this Chapter).
5   Before installing the crankshaft, check the seal journal very carefully for scratches and nicks that could damage the new seal lip

and cause oil leaks. If the crankshaft is damaged, the only alternative is a new or different crankshaft.
6   The old seal can be removed from the housing with a hammer and punch by driving it out from the back side (see Part C of this Chapter). Be sure to note how far it's recessed into the housing bore before removing it; the new seal will have to be recessed an equal amount. Be very careful not to scratch or otherwise damage the bore in the housing or oil leaks could develop.
7   Make sure the housing is clean, then apply a thin coat of engine oil to the outer edge of the new seal. The seal must be pressed squarely into the housing bore (see Part C of this Chapter). Work slowly and make sure the seal enters the bore squarely.
8   The seal lips must be lubricated with moly-base grease or engine assembly lube before the seal/housing is slipped over the crankshaft and bolted to the block. Use a new gasket - no sealant is required - and make sure the dowel pins are in place before installing the housing.
9   Tighten the bolts a little at a time until they're all at the torque listed in the Part C Specifications.

27   Pistons and connecting rods - installation and rod bearing oil clearance check

Refer to illustrations 27.5, 27.9, 27.11, 27.13 and 27.17
1   Before installing the piston/connecting rod assemblies, the cylinder walls must be perfectly clean, the top edge of each cylinder must be chamfered, and the crankshaft must be in place.
2   Remove the cap from the end of the number one connecting rod (check the marks made during removal). Remove the original bearing inserts and wipe the bearing surfaces of the connecting rod and cap with a clean, lint-free cloth. They must be kept spotlessly clean.

Connecting rod bearing oil clearance check

Note: *Don't touch the faces of the new bearing inserts with your fingers. Oil and acids from your skin can etch the bearings.*
3   Clean the back side of the new upper bearing insert, then lay it in place in the connecting rod. Make sure the tab on the bearing fits into the recess in the rod. Don't hammer the bearing insert into place and be very careful not to nick or gouge the bearing face. Don't lubricate the bearing at this time.
4   Clean the back side of the other bearing insert and install it in the rod cap. Again, make sure the tab on the bearing fits into the recess in the cap, and don't apply any lubricant. It's critically important that the mating surfaces of the bearing and connecting rod are perfectly clean and oil free when they're assembled.
5   Position the piston ring gaps at 120-degree intervals around the piston (see illustration).

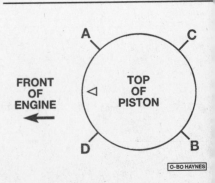

27.5 Ring end gap positions

A   Oil ring side rail gap - lower
B   Oil ring side rail gap - upper
C   Top compression ring gap
D   Second compression ring gap and oil ring spacer gap

27.9 The notch or arrow on each piston must face the front (timing chain) end of the engine as the pistons are installed

27.11 Drive the piston gently into the cylinder bore with the end of a wooden or plastic hammer handle

Slip a section of plastic or rubber hose over each connecting rod cap bolt.

Lubricate the piston and rings with clean engine oil and attach a piston ring compressor to the piston. Leave the skirt protruding about 1/4-inch to guide the piston into the cylinder. The rings must be compressed until they're flush with the piston.

Rotate the crankshaft until the number one connecting rod journal is at BDC (bottom dead center) and apply a coat of engine oil to the cylinder walls.

With the mark or notch on top of the piston (see illustration) facing the front of the engine, gently insert the piston/connecting rod assembly into the number one cylinder bore and rest the bottom edge of the ring compressor on the engine block. If you're working on a Quad-4 engine, make sure the hole in the lower end of the connecting rod is facing the right (exhaust manifold) side of the engine.

Tap the top edge of the ring compressor to make sure it's contacting the block around its entire circumference.

11 Gently tap on the top of the piston with the end of a wooden or plastic hammer handle (see illustration) while guiding the end of the connecting rod into place on the crankshaft journal. The piston rings may try to pop out of the ring compressor just before entering the cylinder bore, so keep some downward pressure on the ring compressor. Work slowly, and if any resistance is felt as the piston enters the cylinder, stop immediately. Find out what's hanging up and fix it before proceeding. Do not, for any reason, force the piston into the cylinder - you might break a ring and/or the piston.

12 Once the piston/connecting rod assembly is installed, the connecting rod bearing oil clearance must be checked before the rod cap is permanently bolted in place.

13 Cut a piece of the appropriate size Plastigage slightly shorter than the width of the connecting rod bearing and lay it in place on the number one connecting rod journal, parallel with the journal axis (see illustration).

14 Clean the connecting rod cap bearing face, remove the protective hoses from the connecting rod bolts and install the rod cap. Make sure the mating mark on the cap is on the same side as the mark on the connecting rod.

15 Install the nuts and tighten them to the torque listed in this Chapter's Specifications. Work up to it in three steps. Note: Use a thin-wall socket to avoid erroneous torque readings that can result if the socket is wedged between the rod cap and nut. If the socket tends to wedge itself between the nut and the cap, lift up on it slightly until it no longer contacts the cap. Do not rotate the crankshaft at any time during this operation.

16 Remove the nuts and detach the rod cap, being very careful not to disturb the Plastigage.

17 Compare the width of the crushed Plastigage to the scale printed on the Plastigage envelope to obtain the oil clearance (see illustration). Compare it to this Chapter's Specifications to make sure the clearance is correct.

18 If the clearance is not as specified, the bearing inserts may be the wrong size (which means different ones will be required). Before

27.13 Lay the Plastigage strips on each rod bearing journal, parallel to the crankshaft centerline

27.17 Measuring the width of the crushed Plastigage to determine the rod bearing oil clearance (be sure to use the correct scale - standard and metric ones are included)

deciding different inserts are needed, make sure no dirt or oil was between the bearing inserts and the connecting rod or cap when the clearance was measured. Also, recheck the journal diameter. If the Plastigage was wider at one end than the other, the journal may be tapered (refer to Section 20).

Final connecting rod installation

19   Carefully scrape all traces of the Plastigage material off the rod journal and/or bearing face. Be very careful not to scratch the bearing - use your fingernail or the edge of a credit card.

20   Make sure the bearing faces are perfectly clean, then apply a uniform layer of clean moly-base grease or engine assembly lube to both of them. You'll have to push the piston into the cylinder to expose the face of the bearing insert in the connecting rod - be sure to slip the protective hoses over the rod bolts first.

21   Slide the connecting rod back into place on the journal, remove the protective hoses from the rod cap bolts, install the rod cap and tighten the nuts to the torque listed in this Chapter's Specifications. Again, work up to the torque in three steps.

22   Repeat the entire procedure for the remaining pistons/connecting rods.

23   The important points to remember are . . .
a) Keep the back sides of the bearing inserts and the insides of the connecting rods and caps perfectly clean when assembling them.

b) Make sure you have the correct piston/rod assembly for each cylinder.
c) The arrow or mark on the piston must face the front (timing chain end) of the engine.
d) Lubricate the cylinder walls with clean oil.
e) Lubricate the bearing faces when installing the rod caps after the oil clearance has been checked.

24   After all the piston/connecting rod assemblies have been properly installed, rotate the crankshaft a number of times by hand to check for any obvious binding.

25   As a final step, the connecting rod endplay must be checked. Refer to Section 14 for this procedure.

26   Compare the measured endplay to this Chapter's Specifications to make sure it's correct. If it was correct before disassembly and the original crankshaft and rods were reinstalled, it should still be right. If new rods or a new crankshaft were installed, the endplay may be inadequate. If so, the rods will have to be removed and taken to an automotive machine shop for resizing.

28   Initial start-up and break-in after overhaul

Warning: *Have a fire extinguisher handy when starting the engine for the first time.*

1   Once the engine has been installed in the vehicle, double-check the oil and coolant levels.

2   With the spark plugs out of the engine and the ECM fuse removed, crank the engine until oil pressure registers on the gauge o the light goes out.

3   Install the spark plugs, hook up the plu wires (except Quad-4 engine) and install th ECM fuse.

4   Start the engine. It may take a fe moments for the fuel system to build up pres sure, but the engine should start without great deal of effort. **Note:** *If the engine keep backfiring, recheck the valve timing (an spark plug wires, where applicable).*

5   After the engine starts, it should b allowed to warm up to normal operating ten perature. While the engine is warming up make a thorough check for fuel, oil an coolant leaks.

6   Shut the engine off and recheck th engine oil and coolant levels.

7   Drive the vehicle to an area with min mum traffic, accelerate at full throttle from 3 to 50 mph, then allow the vehicle to slow 30 mph with the throttle closed. Repeat th procedure 10 or 12 times. This will load th piston rings and cause them to seat proper against the cylinder walls. Check again for and coolant leaks.

8   Drive the vehicle gently for the first 5C miles (no sustained high speeds) and keep constant check on the oil level. It's n unusual for an engine to use oil during th break-in period.

9   At approximately 500 to 600 mile change the oil and filter.

10   For the next few hundred miles, dri the vehicle normally. Don't pamper it abuse it.

11   After 2000 miles, change the oil and ter again and consider the engine broken in

Chapter 3
Cooling, heating and air conditioning systems

Contents

Specifications

General

Radiator cap pressure rating	See Chapter 1
Thermostat rating	195-degrees F
Coolant temperature fan switch rating	230-degrees F (110-degrees C)
Drivebelt tension	See Chapter 1

Torque specifications

Ft-lbs (unless otherwise specified)

Thermostat cover bolts

2.2 liter four-cylinder engines	89 in-lbs
2.5 liter four-cylinder engines	17
2.3 liter four-cylinder (Quad-4) engine	19
3800 V6	20
Other V6 engines	18

Water pump mounting bolts/nuts

2.2 liter four-cylinder engines	18
2.5 liter four-cylinder engines	24
2.3 liter four-cylinder (Quad-4) engine	19
3800 V6	
Short bolts	13
Long bolts	22
Other V6 engines	89 in-lbs

1 General information

Engine cooling system

All vehicles covered by this manual employ a pressurized engine cooling system with thermostatically controlled coolant circulation. An impeller type water pump mounted on the engine block pumps coolant through the engine and radiator. The coolant flows around each cylinder and back to the radiator. Cast-in coolant passages direct coolant around the intake and exhaust ports, near the spark plug areas and the exhaust valve guides.

A wax pellet type thermostat is located in a housing connected to the upper radiator hose. During warm up the closed thermostat prevents coolant from circulating through the radiator. As the engine nears normal operating temperature, the thermostat opens and allows hot coolant to travel through the radiator, where it's cooled before returning to the engine.

The cooling system is sealed by a pressure-type radiator cap, which raises the boiling point of the coolant and increases the cooling efficiency of the radiator. If the system pressure exceeds the cap pressure relief value, the excess pressure in the system forces the spring-loaded valve inside the cap off its seat and allows the coolant to escape through a hose into a coolant reservoir. When the system cools, the excess coolant is automatically drawn from the reservoir back into the radiator.

The coolant reservoir does double duty as both the point at which fresh coolant is added to the cooling system to maintain the proper level and as a holding tank for expelled coolant.

This type of cooling system is known as a closed design because coolant that escapes past the pressure cap is saved and reused.

Heating system

The heating system consists of a blower fan and heater core located in the heater box, the hoses connecting the heater core to the engine cooling system and the heater/air conditioning control head on the dashboard. Hot engine coolant is circulated through the heater core. When the heater mode is activated, a trap door opens to expose the heater box to the passenger compartment. A fan switch on the control head activates the blower motor, which forces air through the core, heating the air.

Air conditioning system

The air conditioning system consists of a condenser mounted in front of the radiator, an evaporator mounted adjacent to the heater core, a compressor mounted on the engine, a filter-drier (accumulator), which contains a high pressure relief valve, and the hoses and lines connecting all of the above components.

A blower fan forces the warmer air of the passenger compartment through the evaporator core (sort of a radiator-in-reverse), transferring the heat from the air to the refrigerant. The liquid refrigerant boils off into low pressure vapor, taking the heat with it when it leaves the evaporator.

2 Antifreeze - general information

Warning: *Don't allow antifreeze to come in contact with your skin or painted surfaces of the vehicle. Rinse off spills immediately with plenty of water. Antifreeze is highly toxic if ingested. Never leave antifreeze lying around in an open container or in puddles on the floor; children and pets are attracted by its sweet smell and may drink it. Check with local authorities about disposing of used antifreeze. Many communities have collection centers, which will see that antifreeze is disposed of safely. Antifreeze is also combustible, so don't store or use it near open flames.* **Caution:** *Never mix green-colored ethylene glycol anti-freeze and orange-colored "DEX-COOL" silicate-free coolant because doing so will destroy the efficiency of the "DEX-COOL" coolant which is designed to last for 100,000 miles or five years.*

The cooling system should be filled with a water/ethylene glycol based antifreeze solution, which will prevent freezing down to at least -20-degrees F, or lower if local climate requires it. It also provides protection against corrosion and increases the coolant boiling point.

The cooling system should be drained, flushed and refilled at least every other year (see Chapter 1). The use of antifreeze solutions for periods of longer than two years is likely to cause damage and encourage the formation of rust and scale in the system. However, 1996 and later models are filled with a new, long-life coolant called "DEX-COOL", with a manufacturer recommended maintenance interval of five years.

Before adding antifreeze, check all hose connections, because antifreeze tends to leak through very minute openings. Engines don't normally consume coolant, so if the level goes down, find the cause and correct it.

The exact mixture of antifreeze-to-water you should use depends on the relative weather conditions. The mixture should contain at least 50 percent antifreeze, but should never contain more than 70 percent antifreeze. Consult the mixture ratio chart on the antifreeze container before adding coolant. Hydrometers are available at most auto parts stores to test the coolant. Use antifreeze that meets the vehicle manufacturer's specifications.

3 Thermostat - check and replacement

Warning: *DO NOT remove the radiator cap, drain the coolant or replace the thermostat until the engine has cooled completely.*

Check

1   Before assuming the thermostat is to blame for a cooling system problem, check the coolant level, drivebelt tension (see Chapter 1) and temperature gauge (or light) operation.
2   If the engine seems to be taking a long time to warm up (based on heater output or temperature gauge operation), the thermostat is probably stuck open. Replace the thermostat with a new one.
3   If the engine runs hot, use your hand to check the temperature of the upper radiator hose. If the hose isn't hot, but the engine is, the thermostat is probably stuck closed, preventing the coolant inside the engine from escaping to the radiator. Replace the thermostat. **Caution:** *Don't drive the vehicle without a thermostat. The computer may stay in open loop and emissions and fuel economy will suffer.*
4   If the upper radiator hose is hot, it means the coolant is flowing and the thermostat is open. Consult the Troubleshooting Section at the front of this manual for cooling system diagnosis.

Replacement

Refer to illustrations 3.8, 3.13a, 3.13b, 3.14a, 3.14b, 3.15, 3.16 and 3.17

5   Disconnect the negative battery cable from the battery and drain the cooling system (see Chapter 1). If the coolant is relatively new or in good condition, save it and reuse it. **Caution:** *On models equipped with the Theft-lock audio system, be sure the lockout feature is turned off before performing any procedure which requires disconnecting the battery.*
6   Remove the air cleaner assembly, if necessary for access (see Chapter 4).
7   On the 3.4L V6, remove the torque strut at the engine.
8   Follow the upper radiator hose to the engine to locate the thermostat cover **(see illustration)**.
9   On all models except those equipped with a 2.5L four-cylinder engine, loosen the hose clamp, then detach the hose from the

**3.8 Thermostat details -
3.4 liter V6 engine**

1   *Upper radiator hose*
2   *Thermostat outlet pipe*

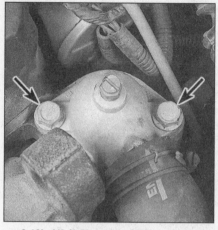

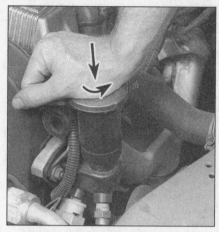

3.13a Quad-4 (2.3L) thermostat housing mounting bolt locations (arrows) - the coolant temperature sensor is mounted on the right side of the assembly

3.13b V6 (2.8L and 3.1L) thermostat housing mounting bolt locations (arrows)

3.14a 2.5L four-cylinder thermostat housing components - first, twist the cap off . . .

...ting. If the hose sticks, grasp it near the end ...ith a pair of adjustable pliers and twist it to ...reak the seal, then pull it off. If the hose is ...d or deteriorated, cut it off and install a new ...e. **Note:** *2.3L Quad-4 engines have two ...oses and a wire harness connected to the ...ermostat cover - remove all of these.*

...) If the outer surface of the large fitting ...at mates with the hose is deteriorated (cor- ...ded, pitted, etc.) it may be damaged further ... hose removal. If it is, the thermostat cover ...ll have to be replaced.

... On the 3.4L V6, relieve the fuel system ...essure (see Chapter 4), then disconnect the ...el lines at the fuel rail.

...2 On the 3.4L V6, remove the heater hose ...acket at the thermostat housing stud, and ...e heater hose at the throttle body.

...3 On all models except those equipped ...th a 2.5L four-cylinder engine, remove the ...lts/nuts and detach the thermostat cover ...ee illustrations). If the cover is stuck, tap it ...th a soft-face hammer to jar it loose. Be ...epared for some coolant to spill as the gas- ...t seal is broken.

... Note how it's installed (which end is fac- ...g up), then remove the thermostat. **Note:** ...n 2.5L engines, the thermostat is removed ...simply twisting the cap off the thermostat ...ousing and pulling up on the bail that ...aches to the thermostat (see illustrations).

... Remove all traces of old gasket material ...d sealant from the housing and cover with ...gasket scraper (see illustration). Clean the ...sket mating surfaces with lacquer thinner ...acetone.

... Some models are equipped with a rub- ...r gasket around the circumference of the ...ermostat (see illustration). If the vehicle ...u are working on is equipped with this type ...thermostat, replace the rubber gasket.

... Install the new thermostat in the hous- ...g. Make sure the correct end faces up - the ...ring end is normally directed into the ...gine (see illustration).

... On 2.3L Quad-4 models, apply a thin, uni- ...m layer of RTV sealant to both sides of the

new gasket and position it on the housing.
19 Install the cover and bolts/nuts. On the 3.4L and 3800 V6 engines, put RTV sealer around the bolt threads. Tighten the bolts to the torque figure listed in this Chapter's Specifications.
20 The remaining steps are the reverse of removal.

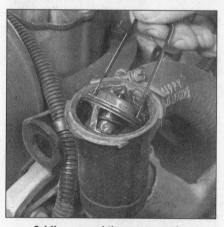

3.14b . . . and then remove the thermostat by the bail

3.16 Some models use a rubber gasket around the circumference of the thermostat

21 Refill the cooling system (see Chapter 1). **Note:** *Be sure to bleed the system of air.*
22 Start the engine and allow it to reach normal operating temperature, then check for leaks and proper thermostat operation (as described in Steps 2 through 4).

3.15 Remove all traces of gasket material (Quad-4 shown)

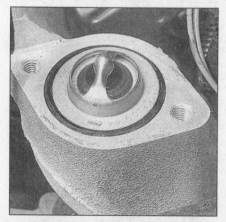

3.17 Position the thermostat in the housing as shown here (V6 shown, others similar)

3

4.1 The motor may be tested by running fused jumper wires directly from the battery to the motor terminals (arrow)

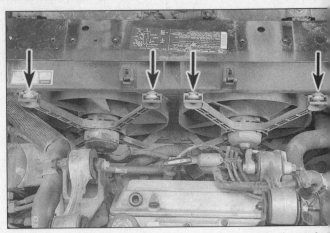

4.6 Cooling fan bolt locations (arrows) - (early models shown, bo locations on later models will vary slightly)

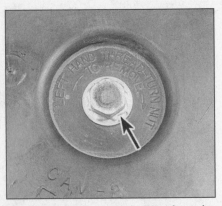

4.7 Remove the motor shaft nut (arrow) - note the left-hand threads

5.2 Do not strip or ruin the coolant drain plug (arrow) - it's made of plastic

5.4a The transaxle cooler lines (arrow) a located in the upper . . .

4   Engine cooling fan - check and replacement

Check

Refer to illustration 4.1

1   To test the fan motor, unplug the electrical connector at the motor and use jumper wires to connect the fan directly to the battery **(see illustration)**. If the fan still doesn't work, replace the motor.

2   If the motor tested OK, the fault lies in the coolant temperature switch, the relay or the wiring which connects the components. Carefully check all wiring and connections. If no obvious problems are found, further diagnosis should be done by a dealer service department or a repair shop.

Replacement

Refer to illustrations 4.6 and 4.7

3   Disconnect the negative battery cable from the battery. **Caution:** *On models equipped with the Theftlock audio system, be sure the lockout feature is turned off before performing any procedure which requires disconnecting the battery.*

4   Remove the air cleaner assembly (see Chapter 4). Remove the engine torque struts (see Chapter 2).

5   Insert a small screwdriver into the connector to lift the lock tab and unplug the fan wire harness.

6   Unbolt the fan assembly **(see illustration)**, then carefully lift it out of the engine compartment.

7   To detach the fan from the motor, remove the motor shaft nut **(see illustration)**. **Note:** *This nut may have left-hand threads. Check the hub of the fan for marks that may indicate which way to loosen the nut.*

8   To remove the fan motor from the bracket, remove the mounting bolts/nuts.

9   Installation is the reverse of removal.

5   Radiator - removal and installation

Refer to illustrations 5.2, 5.4a, 5.4b, 5.6, 5.7 and 5.9

Warning: *Wait until the engine is completely cool before beginning this procedure.*

Removal

1   Disconnect the negative battery cable from the battery. **Caution:** *On models equipped with the Theftlock audio system, be sure the lockout feature is turned off before performing any procedure which requires disconnecting the battery.*

2   Drain the cooling system (see Chapter 1). If the coolant is relatively new or good condition, save it and reuse it. **Note:** *careful not to ruin the plastic drain plug wh removing it* **(see illustration)**.

3   Remove the air cleaner assembly (s Chapter 4).

4   If the vehicle is equipped with an aut matic transaxle, disconnect the cooler lin from the radiator **(see illustrations)**, then c the ends to prevent excessive fluid loss a

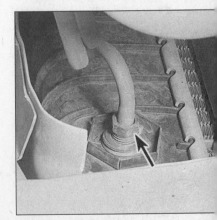

5.4b . . . and lower left corners of the radiator (viewed from below)

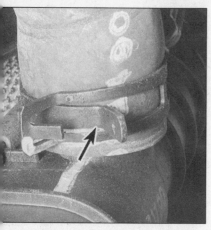

5.6 Use a pair of adjustable pliers to squeeze the ends of the hose clamp (arrow) and slide the clamp down the hose

5.7 The coolant reservoir hose (arrow) connects to the right side of the radiator

mounting panel **(see illustration)**.

10   Remove the mounting panel and carefully lift out the radiator. Don't spill coolant on the vehicle or scratch the paint.

11   With the radiator removed, it can be inspected for leaks and damage. If it needs repair, have a radiator shop or dealer service department perform the work, as special techniques are required.

12   Bugs and dirt can be removed from the radiator with compressed air and a soft brush. Don't bend the cooling fins as this is done.

13   Check the radiator mounts for deterioration and make sure there's nothing in them when the radiator is installed.

Installation

14   Installation is the reverse of the removal procedure.

15   After installation, fill the cooling system with the proper mixture of antifreeze and water. Be sure to bleed the system of air (see Chapter 1).

16   Start the engine and check for leaks. Allow the engine to reach normal operating temperature, indicated by the upper radiator hose becoming hot. Recheck the coolant level and add more if required.

17   If you're working on an automatic transaxle equipped vehicle, check and add fluid as needed.

5.9 Radiator mounting bolt locations (arrows) (early models shown, bolt locations on later models will vary slightly)

ntamination. Use a drip pan to catch pilled fluid.

Remove the engine cooling fan assem- y (see Section 4).

Loosen the hose clamps, then detach e radiator hoses from the fittings **(see illus- ation)**. If they're stuck, grasp each hose ear the end with a pair of adjustable pliers d twist it to break the seal, then pull it off -

be careful not to distort the radiator fittings! If the hoses are old or deteriorated, cut them off and install new ones.

7   Disconnect the reservoir hose from the radiator neck **(see illustration)**.

8   Plug the lines and fittings to prevent the spillage of coolant or ATF as the radiator is removed.

9   Remove the bolts from the radiator

6   Coolant reservoir - removal and installation

Refer to illustrations 6.3 and 6.4

1   Detach the hose from the windshield washer fluid reservoir cap.

2   Disconnect the radiator overflow hose from the top of the radiator.

3   Remove the front fender upper rail, if necessary **(see illustration)**.

4   Remove the mounting bolts and lift the coolant reservoir from the vehicle **(see illustration)**.

5   Installation is the reverse of removal.

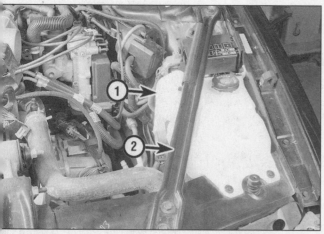

6.3 3.4 liter V6 engine coolant reservoir details

1   *Coolant reservoir*   2   *Front fender upper rail*

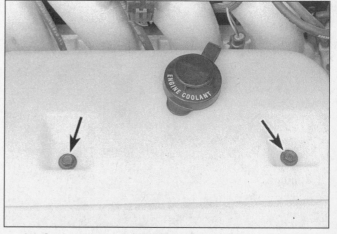

6.4 Coolant reservoir (typical) - remove the mounting bolts (arrows) and lift the reservoir out

7.4 If the pump is leaking through the vent hole, stains will form below the shaft (arrow) - pump removed for clarity

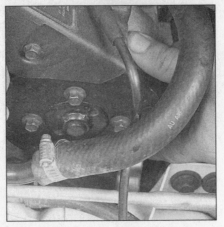

7.5 Rock the pulley back and forth to check for bearing play

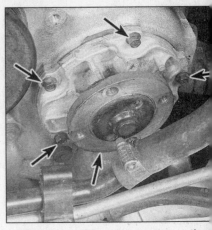

8.10a Typical V6 water pump mounting details (except 3800) - the mounting bolts are located around the perimeter of the pump

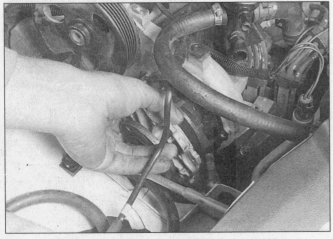

8.10b Pull the pump off - if it's stuck, tap it with a soft-face hammer

8.10c Water pump mounting details - 3800 V6

7 Water pump - check

Refer to illustrations 7.4 and 7.5

1   A failure in the water pump can cause serious engine damage due to overheating.
2   There are three ways to check the operation of the water pump while it's installed on the engine. If the pump is defective, it should be replaced with a new or rebuilt unit.
3   With the engine running at normal operating temperature, squeeze the upper radiator hose. If the water pump is working properly, a pressure surge should be felt as the hose is released. **Warning:** *Keep your hands away from the fan blades!*
4   Water pumps are equipped with weep or vent holes. If a failure occurs in the pump seal, coolant will leak from the hole **(see illustration)**. In most cases you'll need a flashlight and mirror to find the hole on the under side of the water pump to check for leaks.
5   If the water pump shaft bearings fail there may be a howling sound coming from the drivebelt area while the engine is running.

Shaft wear can be felt if the water pump pulley is rocked up-and-down **(see illustration)**. Don't mistake drivebelt slippage, which causes a squealing sound, for water pump bearing failure.

8 Water pump - removal and installation

Refer to illustrations 8.10a, 8.10b, 8.10c and 8.15
Warning: *Wait until the engine is completely cool before beginning this procedure.*

Removal

1   Disconnect the negative battery cable from the battery. **Caution:** *On models equipped with the Theftlock audio system, be sure the lockout feature is turned off before performing any procedure which requires disconnecting the battery.*
2   Drain the cooling system (see Chapter 1). If the coolant is relatively new or in good condition, save it and reuse it. **Note:** *On*

Quad-4 models, disconnect the heater hos from the thermostat housing for more con plete coolant drain.
3   Remove the air cleaner if necessary.
4   Remove the windshield washer reservo and the coolant recovery reservoir (see Se tion 6), as necessary.
5   Remove the serpentine belt and te sioner (see Chapter 1).
6   Remove the alternator and its bracke and the air conditioner compressor, as ne essary.

Quad-4 equipped models only

7   Remove the exhaust manifold he shields (see Chapter 2C) and the exhau manifold.
8   Remove the radiator outlet pipe-to-rad ator pump cover bolts, leaving the lower rad ator hose attached. Pull down on the radiat outlet pipe to disengage it from the wat pump and detach the pipe from the oil pa and transaxle.
9   Remove the water pump cover-to-cyli der block bolts.

8.15 On Quad-4 models, lubricate the splines (arrow) with grease

9.1a Coolant temperature sending unit location (arrow) - 2.5 liter four cylinder engine - the sending unit on the 2.2 liter engine is located on the engine's coolant outlet

9.1b On 2.8, 3.1 and 3.4 liter V6 engines, the coolant temperature sensor (arrow) is adjacent to the EGR pipe (On the 3100 V6, the coolant temperature sensor is located next to the ignition control module)

All models

10   Remove the bolts/nuts and detach the water pump from the engine (see illustrations).

11   Clean the fastener threads and any threaded holes in the engine to remove corrosion and sealant.

12   Compare the new pump to the old one to make sure they're identical.

13   Remove all traces of old gasket material from the engine with a gasket scraper.

14   Clean the engine and water pump mating surfaces with lacquer thinner or acetone.

Installation

15   Carefully attach the pump and gasket to the engine and start the bolts/nuts finger tight. **Note:** *On Quad-4 models, lubricate the splines of the water pump drive* (see illustration) *with chassis grease prior to installation. Lubricate the O-ring on the radiator outlet pipe with antifreeze solution before installing.*

16   Tighten the fasteners in 1/4-turn increments to the torque figure listed in this Chapter's Specifications. Don't overtighten them or the pump may be distorted. **Note:** *On Quad-4 models, tighten the fasteners in this order:*

a) *Pump-to-chain housing*
b) *Pump cover-to-pump assembly*
c) *Cover-to-block, bottom bolt first*
d) *Radiator outlet pipe-to-water pump cover*

17   Reinstall all parts removed for access to the pump.

18   Refill the cooling system (see Chapter 1). Run the engine and check for leaks.

9   Coolant temperature sending unit or Low Coolant sensor - check and replacement

Refer to illustrations 9.1a, 9.1b and 9.1c

Warning: *Wait until the engine is completely cool before beginning this procedure.*

1   The coolant temperature indicator sys-

9.1c Location of the sending unit (arrow) on the 3800 engine

tem is composed of a light or temperature gauge mounted in the instrument panel and a coolant temperature sending unit mounted on the engine (see illustrations). (Some vehicles have more than one sending unit, but the one used for the indicator system has only one wire). **Note:** *Refer to* **illustration 3.13a** *for the location of the coolant temperature sending unit on the Quad-4 engine.* On some models a low coolant warning system is also used. This system is composed of a warning light in the instrument panel and a low coolant sensor mounted on the side of the radiator.

2   If the light or gauge indicates the engine is overheating or that the coolant is low, check the coolant level in the system and then make sure the wiring between the light or gauge and the sending unit is secure and all fuses are intact.

3   When the ignition switch is turned on and the starter motor is turning, the indicator lights (if equipped) should be on (overheated engine or low coolant indication).

4   If either of the lights isn't on, the bulb may be burned out, the ignition switch may be faulty or the circuit may be open. Test the circuit by grounding the wire to the sending unit

while the ignition is on (engine not running for safety). If the gauge deflects full scale or the light comes on, replace the sending unit.

5   As soon as the engine starts, the light should go out and remain out unless the engine overheats or, in the case of the low coolant sensor, the coolant level becomes low. Failure of the light to go out may be due to a grounded wire between the light and the sending unit, a defective sending unit or a faulty ignition switch. If the problem is with the temperature sender, check the coolant to make sure it's the proper type. Plain water may have too low a boiling point to activate the sending unit.

6   If the temperature sending unit must be replaced, simply unscrew it from the engine and install the replacement. Use a light coat of sealant on the threads. Make sure the engine is cool before removing the defective sending unit. If the low coolant sensor must be replaced, drain some coolant from the radiator (see Chapter 1), disconnect the connector then unsnap the sensor from it's mount on the radiator.

7   Check the coolant level after the replacement has been installed and refill as necessary.

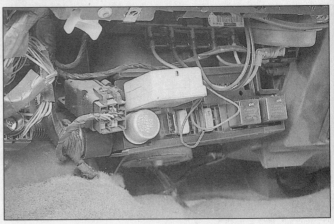

10.3 Detach the electrical connector and remove the mounting screws from the blower assembly

1   Electrical connector   2   Blower mounting screw

10.4 The convenience center is located at the passenger side knee bolster (1996 and earlier models only)

10   Heater and air conditioner blower motor - removal and installation

Refer to illustrations 10.3, 10.4 and 10.6

1   Disconnect the cable from the negative battery terminal. **Caution:** *On models equipped with the Theftlock audio system, be sure the lockout feature is turned off before*

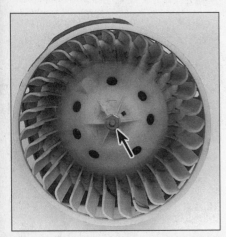

10.6 Remove the nut (arrow) and slip the fan off the motor shaft

performing any procedure which requires disconnecting the battery.

2   Working in the passenger compartment, remove the panel below the glove compartment.

3   Pull down the carpet as necessary so that you can see the blower motor, then disconnect the wires from the blower motor **(see illustration)**.

4   Remove the convenience center rear screws, loosen the front screw and slide the convenience center out **(see illustration)**.

5   Remove the blower motor mounting bolts and separate the motor/fan assembly from the housing.

6   Remove the retaining nut, if so equipped **(see illustration)** and slide the blower fan off the motor shaft. **Note:** *If the blower fan is pressed onto the shaft, it can be removed by cutting through the plastic shaft sleeve with a "hot-knife" until it splits, or by pressing it off.*

7   Installation is the reverse of the removal procedure.

11   Heater core - removal and installation

Refer to illustrations 11.6, 11.7, 11.9a and 11.9b

1   Disconnect the cable from the negative

terminal of the battery. **Caution:** *On models equipped with the Theftlock audio system, be sure the lockout feature is turned off before performing any procedure which requires disconnecting the battery.*

2   Raise the front of the vehicle and support it securely on jackstands. Apply the parking brake and block the rear wheels to keep the vehicle from rolling off the jackstands.

3   Drain the cooling system (see Chapter 1).

4   Remove the drain hose from the bottom of the heater case.

5   Working in the passenger compartment, remove the panels from under the dash and the steering column trim cover. Also remove the center console if equipped.

6   Remove the heater outlet duct and glove box **(see illustration)**.

7   Remove the heater core cover **(see illustration)**.

8   Disconnect the heater hoses from the heater core. **Note:** *The heater hoses and clamps are accessible from inside the engine compartment on the firewall. On the 3100, you may have to rotate the engine forward to gain access (see Chapter 1). On the 3.4 liter, you may have to remove the plenum, the fuel lines, the exhaust crossover pipe and the transmission dipstick tube (see Chapter 4).*

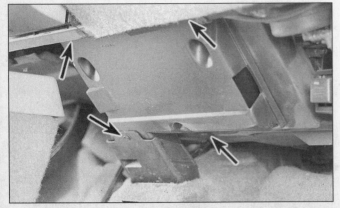

11.6 Heater/air conditioning duct mounting screws (arrows)

11.7 Heater core cover mounting screws (arrows)

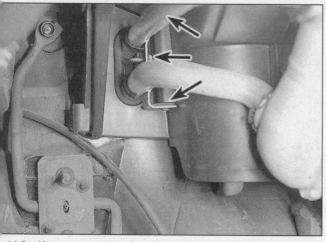

11.9a Heater core mounting clip and screw locations (arrows)

11.9b Handle the heater core by the upper and lower edges (arrows)

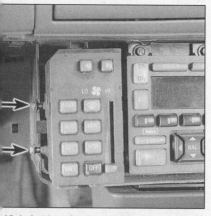

12.4 Automatic air conditioning control assembly mounting screw locations (arrows)

Remove the heater core (see illustra-
ons).
) Installation is the reverse of removal.
efill the cooling system (see Chapter 1).

2 Heater and air conditioner control assembly - removal and installation

efer to illustration 12.4
Disconnect the cable from the negative
rminal of the battery. Caution: On models
quipped with the Theftlock audio system, be
re the lockout feature is turned off before
erforming any procedure which requires dis-
nnecting the battery.
Remove the trim plates from the dash
ee Chapter 11).
If equipped, pull the control knobs off
e face of the control assembly.
Remove the screws on each side of the
ntrol assembly (see illustration).
Disconnect the electrical connectors at
e rear of the control assembly.
Installation is the reverse of removal.

13 Air conditioning system - check and maintenance

Warning: The air conditioning system is
under high pressure. DO NOT loosen any
hose or line fittings or remove any components until after the system has been discharged. Air conditioning refrigerant should
be properly discharged into an approved
container at a dealer service department or an
automotive air conditioning repair facility.
Always wear eye protection when disconnecting air conditioning system fittings.

Check

1   The following maintenance checks
should be performed on a regular basis to
ensure the air conditioner continues to operate at peak efficiency.

  a) Check the compressor drivebelt. If it's
     worn or deteriorated, replace it (see
     Chapter 1).
  b) Check the system hoses. Look for
     cracks, bubbles, hard spots and deterio-
     ration. Inspect the hoses and all fittings
     for oil bubbles and seepage. If there's
     any evidence of wear, damage or leaks,
     replace the hose(s). Warning: Do not
     replace air conditioning hoses until the
     system has been discharged by the
     dealer or an air conditioning shop.
  c) Inspect the condenser fins for leaves,
     bugs and other debris. Use a "fin comb"
     or compressed air to clean the con-
     denser.
  d) Check to make sure the radiator/con-
     denser cooling fan comes on with the air
     conditioning on.
  e) Make sure the system has the correct
     refrigerant charge.

2   It's a good idea to operate the system
for about 10 minutes at least once a month,
particularly during the winter. Long term non-
use can cause hardening, and subsequent
failure, of the seals.
3   Because of the complexity of the air
conditioning system and the special equip-

ment necessary to service it, in-depth troubleshooting and repairs are not included in
this manual (refer to the Haynes Automotive
Heating & Air Conditioning manual), However,
simple checks and component replacement
procedures are provided in this Chapter.
4   The most common cause of poor cooling is simply a low system refrigerant charge.
If a noticeable loss of cool air output occurs,
one of the following quick checks may help
you determine if the refrigerant level is low.
5   Warm the engine up to normal operating
temperature.
6   Place the air conditioning temperature
selector at the coldest setting and put the
blower at the highest setting. Open the doors
(to make sure the air conditioning system
doesn't cycle off as soon as it cools the passenger compartment).
7   With the compressor engaged - the
compressor clutch will make an audible click
and the center of the clutch will rotate - feel
the orifice tube located adjacent to the right
front frame rail near the radiator.
8   If a significant temperature drop is
noticed, the refrigerant level is probably okay.
Further inspection of the system is beyond
the scope of the home mechanic and should
be left to a professional.
9   If the inlet line has frost accumulation or
feels cooler than the accumulator surface,
the refrigerant charge is low.
10   If a low refrigerant charge is suspected,
take your vehicle to a dealer or automotive air
conditioning shop for servicing by a certified
air conditioning technician.

14 Air conditioning accumulator - removal and installation

Refer to illustrations 14.4 and 14.6
Warning: The air conditioning system is
under high pressure. DO NOT loosen any
hose or line fittings or remove any components until after the system has been discharged. Air conditioning refrigerant should

3

14.4 The accumulator is usually located in the left front corner of the engine compartment below the air cleaner

14.6 Remove the air cleaner housing to access the A/C accumulator mounting bolt (arrow)

be properly discharged into an approved container at a dealer service department or an automotive air conditioning repair facility. Always wear eye protection when disconnecting air conditioning system fittings.

1   Have the system discharged (see Warning above).

2   Disconnect the negative battery cable from the battery. **Caution:** *On models equipped with the Theftlock audio system, be sure the lockout feature is turned off before performing any procedure which requires disconnecting the battery.*

3   Remove the air cleaner and intake tube assembly (see Chapter 4).

4   Disconnect the refrigerant lines from the accumulator **(see illustration)**. On the 3800, remove the pressure switch from the accumulator if necessary. **Caution:** *Use a back-up wrench when removing the refrigerant lines to prevent twisting the tubing.*

5   Plug the open fittings to prevent entry of dirt and moisture.

6   Loosen or remove the mounting bracket and bolts **(see illustration)** and remove the accumulator.

7   If a new accumulator is being installed, remove the Schrader valve and pour the oil out into a measuring cup, noting the amount. Add fresh refrigerant oil to the new accumulator equal to the amount removed from the old unit, plus one ounce. **Caution:** *Air conditioner systems equipped with R-134a refrigerant use a special oil designed specifically for use in such systems.*

8   Installation is the reverse of removal.

9   Have the system evacuated, recharged and leak tested by the shop that discharged it.

---

15   Air conditioning compressor - removal and installation

Refer to illustrations 15.7, 15.9a and 15.9b
Warning: *The air conditioning system is*

under high pressure. DO NOT loosen any hose or line fittings or remove any components until after the system has been discharged. Air conditioning refrigerant should be properly discharged into an approved container at a dealer service department or an automotive air conditioning repair facility. Always wear eye protection when disconnecting air conditioning system fittings.

Note: *The accumulator (see Section 14) should be replaced whenever the compressor is replaced.*

1   Have the system discharged (see Warning above).

2   Disconnect the negative battery cable from the battery. **Caution:** *On models equipped with the Theftlock audio system, be sure the lockout feature is turned off before performing any procedure which requires disconnecting the battery.*

3   Remove the coolant recovery reservoir (see Section 6).

4   Set the parking brake and block the rear wheels. Raise the front of the vehicle and support it securely on jackstands.

5   Remove the right lower splash shield.

6   Refer to Chapter 1 and remove th drivebelt (and oil filter, if necessary for clea ance). On the 3.4 liter, remove the engin torque strut and bracket. On the 3.4 liter the 3800, remove the cooling fan assembly.

7   Disconnect the compressor clutc wiring harness **(see illustration)**.

8   Disconnect the refrigerant lines from th rear of the compressor. Plug the open fitting to prevent entry of dirt and moisture.

9   Unbolt the compressor from the moun ing brackets and lift it out of the vehicle **(s illustrations)**.

10   If a new compressor is being installe follow the directions with the compress regarding the draining of excess oil prior installation.

11   The clutch may have to be transferre from the original to the new compressor.

12   Installation is the reverse of remov Replace all O-rings with new ones speci cally made for air conditioning system in yc car and lubricate them with the appropria refrigerant oil.

13   Have the system evacuated, recharg and leak tested by the shop that discharged

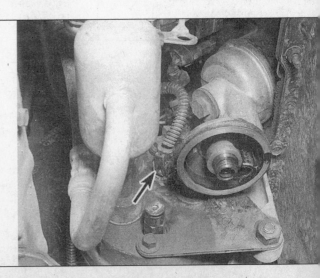

15.7 Disconnect the compressor clutch wiring harness (arrow) - oil filter removed for clarity (V6 shown)

15.9a A/C compressor front and rear mounting bolts

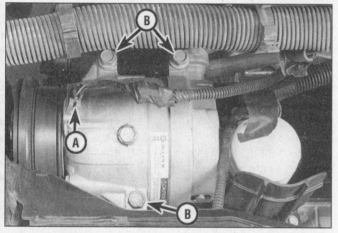

15.9b On 3100 engines disconnect the electrical connector (A) from the air conditioning compressor and remove the mounting bolts (B)

16.8 Remove the bolt (arrow) to disconnect the refrigerant lines

16 Air conditioning condenser - removal and installation

Refer to illustrations 16.8 and 16.10
Warning: *The air conditioning system is under high pressure. DO NOT loosen any hose or line fittings or remove any components until after the system has been discharged. Air conditioning refrigerant should be properly discharged into an approved container at a dealer service department or an automotive air conditioning repair facility. Always wear eye protection when disconnecting air conditioning system fittings.*
Note: *The accumulator (see Section 14) should be replaced whenever the condenser is replaced.*

Removal

1   Have the air conditioning system discharged (see Warning above).
2   Disconnect the negative cable from the battery. **Caution:** *On models equipped with the Theftlock audio system, be sure the lockout feature is turned off before performing*

any procedure which requires disconnecting the battery.
3   Remove the air cleaner assembly and the air cleaner duct(s) (see Chapter 4). **Note:** *On the 2.5L engine, remove the air intake resonator mounting bolt.*
4   Remove the torque struts (upper engine mounts) from the engine (see Chapter 2).
5   Remove the coolant reservoir (see Section 6).
6   Remove the cooling fan (see Section 4).
7   Remove the radiator (see Section 5).
8   Disconnect the refrigerant lines from the condenser **(see illustration)**. Plug the lines to keep dirt and moisture out.
9   Remove the radiator mounting panel from above the condenser **(see illustration 5.9)**.
10 Remove the bolts and clamps and lift out the condenser **(see illustration)**.
11 If the original condenser will be reinstalled, plug the line fittings to prevent oil from draining out.

Installation

12 If a new condenser is being installed, pour one ounce of refrigerant oil into it prior to installation.

13 Reinstall the components in the reverse order of removal. Be sure the rubber pads are in place under the condenser.
14 Have the system evacuated, recharged and leak tested by the shop that discharged it.

16.10 The condenser has rubber mounts at the corners (arrow) - be sure they're in place properly during reassembly

Notes

Chapter 4
Fuel and exhaust systems

Contents

Specifications

Fuel pressure

Throttle Body Injection (TBI)	26 to 32 psi
Port Fuel Injection (PFI)/Sequential Fuel Injection (SFI)	
3800 engine	
1998 and later	48 to 55 psi
All other engines	40 to 47 psi

Torque specifications

	Ft-lbs (unless otherwise indicated)
TBI throttle body mounting bolts	18
PFI/SFI throttle body mounting bolts	
1988	11
1989 on	18
Plenum-to-intake manifold bolts	
3800	
1995 through 1997	132 in-lbs
1998 and later	89 in-lbs
All other V6 engines	16 to 18
Injector retainer screws (2.2L 4-cylinder)	31 in-lbs

1 General information

The fuel system consists of a fuel tank, an electric fuel pump, a fuel pump relay, an air cleaner assembly and either a Throttle Body Injection (TBI) system, or a Port Fuel Injection (PFI) system or a Sequential Fuel Injection (SFI) system. The basic differences between the three types of fuel injection system are in the number of injectors, the location of injectors and the sequence in which they inject fuel.

Throttle Body Injection (TBI) system

The throttle body system utilizes one injector, centrally mounted in a carburetor-like housing. The injector is an electrical solenoid, with fuel delivered to the injector at a constant pressure level. To maintain the fuel pressure at a constant level, excess fuel is returned to the fuel tank.

A signal from the ECM opens the solenoid, allowing fuel to spray through the injector into the throttle body. The amount of time the injector is held open by the ECM determines the fuel/air mixture ratio. This system is used on some Lumina models through 1992.

Port Fuel Injection (PFI) system

The port system utilizes four or six injectors similar in operation to the throttle body injector and the fuel/air ratio is controlled in the same manner. Instead of a single injector mounted in a centrally located throttle body, one injector is installed above each intake port. The throttle body serves only to control the amount of air passing into the system. Because each cylinder is equipped with an injector mounted immediately adjacent to the intake valve, much better control of the fuel/air mixture ratio is possible.

This system is used on various Lumina, Regal, Cutlass and Monte Carlo models through 1994.

Sequential Fuel Injection (SFI) system

The sequential system is like the port fuel injection system in that it has one injector per cylinder. However, in the sequential system the opening of each injector is synchronized with the opening of each intake valve.

This system is used on all Regal models with the 3800 engine, on 1993 California-specification Regals with the 3.1 liter engine, all 1994 models with the 3100 or 3.4 liter engines, and all 1995 and later models.

2   Fuel pressure relief procedure

Refer to illustration 2.4

Warning: *Gasoline is extremely flammable, so take extra precautions when you work on any part of the fuel system. Don't smoke or allow open flames or bare light bulbs near the work area, and don't work in a garage where a natural gas-type appliance (such as a water heater or a clothes dryer) with a pilot light is present. Since gasoline is carcinogenic, wear latex gloves when there's a possibility of being exposed to fuel, and, if you spill any fuel on your skin, rinse it off immediately with soap and water. Mop up any spills immediately and do not store fuel-soaked rags where they could ignite. When you perform any kind of work on the fuel system, wear safety glasses and have a Class B type fire extinguisher on hand.*

Caution: *On models equipped with the Theftlock audio system, be sure the lockout feature is turned off before performing any procedure which requires disconnecting the battery.*

Note: *After the fuel pressure has been relieved, it's a good idea to use a shop towel around any fuel connection to absorb the residual fuel that may leak out when servicing the fuel system.*

1   Before servicing any fuel system component, you must relieve the fuel pressure to minimize the risk of fire or personal injury.

2   Remove the fuel filler cap - this will relieve any pressure built up in the tank.

3   On models with Throttle Body Injection, remove the fuel pump fuse and run the engine until it stalls, then crank the engine for three seconds. Disconnect the cable from the negative terminal of the battery.

4   On models with Port or Sequential Fuel Injection, use one of the two following methods:

a) *Disconnect the fuel pump electrical connector at the fuel tank (lower the tank if necessary). Run the engine until it stops, then engage the starter again for another three seconds. With the ignition turned Off, reconnect the fuel tank electrical connector, then disconnect the cable from the negative terminal of the battery.*

b) *Loosen the fuel filler cap. Attach a fuel pressure gauge to the Schrader valve on the fuel rail (see illustration). Place the gauge bleeder hose in an approved fuel container. Open the valve on the gauge to relieve pressure, then disconnect the cable from the negative terminal of the battery.*

5   Unless this procedure is followed before servicing fuel lines or connections, fuel spray (and possible injury) may occur.

3   Fuel pump/fuel pressure - testing

Warning: *Gasoline is extremely flammable, so extra precautions must be taken when working on any part of the fuel system. See*
the **Warning** *in Section 2.*

Note: *In order to perform the fuel pressure test, you will need to obtain a fuel pressure gauge and adapter set for the fuel injection system being tested.*

Preliminary inspection (all vehicles)

1   Should the fuel system fail to deliver the proper amount of fuel, or any fuel at all, to the fuel injection system, inspect it as follows.

2   Always make certain there is fuel in the tank.

3   With the engine running, inspect for leaks at the threaded fittings at both ends of the fuel line (see Chapter 1). Tighten any loose connections. Inspect all hoses for flattening or kinks which would restrict the flow of fuel.

Pressure check

4   Relieve the fuel system pressure (see Section 2).

Models with Port Fuel Injection (PFI) or Sequential Fuel Injection (SFI)

5   Install a fuel pressure gauge at the Schrader valve on the fuel rail **(see illustration 2.4).**

6   Turn the ignition switch ON with the air conditioning off. The fuel pump should run for about two seconds - note the reading on the gauge. After the pump stops running the pressure should hold steady. It should be within the range listed in this Chapter's Specifications.

7   Start the engine and let it idle at normal operating temperature. The pressure should be lower by 3 to 10 psi. If all the pressure readings are within the limits listed in this Chapter's Specifications, the system is operating properly.

8   If the pressure did not drop by 3 to 10 psi after starting the engine, apply 10 inches of vacuum to the pressure regulator. If the pressure drops, repair the vacuum source to the regulator. If the pressure does not drop, replace the regulator.

9   If the fuel pressure is not within specifications, check the following:

a) *If the pressure is higher than specified, check for a faulty regulator or a pinched or clogged fuel return hose or pipe.*

b) *If the pressure is lower than specified:*

c) *Inspect the fuel filter - make sure it's not clogged.*

d) *Look for a pinched or clogged fuel hose between the fuel tank and the fuel rail.*

e) *Check the pressure regulator for a malfunction.*

f) *Look for leaks in the fuel line.*

g) *Look for a pinched, broken or disconnected regulator vacuum hose.*

h) *Check for leaking injectors.*

i) *Check the in-tank fuel pump check valve.*

10   After the testing is done, relieve the fuel

2.4 Remove the cap to the Schrader valve for connecting a fuel pressure gauge

pressure (see Section 2) and remove the fuel pressure gauge.

11   If there are no problems with any of the above-listed components, check the fuel pump (see below).

Models with Throttle Body Injection (TBI)

Refer to illustration 3.14

12   Relieve fuel system pressure (see Section 2).

13   Install a fuel pressure gauge between the fuel feed hose and the inlet fitting of the throttle body. The fuel hoses will be under high pressure during this check, so make your connections secure, with no possibility of leaks.

14   With the ignition OFF, use a fused jumper wire from a 12 volt source, jump the fuel pump test terminal and note the pressure reading **(see illustration)**.

15   If the pressure is within the limits listed in this Chapter's Specifications, no further testing is necessary.

16   If the pressure was higher than specified, check for a restricted fuel return line. If the line is OK, then replace the pressure regulator.

17   If the pressure was less than specified, slowly pinch the hose between the gauge and the TBI unit and note the pressure. If the pressure goes above 9 psi, then replace the pressure regulator. If there is not any pressure, then check for a plugged fuel filter, plugged fuel pump inlet filter or a restricted fuel line.

18   After testing is completed, relieve the fuel pressure and remove the fuel pressure gauge.

19   If no problems are found with any of the above listed components, check the fuel pump (see below).

Fuel pump check

Refer to illustration 3.22a and 3.22b

20   If you suspect a problem with the fuel pump, verify the pump actually runs. Have an assistant turn the ignition switch to On - you should hear a brief whirring noise as the

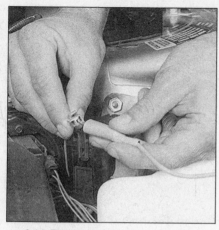

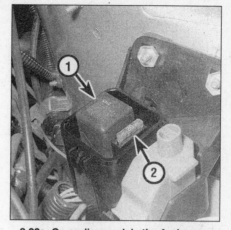

3.14 The fuel pump test terminal is located near the left front strut tower

3.22a On earlier models the fuel pump fuse and fuel pump relay are in the holder near the left front strut tower

1   Fuel pump relay
2   Fuel pump fuse

3.22b On later models the ECM or PCM BAT (fuel pump) fuse (A) and the fuel pump relay (B) locations are clearly marked on the cover of the fuse/relay control box in the engine compartment

pump comes on and pressurizes the system. Have the assistant start the engine. This time you should hear a constant whirring sound from the pump (but it's more difficult to hear with the engine running).

21   If the pump does not come on (makes no sound), proceed to the next Step.

22   Check the fuel pump fuse located in the engine compartment **(see illustration)**. If the fuse is good, see fuel pump relay check. If the fuse is blown, replace the fuse and see if the pump works. If the pump still does not work, go to the next step.

23   With the ignition OFF, apply 12 volts to the fuel pump test terminal **(see illustration 3.14)** and listen for the fuel pump running.

24   If the pump runs, then check the fuel pump relay. If the pump does not run, check for an open circuit between the relay and the fuel pump.

Fuel pump relay check

25   Make sure the fuel pump is good. See fuel pump check.

26   With the ignition off, disconnect the fuel pump relay **(see illustration 3.22a and 3.22b)**.

27   Connect a test light between the orange wire at the connector and the ground wire

(black, black/white, or tan/white, depending on the model). Turn the ignition on. The test light should come on. If it doesn't, check for an open or a bad connection in either the orange or the ground wire, and check that the ground wire is properly grounded.

28   With the ignition off, connect the test light between the dark green/white wire and ground. Make sure the ignition has been off for at least ten seconds, then turn the ignition on. The test light should come on for two seconds. If it does, the relay is bad. If it does not, the problem is in the dark green white wire, it's connection to the PCM, or the PCM itself.

4   Fuel lines and fittings - repair and replacement

Refer to illustrations 4.4a, 4.4b, 4.5, 4.6, 4.16, 4.18 and 4.19

Warning: *Gasoline is extremely flammable, so extra precautions must be taken when working on any part of the fuel system. See*

the **Warning** in Section 2.

1   Always relieve the fuel pressure before servicing fuel lines or fittings (see Section 2).

2   The fuel feed, return and vapor lines extend from the fuel tank to the engine compartment. The lines are secured to the underbody with clip and screw assemblies. These lines must be occasionally inspected for leaks, kinks and dents.

3   If evidence of dirt is found in the system or fuel filter during disassembly, the line should be disconnected and blown out. Check the fuel strainer on the fuel gauge sending unit (see Section 7) for damage and deterioration.

Quick-connect fittings

4   Beginning in model year 1989, "quick-connect fittings" were adopted for certain fuel line connections. Quick-connect fittings come in two varieties. Some have a plastic collar while others have a metal collar **(see illustrations)**. The kind your car is equipped with will depend on the model.

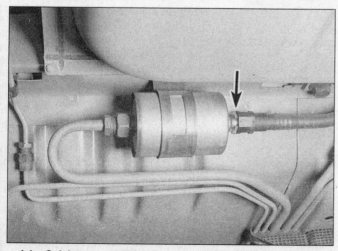

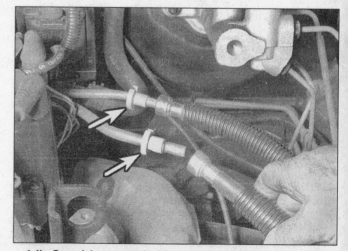

4.4a Quick-connect fitting (arrow) - plastic collar-type shown

4.4b On quick-connect fuel lines, use a special tool and push them into the connector (arrows) to separate the fuel lines

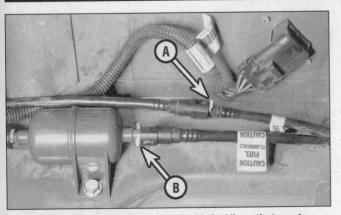

4.5 Some models are equipped with fuel lines that can be disconnected by pinching the tabs and separating each connector

A   *Fuel return line*   B   *Fuel feed line*

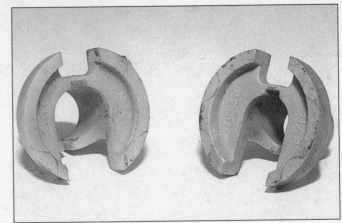

4.6 3/8 and 5/16-inch fuel line disconnect tools

5   To disconnect a plastic quick-connect fitting, simply squeeze the retaining tabs and pull the fitting apart **(see illustration)**.
6   To disconnect a metal quick-connect fitting, a special tool (fuel line separator) is required **(see illustration)**. Slide back the quick-connect fitting's dust cover, then twist the fitting to dislodge any dirt. Use compressed air to blow any dirt away. Install the special tool on the male pipe end and push it in to release the fitting.
7   To reconnect the quick-connect fitting, apply a few drops of clean engine oil to the male pipe end, then push both sides of the fitting together until the retaining tabs snap into place. Make sure the fitting is securely snapped into place by tugging on each end.

Steel tubing

8   If replacement of a fuel line or emission line is called for, use original equipment replacement parts from the dealer parts department.
9   Don't use copper or aluminum tubing to replace steel tubing. These materials cannot withstand normal vehicle vibration.
10   Because fuel lines used on fuel-injected vehicles are under high pressure, they require special consideration.

11   Most fuel lines have threaded fittings with O-rings. Any time the fittings are loosened to service or replace components:

a) Use a backup wrench while loosening and tightening the fittings.
b) Check all O-rings for cuts, cracks and deterioration. Replace any that appear worn or damaged.
c) If the lines are replaced, always use original equipment parts, or parts that meet the GM standards specified in this Section.

Nylon fuel lines

12   Steel or rubber fuel lines can sometimes be replaced with nylon fuel lines. These are designed so that they can be used in fuel injection systems and typically come with a 3/8" inner diameter (for fuel feed) or a 5/16" inner diameter (for fuel return).
13   New nylon fuel lines are flexible enough to be formed around slight bends. If bent too sharply, however, they will kink. Old nylon lines that are to be re-used are less pliable that new ones and are liable to kink more easily, so beware.

Rubber hose

14   When rubber hose is used to replace a

metal line, use reinforced, fuel resistant hose with the word "Fluoroelastomer" imprinted on it. Ask a dealer parts department for replacement parts. Hose(s) not clearly marked like this could fail prematurely and could fail to meet Federal emission standards. Hose inside diameter must match line outside diameter.
15   Don't use rubber hose within four inches of any part of the exhaust system or within ten inches of the catalytic converter. Metal lines and rubber hoses must never be allowed to chafe against the frame. A minimum of 1/4-inch clearance must be maintained around a line or hose to prevent contact with the frame.

Removal and installation

Note: *The following procedure and accompanying illustrations are typical for vehicles covered by this manual.*
16   Relieve the fuel pressure (see Section 2) and disconnect the fuel feed, return or vapor line at the fuel tank **(see illustration)**.
17   Remove all fasteners attaching the lines to the vehicle body.
18   Detach the fitting(s) that attach the metal lines to the engine compartment fuel hoses **(see illustration)**.

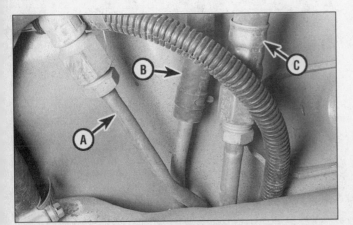

4.16 The fuel return (A), vapor (B), and feed lines (C) at the fuel tank

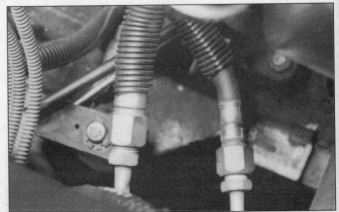

4.18 Typical feed and return line threaded fittings in the engine compartment

19 Installation is the reverse of removal. Be sure to use new O-rings at the threaded fittings **(see illustration)**.

5 Fuel tank - removal and installation

Refer to illustration 5.9
Warning: *Gasoline is extremely flammable, so extra precautions must be taken when working on any part of the fuel system. See the* **Warning** *in Section 2.*
Note: *Don't begin this procedure until the fuel gauge indicates the tank is empty or nearly empty. If the tank must be removed when it's full (for example, if the fuel pump malfunctions), siphon any remaining fuel from the tank prior to removal.*
1   Unless the vehicle has been driven far enough to completely empty the tank, it's a good idea to siphon the residual fuel out before removing the tank from the vehicle. **Warning:** *DO NOT start the siphoning action by mouth! Use a siphoning kit, available at most auto parts stores.*
2   Relieve the fuel pressure (see Section 2).
3   Detach the cable from the negative terminal of the battery. **Caution:** *On models equipped with the Theftlock audio system, be sure the lockout feature is turned off before performing any procedure which requires disconnecting the battery.*
4   Raise the vehicle and place it securely on jackstands.
5   Locate the electrical connector for the electric fuel pump and fuel gauge sending unit in front of the tank, and unplug it. If the vehicle doesn't have a connector, see Step 9 below.
6   Disconnect the fuel feed and return lines (see Section 4), the vapor return line and the filler neck and vent tubes.
7   If necessary, remove the rubber exhaust pipe hangers to allow the exhaust to drop slightly, then remove the exhaust pipe and heat shield.
8   Support the fuel tank with a floor jack.
9   Disconnect both fuel tank retaining straps **(see illustration)**.
10   Lower the tank enough to disconnect

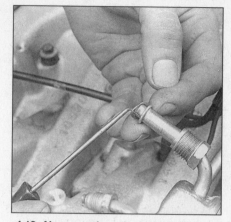

4.19 Always replace the fuel line O-rings

the wires and ground strap from the fuel pump/fuel gauge sending unit, if you haven't already done so.
11   Remove the tank from the vehicle.
12   Installation is the reverse of removal.

6 Fuel tank cleaning and repair - general information

1   All repairs to the fuel tank or filler neck should be carried out by a professional who has experience in this critical and potentially dangerous work. Even after cleaning and flushing of the fuel system, explosive fumes can remain and ignite during repair of the tank.
2   If the fuel tank is removed from the vehicle, it should not be placed in an area where sparks or open flames could ignite the fumes coming out of the tank. Be especially careful inside garages where a natural gas-type appliance is located, because the pilot light could cause an explosion.

7 Fuel pump - removal and installation

Refer to illustrations 7.5a, 7.5b and 7.8
Warning: *Gasoline is extremely flammable, so extra precautions must be taken when*

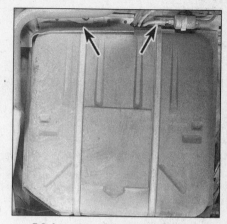

5.9 Location of the fuel tank strap mounting bolts (arrows)

working on any part of the fuel system. See the **Warning** *in Section 2.*

Removal

1   Relieve the fuel pressure (see Section 2).
2   Remove the cable from the negative battery terminal. **Caution:** *On models equipped with the Theftlock audio system, be sure the lockout feature is turned off before performing any procedure which requires disconnecting the battery.*
3   On 1997 and earlier models remove the fuel tank (see Section 5). On 1998 and later models remove the trunk liner and the fuel pump cover from the trunk floor pan.
4   The fuel pump/sending unit assembly is located inside the fuel tank. It's held in place by a cam lock ring mechanism consisting of an inner ring with three locking cams and an outer ring with three retaining tangs.
5   To unlock the fuel pump/sending unit assembly, turn the inner ring clockwise until the locking cams are free of the retaining tangs **(see illustrations)**. **Note:** *If the rings are locked together too tightly to release by hand, tap them gently with a rubber or brass hammer.* **Warning:** *Do not use a steel hammer - a spark could cause an explosion!*
6   Pull the fuel pump/sending unit assembly out of the tank. **Caution:** *The fuel level*

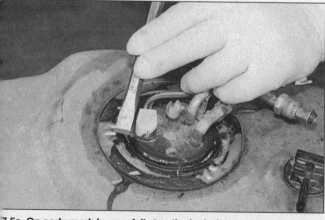

7.5a On early models, carefully tap the lock ring counterclockwise until the locking tabs align with the slots in the fuel tank

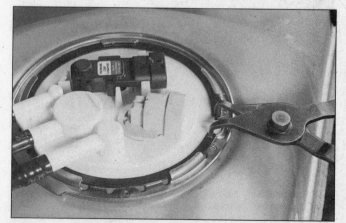

7.5b On later models, use a pair of snap-ring pliers to remove the retaining collar from the fuel pump assembly

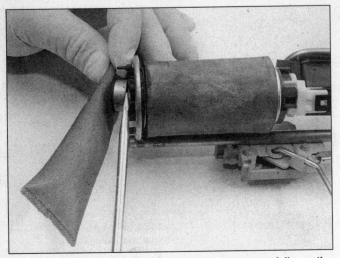

7.8 Inspect the fuel strainer for dirt; if necessary, carefully pry the strainer from the inlet pipe using a screwdriver

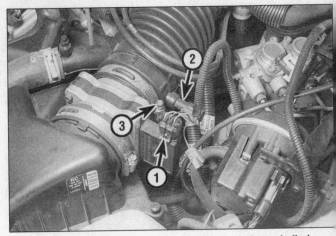

8.4 Air intake details - 3.4 liter V6 shown (others similar)

1	Mass air flow sensor	3	Air intake hose clamp
2	Intake air temperature sensor		

float and sending unit are delicate. Don't bump them into the lock ring during removal or the accuracy of the sending unit may be affected, and don't lift or support the fuel gauge sending unit assembly by the fuel pipes as this could damage the solder joints.

7   Check the condition of the rubber O-ring around the mouth of the lock ring mechanism. If it's dried out, cracked or deteriorated, replace it.

8   Inspect the filter on the lower end of the fuel pump **(see illustration)**. If it's dirty, remove it, clean it with solvent and blow it out with compressed air. If it's too dirty to be cleaned, replace it.

9   If you have to separate the fuel pump and sending unit, pull the fuel pump assembly into the rubber connector and slide the pump away from the bottom support. Care should be taken to prevent damage to the rubber insulator and fuel strainer during removal. After the pump is clear of the bottom support, pull it out of the rubber connector.

Installation

10   Position the rubber gasket around the opening in the fuel tank and guide the fuel pump/sending unit assembly into the tank.

11   Turn the inner lock ring counterclockwise until the locking cams are fully engaged by the retaining tangs. **Note:** *If you've installed a new O-ring, it may be necessary to push down on the inner lock ring until the locking cams slide under the retaining tangs.*

12   Install the fuel tank (see Section 5).

8   Air cleaner housing - removal and installation

Refer to illustrations 8.4, 8.8a and 8.8b

1   Detach the cable from the negative terminal of the battery. **Caution:** *On models equipped with the Theftlock audio system, be sure the lockout feature is turned off before*

performing any procedure which requires disconnecting the battery.

2   Remove the battery and the engine compartment support brace from the vehicle if necessary (see Chapter 5).

3   Remove the air cleaner element (see Chapter 1).

4   Unplug the connector from the intake air temperature sender if necessary **(see illustration)**.

5   Unplug the electrical connector and unclamp the air duct from the Mass Air Flow (MAF) sensor, if equipped.

6   Unclamp the air duct from the housing and detach any other fasteners retaining the duct.

7   Remove the crankcase vent tube (if equipped).

8   Remove the housing mounting bolts and lift the housing from the vehicle **(see illustrations)**.

9   Installation is the reverse of removal.

8.8a Removing the upper air cleaner housing on a 2.3L (Quad 4) engine

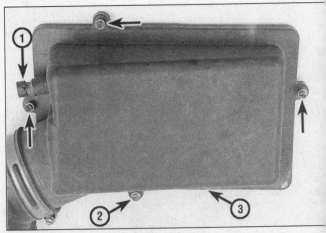

8.8b Upper air cleaner housing mounting details (typical V6 engine)

1	IAT sending unit	3	Upper air filter housing
2	Housing mounting screws		

11.4a Typical location of the IAC valve and the fuel injector electrical connectors (arrows) on a Model 300 TBI unit

1   *Fuel injector electrical connector*
2   *Idle air control valve electrical connector*

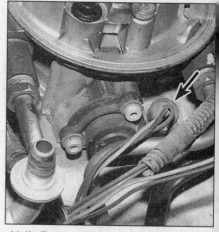

11.4b Typical location of the TPS (arrow) on a Model 700 TBI unit

9   Fuel injection system - general information

Electronic fuel injection provides optimum fuel/air mixture ratios at all stages of combustion and offers immediate throttle response characteristics. It also enables the engine to run at the leanest possible fuel/air mixture ratio, reducing exhaust gas emissions.

On models equipped with a 2.5L four-cylinder engine, the Throttle Body Injection (TBI) unit replaces a conventional carburetor atop the intake manifold. All other models are fitted with either a Port Fuel Injection (PFI) system or a Sequential Fuel Injection (SFI) system. All three systems are controlled by an Electronic Control Module (ECM), which monitors engine performance and adjusts the air/fuel mixture accordingly (see Chapter 6 for a complete description of the fuel control system).

An electric fuel pump located in the fuel tank with the fuel gauge sending unit pumps fuel to the fuel injection system through the fuel feed line and an in-line fuel filter. A pressure regulator keeps fuel available at a constant pressure. Fuel in excess of injector needs is returned to the fuel tank by a separate line.

The basic TBI unit is made up of two major casting assemblies - a throttle body with an Idle Air Control (IAC) valve controls air flow and a Throttle Position Sensor (TPS) monitors throttle angle. The fuel body consists of a fuel meter with a built-in pressure regulator and a fuel injector to supply fuel to the engine.

The fuel injector is a solenoid operated device controlled by the ECM. The ECM turns on the solenoid, which lifts a normally closed ball valve off its seat. The fuel, which is under pressure, is injected in a conical spray pattern at the walls of the throttle body bore above the throttle valve. The fuel which is not used by the injector passes through the pressure regulator before being returned to the fuel tank.

On Port Fuel Injection (PFI) and Sequential Fuel Injection (SFI) systems, the throttle body has a throttle valve to control the amount of air delivered to the engine. The Throttle Position Sensor (TPS) and Idle Air Control (IAC) valves are located on the throttle body.

The fuel rail is mounted on the top of the engine. It distributes fuel to the individual injectors.

Fuel is delivered to the input end of the rail by the fuel feed line, goes through the rail and then to the pressure regulator. The regulator keeps the pressure to the injectors at a constant level.

The remaining fuel is returned to the fuel tank.

10   Fuel injection system - check

Warning: *Gasoline is extremely flammable, so extra precautions must be taken when working on any part of the fuel system. See the* **Warning** *in Section 2.*
Note: *The following procedure is based on the assumption that the fuel pump is working and the fuel pressure is adequate (see Section 3).*

Preliminary checks

1   Check all electrical connectors that are related to the system. Loose connectors and poor grounds can cause many problems that resemble more serious malfunctions.
2   Check to see that the battery is fully charged, as the control unit and sensors depend on an accurate supply of voltage in order to properly meter the fuel.
3   Check the air filter element - a dirty or partially blocked filter will severely impede performance and economy (see Chapter 1).
4   If a blown fuse is found, replace it with a fuse of the same amp rating and see if it blows again. If it does, search for a grounded wire in the harness to the fuel pump.

Port Fuel and Sequential Injection only

5   Check the air intake duct from the mass airflow sensor (if equipped) to the intake manifold for leaks, which will result in an excessively lean mixture. Also check the condition of the vacuum hoses connected to the intake manifold.
6   Remove the air intake duct from the throttle body and check for dirt, carbon or other residue build-up. If it's dirty, clean it with carburetor cleaner and a toothbrush.
7   With the engine running, place a screwdriver against each injector, one at a time, and listen through the handle for a clicking sound, indicating operation.
8   The remainder of the system checks should be left to a GM service department or other qualified repair shop, as there is a chance that the control unit may be damaged if the checks are not performed properly.

11   Throttle Body Injection (TBI) assembly - removal and installation

Refer to illustrations 11.4a and 11.4b
Warning: *Gasoline is extremely flammable, so extra precautions must be taken when working on any part of the fuel system. See the* **Warning** *in Section 2.*
Note: *The fuel injector, pressure regulator, throttle position sensor and the idle air control valve can be replaced without removing the throttle body assembly.*
1   Relieve the fuel system pressure (see Section 2).
2   Disconnect the cable from the negative battery terminal. **Caution:** *On models equipped with the Theftlock audio system, be sure the lockout feature is turned off before performing any procedure which requires disconnecting the battery.*
3   Remove the air cleaner housing (see Section 8).
4   Unplug the electrical connectors from the idle air control valve, throttle position sensor and fuel injector **(see illustrations)**.
5   Remove the wiring harness and insulating grommet from the throttle body.
6   Remove the throttle linkage and return spring, transmission control and cruise control cables (if applicable).

12.4 The best way to remove the fuel injector is to pry on it with a screwdriver, using a second screwdriver as a fulcrum

12.25 The idle air control (IAC) valve pintle must not extend more than 1-1/8 inch - also, replace the O-ring if it is brittle

    A   Distance of pintle extension
    B   O-ring

12.26 To reduce the IAC valve pintle extension, grasp the valve and depress the pintle using a slight side-to-side motion

7   Using pieces of numbered tape, mark all of the vacuum hoses to the throttle body and disconnect them.

8   Disconnect the fuel inlet and return lines. Use a backup wrench on the inlet and return fitting nuts to prevent damage to the throttle body and fuel lines. Remove the fuel fitting O-rings and discard them.

9   Remove the TBI assembly mounting bolts and lift the unit from the intake manifold. It is a good idea to stuff a rag into the intake manifold opening to prevent foreign matter from falling in.

10   Installation is the reverse of the removal procedure. Be sure to install a new throttle body-to-intake manifold gasket, new fuel line O-rings and tighten the mounting bolts to the torque listed in this Chapter's Specifications.

11   Turn the ignition switch to the "ON" position without starting the engine and check for fuel leaks.

12   Check to see if the accelerator pedal is free by depressing the pedal to the floor and releasing it with the ignition switch in the "OFF" position.

12   Throttle Body Injection (TBI) - component replacement

Warning: *Gasoline is extremely flammable, so extra precautions must be taken when working on any part of the fuel system. See the* **Warning** *in Section 2.*

Fuel injector

Refer to illustration 12.4

1   Disconnect the negative battery cable. **Caution:** *On models equipped with the Theft-lock audio system, be sure the lockout feature is turned off before performing any procedure which requires disconnecting the battery.*

2   Unplug the electrical connector at the injector.

3   Remove the injector retainer screw and the retainer **(refer to illustration 11.4a)**.

4   Using one screwdriver as a fulcrum on

the fuel meter body, place another screwdriver tip under the ridge on the fuel injector opposite the electrical connector end and gently pry the injector out **(see illustration)**.

5   If the injector is to be reused, replace the upper and lower O-rings on the injector and in the fuel injector cavity. Install the upper O-ring in the groove on the injector and the lower O-ring flush against the filter element.

6   Install the injector assembly in the fuel meter body by pushing it straight down. Make sure the connector end is facing in the direction of the opening in the fuel meter body for the wire harness grommet.

7   Install the injector retainer and screw. Use a thread locking compound or Loctite 262 on the retainer screw.

8   Reconnect the negative battery cable. Pressurize the fuel system by turning the ignition key to the On position and inspect the area around the injector for leaks.

9   Plug the electrical connector into the injector and start the engine to check for correct operation.

Fuel Pressure Regulator Diaphragm

10   Disconnect the battery negative cable. **Caution:** *On models equipped with the Theft-lock audio system, be sure the lockout feature is turned off before performing any procedure which requires disconnecting the battery.*

11   Relieve the fuel system pressure.

12   Remove the air intake duct.

13   Remove the pressure regulator cover screws. **Caution:** *Apply pressure to the cover while you're removing the screws or the spring inside will force the cover off before you're ready for it.*

14   Carefully remove the regulator cover, then the spring, spring seat and diaphragm assembly.

15   Inspect the regulator valve seat for damage. The throttle body will have to be replaced if damage exists.

16   On reassembly, install a new diaphragm. **Note:** *The diaphragm must be replaced whenever the regulator cover is removed.*

17   Check that the regulator cover screws still retain some thread-locking material on the threads. If necessary apply Loctite 262 to the threads before installing the screws.

18   Reassembly is the reverse of disassembly. Be sure to replace all gaskets and O-ring seals, otherwise a dangerous fuel leak may develop.

19   Turn the ignition switch to the "ON" position without starting the engine and check for fuel leaks.

Idle Air Control (IAC) valve

Refer to illustrations 12.25 and 12.26

20   Unplug the electrical connector from the Idle Air Control (IAC) valve assembly.

21   Remove the two IAC valve attaching screws and withdraw the valve. **Note:** *You may need a Torx bit to remove the screws.*

22   Remove the IAC valve assembly. Check the rubber O-ring for damage and replace it if necessary. **Caution:** *If the IAC valve is to be reused, do not push or pull the IAC valve pintle, as this may damage the threads of the worm drive.*

23   Clean the sealing surface and the bore of the idle air/vacuum signal housing assembly to ensure a good seal. **Caution:** *The IAC valve assembly itself is an electrical component and must not be soaked in any liquid cleaner or solvent or damage may result.*

24   Before installing a new IAC valve assembly, the position of the pintle must be checked. If the pintle is extended too far, damage to the assembly may occur.

25   Measure the distance from the gasket mounting surface of the IAC valve assembly to the tip of the pintle **(see illustration)**.

26   If the distance is greater than 1-1/8 inch, reduce it by applying a firm hand pressure on the pintle to retract it **(see illustration)**.

27   Position the O-ring seal on the IAC valve assembly. Lubricate the O-ring with clean engine oil.

12.33 The throttle position sensor (TPS) mounts to the side of the throttle body with two screws (arrows) and is not adjustable

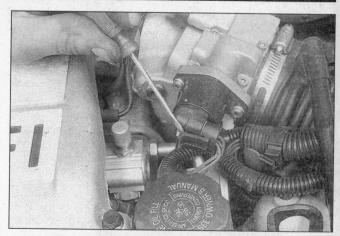

13.2 Using a small screwdriver, release the locking tab and remove the electrical connector

28 Install the IAC valve in the idle air/vacuum signal housing assembly and tighten it securely.

29 Plug in the electrical connector at the IAC valve assembly. **Note:** *The IAC resetting is controlled by the ECM. To initiate this process, turn the ignition on for five seconds, turn it off for ten seconds, then start the engine and check for proper idle.*

Throttle Position Sensor (TPS)

Refer to illustration 12.33

30 Disconnect the cable from the negative battery terminal. **Caution:** *On models equipped with the Theftlock audio system, be sure the lockout feature is turned off before performing any procedure which requires disconnecting the battery.*

31 Remove the air cleaner housing.

32 Unplug the electrical connector from the throttle position sensor.

33 Remove the two sensor mounting screws and pull the sensor from the throttle body **(see illustration)**.

34 To install the TPS, align the slot in the rear of the sensor with the throttle shaft and insert the sensor into the throttle body. Install the mounting screws. This style TPS is not adjustable.

35 The remainder of installation is the reverse of the removal procedure.

---

13 Port Fuel Injection (PFI) and Sequential Fuel Injection (SFI) - component removal and installation

---

Warning: *Gasoline is extremely flammable, so extra precautions must be taken when working on any part of the fuel system. See the* **Warning** *in Section 2.*

Note: *Relieve the fuel system pressure before servicing any fuel system component (see Section 2).*

Throttle Body

Refer to illustrations 13.2, 13.3 and 13.8

Note: *The throttle body on the on 1995 and earlier model 3.4 liter engine is integral with the plenum. See Plenum in this Section.*

1 Disconnect the cable from the negative terminal of the battery. On 1995 and later 3800 engines, remove the acoustic cover from the top of the engine (see Chapter 2F).

Caution: *On models equipped with the Theftlock audio system, be sure the lockout feature*

is turned off before performing any procedure which requires disconnecting the battery.

2 Unplug the Idle Air Control (IAC) valve connector, the Throttle Position Sensor (TPS) connector and, if necessary, the Mass Air Flow (MAF) connector **(see illustration)**.

3 Disconnect the vacuum hoses to the throttle body **(see illustration)**.

4 Disconnect the throttle cable (see Section 14) and, if necessary, the cruise control cable. On 3100 engines, remove the throttle cable bracket.

5 Remove the breather hose.

6 Detach the air inlet duct.

7 Drain the coolant (see Chapter 1) and disconnect the coolant lines.

8 Remove the throttle body bolts and detach the throttle body **(see illustration)**.

9 Install the throttle body and gasket and tighten the bolts to the specified torque.

10 The rest of the procedure is the reverse of removal.

Idle Air Control (IAC) valve

See Section 12.

Throttle Position Sensor (TPS)

11 Most of the models covered by this manual are equipped with non-adjustable TP

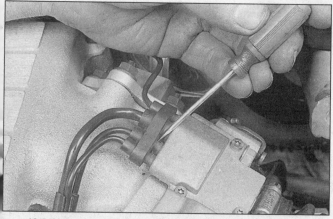

13.3 Using a small screwdriver, remove the vacuum line assembly from the throttle body

13.8 To detach the throttle body from the plenum, remove these two bolts (arrows) (V6 engine is shown)

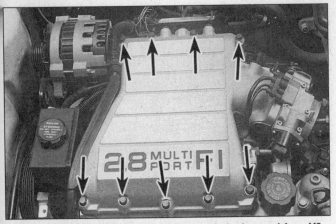

13.18 Location of the plenum mounting bolts (arrows) for a V6 engine (2.8L engine shown, others are similar)

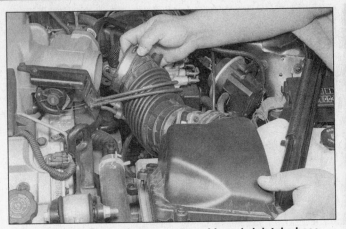

13.27 Remove the air cleaner assembly and air intake hose

sensors, however other models may require adjustment after sensor replacement. The adjustable type of TP sensor is easily identifiable by the elongated slots surounding the sensor mounting holes. Non-adjustable type TP sensors have round mounting holes that are just slighlty larger than the mounting screws that are used to fasten the sensor. To replace non-adjustable type TP sensrs refer to Section 12. On adjustable type TP sensors replacement of this component should be performed by a dealer service department or repair shop equipped with a "scan" tool to properly adjust the TPS.

Plenum (2.2L four-cylinder and 2.8L, 3.1L and 3100 V6)

Refer to illustration 13.18

Note: *Some models may require that the engine be rotated (refer to Chapter 1)*

12   Remove the cable from the negative terminal of the battery. **Caution:** *On models equipped with the Theftlock audio system, be sure the lockout feature is turned off before performing any procedure which requires disconnecting the battery.*

13   Mark and remove all of the vacuum lines that may interfere, then remove the throttle cable bracket nuts.

14   On V6 models, remove the EGR valve (see Chapter 6). On 2.2L four cylinder models, unscrew the EGR tube fitting.

15   On V6 models, remove the throttle body.

16   If necessary, remove the bolts which secure the plastic spark plug wire shield.

17   On the 3100, remove the ignition coil and bracket, the braces to the alternator, and the MAP sensor and bracket.

18   Remove the plenum bolts **(see illustration)**.

19   Remove the plenum and gaskets. If the plenum sticks, use a block of wood and a rubber mallet to dislodge it. Do not pry between the sealing flanges, as this will damage the machined surfaces and vacuum leaks may develop.

20   Remove all traces of old gasket material from the plenum and intake manifold mating surfaces. It is a good idea to stuff rags into the intake manifold openings to prevent debris and old gasket materia! from falling in.

21   Install the new gaskets and set the plenum into position.

22   Install the plenum bolts and tighten them to the torque listed in this Chapter's Specifications.

23   The rest of the procedure is the reverse of removal.

Plenum (3.4 liter V6)

Refer to illustrations 13.27, 13.30 and 13.36

Note: *An eight digit identification number is affixed to the plenum adjacent to the EGR valve flange. Refer to this number if servicing or parts replacement is required.*

24   Remove the cable from the negative terminal of the battery. **Caution:** *On models equipped with the Theftlock audio system, be sure the lockout feature is turned off before performing any procedure which requires disconnecting the battery.*

25   Drain the coolant (see Chapter 1).

26   Remove the throttle cable (see Section 14) and cruise control cable from the throttle lever cam.

27   Remove the air cleaner assembly and air intake hose **(see illustration)**.

28   Remove the vacuum hoses for the brakes and crankcase ventilation valve at the plenum.

29   Disconnect the connectors for the Idle Air Control (IAC) valve and Throttle Position (TP) sensor.

30   Remove the EGR valve and the EGR tube **(see illustration)**. **Caution:** *Avoid bending the EGR tube during removal as this will damage the tube or prevent you from being able to align it during reassembly.*

31   If necessary, remove the bolt that secures the coolant tube to the throttle cable bracket.

32   Remove the fuel pipe clip.

33   Loosen the clamp that secures the coolant hose to the plenum, then slide the clamp down the hose.

34   Disconnect the crankcase vent hose at the front camshaft cover.

35   Remove the vacuum hoses from the vacuum module. **Note:** *Mark the hoses for reassembly.*

36   Remove the engine identification cover **(see illustration)**.

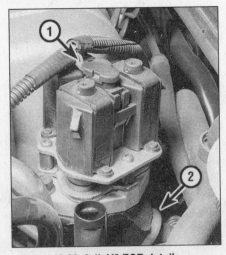

13.30 3.4L V6 EGR details

1   *EGR valve assembly*
2   *EGR Tube*

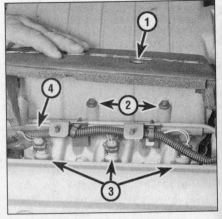

13.36 3.4L V6 plenum details

| 1 | *Engine identification cover* | 3 | *Fuel injectors* |
| 2 | *Plenum bolts* | 4 | *Fuel rail* |

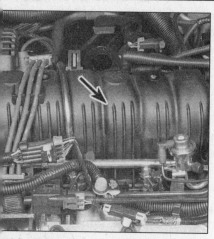

13.54 Remove the upper plenum (arrow) (1995 and later 3800 V6)

13.57 Use a backup wrench when disconnecting the fuel lines (typical V6 engine shown)

13.60a Before removing the injector electrical connector, label the connector according to the cylinder number

47 Remove the wires from the rear spark plugs.

48 Remove the bolts and nuts that secure the plenum.

49 Lift the plenum off the intake manifold. If the plenum sticks, use a block of wood and a rubber mallet to dislodge it. **Caution:** *Do not pry between the sealing flanges, as this will damage the machined surfaces and vacuum leaks may develop. Remove the coolant hose as you lift the plenum.*

50 Remove all traces of old gasket material from the plenum and intake manifold mating surfaces, and remove the remove and discard the coolant O-ring. It is a good idea to stuff rags into the intake manifold openings to prevent debris and old gasket material from falling in.

51 Install the new gaskets and O-ring and set the plenum into position.

52 Install the plenum bolts and tighten them to the torque listed in this Chapter's specifications.

53 The rest of the procedure is the reverse of removal.

Plenum (1995 and later 3800 V6)

Refer to illustration 13.54

44 Disconnect the negative battery cable from the battery. **Caution:** *On models equipped with the Theftlock audio system, be sure the lockout feature is turned off before performing any procedure which requires disconnecting the battery.*

45 Detach the air intake duct (see Section 9).

46 Remove the acoustic cover over the plenum **(see Chapter 2 Part F illustration 24).**

47 Detach the spark plug wires from the spark plugs and set them aside.

48 Unplug the Idle Air Control (IAC) valve, the Throttle Position Sensor (TPS) and the Mass Air Flow sensor (MAF) electrical connectors (see Chapter 6).

49 Mark and disconnect any vacuum hoses connected to the throttle body and the upper plenum. Also detach the breather hose, if equipped.

50 Disconnect the accelerator cable and the cruise control cable (see Section 14) from the throttle lever.

51 Detach the lower throttle body support bracket.

52 Refer to Section 2 to relieve the fuel system pressure then remove the the fuel rail and injectors

53 Refer to Chapter 6 and remove the EGR valve from the manifold.

54 Remove the upper intake plenum mounting bolts and separate the plenum from the engine **(see illustration)**. Do not pry between the manifold and the plenum, as damage to the gasket sealing surfaces may result. If you're installing a new manifold, transfer all fittings and the throttle body to the new manifold.

Fuel rail and related components (All except 2.2L four-cylinder)

Refer to illustrations 13.57, 13.60a, 13.60b, 13.60c, 13.61 and 13.62
Warning: *Before any work is performed on*

the fuel lines, fuel rail or injectors, the fuel system pressure must be relieved (refer to the fuel pressure relief procedure in Section 2).

55 Detach the negative battery cable from the battery. **Caution:** *On models equipped with the Theftlock audio system, be sure the lockout feature is turned off before performing any procedure which requires disconnecting the battery.*

56 Remove the plenum, if equipped (see Plenum).

57 Using a backup wrench, remove the fuel lines at the fuel rail **(see illustration)**. **Note:** *Some models use "Quick-connect" fittings on some fuel line connections. These require a special tool to remove. Tool set J 37088-A includes the tools for the different sizes of quick-connect fittings.*

58 On models with Quad-4 engines, remove the air/oil separator (see Chapter 6).

59 On all models, remove the vacuum line at the fuel pressure regulator.

60 Label and unplug the injector electrical connectors **(see illustrations)**.

61 Remove the fuel rail retaining bolts **(see**

13.60b To remove the connector, push in the retaining clip and pull up

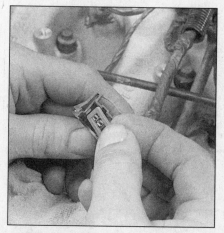

13.60c Sometimes after removing the connector the insulator will pop out - if it does, be sure to reinstall it

4

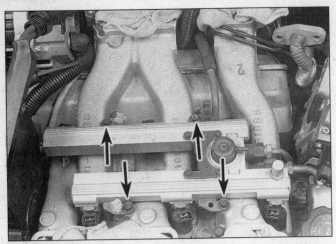

13.61 To remove the fuel rail assembly, remove the retaining bolts (arrows) (typical V6 engine shown)

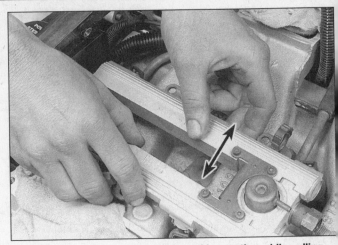

13.62 Use a gentle side-to-side rocking motion while pulling straight up to release the injectors from their bores in the intake manifold (typical V6 shown)

accompanying illustration and illustration 13.36).

62 Carefully remove the fuel rail with the injectors (**see illustration**). **Caution:** *Use care when handling the fuel rail assembly to avoid damaging the injectors.*

63 On reassembly, be sure to replace the fuel pipe O-rings with new ones.

Fuel rail (2.2L four-cylinder)

64 The fuel rail on the 2.2L four-cylinder is integral with the intake manifold (refer to Chapter 2).

Fuel injectors (All except 2.2L four-cylinder)

Refer to illustrations 13.65 and 13.66
Caution: *To prevent dirt from entering the engine, the area around the injectors should be cleaned before servicing.*

65 To remove the fuel injectors, spread open the end of the injector clip slightly and remove it from the fuel rail, then extract the injector (**see illustration**).

66 Inspect the injector O-ring seal(s). These should be replaced whenever the fuel rail is removed (**see illustration**).

67 Install the new O-ring seal(s) on the injector(s) and lubricate them with engine oil.

68 Install the injectors on the fuel rail.

69 Secure the injectors with the retainer clips.

Fuel Injectors (2.2L four-cylinder)

Warning: *Before any work is performed on the fuel lines or injectors, the fuel system pressure must be relieved (refer to the fuel pressure relief procedure in Section 2).*

70 Disconnect the negative battery cable.
Caution: *On models equipped with the Theft-lock audio system, be sure the lockout feature is turned off before performing any procedure which requires disconnecting the battery.*

71 Remove the plenum (see Plenum).

72 Remove the fuel return line retaining bracket and move it away from the fuel pressure regulator.

73 Remove the fuel pressure regulator.

74 Remove the injector retainer bracket.

75 Label then remove the injector connectors.

76 Remove the injectors. **Note:** *Be sure that the small injector O-rings do not remain inside the manifold after the injectors have been removed.*

77 On reassembly, if you're going to reinstall the original injectors be sure to use new O-rings.

78 Apply a light film of clean engine oil to the O-rings.

79 Install the injectors in their holes using twisting motion.

80 Plug in the electrical connectors, the install the injector retainer, making sure that each injector fits properly into its retaining slot.

81 Apply a non-hardening thread locking compound to the retainer screws and tighten the screws to the torque listed in this chapter's Specifications.

82 The remainder of the installation is the reverse of removal. Be sure to check for leaks before returning the car to normal service.

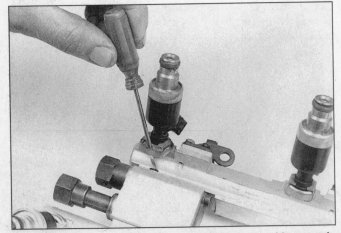

13.65 To remove an injector from the fuel rail assembly, spread the spring clip with a small screwdriver (typical V6 shown)

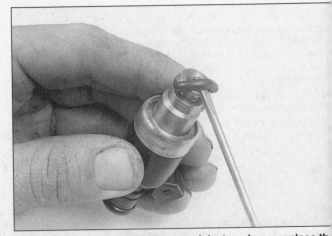

13.66 If you plan to reuse the same injector, always replace the O-ring with a new one

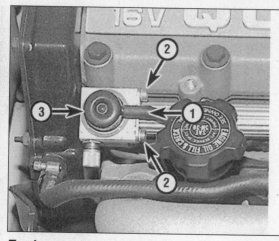

13.84 Details of the Quad-4 engine fuel pressure regulator

1   *Fuel pressure regulator vacuum hose*
2   *Retainer bracket screws*
3   *Fuel pressure regulator*

13.85a A Torx bit or screwdriver is needed to remove the pressure regulator mounting bolts (arrows)

Fuel pressure regulator (Quad-4, 2.8L and 3.1L V6)

Refer to illustrations 13.84, 13.85a, 13.85b and 13.86

83   On V6 models, to remove the fuel pressure regulator from the fuel rail, remove the two fuel line fittings and gaskets.

84   On models with a Quad-4 engine, remove the pressure regulator mounting bracket screws and detach it from the fuel rail **(see illustration)**.

85   On V6 models, detach the pressure regulator mounting bolts and separate the two fuel rails from the pressure regulator assembly **(see illustrations)**.

86   On all models, reassembly is the reverse of disassembly. Be sure to replace all gaskets and O-ring seals **(see illustration)**, otherwise a dangerous fuel leak may develop.

87   Before installing the fuel rail, lubricate all injector O-ring seals with engine oil.

88   Turn the ignition switch to the "ON" position without starting the engine and check for fuel leaks.

Fuel pressure regulator (2.2L four-cylinder and 3.4L, 3100, and 3800 V6)

Refer to illustrations 13.92 and 13.94

89   Disconnect the battery negative cable. **Caution:** *On models equipped with the Theft-lock audio system, be sure the lockout feature is turned off before performing any procedure which requires disconnecting the battery.*

90   Relieve the fuel system pressure.

91   On the 3100, remove the throttle body (see Throttle Body, in this Section). On the 3.4L, remove the plenum (see Plenum, in this Section).

92   Detach the vacuum hose from the regulator **(see illustration)**.

93   On the 3800, clean the area around the snap ring that holds the regulator in place. On others, detach the fuel return line from the regulator. **Caution:** *Be sure to use a back-up wrench.*

94   On the 3800, remove the snap ring that holds the regulator in place **(see illustration)**. On others, remove the screw holding the regulator onto the fuel rail.

95   Remove the regulator using a twisting motion. **Note:** *Fuel is likely to spill when you remove the regulator, so place a towel under the regulator before removing it.*

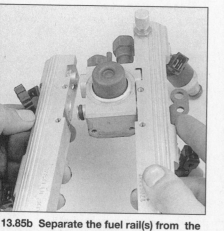

13.85b Separate the fuel rail(s) from the regulator

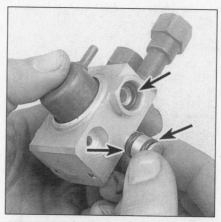

13.86 Always replace all O-rings when replacing the regulator

13.92 Fuel pressure regulator details (typical V6 shown - four-cylinder similar)

A   *Vacuum hose*
B   *Fuel return line fitting*
C   *Mounting bolts*

13.94 On 1996 and later 3.4L engines and all 3800 engines, remove the snap-ring (arrow) from the top of the pressure regulator

13.99 On 1995 and later 3800 engines, remove the retaining screws (arrows) securing the MAF sensor to the throttle body

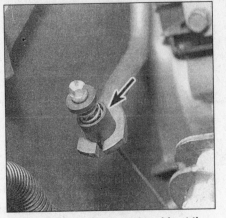

14.2 To detach the throttle cable at the pedal, pull the spring cup (arrow) toward the end of the cable and slide the cable out of the slot

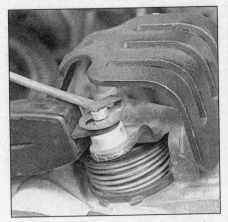

14.7a If your vehicle has this type of retainer, pop off the retaining clip from the throttle lever arm

96 On the 3800, cover the regulator housing to prevent contamination of the fuel system.
97 Reassembly is the reverse of disassembly. Be sure to replace all gaskets and O-ring seals otherwise a dangerous fuel leak may develop.
98 Turn the ignition switch to the "ON" position without starting the engine and check for fuel leaks.

Mass Air Flow (MAF) sensor

Refer to illustration 13.99

99 On 1995 and later 3800 V6 models, the MAF is attached to the throttle body. To remove it, unplug the connector then remove the two MAF sensor screws **(see illustration)**.
100 On others so equipped, the MAF sensor is located between the air intake duct and the air cleaner housing **(see illustration 8.4)**. To remove the sensor, loosen the clamp securing it to the air duct, then loosen the clamp attaching the MAF sensor to the air cleaner housing. Unplug the connector and remove the MAF sensor. **Caution:** *The MAF sensor is*

delicate - if you plan to reinstall the existing unit, handle it carefully.
101 Installation is the reverse of the removal procedure.

---

14 Throttle cable - removal and installation

Refer to illustrations 14.2, 14.7a, 14.7b and 14.8

Removal

1 Detach the screws and the clip retaining the lower instrument panel trim and lower the trim (if necessary).
2 Detach the accelerator cable from the accelerator pedal **(see illustration)**.
3 Squeeze the accelerator cable cover tangs and push the cable through the firewall into the engine compartment.
4 Remove the cable clamp attaching screws and the cable clamp (if equipped).
5 Detach the routing clip and the accelerator cable (if equipped).

6 On 3.4L V6 models, remove the air cleaner duct.
7 Detach the accelerator cable-to-throttle lever retainer and detach the accelerator cable from the throttle body lever **(see illustrations)**.
8 On 3.4L V6 models, turn the throttle body accelerator cam counterclockwise and hold it there while you disconnect the cable **(see illustration)**.
9 Squeeze the accelerator cable retaining tangs and push the cable through the accelerator cable bracket.

Installation

10 Installation is the reverse of removal. **Caution:** *To prevent possible interference, flexible components (hoses, wires, etc.) must not be routed within two inches of moving parts, unless routing is controlled.*
11 Operate the accelerator pedal and check for any binding condition by completely opening and closing the throttle.
12 At the engine compartment side of the firewall, apply sealant around the accelerator cable.

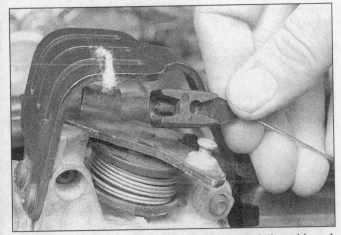

14.7b If your vehicle has this type of retainer, push the cable end forward and lift up to detach it from the throttle lever arm

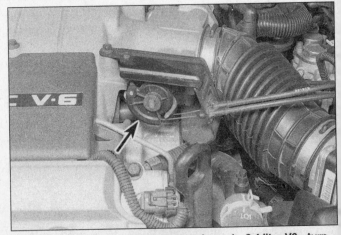

14.8 Throttle body accelerator cam (arrow) - 3.4 liter V6 - turn counterclockwise and hold it there while you disconnect the cable

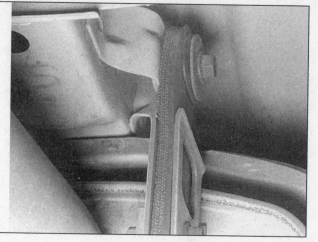

15.1 To detach this rubber block type hanger, remove the bolt

15   Exhaust system servicing - general information

Refer to illustration 15.1
Warning: *Inspection and repair of exhaust system components should be done only after enough time has elapsed after driving the vehicle to allow the system components to cool completely. Also, when working under the vehicle, make sure it is securely supported on jackstands.*

The exhaust system consists of the exhaust manifold(s), the catalytic converter, the muffler, the tailpipe and all connecting pipes, brackets, hangers and clamps. The exhaust system is attached to the body with mounting brackets and rubber hangers **(see illustration)**. If any of the parts are improperly installed, excessive noise and vibration will be transmitted to the body.

Conduct regular inspections of the exhaust system to keep it safe and quiet. Look for any damaged or bent parts, open seams, holes, loose connections, excessive corrosion or other defects which could allow exhaust fumes to enter the vehicle. Deteriorated exhaust system components should not be repaired; they should be replaced with new parts.

If the exhaust system components are extremely corroded or rusted together, welding equipment will probably be required to remove them. The convenient way to accomplish this is to have a muffler repair shop remove the corroded sections with a cutting torch. If, however, you want to save money by doing it yourself (and you don't have a welding outfit with a cutting torch), simply cut off the old components with a hacksaw. If you have compressed air, special pneumatic cutting chisels can also be used. If you do decide to tackle the job at home, be sure to wear safety goggles to protect your eyes from metal chips and work gloves to protect your hands.

Here are some simple guidelines to follow when repairing the exhaust system:
a) *Work from the back to the front when removing exhaust system components.*
b) *Apply penetrating oil to the exhaust system component fasteners to make them easier to remove.*
c) *Use new gaskets, hangers and clamps when installing exhaust systems components.*
d) *Apply anti-seize compound to the threads of all exhaust system fasteners during reassembly.*
e) *Be sure to allow sufficient clearance between newly installed parts and all points on the underbody to avoid overheating the floor pan and possibly damaging the interior carpet and insulation. Pay particularly close attention to the catalytic converter and heat shield.*

16   Turbocharger - general information

A turbocharger is used on some models to increase power. As increased power is required and the throttle is opened, more air/fuel mixture is forced into the combustion chambers by the turbocharger.

A turbine wheel in the turbocharger is driven by the exhaust stream and is connected by a shaft to drive a compressor wheel (fan) to pressurized the intake air stream. The amount of pressurized air allowed into the engine is controlled by an exhaust by-pass valve, or wastegate. In this system, the wastegate is controlled by a solenoid which is controlled by the ECM.

This system also incorporates an air-to-air intercooler (similar to a radiator). The function of the intercooler is to cool the pressurized air from the turbocharger. This cooling of the pressurized air allows for a higher compression ratio (without pre-ignition occurring). This, in turn, will increase power by approximately 15-percent.

17   Turbocharger - inspection

Caution: *Operation of the turbocharger without all the ducts and filters installed can result*

in personal injury, and/or allow foreign objects to damage the turbine blades.
Note: *The turbocharger is not serviceable and must be replaced as a unit.*

1   Every turbocharger has its own noise level when operating. If the noise level changes, suspect a problem. If the sound of the turbocharger goes up and down in pitch, check for heavy dirt buildup in the compressor housing and on the compressor wheel, or for an air inlet restriction. If the noise level is a high pitch or whistling sound, look for an inlet air or exhaust gas leak.
2   With the engine off and the turbocharger stopped, make a visual inspection of the turbocharger and components.
3   Check for loose duct connections from the air cleaner to the turbocharger.
4   Be sure the cross-over duct from the turbocharger-to-intake system is not loose.
5   Visually check the wheels of the turbocharger for damage from foreign objects.
6   Look for evidence of wheel-to-housing contact.
7   Be sure the shaft rotates freely. Rotating stiffness could indicate the presence of sludged oil or coking (hardened oil deposits) from overheating.
8   Push in on one of the shaft wheels while turning it. Be sure the wheels turn freely without contacting the housings, the backplate or the shroud.
9   Be sure the exhaust manifold has no loose connections or cracks.
10   Check the oil drain line for any restrictions.
11   Visually inspect the actuator and wastegate linkage for damage.
12   Check the hose from the throttle body-to-wastegate solenoid and from the wastegate solenoid-to-actuator assembly.
13   Using a pump, apply 3.5 to 4.5 psi to the actuator assembly. The actuator rod end should move 1/64 inch to actuate the wastegate linkage. If not, replace the actuator assembly and recheck.
14   Remove test equipment and reconnect the hose.

18   Turbocharger - removal and installation

Removal

1   Detach the negative cable from the battery. **Caution:** *On models equipped with the Theftlock audio system, be sure the lockout feature is turned off before performing any procedure which requires disconnecting the battery.*
2   Drain the coolant from the radiator.
3   Remove the bolt attaching the intercooler to the intake manifold duct.
4   Detach the intercooler-to-intake manifold duct.
5   At the turbocharger, detach the air cleaner-to-turbocharger duct.
6   Detach the air cleaner inlet duct.

4

7   Remove the air cleaner and duct assembly.

8   At the turbocharger, detach the turbocharger-to-intercooler duct.

9   Remove the turbocharger heat covers.

10   Disconnect the electrical connector for the oxygen sensor.

11   Remove the oxygen sensor.

12   At the turbocharger, detach the coolant return line, the turbocharger oil supply line and the vacuum line.

13   At the actuator, detach the vacuum line. Remove the actuator arm-to-wastegate retaining clip. Detach the actuator arm from the wastegate. Remove the wastegate actuator from the turbocharger.

14   Remove the cruise control servo and place out of the work area.

15   At the turbocharger, detach the downpipe, the coolant supply hose, the oil drain hose (at the drain pipe).

16   Remove the bolt attaching the turbocharger to the exhaust crossover.

17   Remove the turbocharger. **Caution:** *Do not attempt to repair the turbocharger. It is serviced as a unit only.*

Installation

18   Installation is the reverse of removal. Be sure to transfer the oil and water lines to the new turbocharger.

19   Be sure to prime the turbocharger with oil before running the engine. Crank the engine with the fuel pump fuse removed until normal operating oil pressure is reached.

20   Start the engine and check for fluid leaks.

Chapter 5
Engine electrical systems

Contents

1   Ignition system - general information

The engines covered in this manual are equipped with either a Distributorless Ignition System (DIS), on all engines except the Quad-4, or an Integrated Direct Ignition System (IDI) on the Quad-4.

The DIS and IDI systems use a "waste spark" method of spark distribution. Each cylinder is paired with its opposing cylinder in the firing order (1-4, 2-3 on a four cylinder, 1-4, 2-5, 3-6 on a V6) so one cylinder under compression fires simultaneously with its opposing cylinder, where the piston is on the exhaust stroke. Since the cylinder on the exhaust stroke requires very little of the available voltage to fire its plug, most of the voltage is used to fire the plug of the cylinder on the compression stroke.

The DIS system includes a coil pack, an ignition module, a crankshaft reluctor ring, a magnetic sensor, spark plug wires, and the ECM. The IDI system is the same except that it does not have spark plug wires. The ignition module is located under the coil pack and is connected to the ECM.

The magnetic crankshaft sensor is mounted on the bottom of the engine block, just above the oil pan rail, or on the timing chain cover, near the vibration damper. The reluctor ring is a special disc, either cast onto the crankshaft or attached to the vibration damper, which acts as a signal generator for the ignition timing. (Some models also have a camshaft position sensor that is used to determine fuel injection timing.)

The ignition system uses Electronic Spark Timing (EST) and control wires from the ECM, just like conventional distributor systems.

2   Battery - removal and installation

Refer to illustration 2.2
Caution 1: *Always disconnect the negative cable first and hook it up last or the battery may be shorted by the tool being used to loosen the cable clamps.* **Caution 2:** *On models equipped with the Theftlock audio system, be sure the lockout feature is turned off before performing any procedure which requires disconnecting the battery.*

1   Disconnect both cables from the battery terminals.
2   Detach any brackets or braces that would interfere with the removal of the battery, then remove the battery hold-down clamp or strap **(see illustration)**.
3   Lift out the battery. Be careful - it's heavy. **Note:** *On some models it may be necessary to remove the air cleaner housing and to position the engine compartment fuse/junction block aside.*
4   While the battery is out, inspect the carrier (tray) for corrosion (see Chapter 1).
5   If you are replacing the battery, make sure that you get one that's identical, with the same dimensions, amperage rating, cold cranking rating, etc.
6   Installation is the reverse of removal.

3   Battery - emergency jump starting

Refer to the *Booster battery (jump) starting* procedure at the front of this manual.

4   Battery cables - check and replacement

Refer to illustration 4.2
1   Periodically inspect the entire length of each battery cable for damage, cracked or burned insulation and corrosion. Poor battery cable connections can cause starting problems and decreased engine performance.
2   Check the cable-to-terminal connec-

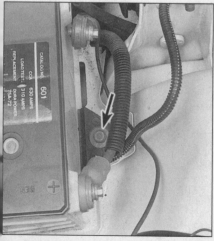

2.2 Remove the battery hold-down clamp bolt (arrow) from the battery carrier

Terminal end corrosion or damage.

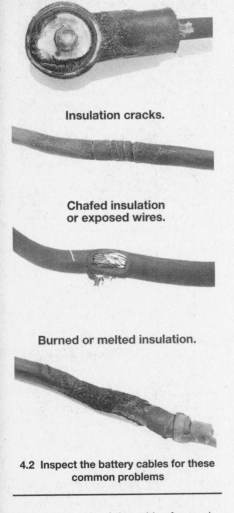

Insulation cracks.

Chafed insulation or exposed wires.

Burned or melted insulation.

4.2 Inspect the battery cables for these common problems

tions at the ends of the cables for cracks, loose wire strands and corrosion (see illustration). The presence of white, fluffy deposits under the insulation at the cable ter-minal connection is a sign that the cable is corroded and should be replaced. Check the terminals for distortion, missing mounting bolts and corrosion.

3   When removing the cables, always dis-connect the negative cable first and hook it up last or the battery may be shorted by the tool used to loosen the cable clamps. Even if only the positive cable is being replaced, be sure to disconnect the negative cable from the battery first (see Chapter 1 for further information regarding battery cable removal). **Caution:** *On models equipped with the Theft-lock audio system, be sure the lockout feature is turned off before performing any procedure which requires disconnecting the battery.*

4   Disconnect the old cables from the bat-tery, then trace each of them to their opposite ends and detach them from the starter solenoid and ground terminals. Note the rout-ing of each cable to ensure correct installation.

5   If you are replacing either or both of the old cables, take them with you when buying new cables. It is vitally important that you replace the cables with identical parts. Cables have characteristics that make them easy to identify: positive cables are usually red, larger in cross-section and have a larger diameter battery post clamp; ground cables are usually black, smaller in cross-section and have a slightly smaller diameter clamp for the negative post.

6   Clean the threads of the solenoid or ground connection with a wire brush to remove rust and corrosion. Apply a light coat of battery terminal corrosion inhibitor, or petroleum jelly, to the threads to prevent future corrosion.

7   Attach the cable to the solenoid or ground connection and tighten the mounting nut/bolt securely.

8   Before connecting a new cable to the battery, make sure that it reaches the battery post without having to be stretched.

9   Connect the positive cable first, fol-lowed by the negative cable.

5   Ignition system - check

Warning: *Because of the very high voltage generated by the ignition system, extreme care should be taken whenever an operation is performed involving ignition components. This not only includes the coils, control mod-ule and spark plug wires (all engines, except Quad-4), but related items connected to the system as well, such as the plug connections, tachometer and any test equipment.*

1   If the engine turns over, but won't start, check for spark at the spark plug by installing a calibrated ignition system tester to one of the spark plug wire/boot terminals. Cali-brated ignition testers are available at most auto parts stores in several different types. One type of spark tester has a short center electrode extending from the insulator and is intended for use on breaker-point and older electronic ignitions. The other type does not have a visible center electrode and is for use on high-voltage ignition systems. Be sure to use the high-voltage ignition tester on these models.

All engines (except Quad-4)
Refer to illustrations 5.2, 5.5, 5.6, 5.7a and 5.7b

2   Insert the ignition tester into the number one spark plug boot terminal. Connect the clip on the tester to a ground such as a metal bracket or valve cover bolt, crank the engine and watch the end of the tester for bright blue, well defined sparks (see illustration).

3   If sparks occur, sufficient voltage is reaching the plugs to fire the engine. How-ever the plugs themselves may be fouled, so remove and check them as described in Chapter 1 or replace them with new ones.

4   If no sparks or intermittent sparks occur, check for a bad spark plug wire by swapping wires or by using an ohmmeter to measure the resistance between the spark plug wire terminal ends. There should be approxi-

5.2 To use a calibrated ignition tester, simply disconnect a spark plug wire, clip the tester to a convenient ground (like a rocker arm cover bolt) and operate the starter - if there's enough power to fire the plug, sparks will be visible between the electrode tip and the tester body

5.5 Check for battery voltage to the ignition module from the ignition key (all engines, except Quad-4)

5.6 On all engines except Quad-4 remove the coil pack from the module assembly and check for a trigger signal between the module terminals with a test light while an assistant cranks the engine over

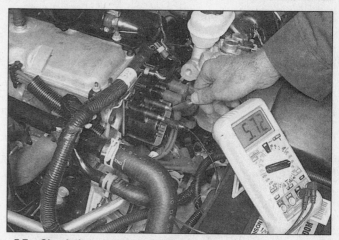

5.7a Check the secondary coil resistance across the towers on each coil pack (all engines, except Quad-4)

...nately 5000 ohms of resistance per foot.

Check the primary wire connections at the ignition control module to make sure they are clean and tight. Refer to the wiring diagrams (see Chapter 12) and check for battery voltage to the ignition control module from the IGN-MOD fuse **(see illustration)**. Battery voltage should be available to the ignition control module with the ignition switch ON. **Note:** *If no voltage is available check the ignition system fuses (see Chapter 12). If the reading is 7 volts or less, repair the primary circuit from the ignition switch to the ignition control module.*

If voltage is available at the ignition control module and there's still no spark, check for a trigger signal from the ignition module. Remove the coil pack from the ignition module to expose the module terminals (see Section 7). Connect a test light between each of the module terminals and have an assistant crank the engine over **(see illustration)**. The test light should blink quickly and constantly as each coil pack is triggered to fire by the switching signal from the ignition module. This test checks for the trigger signal (ground) from the computer and Ignition Control Module. **Caution:** *Use only an LED test light to avoid damaging the PCM.*

If a trigger signal is present at the coil, the computer and the Ignition Control Module are functioning properly and the problem lies in the ignition coils. Check the primary and secondary resistance of the ignition coils and compare it to the Specifications at the beginning of this Chapter **(see illustrations)**.

If the test light does not flash, the ignition module is most likely the problem but not always. **Note:** *Refer to Chapter 6 for additional information on the crankshaft and the camshaft sensors. It will be necessary to verify that the crankshaft sensors and camshaft sensor is operating correctly before changing the ignition module. A defective ignition module can only be diagnosed by process of elimination.*

Have the crankshaft sensor(s), the camshaft sensor and the PCM, checked by a dealer service department or other qualified automotive repair shop.

Quad-4 engines

Refer to illustrations 5.12, 5.15, 5.16a, 5.16b, 5.17a and 5.17b

Note: *This procedure is difficult to perform on Quad-4 engines but it is possible with the correct set-up.*

10   Remove the ignition coil/module housing from the cylinder head (see Section 6). Turn the ignition coil/module assembly upside down and bolt it (module cover) to the engine block using the bolt hole in the cover that the ground strap is attached too, then reconnect the ignition module connector. This will ground the assembly to the engine for testing purposes.

11   Insert the ignition tester into the number one spark plug terminal on the coil assembly, then connect a ground wire from the ignition tester to the engine block.

12   Construct three makeshift spark plug wires from a spare wire set that can be connected to the remaining spark plugs and to each of their corresponding spark plug terminals on the coil assembly **(see illustration)**.

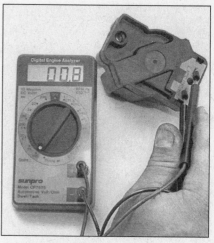

5.7b Checking the coil primary resistance (all engines, except Quad-4)

5

13   Disable the fuel pump (see Chapter 4) and crank the engine over to check for spark at the ignition tester. Test each cylinder one-by-one making sure to ground the remaining spark plugs not being tested before cranking the engine over. Failure to ground the three spark plugs not being tested may result in ignition control module damage!

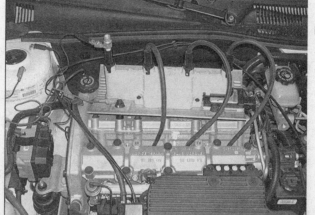

5.12 To use a calibrated ignition tester on Quad-4 engines, it will be necessary to flip the coil assembly over and bolt it to engine while grounding the ignition tester - then fabricate three spark plug wires which connect from the remaining spark plug boot terminals on the coil assembly and to spark plugs not being tested

5.15 Check for battery voltage to the ignition module connector (arrow) from the ignition key (Quad-4 engines)

5.16a On Quad-4 engines, first disconnect the ignition coil to ignition module connector and check for battery voltage to the coil with the ignition key On . . .

5.16b . . . then connect an LED test light to the positive terminal of the battery and to each of the coil negative (-) terminals on the ignition module and watch for a blinking light when the engine is cranked

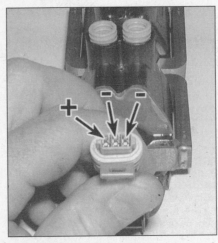

5.17a To check the primary resistance of the ignition coil on Quad-4 engines, connect the positive probe to the positive (+) terminal and the negative probe to each of negative (-) terminals of the coil connector - the resistance should be the same for each check

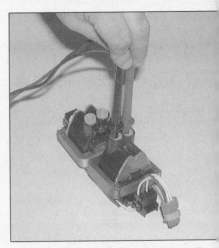

5.17b Checking the secondary coil resistance (Quad-4 engines) - be sure the probes touch the contact in the center of each socket

14   If sparks occur, sufficient voltage is reaching the plugs to fire the engine. However the plugs themselves may be fouled, so remove and check them as described in Chapter 1 or replace them with new ones.

15   If no sparks or intermittent sparks occur, check the primary wire connections at the ignition control module to make sure they are clean and tight. Refer to the wiring diagrams (see Chapter 12) and check for battery voltage to the ignition control module from the IGN-MOD fuse **(see illustration)**. Battery voltage should be available to the ignition control module with the ignition switch ON. **Note:** *If no voltage is available check the ignition system fuses (see Chapter 12). If the reading is 7 volts or less, repair the primary circuit from the ignition switch to the ignition control module.*

16   If voltage is available at the ignition control module and there's still no spark, check for a trigger signal from the ignition module.

Remove the coil housing cover and the ignition coil to ignition module electrical connector to expose the module output terminals (see Section 7). First check for battery voltage to the coils with the ignition key On, then connect a test light to the positive terminal of the battery and to each of the coil/module negative terminals and have an assistant crank the engine over **(see illustrations)**. The test light should blink quickly and constantly as each coil pack is triggered to fire by the switching signal from the ignition module. This test checks for the trigger signal (ground) from the computer and Ignition Control Module. **Caution:** *Use only an LED test light to avoid damaging the PCM*

17   If a trigger signal is present at the coil, the computer and the Ignition Control Module are functioning properly and the problem lies in the ignition coils or the wiring to the ignition coils. Check the primary and secondary

resistance of the ignition coils and compare it to the Specifications at the beginning of this Chapter **(see illustrations)**. **Note:** *On Quad-4 engines, primary resistance can be checked at the coil/ignition module connector or at the bottom of each coil. Always verify that the wires leading to the coils are not defective (have continuity) before replacing an ignition coil.*

18   If the test light does not flash, the ignition module is most likely the problem but not always. **Note:** *Refer to Chapter 6 for additional information on the crankshaft and the camshaft sensors. It will be necessary to verify that the crankshaft sensors and camshaft sensor is operating correctly before changing the ignition module. A defective ignition module can only be diagnosed by process of elimination.*

19   Have the crankshaft sensor(s), the camshaft sensor and the PCM, checked by dealer service department or other qualified automotive repair shop.

6.4 Coils and plug wires are numbered according to their associated cylinders

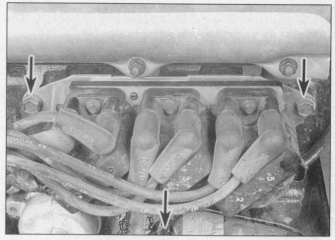

6.5 Mounting bolt locations (arrows) for the module/coil assemblies on a V6 engine

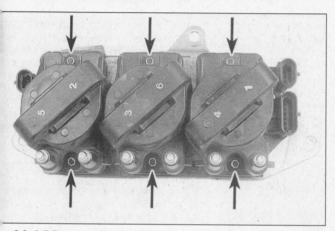

6.6 A 5.5 mm socket is required to remove the coil mounting bolts (arrows)

6.8 When installing a coil, make sure that it plugs into the module completely

6 Ignition coil and module - removal and installation

Refer to illustrations 6.4, 6.5, 6.6, 6.8, 6.11 and 6.13

Direct Ignition System (DIS) (all except Quad-4 engines)

Refer to the previous Section for checking procedures.

Detach the cable from the negative terminal of the battery. **Caution:** *On models equipped with the Theftlock audio system, be sure the lockout feature is turned off before performing any procedure which requires disconnecting the battery.*

Unplug the electrical connectors from the module.

If the plug wires are not numbered, label them and detach the plug wires at the coil assembly **(see illustration)**.

Remove the module/coil assembly mounting bolts and lift the assembly from the vehicle **(see illustrations)**.

Remove the bolts attaching the coils to

6.11 The Quad-4 ignition coil and module retaining bolt locations (arrows)

the module and separate them **(see illustration)**.

7 Installation is the reverse of removal.

8 When installing the coils, make sure they are connected properly **(see illustration)**.

Integrated Direct Ignition (IDI) (Quad 4 engines)

9 Detach the cable from the negative ter-

minal of the battery. **Caution:** *On models equipped with the Theftlock audio system, be sure the lockout feature is turned off before performing any procedure which requires disconnecting the battery.*

10 Disconnect the IDI electrical connector.

11 Remove the ignition system assembly-to-camshaft housing bolts **(see illustration)**.

12 Lift the ignition system assembly from

the engine **(see illustration)**.

13   Detach the coil housing-to-cover screws and remove the housing from the cover **(see illustration)**.

14   Detach the coil electrical connectors.

15   Remove the coils.

16   Remove the module-to-cover screws and detach the module from the cover.

17   Installation is the reverse of removal.

7   Charging system - general information and precautions

Caution: *On models equipped with the Theftlock audio system, be sure the lockout feature is turned off before performing any procedure which requires disconnecting the battery.*

The charging system includes the alternator, an internal voltage regulator, a charge indicator, the battery, a fusible link and the wiring between all the components. The charging system supplies electrical power for the ignition system, the lights, the radio, etc. The alternator is driven by a drivebelt at the front of the engine.

The purpose of the voltage regulator is to limit the alternator's voltage to a preset value. This prevents power surges, circuit overloads, etc., during peak voltage output. The fusible link is a short length of insulated wire integral with the engine compartment wiring harness. The link is four wire gauges smaller in diameter than the circuit it protects. Production fusible links and their identification flags are identified by the flag color. See Chapter 12 for additional information regarding fusible links.

The charging system doesn't ordinarily require periodic maintenance. However, the drivebelt, battery and wires and connections should be inspected at the intervals outlined in Chapter 1.

The dashboard warning light should come on when the ignition key is turned to Start, then go off immediately. If it remains on, there is a malfunction in the charging system (see Section 8). Some vehicles are also equipped with a voltmeter. If the voltmeter indicates abnormally high or low voltage, check the charging system (see Section 8).

Be very careful when making electrical circuit connections to a vehicle equipped with an alternator and note the following:

a) *When reconnecting wires to the alternator from the battery, be sure to note the polarity.*

b) *Before using arc welding equipment to repair any part of the vehicle, disconnect the wires from the alternator and the battery terminals.*

c) *Never start the engine with a battery charger connected.*

d) *Always disconnect both battery leads before using a battery charger.*

e) *The alternator is turned by an engine drivebelt which could cause serious injury if your hands, hair or clothes become entangled in it with the engine running.*

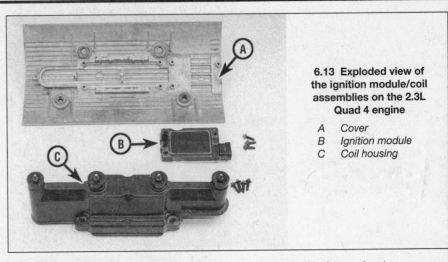

6.13 Exploded view of the ignition module/coil assemblies on the 2.3L Quad 4 engine

A   *Cover*
B   *Ignition module*
C   *Coil housing*

f) *Because the alternator is connected directly to the battery, it could arc or cause a fire if overloaded or shorted out.*

g) *Wrap a plastic bag over the alternator and secure it with rubberbands before steam cleaning the engine.*

8   Charging system - check

1   If a malfunction occurs in the charging circuit, don't automatically assume that the alternator is causing the problem. First check the following items:

a) *Check the drivebelt tension and condition (see Chapter 1). Replace it if it's worn or deteriorated.*

b) *Make sure the alternator mounting and adjustment bolts are tight.*

c) *Inspect the alternator wiring harness and the electrical connectors at the alternator and voltage regulator. They must be in good condition and tight.*

d) *Check the fusible link (if equipped) located between the starter solenoid and the alternator. If it's burned, determine the cause, repair the circuit and replace the link (the vehicle won't start and/or the accessories won't work if the fusible link blows). Sometimes a fusible link may look good, but still be bad. If in doubt, remove it and check for continuity.*

e) *Start the engine and check the alternator for abnormal noises (a shrieking or squealing sound indicates a bad bearing).*

f) *Check the specific gravity of the battery electrolyte. If it's low, charge the battery (doesn't apply to maintenance free batteries).*

g) *Make sure the battery is fully charged (one bad cell in a battery can cause overcharging by the alternator).*

h) *Disconnect the battery cables (negative first, then positive). Inspect the battery posts and the cable clamps for corrosion. Clean them thoroughly if necessary (see Chapter 1). Reconnect the cable to the positive terminal. Caution: On models equipped with the Theftlock audio system, be sure the lockout feature is*

turned off before performing any procedure which requires disconnecting the battery.

i) *With the key off, connect a test light between the negative battery post and the disconnected negative cable clamp.*

1) *If the test light does not come on, reattach the clamp and proceed to the next Step.*

2) *If the test light comes on, there is a short (drain) in the electrical system of the vehicle. The short must be repaired before the charging system can be checked.*

3) *Disconnect the alternator wiring harness.*

(a) *If the light goes out, the alternator is bad.*

(b) *If the light stays on, pull each fuse until the light goes out (this will tell you which component is shorted).*

2   Using a voltmeter, check the battery voltage with the engine off. It should be approximately 12 volts.

3   Start the engine and check the battery voltage again. It should now be approximately 14 to 15 volts.

4   Turn on the headlights. The voltage should drop, and then come back up, if the charging system is working properly.

5   If the voltage reading is more or less than the specified charging voltage, alternator replacement is necessary (refer to Section 9).

9   Alternator - removal and installation

All except 3.4 liter V6

Refer to illustrations 9.4a and 9.4b

1   Detach the cable from the negative terminal of the battery. **Caution:** *On models equipped with the Theftlock audio system, be sure the lockout feature is turned off before performing any procedure which requires disconnecting the battery.*

2   Detach the electrical connectors from the alternator.

9.4a Alternator mounting bolts (arrows)
(typical four-cylinder engine)

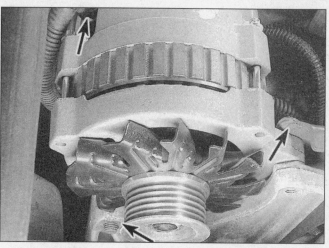

9.4b Alternator mounting bolts (arrows) (typical V6 engine)

On some models it will be necessary to loosen the adjustment and pivots to remove the drivebelt. On other models it will be necessary to rotate the drivebelt tensioner to remove the drivebelt (see Chapter 1).

Remove the alternator support brackets if equipped, then remove the alternator mounting bolts **(see illustrations)**.

3.4 liter V6

Remove the air cleaner and duct assembly (see Chapter 4).

Detach the cable from the negative terminal of the battery. **Caution:** *On models equipped with the Theftlock audio system, be sure the lockout feature is turned off before performing any procedure which requires disconnecting the battery.*

Remove the coolant recovery tank (see Chapter 3).

Remove the drivebelt (see Chapter 1).

Loosen the wheel lug nuts and the right-side driveaxle/hub nut. Raise the front of the vehicle and support it securely on jackstands, then remove the wheels.

Remove the under-vehicle splash shield. Also remove the right inner fender splash shield (see Chapter 11, Section 12).

Remove the front exhaust pipe and catalytic converter (see Chapter 4).

Remove the right driveaxle (see Chapter 8).

Also remove the driveaxle shield.

Remove the steering gear heat shield.

Unplug the electrical connector from the knock sensor (see Chapter 6, Section 7).

Remove the alternator cooling duct, if so equipped.

Remove the alternator rear brace bolt and nut.

Unscrew the nut and detach the power steering line from the alternator upper mounting stud.

Detach the electrical connector and battery terminal from the rear of the alternator.

Unscrew the alternator lower mounting bolt. Be careful not to lose any spacers or washers that may be present.

21 Unscrew the upper mounting bolt/stud and remove the alternator.

All engines

22 If you are replacing the alternator, take the old one with you when purchasing the replacement unit. Make sure the new/rebuilt unit is identical to the alternator. Look at the terminals - they should be the same in number, size and location as those on the old alternator. Finally, look at the identification numbers - they will be stamped into the housing or printed on a tag attached to the housing. Make sure the numbers are the same on both alternators.

23 Many new alternators DO NOT have a pulley installed, so you may have to switch the pulley from the old unit to the new/rebuilt one. When buying an alternator, find out the shop's policy regarding pulleys - some shops will perform this service free of charge.

24 Installation is the reverse of removal.

25 After installation, adjust the drive belt tension (see Chapter 1).

26 Check the charging voltage to verify proper operation of the alternator (see Section 8).

10 Starting system - general information and precautions

The starting system consists of the battery, the starter motor, the starter solenoid and the wires connecting them. The solenoid is mounted directly on the starter motor. The solenoid/starter motor assembly is installed on the lower part of the engine, next to the transmission bellhousing.

When the ignition key is turned to the "Start" position, the starter solenoid is actuated through the starter control circuit. The starter solenoid then connects the battery to the starter. The battery supplies the electrical energy to the starter motor, which does the actual work of cranking the engine.

The starter motor on a vehicle equipped

with a manual transaxle can only be operated when the clutch pedal is depressed; the starter on a vehicle equipped with an automatic transaxle can only be operated when the selector lever is in Park or Neutral.

Always observe the following precautions when working on the starting system:

a) *Excessive cranking of the starter motor can overheat it and cause serious damage. Never operate the starter motor for more than 15 seconds at a time without pausing to allow it to cool for at least two minutes.*

b) *The starter is connected directly to the battery and could arc or cause a fire if mishandled, overloaded or shorted out.*

c) *Always detach the cable from the negative terminal of the battery before working on the starting system.* **Caution:** *On models equipped with the Theftlock audio system, be sure the lockout feature is turned off before performing any procedure which requires disconnecting the battery.*

11 Starter motor - testing in vehicle

Note: *Before diagnosing starter problems, make sure the battery is fully charged.*

1 If the starter motor does not turn at all when the switch is operated, make sure that the shift lever is in Neutral or Park (automatic transmission) or that the clutch pedal is depressed (manual transmission).

2 Make sure that the battery is charged and that all cables, both at the battery and starter solenoid terminals, are clean and secure.

3 If the starter motor spins but the engine is not cranking, the overrunning clutch in the starter motor is slipping and the starter motor must be replaced.

4 If, when the switch is actuated, the starter motor does not operate at all but the solenoid clicks, then the problem lies with either the battery, the main solenoid contacts or the starter motor itself (or the engine is seized).

5

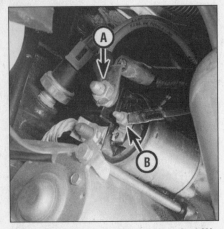

12.8 Disconnect the battery terminal (A) and the switch terminal (B) from the starter solenoid

12.9 Typical starter motor mounting bolt locations (arrows)

13.3 Disconnect the solenoid-to-starter motor strap by removing the retaining bolt (arrow)

5   If the solenoid plunger cannot be heard when the switch is actuated, the battery is bad, the fusible link is burned (the circuit is open) or the solenoid itself is defective.

6   To check the solenoid, connect a jumper lead between the battery (+) and the ignition switch wire terminal (the small terminal) on the solenoid. If the starter motor now operates, the solenoid is OK and the problem is in the ignition switch, neutral start switch or the wiring.

7   If the starter motor still does not operate, remove the starter/solenoid assembly for disassembly, testing and repair.

8   If the starter motor cranks the engine at an abnormally slow speed, first make sure that the battery is charged and that all terminal connections are tight. If the engine is partially seized, or has the wrong viscosity oil in it, it will crank slowly.

9   Run the engine until normal operating temperature is reached, then disconnect the coil wire from the distributor cap and ground it on the engine.

10   Connect a voltmeter positive lead to the positive battery post and connect the negative lead to the negative post.

11   Crank the engine and take the voltmeter readings as soon as a steady figure is indicated. Do not allow the starter motor to turn for more than 15 seconds at a time. A reading of 9 volts or more, with the starter motor turning at normal cranking speed, is normal. If the reading is 9 volts or more but the cranking speed is slow, the motor is faulty. If the reading is less than 9 volts and the cranking speed is slow, the solenoid contacts are burned, the starter motor is bad, the battery is discharged or there is a bad connection.

12   Starter motor - removal and installation

Refer to illustrations 12.8 and 12.9
Note: *On some vehicles it may be necessary*

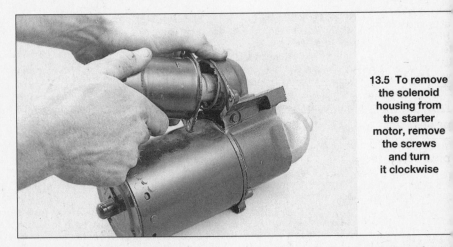

13.5 To remove the solenoid housing from the starter motor, remove the screws and turn it clockwise

to remove the exhaust pipe(s) or frame crossmember to gain access to the starter motor. In extreme cases it may even be necessary to unbolt the mounts and raise the engine slightly to get the starter out.

1   Detach the negative cable from the battery. **Caution:** *On models equipped with the Theftlock audio system, be sure the lockout feature is turned off before performing any procedure which requires disconnecting the battery.*

2   Remove the air cleaner and duct assembly, as necessary.

3   On the 3800, remove the upper mounting bracket, the cooling fan, and the oil cooler lines from the radiator.

4   Raise the vehicle and support it firmly on jackstands.

5   On 3.4 liter, 3800 and 3100 V6 engines with 4T60-E/4T65-E transaxles, remove the flywheel inspection cover.

6   Remove the oil filter splash shield, if necessary.

7   On the 3800, remove the engine harness retainers and position the harness away from the starter.

8   Clearly label, then detach, the electrical connectors from the starter (see illustration).

9   Remove the mounting bolts and nuts equipped), then remove the starter (see illustration). Note the location of any shims use

10   Installation is the reverse of removal.

13   Starter solenoid - removal and installation

Refer to illustrations 13.3 and 13.5

1   Disconnect the cable from the negati terminal of the battery. **Caution:** *On mode equipped with the Theftlock audio system, sure the lockout feature is turned off befo performing any procedure which requires d connecting the battery.*

2   Remove the starter motor (see Se tion 12).

3   Disconnect the strap from the soleno to the starter motor terminal (see illustr tion).

4   Remove the screws which secure t solenoid to the starter motor.

5   Twist the solenoid in a clockwise dire tion to disengage the flange from the star body (see illustration).

6   Installation is the reverse of removal.

Chapter 6
Emissions and engine control systems

Contents

Specifications

Torque specifications

Crankshaft sensor bolts	
1995 and earlier	
3X or 7X sensors	
2.5L	20 in-lbs
All others	70 to 85 in-lbs
24X sensor	96 in-lbs
1996 and later	
3100 and 3.4L engines	
7X sensors	70 in-lbs
24X sensor	96 in-lbs
3800 engine	14 to 28 ft-lbs
Camshaft sensor bolts	
3100 and 3.4L engines	96 in-lbs
3800 engine	35 to 53 in-lbs

General information

Refer to illustration 1.5

To prevent pollution of the atmosphere from burned and evaporating gases, a number of emissions control systems are incorporated on the vehicles covered by this manual. The combination of systems used depends on the year in which the vehicle was manufactured, the locality to which it was originally delivered and the engine type. The major systems incorporated on the vehicles with which this manual is concerned include the:

Fuel Control System
Exhaust Gas Recirculation (EGR) system
Evaporative Emissions Control (EVAP) system
Transmission Converter Clutch (TCC) system
Positive Crankcase Ventilation (PCV) system
Catalytic converter

All of these systems are linked, directly or indirectly, to the On Board Diagnostic (OBD) system. The Sections in this Chapter include general descriptions, checking procedures (where possible) and component replacement procedures (where applicable) for each of the systems listed above.

Before assuming that an emissions control system is malfunctioning, check the fuel and ignition systems carefully. In some cases special tools and equipment, as well as specialized training, are required to accurately diagnose the causes of a rough running or difficult to start engine. If checking and servicing become too difficult, or if a procedure is beyond the scope of the home mechanic, consult your dealer service department or other qualified repair shop. This does not necessarily mean, however, that the emissions control systems are particularly difficult to maintain and repair. You can quickly and easily perform many checks and do most (if not all) of the regular maintenance at home with common tune-up and hand tools. **Note:** *The most frequent cause of emissions system problems is simply a loose or broken vacuum hose or wiring connection. Therefore, always check the hose and wiring connections first.*

Pay close attention to any special precautions outlined in this Chapter. It should be noted that the illustrations of the various systems may not exactly match the system installed on your particular vehicle due to changes made by the manufacturer during production or from year to year.

A Vehicle Emissions Control Information (VECI) label is located in the engine compart-

6

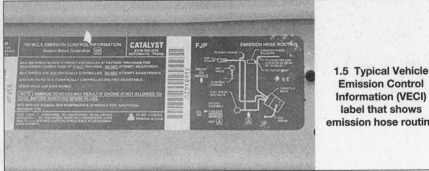

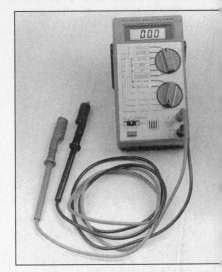

1.5 Typical Vehicle Emission Control Information (VECI) label that shows emission hose routing

ment of all vehicles with which this manual is concerned **(see illustration)**. This label contains important emissions specifications and setting procedures, as well as a vacuum hose schematic with emissions components identified. When servicing the engine or emissions systems, the VECI label in your particular vehicle should always be checked for up-to-date information. **Note:** *Because of a federally mandated extended warranty which covers the emission control system components (and any components which have a primary purpose other than emission control but have significant effects on emissions), check with your dealer about warranty coverage before working on any emission related systems.*

The number of emissions control system components on later model fuel-injected vehicles has actually decreased due to the high efficiency of the new fuel injection and ignition systems. These models are equipped with a three way catalytic converter containing beads which are coated with a catalyst material containing platinum, palladium and rhodium to reduce the level of nitrogen oxides.

2   On Board Diagnostic (OBD) system and trouble codes

Note: *All 1995 and earlier models (except 1994 and later 3.4L and 1995 3800 models) are equipped with OBD I self diagnosis sys-* *tem, while 1994 and later 3.4L and 1995 3800 models and all 1996 and later models are equipped with OBD II self diagnosis system. Both systems require the use of a Scan tool to access trouble codes. However, many of the information sensor checks and replacement procedures do apply to both systems. Because OBD I and OBD II systems require a special SCAN tool to access the trouble codes, have the vehicle diagnosed by a dealer service department or other qualified automotive repair facility if the proper SCAN tool is not available. The codes indicated in the text are designed and mandated by the EPA for all OBD I and OBD II vehicles produced by automobile manufacturers. These generic trouble codes do not include the manufacturer's specific trouble codes. Consult a dealer service department or other qualified repair shop for additional information. Refer to the troubleshooting tips in the beginning of this manual to gain some insight to the most likely causes of a problem.*

Diagnostic tool information

Refer to illustrations 2.1, 2.2 and 2.4

1   A digital multimeter is a necessary tool for checking fuel injection and emission related components **(see illustration)**. A digital volt-ohmmeter is preferred over the older style analog multimeter for several reasons. The analog multimeter cannot display the volts-ohms or amps measurement in hundredths and thousandths increments. When

2.1 Digital multimeters can be used for testing all types of circuits; because of their high impedance, they are much more accurate than analog meters for measuring millivolts in low-voltage computer circuits

working with electronic circuits which are often very low voltage, this accurate reading is most important. Another good reason for the digital multimeter is the high impedance circuit. The digital multimeter is equipped with a high resistance internal circuitry (1 million ohms). Because a voltmeter is hooked up in parallel with the circuit when testing, it is vital that none of the voltage being measured should be allowed to travel the parallel path set up by the meter itself. This dilemma does not show itself when measuring large amounts of voltage (9 to 12 volt circuits) but if you are measuring a low voltage circuit such as the oxygen sensor signal voltage, a fraction of a volt may be a significant amount when diagnosing a problem.

2   Hand-held scanners are the most powerful and versatile tools for analyzing engine management systems used on later model vehicles **(see illustration)**. Early model scan

2.2 Scanners like the Actron Scantool and the AutoXray XP240 are powerful diagnostic aids - programmed with comprehensive diagnostic information, they can tell you just about anything you want to know about your engine management system

2.4 Trouble code tools simplify the task of extracting the trouble codes on OBD I systems

2.5a Typical engine control components – Quad-4 engine

1 Vehicle speed sensor (on transaxle)	5 IAC valve and TPS sensor (on throttle body)
2 Oxygen sensor (on exhaust manifold)	6 MAT sensor
3 Coolant temperature sensor	7 MAP sensor
4 Power steering pressure switch (on steering gear)	

ners handle codes and some diagnostics for many OBD I systems. Each brand scan tool must be examined carefully to match the year, make and model of the vehicle you are working on. Often interchangeable cartridges are available to access the particular manufacturer; Ford, GM, Chrysler, etc.). Some manufacturers will specify by continent; Asia, Europe, USA, etc. Seek the advice of your local parts retailer.

3   With the arrival of the federally mandated emission control system (OBD II), a specially designed scanner must also be developed. At this time, several manufacturers plan to release OBD II scan tools for the home mechanic. Ask the parts salesperson at a local auto parts store for additional information concerning dates and costs. **Note:** *Although OBD II codes cannot be accessed without a Scan tool, follow the simple component checks in Section 4.*

4   Another type of code reader and less expensive is available at parts stores **(see illustration)**. These tools simplify the procedure for extracting codes from the engine management computer by simply "plugging in" to the diagnostic connector on the vehicle wiring harness.

OBD system general description

Refer to illustrations 2.5a, 2.5b, 2.5c and 2.5d

5   The electronically controlled fuel and emissions system is linked with many other related engine management systems. It consists mainly of sensors, output actuators and a Powertrain Control Module (PCM) **(see illustrations)**. Completing the system are various other components which respond to

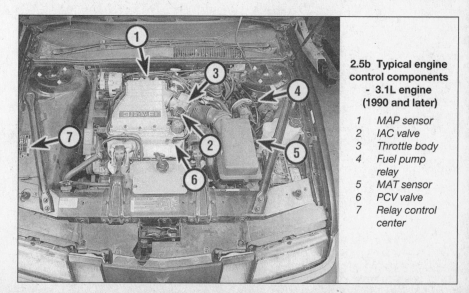

2.5b Typical engine control components - 3.1L engine (1990 and later)

1	MAP sensor
2	IAC valve
3	Throttle body
4	Fuel pump relay
5	MAT sensor
6	PCV valve
7	Relay control center

2.5c Typical engine control components - 3100 engine (1994 and later)

1	Manifold Absolute Pressure sensor	5	Intake Air Temperature sensor	8	PCV valve
2	Digital EGR valve	6	Mass Air Flow sensor	9	Powertrain Control Module
3	Idle Air Control valve	7	Engine Coolant Temperature sensor	10	Camshaft position sensor (not visible)
4	Throttle Position Sensor (not visible)		(not visible)		

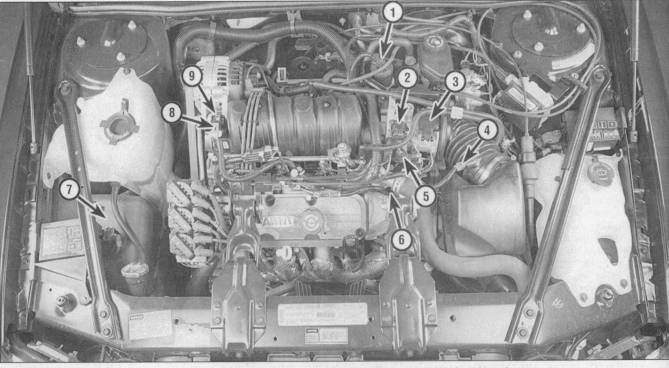

2.5d Typical engine control components - 3800 engine (1995 and later)

1	Digital EGR valve	4	Intake Air Temperature sensor	7	Powertrain Control Module
2	Idle Air Control valve	5	Throttle Position Sensor	8	PCV valve (underneath MAP sensor)
3	Mass Air Flow sensor	6	Engine Coolant Temperature sensor	9	Manifold Absolute Pressure sensor
			(not visible)		

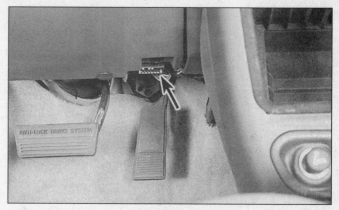

2.11a The 12-pin Data Link Connector (DLC) found on models equipped with OBD I engine control system is located under the dash

A   Ground
B   Diagnostic test terminal (not present on all models)

2.11b The 16-pin Data Link Connector (DLC) found on models equipped with OBD II - the DLC (arrow) is located to the right of the steering column

commands from the PCM.

6   In many ways, this system can be compared to the central nervous system in the human body. The sensors (nerve endings) constantly gather information and send this data to the PCM (brain), which processes the data and, if necessary, sends out a command for some type of vehicle change (limbs).

7   Here's a specific example of how one portion of this system operates: An oxygen sensor, mounted in the exhaust manifold and protruding into the exhaust gas stream, constantly monitors the oxygen content of the exhaust gas as it travels through the exhaust pipe. If the percentage of oxygen in the exhaust gas is incorrect, an electrical signal is sent to the PCM. The PCM takes this information, processes it and then sends a command to the fuel injectors, telling it to change the fuel/air mixture. To be effective, all this happens in a fraction of a second, and it goes on continuously while the engine is running. The end result is a fuel/air mixture which is constantly kept at a predetermined ratio, regardless of driving conditions.

Obtaining trouble codes

Refer to illustrations 2.11a and 2.11b

8   One might think that a system which uses exotic electrical sensors and is controlled by an on-board computer would be difficult to diagnose. This is not necessarily the case.

9   The On Board Diagnostic (OBD) system has a built-in self-diagnostic system, which indicates a problem by turning on a "SERVICE ENGINE SOON" light on the instrument panel when a fault has been detected. **Note:** Since some of the trouble codes do not set the 'SERVICE ENGINE SOON' light, it is a good idea to access the OBD system and look for any trouble codes that may have been recorded and need tending.

10   Perhaps more importantly, the PCM will recognize this fault, in a particular system monitored by one of the various information sensors, and store it in its memory in the form of a trouble code. Although the trouble code cannot reveal the exact cause of the malfunc-

tion, it greatly facilitates diagnosis as you or a dealer mechanic can "tap into" the PCM's memory and be directed to the problem area.

11   To retrieve this information from the PCM on most OBD I and all OBD II systems, a SCAN tool must be connected to the Data Link Connector (DLC) **(see illustrations)**. The SCAN tool is a hand-held digital computer scanner that interfaces with the on-board computer. The SCAN tool is a very powerful tool; it not only reads the trouble codes but also displays the actual operating conditions of the sensors and actuators. SCAN tools are expensive, but they are necessary to accurately diagnose a modern computerized fuel-injected engine. SCAN tools are available from automotive parts stores and specialty tool companies. **Note:** On some OBD I models with a 12-pin Data Link Connector, trouble codes can be access by connecting terminal B to terminal A with a jumper wire with the ignition key in the ON position, and the codes can be read by watching the flashes of the SERVICE ENGINE SOON light on the instrument panel (for example, one flash, pause, followed by four flashes would indicate a code 14). It should be noted, however, that some models with a 12-pin diagnostic connector do not have a terminal B present in the connector. On these models a scan tool is required to access trouble codes. **Caution:** Don't crank or start the engine when terminals A and B are connected by the jumper wire.

12   The self-diagnosis feature built into this system does not detect all possible faults. If you suspect a problem with the On Board Diagnostic (OBD) system, but the SERVICE ENGINE SOON light has not come on and no trouble codes have been stored, take the vehicle to a dealer service department or other qualified repair shop for diagnosis.

13   Furthermore, when diagnosing an engine performance, fuel economy or exhaust emissions problem (which is not accompanied by a SERVICE ENGINE SOON light) do not automatically assume the fault lies in this system. Perform all standard troubleshooting procedures, as indicated elsewhere in this manual, before turning to the On Board Diagnostic (OBD) system.

14   Finally, since this is an electronic system, you should have a basic knowledge of automotive electronics before attempting any diagnosis. Damage to the PCM, Programmable Read Only Memory (PROM) calibration unit or related components can easily occur if care is not exercised.

Clearing trouble codes

15   To clear the codes from the PCM memory, install the SCAN tool, scroll the menu for the function that describes "CLEARING CODES" and follow the prescribed method for that particular SCAN tool or momentarily remove the PCM/IGN fuse from the fuse box for 30 seconds. Clearing codes may also be accomplished by removing the fusible link (main power fuse) located near the battery positive terminal (see Chapter 12) or by disconnecting the cable from the positive terminal (+) of the battery. **Caution:** On models equipped with a Theftlock audio system, be sure the lockout feature is turned off before disconnecting the battery cable. Disconnecting the power to the PCM to clear the memory can be an important diagnostic tool, especially on intermittent problems. **Caution:** To prevent damage to the PCM, the ignition switch must be OFF when disconnecting or connecting power to the PCM. **Note:** Disconnecting the negative battery terminal will erase any radio preset codes that have been stored.

Trouble Code Identification

16   Following is a list of the typical Trouble Codes which may be encountered while diagnosing the On Board Diagnostic (OBD I and OBD II) system. Also included are simplified troubleshooting procedures. If the problem persists after these checks have been made, the vehicle must be diagnosed by a professional mechanic who can use specialized diagnostic tools and advanced troubleshooting methods to check the system. Procedures marked with an asterisk (*) indicate component replacements which may not cure the problem in all cases. For this reason, you may want to seek professional advice before purchasing replacement parts.

OBD I Trouble Codes

Trouble codes	Circuit or system	Probable cause
Code 12 (1 flash, pause, 2 flashes)	Diagnostic	This code will flash whenever the diagnostic terminal is grounded with the ignition turned On and the engine not running. If additional trouble codes are stored in the ECM they will appear after this code has flashed three times.
Code 13 (1 flash, pause, 3 flashes)	Oxygen sensor circuit (open circuit)	Check the wiring and connectors from the oxygen sensor. Replace the oxygen sensor.*
Code 14 (1 flash, pause, 4 flashes)	Coolant sensor circuit (high temperature)	If the engine is experiencing overheating problems, the problem must be rectified before continuing. Check all wiring and connectors associated with the coolant temperature sensor. Replace the coolant temperature sensor.*
Code 15 (Except 2.2L) (1 flash, pause, 5 flashes)	Coolant sensor circuit (low temperature)	See above, then check the wiring connections at the ECM.
Code 16 (3100, '93 3.1 Calif.) (1 flash, pause 6 flashes)	System voltage low	This code will set when the voltage on the ignition feed is 8 volts or less. Check the wiring and ground connections at the ECM, and check for open or chafed wiring.
Code 17 (3100 V6, '93 3.1 Calif.) (1 flash, pause, 7 flashes)	Cam position sensor circuit	Check the wiring and connections between the camshaft position sensor and the ECM.
Code 18 ('93 3800 V6) (1 flash, pause, 8 flashes)	Cam/Crank sensor circuit	Check the connections at the ECM, the crank position flash, sensor and the ignition control module. Also check for opens or shorts in the sensor wiring. Check for bent vanes in the interruptor ring on the vibration damper, and check that the crank sensor is not damaged or misaligned.
Code 21 (2 flashes, pause, 1 flash)	Throttle position sensor	Check for a sticking or misadjusted TPS plunger. Check all wiring and connections between the TPS and the ECM. Adjust or replace the TPS (see Chapter 4).*
Code 22 (Except 2.2L) (2 flashes, pause, 2 flashes)	Throttle position sensor	Check the TPS adjustment (Chapter 4). Check the ECM connector. Replace the TPS (Chapter 4).*
Code 23 (2 flashes, pause, 3 flashes)	Manifold (Intake) air temperature	Check the MAT/IAT sensor, wiring and connectors for an open sensor circuit. Replace the MAT/IAT sensor.*
Code 24 (Except '93 3.1 Calif.) (2 flashes, pause, 4 flashes)	Vehicle speed sensor (3100: No voltage)	A fault in this circuit should be indicated only when the vehicle is in motion. Disregard Code 24 if it is set when the drive wheels are not turning. Check the connections at the ECM. Check the TPS setting.
Code 25 (Except 2.2L) (2 flashes, pause, 5 flashes)	Manifold (Intake) air temperature	Check the voltage signal from the MAT/IAT sensor to the ECM. It should be above 4 volts.
Code 26 (Except 2.2, 2.5, 3100, '93 3.1 Calif.) (2 flashes, pause, 6 flashes)	Quad driver circuit	Have the vehicle checked by a dealer service department.
Code 27 ('91/'92 3800 V6) (2 flashes, pause, 7 flashes)	Gear switch diagnosis	Check for poor connections at the ECM, or for mis-routed or damaged wiring.
Code 28 ('91/'92 3800 V6) (2 flashes, pause, 8 flashes)	Gear switch diagnosis	Check for poor connections at the ECM, or for mis-routed or damaged wiring.
Code 29 ('91/'92 3800 V6) (2 flashes, pause, 9 flashes)	Gear switch diagnosis	Check for poor connections at the ECM, or for mis-routed or damaged wiring.
Code 31 (Turbo models) (3 flashes, pause, 1 flash)	Turbo overboost	Wastegate system malfunction. Have the vehicle checked by a dealer service department.
Code 32 (Except '93 3.1L for Calif.) (3 flashes, pause, 2 flashes)	EGR system	Check vacuum hoses and connections for leaks and restrictions. Replace the EGR valve.*

OBD I Trouble Codes (continued)

Trouble codes	Circuit or system	Probable cause
Code 33 (1988 models) (3 flashes, pause, 3 flashes)	MAF sensor	Replace MAF sensor.*
Code 33 (1989 On) (3 flashes, pause, 3 flashes)	MAP sensor	Check the vacuum hoses from the MAP sensor. Check the electrical connections at the ECM. Replace the MAP sensor.*
Code 34 (1988 models and 3800 through 1993) (3 flashes, pause, 4 flashes)	MAF sensor	Check for loose or damaged air duct, misadjusted minimum idle speed and vacuum leaks. Inspect the MAF sensor and the electrical connections.
Code 34 (1989 on, except 2.2L) (3 flashes, pause, 4 flashes)	MAP sensor	Code 34 will set when the signal voltage from the MAP sensor is too low. Instead the ECM will substitute a fixed MAP value and use the TPS to control fuel delivery. Replace the MAP sensor.*
Code 35 (Except 2.2L) (3 flashes, pause, 5 flashes)	Idle speed error	Code 35 will set when the closed throttle speed is 300 rpm above or below the correct idle speed for 5 seconds. Replace the IAC.*
Code 36 (3100, '93 3.1 Calif.) (3 flashes, pause, 6 flashes)	Ignition control 24X reference	Check the connections between the ECM and the crankshaft position sensor.
Code 36 ('93 3800 V6 only) (3 flashes, pause, 6 flashes)	Transaxle shift control	Check wiring for opens or shorts. Check connectors at transaxle as well as the solenoids. If these are all good, the problem may be an internal hydraulic or mechanical failure.
Code 37 (3100 V6) (3 flashes, pause, 7 flashes)	TCC/Brake switch	Check the connections at the stop lamp/torque converter clutch/cruise control switch, and check the adjustment of the switch.
Code 39 ('92/'93 3.4L V6 only) (3 flashes, pause, 9 flashes)	Transaxle clutch switch	Check for a short to ground between the ECM and the clutch switch, bad connections at the ECM, and make sure that the clutch switch is securely fastened.
Code 41 (V6 through '93 except 3800 and '93 3.1 Calif.) (4 flashes, pause, 1 flash)	Cylinder select error	Check for faulty connections at MEM-CAL or incorrect MEM-CAL installed in ECM. Replace the MEM-CAL.*
Code 41 (Quad-4 engine) (4 flashes, pause, 1 flash)	1X reference	Check the ignition module and ECM connections.
Code 41 (3100 V6) (4 flashes, pause, 1 flash)	Ignition control	Look for an open circuit between the ignition control module and the powertrain control module.
Code 41 (3800 through '93) (4 flashes, pause, 1 flash)	Camshaft position sensor	Check the connections and wiring at the camshaft position sensor and ignition control module. Check for a missing camshaft magnet, or a damaged or faulty sensor.
Code 42 (4 flashes, pause, 2 flashes)	Electronic Spark Timing (Ignition control circuit)	Faulty connections or ignition module.
Code 43 (Except 2.2L, 2.5L) (4 flashes, pause, 3 flashes)	Electronic spark control (Knock sensor circuit)	Faulty knock sensor or MEM-CAL.*
Code 44 (4 flashes, pause, 4 flashes)	Lean exhaust	Check the ECM wiring connections. Check for vacuum leaks at the hoses and intake manifold gasket.*
Code 45 (4 flashes, pause, 5 flashes)	Rich exhaust	Check the evaporative charcoal canister and its components for the presence of fuel.
Code 46 (3100 V6) (4 flashes, pause, 6 flashes)	Security system error	Check for an open or short circuit between the theft deterrent module and the powertrain control module.
Code 51 (5 flashes, pause, 1 flash)	MEM-CAL or PROM	Make sure the MEM-CAL or PROM is properly installed in the ECM. Replace the MEM-CAL or PROM.*

OBD I Trouble Codes (continued)

Trouble codes	Circuit or system	Probable cause
Code 53 (Except 2.2L and '93 3800 V6) (5 flashes, pause, 3 flashes)	System over-voltage	Code 53 will set if the voltage at ECM terminal B2 is greater than 17.1 volts for 2 seconds. Check the charging system.
Code 53 ('93 3800 V6) (5 flashes, pause, 3 flashes)	EGR system	If the EGR valve shows signs of excessive heat, the exhaust system may have a blockage, such as a plugged converter. Also, check the wiring between the EGR valve and the ECM. Check for a blockage in the EGR passage too.
Code 54 (V6 except '93 3800) (5 flashes, pause, 4 flashes)	Fuel pump	Check for a faulty fuel pump relay.
Code 54 ('93 3800 V6) (5 flashes, pause, 4 flashes)	EGR system	If the EGR valve shows signs of excessive heat, the exhaust system may have a blockage, such as a plugged converter. Also, check the wiring between the EGR valve and the ECM. Check for a blockage in the EGR passage too.
Code 55 (Except 2.2, 2.5, '93 3800, and '93 3.1 Calif.) (5 flashes, pause, 5 flashes)	ECM	Check the ECM grounds and the MEM-CAL or PROM seating. Replace the ECM *
Code 55 ('93 3800 V6) (5 flashes, pause, 5 flashes)	EGR system	If the EGR valve shows signs of excessive heat, the exhaust system may have a blockage, such as a plugged converter. Also, check the wiring between the EGR valve and the ECM. Check for a blockage in the EGR passage too.
Code 56 ('93 3800 V6 only) (5 flashes, pause, 6 flashes)	Quad driver	Have the vehicle checked by a dealer service department.
Code 58 (3100 V6) (5 flashes, pause, 8 flashes)	Transaxle fluid temperature	A low voltage signal indicates high transaxle fluid temperature, or a circuit (low voltage) failure in the sensor circuit. Check for a short circuit between the transaxle fluid temperature sensor and the control unit, or a bad control unit.
Code 59 (3100 V6) (5 flashes, pause, 9 flashes)	Transaxle fluid temperature circuit (high voltage)	A high voltage signal indicates low transaxle fluid temperature, or a failure in the circuit. Check for an open circuit between the transaxle fluid temperature sensor and the control unit, a bad sensor ground, or a bad control unit.
Code 61 (V6 except '93 3800 and 3.1 Calif.) (6 flashes, pause, 1 flash)	Degraded oxygen sensor	Check for contaminated fuel. Replace the sensor.*
Code 61 ('93 3800 V6 only) (6 flashes, pause, 1 flash)	Cruise control vent circuit	This particular trouble code will not illuminate the Service Engine Soon light. If code 61 is displayed during the course of troubleshooting, check the wiring and connections between the cruise servo and the ECM. If necessary, replace the cruise servo.*
Code 62 (Quad-4 and 3.4L V6) (6 flashes, 2 flashes)	Trans. gear switch circuit	Check electrical switching circuits of transaxle.
Code 62 ('93 3800 V6) (6 flashes, pause, 2 flashes)	Cruise control VAC solenoid	This particular trouble code will not illuminate the Service Engine Soon light. If code 62 is displayed during the course of troubleshooting, check the wiring and connections between the cruise servo and the ECM. If necessary, replace the cruise servo.*
Code 63 ('93 3800 V6) (6 flashes, pause, 3 flashes)	Cruise control system problem	This particular trouble code will not illuminate the Service Engine Soon light. If code 63 is displayed during the course of troubleshooting, check the wiring and connections between the cruise servo and the ECM. Also check for a binding cruise control throttle cable or leaking vacuum lines. If necessary, replace the cruise servo.*

OBD I Trouble Codes (continued)

Trouble codes	Circuit or system	Probable cause
Code 65 (Quad-4 engine) (6 flashes, pause, 5 flashes)	Fuel injector	Check the ECM and injector wiring and connectors.
Code 65 ('93 3800 V6) (6 flashes, pause, 5 flashes)	Cruise servo position sensor (SPS low)	This particular trouble code will not illuminate the Service Engine Soon light. If code 65 is displayed during the course of troubleshooting, check the wiring and connections between the cruise servo and the ECM. If necessary, replace the cruise servo.*
Code 66 (Quad-4, 3.1, 3.4 and 3100 V6) (6 flashes, pause, 6 flashes)	A/C pressure sensor	Check the sensor connections.
Code 66 (3800 V6) (6 flashes, pause, 6 flashes)	Excessive A/C cycling (Low refrigerant charge)	If the refrigerant pressure is low the ECM can cause the compressor to cycle too often. If excessive cycling is detected, the ECM protects the compressor by disabling the relay, and storing a code 66, until the ignition is turned off and on again. If this happens three times, the ECM will disable the A/C clutch until code 66 is cleared. If the refrigerant charge is low, check all refrigerant lines and connections for leaks. If refrigerant charge is OK, check the A/C wiring and connections. Replace the pressure switch if necessary.*
Code 67 ('93 3800 V6) (6 flashes, pause, 7 flashes)	Cruise control switches	Check the wiring and connections between the cruise engage switch and the ECM.
Code 68 ('93 3800 V6) (6 flashes, pause, 8 flashes)	Cruise servo position sensor (SPS low)	This particular trouble code will not illuminate the Service Engine Soon light. If code 68 is displayed during the course of troubleshooting, check the wiring and connections between the cruise servo and the ECM. Also check for a binding cruise control throttle cable or a blockage in the vacuum dump hose. Check that the vacuum dump valve is properly adjusted. If necessary, replace the cruise servo.*
Code 69 (3800 V6) (6 flashes, pause, 9 flashes)	A/C head pressure high	This particular trouble code will not illuminate the Service Engine Soon light. If code 69 is displayed during the course of troubleshooting, check the wiring and connections between the A/C system and the ECM.
Code 70 (3100, '93 3.1Calif.) (7 flashes)	A/C pressure sensor (high pressure)	Code 70 sets when the voltage signal from the sensor is beyond the normal range, but is not a problem in the refrigerant system. Check for an open sensor ground or a bad wiring connection. If these are good, the sensor may have to be replaced.*
Code 72 (3100 V6) (7 flashes, pause, 2 flashes)	Vehicle speed sensor (Signal error)	This code sets when the voltage signal is lost while the car is traveling at road speeds. Check for a short in the speed sensor wires. If these are good, the speed sensor or the ECM may have to be replaced.*
Code 75 (3100, '93 3.1 Calif.) (7 flashes, pause, 5 flashes)	Digital EGR # 1 solenoid	Check the wires and connections between the EGR valve and the ECM. Also check for plugged passages in the intake plenum. If these are good the EGR valve or the ECM may have to be replaced.*
Code 76 (3100, '93 3.1 Calif.) (7 flashes, pause, 6 flashes)	Digital EGR # 2 solenoid	Check the wires and connections between the EGR valve and the ECM. Also check for plugged passages in the intake plenum. If these are good the EGR valve or the ECM may have to be replaced.*
Code 77 (3100, '93 3.1 Calif.) (7 flashes, pause, 7 flashes)	Digital EGR # 3 solenoid	Check the wires and connections between the EGR valve and the ECM. Also check for plugged passages in the intake plenum. If these are good the EGR valve or the ECM may have to be replaced.*
Code 79 (3100 V6) (7 flashes, pause, 9 flashes)	Transaxle fluid (temperature high)	Certain driving conditions can cause the transaxle fluid to get too hot. For example, trailer towing and driving on steep grades.

6

OBD I Trouble Codes (continued)

Trouble codes	Circuit or system	Probable cause
Code 79 ('93 3.1 Calif.) (7 flashes, pause, 9 flashes)	Vehicle speed sensor (High signal voltage)	A misadjusted park/neutral switch can cause a false code 79. Also check for short circuits in the wiring between the speed sensor and the ECM.
Code 80 (3100 V6) (8 flashes)	Transaxle component error	Check torque converter clutch (TCC) fuse; if blown, check for a short circuit in the wiring between the fuse and the transaxle. If the fuse and wiring are good check for faulty terminal connections at the ECM and transaxle.
Code 80 ('93 3.1 Calif.) (8 flashes)	Vehicle speed sensor (Low signal voltage)	A misadjusted park/neutral switch can cause a false code 79. Also check for short circuits in the wiring between the speed sensor and the ECM. If necessary the vehicle speed sensor or ECM may have to be replaced.
Code 81 ('93 3.1 Calif.) (8 flashes, pause, 1 flash)	Brake switch error	Check the adjustment of the brake switch, as well as the wiring between the brake switch and the ECM.
Code 82 (3100, '93 3.1 Calif.) (8 flashes, pause, 2 flashes)	Ignition control 3X error	An open or short in the 3X signal will cause the symptom cranks but won't run. However, an intermittent problem, such as poor connections, can also set the code. If the wiring is good the problem may be in the ignition control module or the crankshaft position sensor.
Code 85 (3100, '93 3.1 Calif.) (8 flashes, pause, 5 flashes)	PROM error	Check all wiring and connections at the ECM. If these are good, the ECM may need to be reprogrammed or replaced. A replacement ECM must be reprogrammed also.
Code 86 (3100, '93 3.1 Calif.) (8 flashes, pause, 6 flashes)	Analog/Digital (A/D) error	The A/D multiplexer is a chip inside the ECM. If code 86 sets, check for and repair any other codes first. Also check the ECM terminals and wiring.
Code 87 (3100, '93 3.1 Calif.) (8 flashes, pause, 7 flashes)	EEPROM error	Check all wiring and connections at the ECM. If these are good, the ECM may need to be reprogrammed or replaced. A replacement ECM must be reprogrammed also. Reprogramming is a job for your dealer to carry out.
Code 90 (3100 V6) (9 flashes)	Torque Converter Clutch (TCC) error	Check torque converter clutch (TCC) fuse; if blown, check for a short circuit in the wiring between the fuse and the transaxle. If the fuse and wiring are good check for faulty terminal connections at the ECM and transaxle.
Code 96 (3100 V6) (9 flashes, pause, 6 flashes)	Transaxle low voltage	Check the charging system, as well as the wiring and connections at the ECM.
Code 98 (3100 V6) (9 flashes, pause, 8 flashes)	Invalid ECM program	The ECM may have to be reprogrammed.
Code 99 (3100, '93 3800 V6) (9 flashes, pause, 9 flashes)	Invalid ECM program	The ECM may have to be reprogrammed.

Component replacement may not cure the problem in all cases. For this reason, you may want to seek professional advice before purchasing replacement parts.

OBD II Trouble Codes

Code	Code Definition	Location
P0101	Mass Air Flow (MAF) sensor error	See Section 4
P0107	Manifold Absolute Pressure (MAP) sensor circuit low input	See Section 4
P0108	Manifold Absolute Pressure (MAP) sensor circuit high input	See Section 4
P0112	Intake Air Temperature (IAT) sensor circuit low input	See Section 4
P0113	Intake Air Temperature (IAT) sensor circuit high input	See Section 4
P0117	Electronic Coolant Temperature (ECT) sensor circuit low input	See Section 4
P0118	Electronic Coolant Temperature (ECT) sensor circuit high input	See Section 4
P0121	Throttle Position Sensor (TPS) range/performance fault	See Section 4
P0122	Throttle Position Sensor (TPS) circuit low input	See Section 4
P0123	Throttle Position Sensor (TPS) circuit high input	See Section 4
P0131	Upstream heated O2 sensor circuit low voltage (Bank 1, Sensor 1)	See Section 4
P0133	Upstream heated O2 sensor circuit high voltage (Bank 1, Sensor 1)	See Section 4
P0137	Downstream heated O2 sensor circuit low voltage (Bank 1, Sensor 2)	See Section 4
P0138	Downstream heated O2 sensor circuit high voltage (Bank 1, Sensor 2)	See Section 4
P0141	O2 sensor heater circuit fault (Bank 1, Sensor 2)	See Section 4
P0171	System Adaptive fuel too lean	See Chapter 4
P0172	System Adaptive fuel too rich	See Chapter 4
P0191	Injector Pressure sensor system performance	See Chapter 4
P0192	Injector Pressure sensor circuit low input	See Chapter 4
P0193	Injector Pressure sensor circuit high input	See Chapter 4
P0301	Cylinder number 1 misfire detected	See Chapter 5
P0302	Cylinder number 2 misfire detected	See Chapter 5
P0303	Cylinder number 3 misfire detected	See Chapter 5
P0304	Cylinder number 4 misfire detected	See Chapter 5
P0325	Knock sensor circuit 1 fault	See Section 4
P0326	Knock sensor circuit performance	See Section 4

OBD II Trouble Codes (continued)

Code	Code Definition	Location
P0351	COP ignition coil 1 primary circuit fault	See Chapter 5
P0352	COP ignition coil 2 primary circuit fault	See Chapter 5
P0353	COP ignition coil 3 primary circuit fault	See Chapter 5
P0354	COP ignition coil 4 primary circuit fault	See Chapter 5
P0400	EGR flow fault	See Section 8
P0401	EGR insufficient flow detected	See Section 8
P0402	EGR excessive flow detected	See Section 8
P0420	Catalyst system efficiency below threshold (Bank 1)	See Section 11
P0421	Catalyst system efficiency below threshold (Bank 1)	See Section 11
P0430	Catalyst system efficiency below threshold (Bank 2)	See Section 11
P0431	Catalyst system efficiency below threshold (Bank 2)	See Section 11
P0441	EVAP incorrect purge flow	See Section 6
P0443	EVAP VMV circuit fault	See Section 6
P0452	EVAP fuel tank pressure sensor low input	See Section 4
P0453	EVAP fuel tank pressure sensor high input	See Section 4
P0502	VSS circuit low input	See Section 4
P0503	VSS circuit range performance	See Section 4
P0506	IAC system rpm lower than expected	See Chapter 4
P0507	IAC system rpm higher than expected	See Chapter 4
P0602	PCM control module programming error	See Section 4
P0705	Transaxle Range sensor circuit malfunction	See Section 4
P1171	Inadequate fuel flow	See Chapter 4

*Component replacement may not cure the problem in all cases. For this reason, you may want to seek professional advice before purchasing replacement parts.

3 Powertrain Control Module (PCM/ECM) - replacement

Note 1: All vehicles equipped with OBD I diagnostic system have a replaceable PROM or MEM-CAL in the PCM that must be installed into the new PCM when exchanged. All vehicles equipped with OBD II diagnostic system are equipped with (non-replaceable) EEPROM that must be recalibrated with a special factory SCAN tool (TECH 1) after replacement of the PCM.

Note 2: 1995 3100 engines and all 1996 and 1997 models are equipped with a replaceable Knock sensor (KS) module in the PCM that must be installed into the new PCM when exchanged.

Note 3: The Powertrain Control Module (PCM) is often referred to as the Electronic Control Module (ECM) in several instances throughout this manual. For clarification purposes the PCM is the same component as the ECM. When electronic engine control systems where first introduced, the manufacturers named this control unit the ECM, as the engine control systems became more advanced with the additional sensors and related components the manufacturers renamed this control unit the PCM.

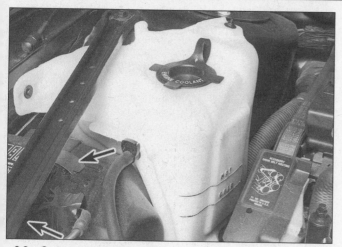

3.3a On some models it will be necessary to detach the clasps (arrows) from the PCM cover

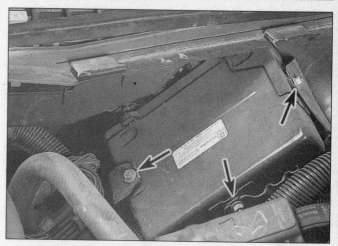

3.3b Location of the ECM mounting screws (arrows)

PCM/ECM replacement

Refer to illustrations 3.3a and 3.3b

Caution: *The ignition switch must be turned off when disconnecting or plugging in the electrical connectors to prevent damage to the PCM.*

1   The Powertrain Control Module (PCM) on 1997 and earlier models is located in the engine compartment in front of the right front strut tower. The Powertrain Control Module (PCM) on 1998 and later models is located at the left front corner of the engine compartment in air cleaner housing.

2   Disconnect the negative battery cable from the battery. **Caution:** *On models equipped with the Theftlock audio system, be sure the lockout feature is turned off before performing any procedure which requires disconnecting the battery.*

3   On 1997 and earlier models remove the retaining nuts **(see illustrations)**. On 1998 and later models remove the air intake duct, the upper air cleaner housing and the air filter (see Chapter 4).

4   Detach the PCM from the inner fender well (1997 and earlier) or air cleaner housing (1998 and later) and unplug the electrical connectors from the PCM.

5   Installation is the reverse of removal.

MEM-CAL or PROM replacement (OBD I diagnostic system)

Refer to illustration 3.6

6   To allow one model of PCM to be used for many different vehicles, a device called a MEM-CAL (MEMory and CALibration) or PROM (Programmable Read Only Memory) is used **(see illustration)**. This device is located inside the PCM and contains information on the vehicle's weight, engine, transaxle, axle ratio, etc. One PCM part number can be used by many GM vehicles but the MEM-CAL or PROM is very specific and must be used only in the vehicle for which it was designed. For this reason, it's essential to check the latest parts book and Service Bulletin information

for the correct part number when replacing a MEM-CAL or PROM. A PCM purchased at a dealer doesn't come with a MEM-CAL or PROM. The MEM-CAL or PROM from the old PCM must be carefully removed and installed in the new PCM.

MEM-CAL

Refer to illustrations 3.8 and 3.11

7   Remove the MEM-CAL access cover.

8   Using two fingers, push both retaining clips back away from the MEM-CAL **(see illustration)**. At the same time, grasp the MEM-CAL at both ends and lift it up out of the socket. Don't remove the MEM-CAL cover itself. **Caution:** *Use of unapproved removal or installation methods may damage the MEM-CAL or socket.*

9   Verify that the numbers on the old PCM and new PCM match up (or that the numbers of the old and new MEM-CALs match up, depending on the component(s) being replaced).

10   To install the MEM-CAL, press only on the ends.

3.6 Two typical Electronic Control Modules (ECM)

3.8 Using two fingers, push the retaining clips (arrows) away from the MEM-CAL and simultaneously grasp it at both ends and lift it up, out of the socket

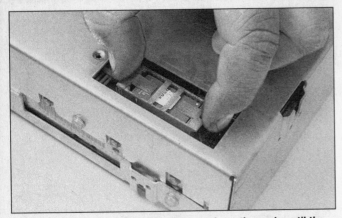

3.11 To install the MEM-CAL, press only on the ends until the retaining clips snap into the ends of the MEM-CAL - make sure the notches in the MEM-CAL are aligned with the small notches in the socket

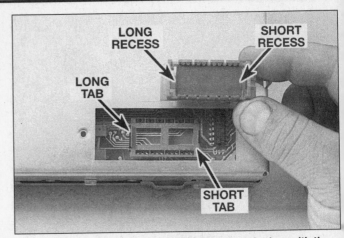

3.14 Note how the notch in the PROM is matched up with the smaller notch in the carrier

11   The small notches in the MEM-CAL must be aligned with the small notches in the MEM-CAL socket. Press on the ends of the MEM-CAL until the retaining clips snap into the ends of the MEM-CAL. Don't press on the middle of the MEM-CAL - press only on the ends **(see illustration)**.

12   The remainder of the installation is the reverse of removal.

PROM

Refer to illustrations 3.14 and 3.15

13   To remove the PROM a special tool should be used. These are usually supplied when a replacement PCM is purchased. **Caution:** *Removal without this tool or with any other tool may cause damage.* Grasp the PROM carrier at the narrow ends. Gently rock the carrier from end-to-end while carefully pulling up.

14   Note the reference end of the PROM carrier **(see illustration)** before setting it aside.

15   Position the PROM and carrier assembly squarely over the socket with the small notched end of the carrier aligned with the

small notch in the socket **(see illustration)**. **Caution:** *Don't press on the PROM; press only on the carrier. Also, if the unit is installed backwards it will be destroyed when the ignition is turned on.*

Final Installation

16   Once the new MEM-CAL or PROM is installed in the old PCM (or the old MEM-CAL or PROM is installed in the new PCM), check the installation to verify its been installed properly by doing the following test:

a) *Turn the ignition switch on.*
b) *Enter the diagnostics mode at the ALDL (data link connector) (see Section 2).*
c) *Allow Code 12 to flash four times to verify that no other codes are present. This indicates the MEM-CAL or PROM is installed properly and the PCM is functioning properly.*

17   If trouble codes 41, 42, 43, 51 or 55 (41, 43, or 51 on Quad-4 engine) occur, or if the Service Engine Soon or Check Engine light is on constantly but isn't flashing any codes, the MEM-CAL or PROM is either not completely seated or it's defective. If it's not

seated, press firmly on the ends of the MEM-CAL or PROM. If it's necessary to remove the MEM-CAL or PROM, follow the above Steps again.

EEPROM reprogramming (OBD II diagnostic system)

18   The 1995 models equipped with 3100 engines and all 1996 and later models are equipped with Electrical Erasable Programmable Read Only Memory (EEPROM) chip that is permanently soldered to the PCM circuit board. The EEPROM can be reprogrammed using the GM TECH 1 scan tool. Do not attempt to remove this component from the PCM. Have the EEPROM reprogrammed at a dealer service department or other qualified repair shop.

Knock Sensor (KS) module replacement

Refer to illustrations 3.21 and 3.22

19   1995 3100 engine, and all 1996 and 1997 models are equipped with a knock sensor (KS) module. New PCMs are not supplied

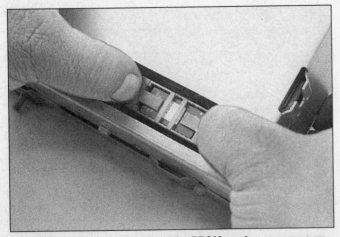

3.15 Press only on the ends of the PROM carrier - pressure on the area in between could result in bent or broken pins or damage to the PROM

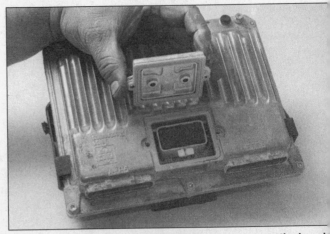
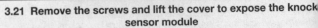

3.21 Remove the screws and lift the cover to expose the knock sensor module

3.22 Pinch the tabs and lift the knock sensor module from the PCM

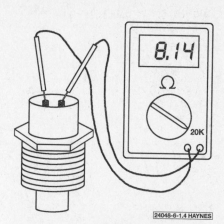

4.1a Be sure the ohmmeter probes make clean contact with the terminals of the sensor when checking the resistance (thermistor type sensor shown)

Temperature (degrees-F)	Resistance (ohms)
212	176
194	240
176	332
158	458
140	668
122	972
112	1182
104	1458
95	1800
86	2238
76	2795
68	3520
58	4450
50	5670
40	7280
32	9420

4.1b Coolant temperature and intake air temperature sensors approximate temperature vs. resistance relationships

with a KS module, therefore it is necessary to remove the original and install it into the new PCM. Use care not to damage the KS module when working with these delicate electrical components.

20 Remove the PCM from the engine compartment (see Step 4).

21 Remove the screws and detach the access cover from the PCM **(see illustration)**.

22 Using two fingers, carefully squeeze the retaining clips together and lift the KS module straight out of the PCM **(see illustration)**.

23 Installation is the reverse of removal. Be sure to use the alignment notches in the seat area of the PCM when sliding the KS module back in place. Press only on the end of the KS module assembly until it snaps back into place.

4   Information sensors - general information and testing

Caution 1: *When performing the following tests, use only a high-impedance (10 mega ohms) digital multi-meter to prevent damage to the PCM.*

Caution 2: *On models equipped with a Theft-lock audio system, be sure the lockout feature is turned off before disconnecting the battery cable.*

Note 1: *Because this system requires a special TECH 1 SCAN tool to access the self-diagnosis system, have the vehicle codes extracted by a dealer service department or other qualified repair facility if the SCAN tool is not available for diagnostic purposes. There are several checks the home mechanic can perform to test for a NO START condition but in the event the driveability symptoms are intermittent or varied, it will be necessary to monitor the operation of the sensor(s) with a SCAN tool or oscilloscope.*

Note 2: *Refer to Chapter 5 for additional information concerning the ignition system and the crankshaft sensor diagnostics.*

Thermistor (two-wire) sensors (coolant temperature, intake air temperature, etc.)

Refer to illustrations 4.1a and 4.1b

1   Thermistors are variable resistors that sense temperature level changes and convert them to a voltage signal. The engine coolant temperature sensor (ECT) and the intake air temperature sensor (IAT) are thermistor type sensors. As the sensor temperature DECREASES, the resistance will INCREASE. As the sensor temperature INCREASES, the resistance will DECREASE. to check thermistor type sensors select the ohms range on the multi-meter, disconnect the harness connector at the sensor and connect the test probes to the sensor terminals **(see illustration)**. The resistance reading (ohms) should be high when the sensor is cold, and low when the sensor is hot **(see illustration)**. Be sure the tips of the probe make clean contact with the terminals inside the sensor to insure an accurate reading.

2   After the resistance of the sensor has been checked, test the system for the proper reference voltage from the computer. Simply disconnect the harness connector at the sensor, select the voltage range on the multimeter and probe the terminals on the harness for the voltage signal. Reference voltage should be approximately 5.0 volts. The ignition switch must be in the ON position (engine not running). If there is no reference voltage available to the sensor, then the circuit and the computer must be checked.

Potentiometers (three-wire) sensors (TPS, EGR valve position sensor)

Refer to illustration 4.3

3   The potentiometer is a variable resistor that converts the voltage signal by varying the resistance according to the driving situation. The signal the potentiometer generates

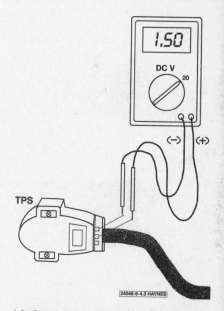

4.3 Carefully backprobe the GROUND (-) and the SIGNAL (+) terminals using straight pins to make contact without disconnecting the harness connector (potentiometer type sensor shown)

is used by the computer to determine position and direction of the device within the component. Although it is possible to measure resistance changes of the sensor, it is best to measure voltage changes as the end result of the potentiometer's performance. Select the DC volts function on the multi-meter, carefully backprobe the harness connector using straight pins inserted into the correct terminals and connect the meter probes to the pins. Connect the negative probe (-) to the ground terminal and the positive probe to the SIGNAL terminal. It will be necessary to refer to the wiring diagrams at the end of this manual for the correct terminals. Observe the meter as the signal arm is moved through its complete range (sweep). The voltage should vary as the arm is moved

VOLTAGE RANGE	ALTITUDE
3.8 - 5.5V	Below 1,000
3.6 - 5.3V	1,000 - 2,000
3.5 - 5.1V	2.000 - 3,000
3.3 - 5.0V	3,000 - 4,000
3.2 - 4.8V	4,000 - 5,000
3.0 - 4.6V	5,000 - 6,000
2.9 - 4.5V	6,000 - 7,000
2.8 - 4.3V	7,000 - 8,000
2.6 - 4.2V	8,000 - 9,000
2.5 - 4.0V	9,000 - 10,000 FEET

LOW ALTITUDE = HIGH PRESSURE = HIGH VOLTAGE

24071-6-4.09 HAYNES

4.5a Typical Manifold Absolute Pressure (MAP) sensor altitude (pressure) vs. voltage values

through its range. These sensors are often used as the throttle position sensor or the EGR valve position sensor. Most of these type sensors will vary from as little as 0.2 volts, up to 5.0 volts (closed to open).

4   After the SIGNAL voltage has been checked, test the system for the proper reference voltage from the computer. Simply disconnect the harness connector at the sensor, select the voltage range on the rotary switch on the volt/ohmmeter and probe the correct terminals on the harness for the voltage signal. Reference voltage should be approximately 5.0 volts. The ignition switch must be in the ON position (engine not running). If there is no reference voltage available to the sensor, then the circuit and the computer must be checked.

Pressure/vacuum (three-wire) sensors (MAP, BARO, vacuum sensors, etc.)

Refer to illustrations 4.5a and 4.5b

5   The pressure/vacuum sensors monitor the intake manifold pressure changes resulting from changes in engine load and speed and converts the information into a voltage output. The PCM uses the sensor to control fuel delivery and ignition timing. The sensor SIGNAL voltage to the PCM varies from below 2.0 volts at idle (high vacuum) to above 4.0 volts with the ignition key ON (engine not running) or wide open throttle (WOT) (low vacuum). These values correspond with the altitude and pressure changes the vehicle experiences while driving **(see illustration)**. Select the DC volts function on the meter, carefully backprobe the harness connector terminals using straight pins and connect the voltmeter probes to the pins **(see illustration)**. Connect the negative probe (-) to the ground terminal and the positive probe to the SIGNAL terminal. It will be necessary to refer to the manufacturer's schematic or wiring diagram for the correct terminals. Observe the meter as the engine idles and then slowly raise the engine rpm to wide open throttle. The voltage should increase from approximately 0.5 to 1.5 volts (high vacuum) to 5.0 volts (low vacuum). If the engine stalls or runs roughly, it is possible to simulate conditions by attaching a hand-held vacuum pump to the pressure/vacuum sensor to simulate running conditions. Also, be sure the manufacturer specifies the type of pressure/vacuum sensor installed on the vehicle. Some manufacturers equip the fuel injection system with a voltage varying sensor while others use a

frequency varying sensor. The latter type must be checked using the frequency (Hz) scale on the meter.

6   If SIGNAL voltage does not exist, test the system for the proper reference voltage from the computer. Simply disconnect the harness connector at the sensor, select the voltage range on the meter and probe the correct terminals on the harness for the voltage signal. Reference voltage should be approximately 5.0 volts. The ignition switch must be in the ON position (engine not running). If there is no reference voltage available to the sensor, then the circuit and the computer must be checked.

Magnetic reluctance (two-wire) sensors (crankshaft, vehicle speed, etc.)

Refer to illustration 4.7

7   Magnetic reluctance type sensors consist of a permanent magnet with a coil wire wound around the assembly. Crankshaft sensors and vehicle speed sensors are common applications of this type sensor. A steel disk mounted on a gear (crankshaft, input shaft, etc.) has tabs that pass between the pole pieces of the magnet causing a break in the magnetic field when passed near the sensor. This break in the field causes a magnetic flux producing reluctance (resistance) thereby changing the voltage signal. This voltage signal is used to determine crankshaft position, vehicle speed, etc. Because magnetic energy is a free-standing source of energy and does not require battery power to produce voltage, this type of sensor must be checked by observing voltage fluctuations with an AC volt meter. Simply switch the voltmeter to the

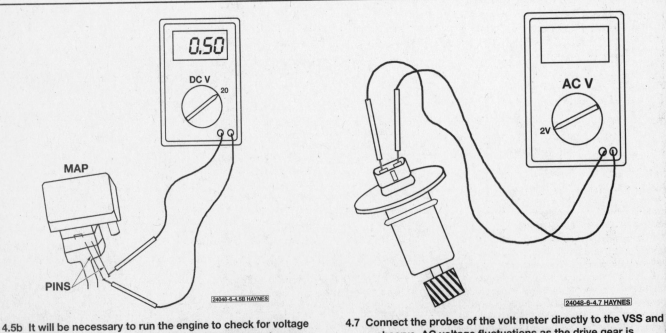

4.5b It will be necessary to run the engine to check for voltage fluctuations in order to create pressure and vacuum changes within the system (pressure/vacuum type sensor shown)

4.7 Connect the probes of the volt meter directly to the VSS and observe AC voltage fluctuations as the drive gear is slowly rotated

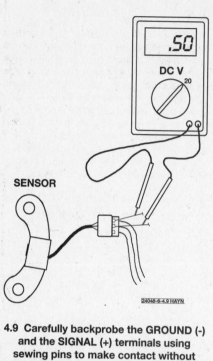

4.9 Carefully backprobe the GROUND (-) and the SIGNAL (+) terminals using sewing pins to make contact without disconnecting the harness connector (Hall effect type sensor shown)

AC scale and connect the probes to the sensor. On vehicle speed sensors it will be necessary to place the transmission in neutral, then with the help of an assistant, hold one tire still while spinning the other tire (approximately 2 MPH or faster) and, observe the voltage fluctuations. This test can also be performed with the sensor removed from the vehicle, by turning the sensor's drive gear **(see illustration)**. **Note:** *Some vehicle speed sensors don't have a drive gear - this type of sensor will have to be checked in place.* On crankshaft position sensors it will be necessary to have an assistant crank the engine over in short bursts at the ignition key while you observe the voltage fluctuations. The meter should register slight voltage fluctuations that are constant and relatively the same range. These small voltage fluctuations indicate that the magnetic portion of the sensor is producing a magnetic field and "sensing" engine parameters for the computer.

Hall effect (three-wire) sensors (crankshaft, camshaft, etc.)

Refer to illustration 4.9

8   These sensors consist of a single hall effect switch, and a magnet which are separated by an air gap. The hall effect switch completes the ground circuit when a magnetic field is present. As the engine rotates, these sensors will produce ON OFF signals to the PCM to determine crankshaft and camshaft position and speed. Hall effect crankshaft sensors generally use an interrupter ring located on the back of the

crankshaft balancer to disrupt the ground signal from the sensor. Always be sure to check the interrupter ring for damage while diagnosing a faulty crankshaft position sensor. Hall effect camshaft sensors typically use a rotating magnet on the camshaft sprocket or in the sensor housing to complete the hall effect switch ground signal. The PCM uses the sensors to control fuel delivery and ignition timing. The camshaft sensor typically produces only one ON OFF signal per camshaft revolution while the crankshaft sensor produces as many as thirty ON OFF signals per crankshaft revolution depending on the make and manufacturer of the vehicle.

9   To check Hall effect type sensors, select the DC volts function on the multi-meter, carefully backprobe the harness connector terminals using pins and connect the voltmeter probes to the pins **(see illustration)**. Connect the negative probe (-) to the ground connection and the positive probe to the SIGNAL terminal. It will be necessary to refer to the wiring diagrams at the back of this manual for the correct terminals. With the ignition key in the ON position and the engine OFF, rotate the engine crankshaft by hand (one full turn) and observe the voltmeter. If the sensor is operating properly it should trigger ON OFF voltage readings on the voltmeter while the engine is being rotated. Depending on the make and model of the vehicle, there should be 5.0 to 12 volts of signal voltage. **Note:** *When testing camshaft sensors it may be necessary to have an assistant crank the engine over in short bursts at the ignition key (detach the primary (low-voltage) wires from the coil to prevent the engine from starting) while observing the voltmeter. It could take as many as two complete crankshaft revolutions before a camshaft sensor signal is detected.*

10   If SIGNAL voltage does not exist, test the system for the proper reference voltage from the computer. Simply disconnect the harness connector at the sensor, select the voltage range on the rotary switch on the volt/ohmmeter and probe the correct terminals on the harness for the reference signal. Depending on the make and model of the vehicle there should be 5.0 to 12 volts of reference voltage (5 volts is the most common). The ignition switch must be in the ON position (engine not running). If there is no reference voltage available to the sensor, the circuit and the computer must be checked.

Dual Hall effect (four-wire) sensors (crankshaft)

11   Dual Hall effect sensors consist of two hall-effect switches and a shared magnet mounted between them. The magnet and each Hall-effect switch are separated by an air gap. Each hall effect switch completes the ground circuit when a magnetic field is present. Dual hall effect sensors are often used as a crankshaft sensor in place of two single hall effect switches described above. Typically the crankshaft balancer is equipped with two "interrupter rings" (each having a differ-

ent number of interrupter blades) which create two separate ON OFF signal patterns that can be used for the timing and the ignition sequence.

12   To check dual hall effect type sensors, select the DC volts function with the rotary switch, carefully backprobe the harness connector terminals using pins and position the voltmeter probes onto the pins **(see illustration 4.9)**. Connect the negative probe (-) to the ground connection and the positive probe to one SIGNAL terminal. It will be necessary to refer to the manufacturer's schematic or wiring diagram for the correct terminals. With the ignition key in the ON position and the engine OFF, rotate the engine crankshaft by hand (one full turn) and observe the volt meter. Next, connect the positive probe to the second SIGNAL terminal. If the sensor is operating properly it should trigger ON OFF voltage readings on the voltmeter from both SIGNAL terminals while the engine is being rotated. Depending on the make and model of the vehicle, there should be 5 to 12 volts of signal voltage at each terminal.

13   If SIGNAL voltage does not exist, test the system for the proper reference voltage from the computer. Simply disconnect the harness connector at the sensor, select the voltage range on the rotary switch on the volt/ohmmeter and probe the correct terminals on the harness for the reference signal. Depending on the make and model of the vehicle there should be 5.0 to 12 volts of reference voltage (5 volts is the most common). The ignition switch must be in the ON position (engine not running). If there is no reference voltage available to the sensor, the circuit and the computer must be checked.

Oxygen (O2) sensors

14   The oxygen sensor(s) monitors the oxygen content of the exhaust gas stream. The oxygen content in the exhaust reacts with the oxygen sensor to produce a voltage output which varies from 0.1-volt (high oxygen, lean mixture) to 0.9-volts (low oxygen, rich mixture). The PCM constantly monitors this variable voltage output to determine the ratio of oxygen to fuel in the mixture. The PCM alters the air/fuel mixture ratio by controlling the pulse width (open time) of the fuel injectors. The PCM and the oxygen sensor(s) attempt to maintain a mixture ratio of 14.7 parts air to 1 part ratio of fuel at all times. The oxygen sensor produces no voltage when it is below its normal operating temperature of about 600-degrees F. During this initial period before warm-up, the PCM operates in OPEN LOOP mode. When checking the oxygen sensor system, it will be necessary to test all oxygen sensors. **Note:** *Because the oxygen sensor(s) are difficult to access, probing the harness electrical connectors for testing purposes will require patience. The exhaust manifolds and pipes are extremely hot and will melt stray electrical probes and leads that touch the surface during testing. If possible, use a SCAN tool that plugs into the DLC*

(diagnostic link). This tool will access the PCM data stream and indicates the millivolt changes for each individual oxygen sensor.

15   Check the oxygen sensor millivolt signal. Locate the oxygen sensor electrical connector and carefully backprobe it using a long pin(s) into the appropriate wire terminals. In most models, connect the positive probe (+) of a voltmeter onto the SIGNAL wire and the negative probe (-) to the ground wire. Consult the wiring diagrams at the end of Chapter 12 for additional information on the oxygen sensor electrical connector wire color designations. **Note:** *Downstream oxygen sensors will produce much slower fluctuating voltage values to reflect the results of the catalyzed exhaust mixture from rich or lean to less presence of CO, HC and Nox molecules. Here the CO_2 and H_2O gaseous forms do not register or react with the oxygen sensors to such a large degree.* Monitor the SIGNAL voltage (millivolts) as the engine goes from cold to warm.

16   The oxygen sensor will produce a steady voltage signal of approximately 0.1 to 0.2 volts (100 to 200 millivolts) with the engine cold (open loop). After a period of approximately two minutes, the engine will reach operating temperature and the oxygen sensor will start to fluctuate between 0.1 to 0.9 volts (100 to 900 millivolts) (closed loop). If the oxygen sensor fails to reach the closed loop mode or there is a very long period of time until it does switch into closed loop mode, replace the oxygen sensor with a new part. **Note:** *Downstream oxygen sensors will not change voltage values as quickly as upstream oxygen sensors. Because the downstream oxygen sensors detect oxygen content after the exhaust has been catalyzed, voltage values should fluctuate much slower and deliberate.*

17   Also inspect the oxygen sensor heater. Disconnect the oxygen sensor electrical connector and working on the oxygen sensor side, connect an ohmmeter between the black wire (-) and brown wire (+). It should measure approximately 5 to 7 ohms. **Note:** *The wire colors often change from the harness connector to the oxygen sensor connector wires according to manufacturer's specifications. Follow the wire colors to the oxygen sensor electrical connector and determine the matching wires and their colors before testing the heater resistance.*

18   Check for proper supply voltage to the heater. Disconnect the oxygen sensor electrical connector and working on the engine side of the harness, measure the voltage between the black wire (-) and brown wire (+) on the oxygen sensor electrical connector. There should be battery voltage with the ignition key ON (engine not running). If there is no voltage, check the circuit between the main relay, the fuse and the sensor. **Note:** *It is important to remember that supply voltage will only reach the O2 sensor with the ignition key ON (engine not running).*

19   If the oxygen sensor fails any of these tests, replace it with a new part.

Mass Air Flow (MAF) sensors

20   The MAF sensor measures the amount of air passing through the sensor body and ultimately entering the engine through the throttle body. The PCM uses this information to control fuel delivery - the more air entering the engine (acceleration), the more fuel needed.

21   A SCAN tool is necessary to check the output of the MAF sensor. The SCAN tool displays the sensor output in grams per second. With the engine idling at normal operating temperature, the display should read 4 to 7 grams per second. When the engine is accelerated the values should raise and remain steady at any given RPM. A failure in the MAF sensor or circuit will also set a diagnostic trouble code.

Knock sensors

22   Knock sensors detect abnormal vibration in the engine. The knock control system is designed to reduce spark knock during periods of heavy detonation. This allows the engine to use maximum spark advance to improve driveability. Knock sensors produce AC output voltage which increases with the severity of the knock. The signal is fed into the PCM and the timing is retarded to compensate for the severe detonation.

23   To check a knock sensor, disconnect the electrical connector and drain the engine coolant as described in Chapter 1. Remove the sensor from the engine block, then reconnect the wiring harness to it. **Note:** *Most knock sensors have a 1/4-inch pipe thread, use a pipe plug of the same size to thread into the engine block as the sensor is removed. This will relieve the task of draining the coolant with minimal coolant loss.* These type of sensors must be checked by observing voltage fluctuations with a voltmeter. Simply switch the voltmeter to the lowest voltage scale and connect the negative probe (-) to the sensor body (ground) and the positive probe to the sensor terminal. With the voltmeter connected to the sensor, gently tap on the bottom of the knock sensor with a hammer or similar device (this simulates the knock from the engine) and observe voltage fluctuations on the meter. If no voltage fluctuations can be detected, the sensor is bad and should be replaced with a new part.

Neutral Start switch

24   The Neutral Start switch or Transaxle range sensor, located on the rear upper part of the automatic transaxle, indicates to the PCM when the transaxle is in Park or Neutral. This information is used for Transaxle Converter Clutch (TCC), Exhaust Gas Recirculation (EGR) and Idle Air Control (IAC) valve operation. **Caution:** *The vehicle should not be driven with the Neutral Start switch disconnected because idle quality will be adversely affected.*

25   For more information regarding the Neutral Start switch, which is part of the Neutral start and back-up light switch assembly, see Chapter 7.

Air conditioning control

26   During air conditioning operation, the PCM controls the application of the air conditioning compressor clutch. The PCM controls the air conditioning clutch control relay to delay clutch engagement after the air conditioning is turned ON to allow the IAC valve to adjust the idle speed of the engine to compensate for the additional load. The PCM also controls the relay to disengage the clutch on WOT (wide open throttle) to prevent excessively high rpm on the compressor. Be sure to check the air conditioning system as detailed in Chapter 3 before attempting to diagnose the air conditioning clutch or electrical system.

Power steering pressure sensor

27   Turning the steering wheel increases power steering fluid pressure and engine load. The pressure switch will close before the load can cause an idle problem. A pressure switch that will not open or an open circuit from the PCM will cause timing to retard at idle and this will affect idle quality. A pressure switch that will not close or an open circuit may cause the engine to die when the power steering system is used heavily. Any problems with the power steering pressure switch or circuit should be repaired by a dealer service department or other qualified repair shop.

Fuel Tank Pressure (FTP) sensor

28   The fuel tank pressure (FTP) sensor is used to monitor the fuel tank pressure or vacuum during the OBD II test portion for emissions integrity. This test scans various sensors and output actuators to detect abnormal amounts of fuel vapors that may not be purging into the canister and/or the intake system for recycling. The FTP sensor helps the PCM monitor this pressure differential (pressure vs. vacuum) inside the fuel tank. Any problems with the fuel tank pressure (FTP) sensor or circuit should be repaired by a dealer service department or other qualified repair shop.

Transaxle Converter Clutch (TCC) system

29   The purpose of the Torque Converter Clutch (TCC) system, equipped in automatic transaxles, is to eliminate the power loss of the torque converter stage when the vehicle is in the cruising mode (usually above 35 mph). This economizes the automatic transaxle to the fuel economy of the manual transaxle. The lock-up mode is controlled by the PCM through the activation of the TCC apply solenoid which is built into the automatic transaxle. When the vehicle reaches a specified speed, the PCM energizes the solenoid and allows the torque converter to lock-up and mechanically couple the engine to the transaxle, under which conditions emissions are at their minimum. However, because of other operating condition

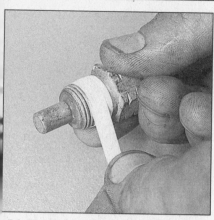

5.2 To prevent coolant leakage, be sure to wrap the temperature sensor threads with Teflon tape before installation

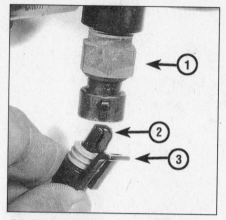

5.3 A typical engine coolant temperature sensor (1) - electrical connector (2) has a locking tab (3) that must be released to unplug the connector

5.7 Typical Mass Air Flow (MAF) sensor installation (arrow)

demands (deceleration, passing, idle, etc.), the transaxle must also function in its normal, fluid-coupled mode. When such latter conditions exist, the solenoid de-energizes, returning the torque converter to normal operation. The converter also returns to normal operation whenever the brake pedal is depressed.
30   Due to the requirement of special diagnostic equipment for the testing of this system, and the possible requirement for dismantling of the automatic transaxle to replace components of this system, Checking and replacing of the components should be handled by a dealer service department or other qualified repair facility.

5   Information sensors - replacement

Engine Coolant Temperature (ECT) sensor
Refer to illustrations 5.2 and 5.3
Warning: Wait until the engine is completely cool before beginning this procedure.
     The coolant temperature sensor is typically located in the thermostat housing or near the thermostat. On the 3.4L engine it is located on the lower intake manifold at the rear of the right cylinder head (facing the firewall). The coolant temperature sensor is a thermistor (a resistor which varies the value of its resistance in accordance with temperature changes). Refer to Section 4 for the checking procedures.
2   Before installing the new sensor, wrap the threads with Teflon sealing tape to prevent leakage and thread corrosion (see illustration).
3   To remove the sensor, release the locking tab, unplug the electrical connector (see illustration), then carefully unscrew the sensor. Caution: Handle the coolant sensor with care. Damage to this sensor will affect the operation of the entire fuel injection system.
4   Installation is the reverse of removal. Check the coolant level and add some, if necessary (see Chapter 1).

Mass Airflow (MAF) sensor
Refer to illustrations 5.7 and 5.8
5   The Mass Airflow Sensor (MAF) is typically located in the air intake duct between the throttle body and the air cleaner housing.

On 1995 and later 3800 engines the MAF sensor is located on the throttle body. The MAF sensor is a hot-wire type sensor and is used to measure the amount of air entering the engine. Refer to Section 4 for the checking procedures. Note : Not all engines are equipped with a MAF sensor, engines not equipped with a MAF sensor will be equipped with a MAP sensor.
6   Disconnect the electrical connector from the MAF sensor.
7   Loosen the clamps securing the MAF sensor to the air intake duct and remove the sensor from the vehicle (see illustration).
8   On 1995 and later 3800 engines, remove the screws and lift the MAF sensor from the throttle body (see illustration).
9   Installation is the reverse of removal.

Manifold Absolute Pressure (MAP) sensor
Refer to illustrations 5.10 and 5.13
10   The Manifold Absolute Pressure (MAP) sensor is typically mounted to the upper intake manifold or to the firewall (see illustration). On the 1995 and later 3800 engines it is located at the front of the engine underneath the fuel injector (acoustic) trim cover. This sensor is a pressure/ vacuum type sensor

5.8 On the 1995 and later 3800 engines, remove the screws (arrows) securing the MAF sensor to the throttle body

5.10 Typical MAP sensor mounting details

1	MAP sensor assembly	3	Vacuum line
2	Mounting screws	4	Electrical connector

6

5.13 On 1995 and later 3800 engines, squeeze the clips and disengage the MAP sensor from the PCV valve housing

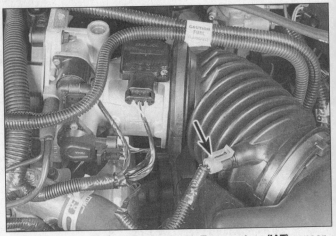

5.15a On some models the Intake Air Temperature (IAT) sensor (arrow) is mounted in the air intake duct

which monitors the intake manifold pressure changes resulting from changes in engine load and speed. Refer to Section 4 for the checking procedures.

11   To replace the sensor, detach the vacuum hose, unplug the electrical connector

5.15b On other models the Intake Air Temperature (IAT) sensor (arrow) is mounted in the air cleaner housing

and remove the mounting screws.

12   To replace the sensor on 3.4L models, unplug the electrical connector and remove the sensor retaining bracket and bolt. Then pull straight up to remove it.

13   To replace the sensor on 1995 and later 3800 engines, unplug the electrical connector and detach the retaining clips to remove it **(see illustration)**.

14   Installation is the reverse of removal.

Intake Air Temperature (IAT) sensor

Refer to illustrations 5.15a and 5.15b

15   The Intake Air Temperature (IAT) sensor is located inside the air duct directly downstream of the air filter housing or in the air cleaner housing **(see illustrations)**. On earlier models this sensor is also referred to as a Manifold Air Temperature (MAT) sensor. The IAT or MAT sensor is a thermistor (a resistor which varies the value of its resistance in accordance with temperature changes). Refer to Section 4 for the checking procedures.

16   To remove an IAT or MAT sensor,

unplug the electrical connector and remove the sensor from the air intake duct or the air cleaner housing. On sensors located in the air intake duct carefully twist the sensor to release it from the rubber boot. Sensors mounted to the air cleaner housing will require a wrench to unscrew them.

17   Installation is the reverse of removal.

Oxygen sensor(s)

Refer to illustrations 5.18a, 5.18b and 5.23

Note: *Because it is installed in the exhaust manifold or pipe, which contracts when cool, the oxygen sensor may be very difficult to loosen when the engine is cold. Rather than risk damage to the sensor (assuming you are planning to reuse it in another manifold or pipe) or the threads, which it screws into, start and run the engine for a minute or two, then shut it off. Be careful not to burn yourself during the following procedure.*

18   Vehicles equipped with OBD-I engine control systems have a single heated oxygen sensor located in the exhaust manifold. Vehicles equipped with OBD-II engine control systems have an upstream O2 sensor (before

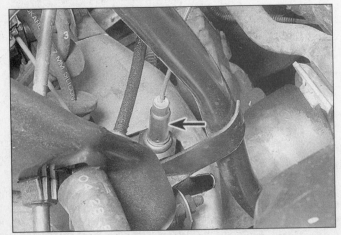

5.18a The oxygen sensor (arrow) is located in the exhaust manifold

5.18b Location of the downstream oxygen sensor – OBD II engine control systems

5.23 Use a special slotted socket to remove the oxygen sensor from the exhaust manifold or exhaust pipe

the catalytic converter in the rear exhaust manifold) and a downstream O_2 sensor (after the catalytic converter) **(see illustration)**. The oxygen sensor(s) monitors the oxygen content of the exhaust gas stream. Oxygen content in the exhaust reacts with the oxygen sensor to produce a voltage output which allows the PCM to change the air fuel ratio in the engine. Refer to Section 4 for the checking procedures.

19   The following is a list of special precautions which must be taken whenever the sensor is serviced.

a)   *The oxygen sensor has a permanently attached pigtail and electrical connector which should not be removed from the sensor. Damage or removal of the pigtail or electrical connector can adversely affect operation of the sensor.*

b)   *Grease, dirt and other contaminants should be kept away from the electrical connector and the louvered end of the sensor.*

c)   *Do not use cleaning solvents of any kind on the oxygen sensor.*

d)   *Do not drop or roughly handle the sensor.*

e)   *The silicone boot must be installed in the correct position to prevent the boot from being melted and to allow the sensor to operate properly.*

f)   *The sensor is designed to allow air circulation to the internal portion of the sensor. Whenever the sensor is removed and installed or replaced, make sure the air passages are not restricted.*

20   Disconnect the cable from the negative terminal of the battery. **Caution:** *On models equipped with a Theftlock audio system, be sure the lockout feature is turned off before disconnecting the battery cable.*

21   Raise the vehicle and place it securely on jackstands.

22   Remove any exhaust heat shields which would interfere with the removal of the oxygen sensor(s), then disconnect the electrical connector from the sensor.

23   Carefully unscrew the sensor from the exhaust manifold or the exhaust pipe **(see illustration)**.

24   Anti-seize compound must be used on the threads of the sensor to facilitate future removal. The threads of new sensors will already be coated with this compound, but if an old sensor is removed and reinstalled, re-coat the threads.

25   Install the sensor and tighten it securely.

26   Reconnect the electrical connector of the pigtail lead to the main engine wiring harness.

27   Lower the vehicle, take it on a test drive and check to see that no trouble codes set.

Throttle Position Sensor (TPS)

28   The Throttle Position Sensor (TPS) is located on the end of the throttle shaft on the throttle body regardless of whether the engine is equipped with a Throttle Body Injection (TBI), Port Fuel Injection (PFI) or a Sequential Fuel Injection (SFI) system. The TPS is a potentiometer type sensor and is used to measure the throttle valve position

and angle. Refer to Section 4 of this Chapter for the checking procedures of the TPS or to Chapter 4 for the removal and installation procedures of the TPS.

Crankshaft position sensor(s)

Refer to illustrations 5.29 and 5.30

29   2.2L, 2.3L (Quad-4) and 2.5L four cylinder engines and 2.8L, 3.1L and 1993 and earlier 3.4L V6 engines are equipped with a single 3X or 7X crankshaft position sensor (CKP) which is mounted on the side of the engine block **(see illustration)**. 3X and 7X sensors are hall effect type sensors that produce 7 ON/OFF signals per crankshaft revolution. This sensor monitors crankshaft position and speed for the Ignition control module which then in turn sends a signal to the PCM to control the ignition and fuel systems. Refer to Section 4 for the checking procedures.

30   3100 and 1994 and later 3.4L V6 engines are equipped with two crankshaft position sensor(s), the first crankshaft sensor (24X sensor) is mounted at the front of the engine behind the crankshaft balancer **(see illustration)**. The 24X sensor is a hall effect type sensor that produces 24 ON/OFF signals per crankshaft revolution. This sensor monitors crankshaft position and speed to improve idle spark control below 1250 RPM. The second crankshaft sensor on 3100 and 1994 and later 3.4L engines (3X sensor on 1995 models or a 7X sensor on 1996 and later models) is mounted on the side of the engine block **(see illustration 5.29)**. 3X and 7X sensors are hall effect type sensors that produce 7 ON/OFF signals per crankshaft revolution and are the same sensors used on four cylinder engines and earlier V6 engines. This sensor monitors crankshaft position and speed for the Ignition control module which then in turn sends a signal to the PCM to control the ignition and fuel systems. Refer to Section 4 for the checking procedures

31   3800 models are equipped with a dual (18X and 3X) crankshaft position sensor which is mounted at the front of the engine

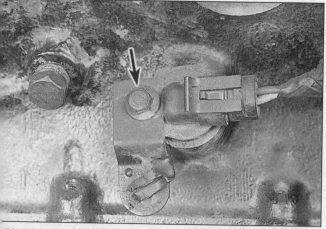

5.29 3X or 7X crankshaft position sensors (A) receive pulses from a magnet on the crankshaft and are mounted to the side of the engine block - do not confuse the CKP sensor with a knock sensor which is threaded into a coolant passage

5.30 24X and dual hall effect crankshaft sensors (arrow) are mounted at the front of the engine behind the crankshaft balancer (3800 engine shown)

6

5.42a On 3100 engines, follow the cam sensor wiring harness (arrows) to locate the cam sensor

5.42b On the 3800 engine, the camshaft sensor (arrow) is located on the front cover (view from below)

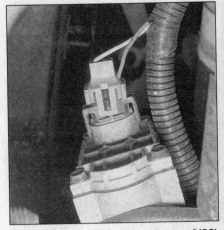

5.47a Typical Vehicle Speed Sensor (VSS) found on 3T40 transaxles - the speed sensor is located at the rear of the transaxle housing near the right driveaxle

behind the crankshaft balancer **(see illustration 5.30)**. This sensor is a dual hall effect type sensor that can produce two signal patterns (outputs) at the same time. The 18X portion of the switch will produce 18 ON/OFF signals per crankshaft revolution while the 3X portion of the switch will produce 3 ON/OFF signals per crankshaft revolution. Refer to Section 4 for the checking procedures.

24X and dual hall effect sensor replacement

32   Disconnect the negative terminal from the battery. **Caution:** *On models equipped with a Theftlock audio system, be sure the lockout feature is turned off before disconnecting the battery cable.*

33   Disconnect the electrical connector from the crankshaft sensor.

34   Remove the crankshaft balancer (see Chapter 2).

35   On 3800 models, carefully pry off the sensor cover. On 3100 and 3.4L models, remove the connector retaining bracket.

36   Remove the bolts securing the sensor to the front cover **(see illustration 5.30)**.

37   Installation is the reverse of removal. Tighten the bolts to the torque listed in this Chapter's Specifications.

3X and 7X sensor replacement

38   Disconnect the negative terminal from the battery. **Caution:** *On models equipped with a Theftlock audio system, be sure the lockout feature is turned off before disconnecting the battery cable.*

39   Disconnect the electrical connector from the crankshaft sensor.

40   Remove the bolt from the crankshaft sensor and remove the sensor from the side of the engine block **(see illustration 5.29)**. **Note:** *On 2.5L four cylinder engines, remove the ignition coil and module assembly from the side of the engine block then detach the crankshaft sensor from the bottom of the ignition module assembly.*

41   Installation is the reverse of removal.

Tighten the bolts to the torque listed in this Chapter's Specifications.

Camshaft position sensor

Refer to illustration 5.42a and 5.42b

42   The camshaft sensor on 3100 models is mounted at the front of the engine block in between the cylinder heads **(see illustration)**. The camshaft sensor on 3.4L models is mounted to the top of the left cylinder head at the rear of the engine. The camshaft sensor on 3800 models is mounted in the front cover of the engine above the crankshaft sensor **(see illustration)**. The camshaft sensor is a hall effect type sensor which produces 1 ON/OFF signal for every two crankshaft revolutions. This signal tells the PCM that the No. 1 piston is on the intake stroke. Refer to Section 4 for the checking procedures.

43   Disconnect the negative terminal from the battery. **Caution:** *On models equipped with a Theftlock audio system, be sure the lockout feature is turned off before disconnecting the battery cable.*

44   Disconnect the electrical connector from the camshaft sensor. **Note:** *On 3100 models it may be necessary remove the power steering pump to allow access to the camshaft sensor.*

45   Remove the bolt from the camshaft sensor and remove the sensor.

46   Installation is the reverse of removal. Tighten the bolts to the torque listed in this Chapter's Specifications.

Vehicle Speed Sensor (VSS)

Refer to illustrations 5.47a and 5.47b

47   The Vehicle Speed Sensor (VSS) is located at the rear the transaxle housing near the right drive axle **(see illustrations)**. This sensor is a magnetic reluctance type sensor which sends a pulsing voltage signal to the PCM, which the PCM converts to miles per hour. The VSS is major component in the Transaxle Converter Clutch (TCC) system. Refer to Section 4 for the checking procedures.

5.47b Typical Vehicle Speed Sensor (VSS) found on 4T60-E and 4T65-E transaxle

48   To replace the VSS, detach the sensor retaining bolt(s) and or bracket, unplug the sensor and remove it from the transaxle.

49   Installation is the reverse of removal.

Park/Neutral switch

Refer to illustration 5.50

50   The Park/Neutral (P/N) switch, located on the upper part of the automatic transaxle **(see illustration)**, indicates to the PCM when the transaxle is in Park or Neutral. This information is used for Torque Converter Clutch (TCC), Exhaust Gas Recirculation (EGR) and Idle Air Control (IAC) valve operation. **Caution:** *The vehicle should not be driven with the Park/Neutral switch disconnected because idle quality will be adversely affected and a trouble code may be set.*

51   For more information regarding the P/N switch, which is part of the Neutral/start and back-up light switch assembly, see Chapter Part B.

Power Steering Pressure Switch

52   This sensor located in the steering gear

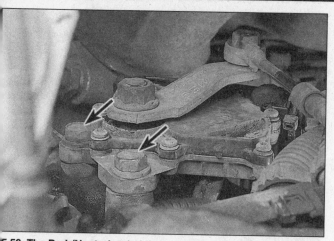

5.50 The Park/Neutral switch is located on the upper rear part of the automatic transaxle - arrows are pointing to mounting bolts

7.3 The Electronic Spark Control (ESC) knock sensor (arrow) is located on the engine block

ells the PCM when the pressure in the steering hydraulic system exceeds a certain pressure. If the engine is at idle the air conditioner will be shut off and the idle speed will increase. to replace this sensor, disconnect the electrical connector and unscrew the sensor from the steering gear.

Electronic Spark Timing (EST)

To provide improved engine performance, fuel economy and control of exhaust missions, the Electronic Control Module (CM) controls spark advance (ignition timing) with the Electronic Spark Timing (EST) ystem.

The PCM receives a reference pulse om the crankshaft sensor (see illustrations 29 and 5.30), and/or cam sensor (if quipped) which indicate engine rpm, ankshaft position and/or camshaft position. e PCM then determines the proper spark vance for the engine operating conditions d sends an EST pulse to the ignition mod- . A fault in the EST system will usually set a uble code.

Electronic Spark Control (ESC) system

fer to illustration 7.3

eneral description

Irregular octane levels in modern gaso- e can cause detonation in an engine. Deto- on is sometimes referred to as "spark ck."

The Electronic Spark Control (ESC) sys- is designed to retard spark timing up to degrees to reduce spark knock in the ine. This allows the engine to use maxi- m spark advance to improve driveability fuel economy.

The ESC knock sensor, which is located on the engine block (see illustration), creates and sends an alternating current voltage signal to the PCM. The PCM then retards the timing until spark knock is eliminated.

4   Loss of the ESC signal to the PCM will cause the PCM to constantly retard EST. This will result in sluggish performance and cause the PCM to set a trouble code.

Knock sensor replacement

5   Detach the cable from the negative terminal of the battery. **Caution:** *On models equipped with the Theftlock audio system, be sure the lockout feature is turned off before performing any procedure which requires disconnecting the battery.*
6   Disconnect the electrical connector from the sensor.
7   Unscrew the sensor from the block.
8   Installation is the reverse of the removal procedure.

8   Exhaust Gas Recirculation (EGR) system

General description

1   The Exhaust Gas Recirculation (EGR) system is used to lower NOx (oxides of nitrogen) emission levels caused by high combustion temperatures. It does this by decreasing combustion temperature. The main element of the system is the EGR valve, which feeds small amounts of exhaust gas back into the combustion chamber.
2   The EGR valve is usually open during warm engine operation and anytime the engine is running above idle speed. The amount of gas recirculated is controlled by variations in vacuum and exhaust backpressure.
3   There are three types of EGR valves. Their names refer to the means by which they are controlled:

Digital/Linear EGR valve (Quad-4 and 1990 and later non-turbo V6 engines)

Negative backpressure EGR valve (2.2L and 2.5L engines)
Integrated electronic EGR valve (turbo V6 engine)

Digital/Linear EGR valve

4   The digital EGR valve feeds small amounts of exhaust gas back into the intake manifold and then into the combustion chamber.
5   The digital EGR valve is designed to accurately supply recirculated exhaust gas to an engine, independent of intake manifold vacuum. The valve controls EGR flow from the exhaust to the intake manifold through multiple orifices to produce various combinations of EGR flow. (The exact number of orifices and combinations of EGR flow depends on the year and model of the car.) When a solenoid is energized, the armature, with attached shaft and swivel pintle, is lifted, opening the orifice. The flow accuracy is dependent on metering orifice size only, which results in improved control.
6   The digital EGR valve is opened by the PCM, grounding each solenoid circuit. This activates the solenoid, raises the pintle, and allows exhaust gas flow into the intake manifold. The exhaust gas then moves with the air/fuel mixture into the combustion chamber.

Negative backpressure EGR valve

7   On the negative backpressure EGR valve, the diaphragm on this valve has an internal vacuum bleed hole which is held closed by a small spring when there is no exhaust backpressure. Engine vacuum opens the EGR valve against the pressure of a large spring. When manifold vacuum combines with negative exhaust backpressure, the vacuum bleed hole opens and the EGR valve closes.

Integrated electronic EGR valve

8   The integrated electronic EGR valve functions like a port valve with a remote vacuum regulator, except the regulator and a pintle position sensor are sealed in the black

6

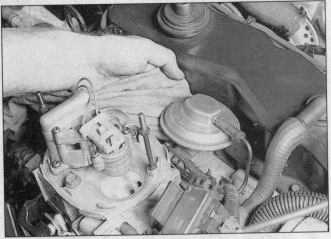

8.12 To check for proper circulation of recirculated exhaust gas, warm up the engine and, using a rag to protect your fingers, push up on the diaphragm - the engine should stumble or stall

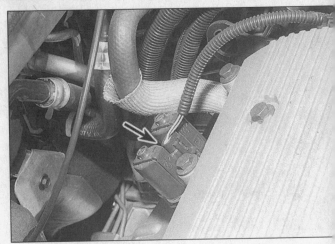

8.23a The EGR valve (arrow) on the Quad-4 engine is near the firewall

8.23b EGR valve mounting bolt locations (arrows) (negative back pressure type)

8.23c EGR valve-to-intake manifold mounting nut locations (arrows) (1995 and earlier V6 engines with digital EGR)

plastic cover. The regulator and the position sensor are not serviceable.

9   This valve has a vacuum regulator, to which the PCM provides variable current. This variable current produces the desired EGR flow using inputs from the MAT sensor (MAF sensor on 1988 models), coolant temperature sensor and engine rpm.

Checking

Refer to illustration 8.12

Negative backpressure EGR valves

10   Hold the top of the EGR valve and try to rotate it back-and-forth. If play is felt, replace the valve.

11   If no play is felt, place the transaxle in Neutral (manual) or Park (automatic), run the engine at idle until it warms up to at least 195-degrees F.

12   Using a rag to protect your hand from the engine heat, push up on the underside of the EGR valve diaphragm **(see illustration)**. If the rpm drops, go to step 14.

13   If there's no change in rpm, clean the

EGR passages and repeat step 12. If there's still no change in rpm, replace the valve.

14   If the rpm drops is step 12, check for movement of the EGR valve diaphragm as the rpm is changed from approximately 2000 rpm to idle. If the diaphragm moves, there is no problem.

15   If the diaphragm doesn't move, check the vacuum signal at the EGR valve as the engine rpm is changed from approximately 2000 rpm to idle.

16   If the vacuum is over six inches, replace the EGR valve. If it's under six inches, check the vacuum hoses for restrictions, leaks and poor connections.

Integrated electronic EGR valves

17   With the ignition off, connect a vacuum pump to the valve - apply vacuum and the valve should not move. Repeat the test with the ignition switch On. The valve should not move.

18   Ground the diagnostic terminal (see Section 2) and repeat the test. The valve should move and should be able to hold vacuum.

19   Start the engine and lift the EG diaphragm using a rag to protect your f gers. The idle should roughen **(see illustra tion 8.12)**.

20   Due to the complexity and the interrel tionship with ECM/PCM, any further chec should be left to a dealer service departme

Digital/Linear EGR valves

21   A special "scan" tool is needed to che this valve and should be left to a dealer s vice department.

Component replacement

Refer to illustrations 8.23a, 8.23b, 8.23c 8.23d and 8.24

EGR valve

22   Disconnect the vacuum hose equipped) from the EGR valve and disce nect the electrical connector (if equipped).

23   Remove the nuts or bolts which sec the valve to the intake manifold or adap **(see illustrations)**.

24   On V6 engines so equipped, disconr

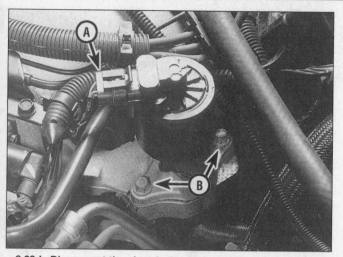

8.23d Disconnect the electrical connector (A) and remove the valve mounting bolts (B) (linear type EGR)

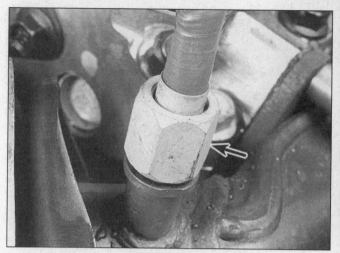

8.24 On V6 engines, disconnect the EGR pipe at the exhaust manifold

the EGR pipe from the exhaust manifold (see illustration).

25   On all engines, separate the EGR valve from the engine.

EGR valve cleaning

26   Inspect the valve pintle for deposits.

27   Depress the valve diaphragm and check for deposits around the valve seat area.

28   Use a wire brush to carefully clean deposits from the pintle.

29   Remove any deposits from the valve outlet with a screwdriver.

30   If EGR passages in the intake manifold have an excessive build-up of deposits, the passages should be cleaned. Care should be taken to ensure that all loose particles are completely removed to prevent them from clogging the EGR valve or from being ingested into the engine. **Note:** *It's a good idea to place a rag in the passage opening to keep debris from entering while cleaning the manifold.*

31   Using a wire wheel, buff the exhaust deposits off the mounting surface.

32   Clean the mounting surfaces of the EGR valve. Remove all traces of old gasket material.

33   Install the new EGR valve, with a new gasket, on the intake manifold or adapter.

34   Installation is the reverse of removal.

9   Evaporative Emission Control System (EECS or EVAP)

Refer to illustrations 9.2, 9.4 and 9.12

General description

1   This system is designed to trap and store fuel vapors that evaporate from the fuel tank, throttle body and intake manifold.

2   The Evaporative Emission Control System (EECS or EVAP) consists of a charcoal-filled canister and the lines connecting the

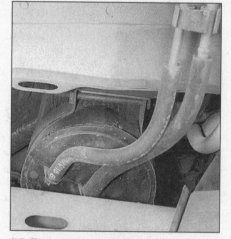

9.2 Typical evaporative canister on a 1989 3.1L engine

canister to the fuel tank and ported vacuum (see illustration).

3   Fuel vapors are transferred from the fuel tank, throttle body and intake manifold to a canister where they're stored when the engine isn't running. When the engine is running, the fuel vapors are purged from the canister by intake air flow and consumed in the normal combustion process.

4   On all engines except the 2.2L and 2.5L, the ECM/PCM operates a solenoid valve which controls vacuum to the charcoal canister. Under cold engine or idle conditions, the solenoid is turned on by the PCM, which closes the valve and blocks vacuum to the canister. The PCM turns off the solenoid valve and allows purge when the engine is warm (see illustration).

5   On 2.2L and 2.5L engines, the purging of the canister is controlled by a purge valve located in the engine compartment in-line between the canister and the intake manifold (2.2L), or, in the case of the 2.5L, on the canister. The purge valve is opened by vacuum when the engine speed is above idle.

9.4 View of canister purge control valve (arrow)

Checking

6   Poor idle, stalling and poor driveability can be caused by an inoperative purge valve, a damaged canister, split or cracked hoses or hoses connected to the wrong tubes.

7   Evidence of fuel loss or fuel odor can be caused by liquid fuel leaking from fuel lines or the TBI, a cracked or damaged canister, an inoperative purge valve, disconnected, mis-routed, kinked, deteriorated or damaged vapor or control hoses or an improperly seated air cleaner or air cleaner gasket.

8   Inspect each hose attached to the canister for kinks, leaks and cracks along its entire length. Repair or replace as necessary.

9   Inspect the canister. If it's cracked or damaged, replace it.

10   Look for fuel leaking from the bottom of the canister. If fuel is leaking, replace the canister and check the hoses and hose routing.

11   Any further testing should be left to a dealer service department.

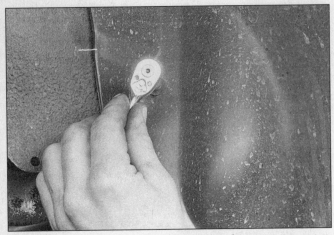

9.12 Remove the access cover in the left rear wheelwell to get at the canister

10.5 Typical engine oil/air separator (arrow)

Component replacement

Canister

12   Remove the access cover at the left rear wheel **(see illustration)**.
13   Clearly label, then detach the vacuum hoses from the canister.
14   Remove the canister.
15   Installation is the reverse of removal.

Purge valve

16   Label and disconnect the vacuum hoses from the valve.
17   If necessary, disconnect the electrical connector.
18   Unscrew the mounting bolts and remove the valve.
19   Installation is the reverse of removal.

10   Positive Crankcase Ventilation (PCV) system

General description

All models except Quad-4 engine

1   The Positive Crankcase Ventilation (PCV) system reduces hydrocarbon emissions by scavenging crankcase vapors. It does this by circulating fresh air from the air cleaner through the crankcase, where it mixes with blow-by gases and is then rerouted through a PCV valve to the intake manifold.
2   The main components of the PCV system are the PCV valve, a fresh air filtered inlet and the vacuum hoses connecting these two components with the engine.
3   To maintain idle quality, the PCV valve restricts the flow when the intake manifold vacuum is high. If abnormal operating conditions arise, the system is designed to allow excessive amounts of blow-by gases to flow back through the crankcase vent tube into the air cleaner to be consumed by normal combustion.
4   Checking and replacement of the PCV valve and filter is covered in Chapter 1.

Quad-4 engine only

Refer to illustration 10.5

5   The Quad-4 engine uses a Crankcase Ventilation (CV) system to provide scavenging of crankcase vapors. Blow-by gases are passed through a crankcase ventilation oil/air separator into the intake manifold **(see illustration)**.
6   The oil/air separator causes oil, which may be suspended in the blow-by gases, to be separated and allows it to drain back to the crankcase through a hose.
7   If the CV system becomes plugged, it must be replaced as a unit.

Oil/air separator replacement (Quad-4 engine only)

8   Label and remove all hoses from the oil/air separator.
9   Remove the mounting bolts and lift it from the vehicle.
10   Installation is the reverse of removal.

11   Catalytic converter

General description

1   The catalytic converter is an emission control device added to the exhaust system to reduce pollutants from the exhaust gas stream. The converter contains a honeycomb mesh.

Checking

2   The test equipment for a catalytic converter is expensive and highly sophisticated. If you suspect the converter is malfunctioning, take it to a dealer service department or authorized emissions inspection facility for diagnosis and repair.
3   Whenever the vehicle is raised for servicing of underbody components, check the converter for leaks, corrosion and other damage. If damage is discovered, the converter should be replaced.

Removal and installation

4   On some models the converter is welded to the exhaust system, so converter replacement requires removal of the exhaust pipe assembly (see Chapter 4). Take the vehicle, or the exhaust system, to a dealer service department or a muffler shop.
5   On other models, raise the car and support it securely on jack stands.
6   Unbolt the converter from the manifold and from the intermediate exhaust pipe, then remove the converter.
7   Installation is the reverse of removal.

12   Air Injection Reactor (AIR) system

General description

1   The AIR system is used on V6 engine models equipped with manual transaxles. The AIR system helps reduce hydrocarbons and carbon monoxide levels in the exhaust by injecting air into the exhaust ports of each cylinder.
2   The AIR system uses an air pump to force the air into the exhaust stream. An Electric Air Control (EAC) valve, controlled by the vehicle's Electronic Control Module (ECM) directs the air to the correct location depending on engine temperature and driving conditions. During certain situations, such as deceleration, the air is diverted to the atmosphere to prevent backfiring from too much oxygen in the exhaust stream.
3   One-way check valves are also used in the AIR system's air lines to prevent exhaust gases from being forced back through the system.
4   The following components are utilized in the AIR system: an engine driven air pump, an Electric Air Control (EAC) valve, air hoses and a check valve.

Checking

5   Because of the complexity of this system it is difficult for the home mechanic to

12.14 Remove the AIR pump pulley bolts and separate the pulley from the pump

12.17 Removing the AIR pump filter (remove as shown - do not insert any tool behind the filter, as damage to the pump may occur)

make a proper diagnosis. If the system is suspected of not operating properly, individual components can be checked.

5   Begin any inspection by carefully checking all hoses and wires. Be sure they are in good condition and that all connections are tight and clean. Also make sure that the pump drivebelt is in good condition and properly adjusted.

7   To check the pump allow the engine to reach normal operating temperature and run it at about 1500 rpm. Locate the hose running from the air pump and squeeze it to feel the pulsations. Have an assistant increase the engine speed and check for an increase in air flow. If this is observed as described, the pump is functioning properly. If it is not operating in this manner, check for proper drive belt tension. If belt tension is okay, a faulty pump is indicated.

8   The check valve can be inspected by first removing it from the air line. Attempt to blow through it from both directions. Air should only pass through it in the direction of normal air flow. If it is either stuck open or stuck closed the valve should be replaced.

9   Disconnect the vacuum signal line at the air control valve. With the engine at idle speed, 10 inches of vacuum should be present in the line.

10   When the EAC is energized (12-volts) and has good manifold vacuum to it, air from the pump should be directed to the exhaust ports. When the EAC is de-energized and good manifold vacuum to it, the air should be diverted to the atmosphere.

Component replacement

Drivebelt

11   Loosen the pump mounting bolt and the pump adjustment bracket bolt.

12   Move the pump until the belt can be removed.

13   Install the new belt and adjust it (refer to Chapter 1).

AIR pump pulley and filter

Refer to illustrations 12.14 and 12.17

14   Compress the drivebelt to keep the pulley from turning and loosen the pulley bolts **(see illustration)**.

15   Remove the drivebelt as described above.

16   Remove the mounting bolts and lift off the pulley.

17   If the fan-like filter must be removed, grasp it firmly with needle-nose pliers, and pull it from the pump **(see illustration)**. **Note:** *Do not insert a screwdriver between the filter and pump housing as the edge of the housing could be damaged. The filter will usually be distorted when it is pulled off. Be sure no fragments fall into the air intake hose.*

18   The new filter is installed by placing it in position on the pump, placing the pulley over it and tightening the pulley bolts evenly to draw the filter into the pump. Do not attempt to install a filter by pressing or hammering it into place. **Note:** *It is normal for the new filter to have an interference fit with the pump housing and, upon initial operation, it may squeal until it has worn in.*

19   Install the drivebelt and, while compressing the belt, tighten the pulley bolts.

20   Adjust the drivebelt tension.

Hoses and tubes

21   To replace any tube or hose always note how it is routed first, either with a sketch or with numbered pieces of tape.

22   Remove the defective hose or tube and replace it with a new one of the same material and size and tighten all connections.

Check valve

23   Disconnect the pump outlet hose at the check valve.

24   Remove the check valve from the pipe assembly, making sure not to bend or twist the assembly.

25   Install a new valve after making sure that it is a duplicate of the part removed, then tighten all connections.

Electric Air Control (EAC) valve

26   Disconnect the negative cable at the battery. **Caution:** *On models equipped with the Theftlock audio system, be sure the lockout feature is turned off before performing any procedure which requires disconnecting the battery.*

27   Disconnect the vacuum signal line from the valve. Also disconnect the air hoses and electrical connectors.

28   If the mounting bolts are retained by tabbed lock washers, bend the tabs back, then remove the mounting bolts and the valve.

29   Installation is the reverse of the removal procedure.

AIR pump

30   If the pulley must be removed from the pump it should be done prior to removing the drivebelt.

31   If the pulley is not being removed, remove the drivebelt.

32   Remove the pump mounting bolts and separate the pump from the engine.

33   Installation is the reverse of the removal procedure. **Note:** *Do not tighten the pump mounting bolts until all components are installed.*

34   Following installation adjust the drivebelt tension as described in Chapter 1.

6

Notes

Chapter 7 Part A
Manual transaxle

Contents

Specifications

Torque specifications

	Ft-lbs
Clutch housing cover bolts	10
Transaxle–to–engine bolts/nuts	55
Transaxle mount–to–frame nut	42
Transaxle mount–transaxle bolts	35
Shift control bolts/nuts	18
Shift linkage bracket bolts	17
Shift cable stud nuts	18

2.2a Details of the shift cables at the retainer bracket on the transaxle

1 Retainer bracket   3 Shift cables
2 Retainer mounting nut

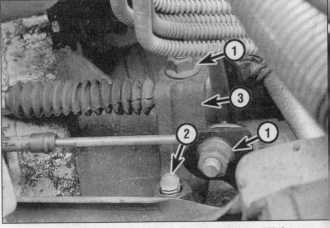

2.2b Details of the shift cables at the transaxle shift levers

1 Shift cable-to-shift lever mounting nut
2 Shift lever-to-transaxle mounting nut
3 Shift lever

1 General information

The vehicles covered by this manual are equipped with either a five–speed manual or a three or four–speed automatic transaxle. Information on the manual transaxle is included in this Part of Chapter 7. Information on the automatic transaxle can be found in Part B of this Chapter.

The five–speed manual transaxle is essentially a transmission coupled together with a differential in one assembly.

Due to the complexity, availability of replacement parts and special tools necessary, internal repair procedures should not be attempted by the home mechanic. The information contained in this manual will be limited to general diagnosis, external adjustments and removal and installation.

Depending on the expense involved in having a faulty transaxle overhauled, it may be a good idea to consider replacing the unit with either a new or rebuilt one. Your local dealer or transmission shop should be able to supply information concerning cost, availability and exchange policy. Regardless of how you decide to remedy a faulty transaxle problem, however, you can still save money by removing and installing it yourself.

2 Manual transaxle shift cables - removal and installation

Refer to illustrations 2.2a and 2.2b

Removal

1 Disconnect the negative cable from the battery. **Caution:** *On models equipped with the Theftlock audio system, be sure the lock-out feature is turned off before performing any procedure which requires disconnecting the battery.*
2 Remove the nuts retaining the selector and shift cables to the transaxle levers, and the nut securing the cable retainer bracket **(see illustrations)**.
3 Remove the console (see Chapter 11).
4 Use a small screwdriver to pry the cable free of the shift control ball sockets.
5 Remove the screws from the carpeting sill plate in the left front corner of the passenger compartment, remove the plate and then pull the carpet back for access to the cables.
6 Remove the cable retainer and grommet nuts at the floor pan and pry the two retaining tabs up.
7 Pull the cables through into the passenger compartment and remove them from the vehicle.

Installation

8 Push the cable assembly through the opening from the passenger compartment into the engine compartment.
9 Install the grommet and cable retainer nuts and bend the two retaining tabs down.
10 Install the carpet and sill plate.
11 Connect the cable ends to the shifter.
12 Install the console.
13 Connect the cable assembly to the transaxle bracket and shift levers.

3 Manual transaxle shifter shaft seal - replacement

1 Disconnect the negative cable from the battery. **Caution:** *On models equipped with the Theftlock audio system, be sure the lock-out feature is turned off before performing any procedure which requires disconnecting the battery.*
2 Disconnect the shift cable(s) from the shift lever(s).
3 Remove the shift lever nut, making sure the shift lever itself doesn't move while the nut is loosened.
4 Remove the shift lever assembly, keep-ing all of the components in order.
5 Pry out the old seal with a screwdriver.
6 Install the new seal in the bore and tap it into place with a deep socket or piece of pipe and a hammer.
7 Installation of the remaining components is the reverse of removal.

4 Manual transaxle shift control - removal and installation

Removal

1 Disconnect the negative cable from the battery. **Caution:** *On models equipped with the Theftlock audio system, be sure the lock-out feature is turned off before performing any procedure which requires disconnecting the battery.*
2 Remove the console, shift boot and knob (see Chapter 11).
3 Disconnect the shift cables from the shifter (see Section 2).
4 Remove the shift cable retaining clips.
5 Remove the retaining nuts/bolts and lift the shift control assembly out of the vehicle.

Installation

6 Place the shift control assembly in position and install the retaining nuts.
7 Connect the shift cables to the shift control assembly.
8 Install the console.
9 Connect the negative battery cable.

5 Manual transaxle - removal and installation

Refer to illustration 5.7

Removal

1 Disconnect the negative cable from the battery. **Caution:** *On models equipped with the Theftlock audio system, be sure the lock-*

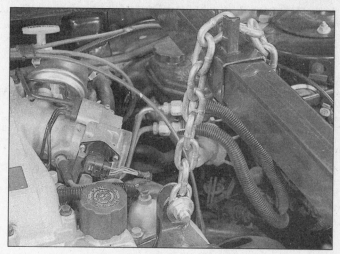

5.7 GM recommends using a special support fixture for the engine - you can also use an engine hoist - the engine can also be supported from below, with a jack and block of wood under the oil pan, but transaxle removal will be more difficult

...ut feature is turned off before performing ...ny procedure which requires disconnecting ...e battery.

Raise the vehicle and support it securely ...jackstands.

Drain the transaxle fluid (Chapter 1).

Disconnect the shift linkage from the ...ansaxle (see Section 2), then remove the ...utch master cylinder pushrod clip (near the ...p of the clutch pedal arm) and slide the ...ushrod off the pedal pin.

Detach the speedometer sensor and ...ire harness connectors from the transaxle. ...emove the clutch release cylinder from the ...ansaxle (see Chapter 8). On 1991 and later ...odels, disconnect the release lever from the ...lease bearing (see Chapter 8).

Remove the exhaust system compo-...nts as necessary for clearance.

Support the engine. This can be done ...om above with an engine hoist or support ...xture, or by placing a jack (with a block of ...ood as an insulator) under the engine oil ...an **(see illustration)**. The engine must ...main supported at all times while the ...ansaxle is out of the vehicle! Be sure the ...gine is supported securely.

Remove any chassis or suspension ...mponents that will interfere with transaxle ...moval (Chapter 10).

Disconnect the driveaxles from the

transaxle (Chapter 8).

10 Support the transaxle with a jack, preferably one designed for removing transaxles, then remove the bolts securing the transaxle to the engine.

11 Remove the transaxle mount nuts and bolts.

12 Make a final check that all wires and hoses have been disconnected from the transaxle, then carefully pull the transaxle and jack away from the engine.

13 Once the input shaft is clear, lower the transaxle and remove it from under the vehicle. **Caution:** *Do not depress the clutch pedal while the transaxle is out of the vehicle.*

14 With the transaxle removed, the clutch components are now accessible and can be inspected. In most cases, new clutch components should be routinely installed when the transaxle is removed.

Installation

15 If removed, install the clutch components (Chapter 8.)

16 With the transaxle secured to the jack with a chain, raise it into position behind the engine, then carefully slide it forward, engaging the input shaft with the clutch plate hub splines. Do not use excessive force to install the transaxle - if the input shaft does not slide into place, readjust the angle of the transaxle

so it is level and/or turn the input shaft so the splines engage properly with the clutch plate hub.

17 Install the transaxle-to-engine bolts. Tighten the bolts securely.

18 Install the transaxle mount nuts or bolts.

19 Install the chassis and suspension components which were removed. Tighten all nuts and bolts securely.

20 Remove the jacks supporting the transaxle and engine.

21 Install or reconnect the various items removed or disconnected previously, referring to Chapter 8 for installation of the driveaxles and Chapter 4 for information regarding the exhaust system components.

22 Make a final check that all wires, hoses, linkages and the speedometer sensor have been connected and that the transaxle has been filled with lubricant to the proper level (see Chapter 1).

23 Connect the negative battery cable. Road test the vehicle for proper operation and check for leaks.

6 Manual transaxle overhaul - general information

Overhauling a manual transaxle is a diffi-cult job for the do-it-yourselfer. It involves

7

the disassembly and reassembly of many small parts. Numerous clearances must be precisely measured and, if necessary, changed with seiect fit spacers and snap–rings. As a result, if transaxle problems arise, it can be removed and installed by a competent do–it–yourselfer, but overhaul should be left to a transmission repair shop. Rebuilt transaxles may be available - check with your dealer parts department and auto parts stores. At any rate, the time and money involved in an overhaul is almost sure to exceed the cost of a rebuilt unit.

Nevertheless, it's not impossible for an inexperienced mechanic to rebuild a transaxle if the special tools are available and the job is done in a deliberate step–by–step manner so nothing is overlooked.

The tools necessary for an overhaul include internal and external snap–ring pliers, a bearing puller, a slide hammer, a set of pin punches, a dial indicator and possibly a hydraulic press. In addition, a large, sturdy workbench and a vise or transaxle stand will be required.

During disassembly of the transaxle, make careful notes of how each piece comes off, where it fits in relation to other pieces and what holds it in place.

Before taking the transaxle apart for repair, it will help if you have some idea what area of the transaxle is malfunctioning. Certain problems can be closely tied to specific areas in the transaxle, which can make component examination and replacement easier. Refer to the Troubleshooting section at the front of this manual for information regarding possible sources of trouble.

Chapter 7 Part B
Automatic transaxle

Contents

Specifications

General

Fluid type and capacity	See Chapter 1

Torque specifications

	Ft-lbs (unless otherwise indicated)
Shift control assembly nuts	18
Starter safety switch-to-case bolts	18
Transaxle-to-engine bolts	55
Torque converter-to-driveplate bolts	
3T40	44
4T60-E and 4T65-E	46
Torque converter cover bolts	
3T40	60 in-lbs
4T60-E and 4T65-E	89 in-lbs
Transaxle mount bracket-to-transaxle bolts	
3T40	
Top (12 mm)	61
Front (10 mm)	35
4T60-E and 4T65-E	70
Transaxle mount	
Mount-to-transaxle bracket nuts	
3T40	22
4T60-E and 4T65-E	35
Mount-to-crossmember bracket nuts	35
Crossmember bracket-to-frame bolts	
3T40	35
4T60-E and 4T65-E	43
Transaxle to engine support brace bolts	35
TV cable-to-transaxle case bolt	
3T40	89 in-lbs

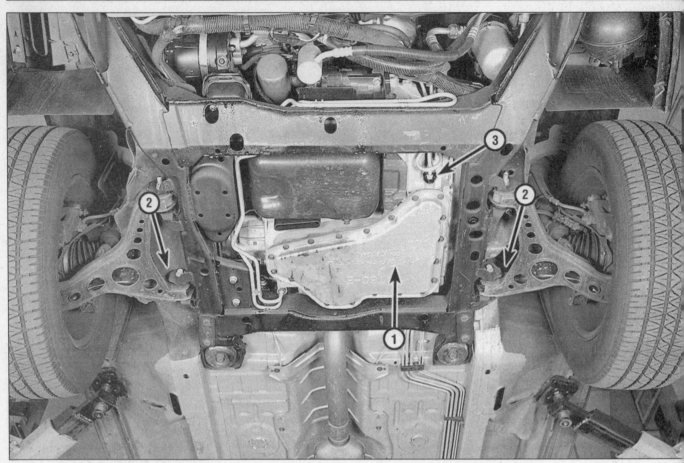

1.1 An underside view of an automatic transaxle and its related components

| 1 | Transaxle fluid pan | 2 | Driveaxles | 3 | Transaxle cooler lines |

1   General information

Refer to illustration 1.1

The vehicles covered by this manual are equipped with either a five-speed manual or three or four-speed Hydra-Matic automatic transaxles **(see illustration)**. Three models of Hydra-Matic automatic transaxles are used on these vehicles: 125C/3T40, 440-T4/4T60-E and the 4T65-E.

Due to the complexity of the clutches and the hydraulic control system, and because of the special tools and expertise required to perform an automatic transaxle overhaul, it should not be undertaken by the home mechanic. Therefore, the procedures in this Chapter are limited to general diagnosis, routine maintenance, adjustment and transaxle removal and installation.

If the transaxle requires major repair work, it should be left to a dealer service department or an automotive or transmission repair shop. You can, however, remove and install the transaxle yourself and save the expense, even if the repair work is done by a transmission shop.

Replacement and adjustment procedures the home mechanic can perform include those involving the throttle valve (TV)

cable and the shift linkage. **Caution:** *Never tow a disabled vehicle with an automatic transaxle at speeds greater than 35 mph or for distances over 50 miles if the front wheels are on the ground.*

2   Diagnosis - general

Note: *Automatic transaxle malfunctions may be caused by five general conditions: poor engine performance, improper adjustments, hydraulic malfunctions, mechanical malfunctions or malfunctions in the computer or its signal network. Diagnosis of these problems should always begin with a check of the easily repaired items: fluid level and condition (see Chapter 1), shift linkage adjustment and throttle linkage adjustment. Next, perform a road test to determine if the problem has been corrected or if more diagnosis is necessary. If the problem persists after the preliminary tests and corrections are completed, additional diagnosis should be done by a dealer service department or transmission repair shop. Refer to the Troubleshooting section at the front of this manual for transaxle problem diagnosis.*

Preliminary checks

1   Drive the vehicle to warm the transaxle to normal operating temperature.

2   Check the fluid level as described in Chapter 1:

a) *If the fluid level is unusually low, add enough fluid to bring the level within the designated area of the dipstick, then check for external leaks.*

b) *If the fluid level is abnormally high, drain off the excess, then check the drained fluid for contamination by coolant. The presence of engine coolant in the automatic transmission fluid indicates that a failure has occurred in the internal radiator walls that separate the coolant from the transmission fluid (see Chapter 3).*

c) *If the fluid is foaming, drain it and refill the transaxle, then check for coolant in the fluid or a high fluid level.*

3   Check the engine idle speed. **Note:** *If the engine is malfunctioning, do not proceed with the preliminary checks until it has been repaired and runs normally.*

4   Check the throttle valve cable for freedom of movement. Adjust it if necessary (see Section 3). **Note:** *The throttle valve cable may function properly when the engine is shut off and cold, but it may malfunction once the*

engine is hot. Check it cold and at normal engine operating temperature.

5   Inspect the shift cable (see Section 5). Make sure that it's properly adjusted and that the linkage operates smoothly.

Fluid leak diagnosis

6   Most fluid leaks are easy to locate visually. Repair usually consists of replacing a seal or gasket. If a leak is difficult to find, the following procedure may help.

7   Identify the fluid. Make sure it's transmission fluid and not engine oil or brake fluid (automatic transmission fluid is a deep red color).

8   Try to pinpoint the source of the leak. Drive the vehicle several miles, then park it over a large sheet of cardboard. After a minute or two, you should be able to locate the leak by determining the source of the fluid dripping onto the cardboard.

9   Make a careful visual inspection of the suspected component and the area immediately around it. Pay particular attention to gasket mating surfaces. A mirror is often helpful for finding leaks in areas that are hard to see.

10   If the leak still cannot be found, clean the suspected area thoroughly with a degreaser or solvent, then dry it.

11   Drive the vehicle for several miles at normal operating temperature and varying speeds. After driving the vehicle, visually inspect the suspected component again.

12   Once the leak has been located, the cause must be determined before it can be properly repaired. If a gasket is replaced but the sealing flange is bent, the new gasket will not stop the leak. The bent flange must be straightened.

13   Before attempting to repair a leak, check to make sure that the following conditions are corrected or they may cause another leak. **Note:** *Some of the following conditions cannot be fixed without highly specialized tools and expertise. Such problems must be referred to a transmission shop or a dealer service department.*

Gasket leaks

14   Check the pan periodically. Make sure the bolts are tight, no bolts are missing, the gasket is in good condition and the pan is flat (dents in the pan may indicate damage to the valve body inside).

15   If the pan gasket is leaking, the fluid level or the fluid pressure may be too high, the vent may be plugged, the pan bolts may be too tight, the pan sealing flange may be warped, the sealing surface of the transaxle housing may be damaged, the gasket may be damaged or the transaxle casting may be cracked or porous. If sealant instead of gasket material has been used to form a seal between the pan and the transaxle housing, it may be the wrong sealant.

Seal leaks

16   If a transaxle seal is leaking, the fluid level or pressure may be too high, the vent may be plugged, the seal bore may be damaged, the seal itself may be damaged or improperly installed, the surface of the shaft protruding through the seal may be damaged or a loose bearing may be causing excessive shaft movement.

17   Make sure the dipstick tube seal is in good condition and the tube is properly seated. Periodically check the area around the speedometer gear or sensor for leakage. If transmission fluid is evident, check the O-ring for damage. Also inspect the side gear shaft oil seals for leakage.

Case leaks

18   If the case itself appears to be leaking, the casting is porous and will have to be repaired or replaced.

19   Make sure the oil cooler hose fittings are tight and in good condition.

Fluid comes out vent pipe or fill tube

20   If this condition occurs, the transaxle is overfilled, there is coolant in the fluid, the case is porous, the dipstick is incorrect, the vent is plugged or the drain back holes are plugged.

3   Throttle valve (TV) cable (3T40) - replacement and adjustment

Refer to illustrations 3.2, 3.4, 3.5 and 3.8

Replacement

1   Disconnect the TV cable from the throttle lever by grasping the connector, pulling it forward to disconnect it and then lifting up and off the lever pin.

2   Disconnect the TV cable housing from the bracket by compressing the tangs and pushing the housing back through the bracket **(see illustration)**.

3   Disconnect any clips or straps retaining the cable to the transaxle.

4   Remove the bolt retaining the cable to the transaxle **(see illustration)**.

5   Pull up on the cover until the end of the cable can be seen, then disconnect it from the transaxle TV link **(see illustration)**. Remove the cable from the vehicle.

6   To install the cable, connect it to the transaxle TV link and push the cover securely over the cable. Route the cable to the top of

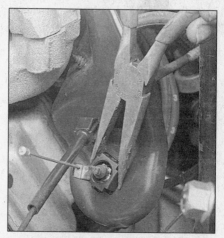

3.2 Use needle-nose pliers to compress the TV cable tangs, then push the housing back through the bracket

3.4 Remove the TV cable bolt and pull up on the cable until it's out of the transaxle

3.5 Hold the transaxle TV link with the needle-nose pliers and slide the cable link off the pin

7

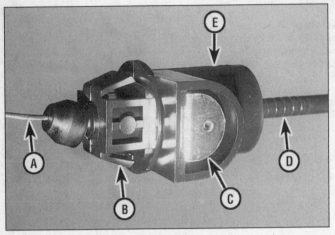

3.8 TV cable adjuster details

A	TV cable	D	Cable casing
B	Locking lugs	E	Slider
C	Release tabs		

4.4 Neutral start switch installation details

1	Neutral start switch	2	Neutral start switch electrical connector

the engine, push the housing through the bracket until it clicks into place, place the connector over the throttle lever pin and pull back to lock it. Secure the cable with any retaining clips or straps.

Adjustment

7   The engine MUST NOT be running during this adjustment.

8   Depress the re-adjust button and push the slider through the fitting (away from the throttle lever) as far as it will go **(see illustration)**.

9   Release the re-adjust button.

10   Manually turn the throttle lever to the "wide open throttle" position until the re-adjust tab makes an audible click, then release the throttle lever. The cable is now adjusted. **Note:** *Don't use excessive force at the throttle lever to adjust the TV cable. If great effort is required to adjust the cable, disconnect the cable at the transaxle end and check for free operation. If it's still difficult, replace the cable. If it's now free, suspect a bent TV link in the transaxle or a problem with the throttle lever.*

4   Neutral Start (Starter Safety) switch – replacement and adjustment

Replacement

Refer to illustration 4.4

1   Disconnect the negative cable from the battery. **Caution:** *On models equipped with the Theftlock audio system, be sure the lock-out feature is turned off before performing any procedure which requires disconnecting the battery.*

2   Shift the transaxle into Neutral.

3   Raise the vehicle and support it securely on jackstands.

4   Trace the wire harness from the Neutral start switch to the connector **(see illustration)** and unplug it. Detach the wire harness from the two retention clips.

5   Lower the vehicle.

6   Disconnect the electrical and vacuum connectors from the cruise control servo (if equipped). Remove the servo and set it out of the way on the cowl.

7   Remove the nut and detach the shift lever from the transaxle.

8   Remove the bolts and detach the switch.

9   To install the switch, line up the flats on the shift shaft with the flats in the switch and lower the switch onto the shaft.

10   Install the bolts. If the switch is new and the shaft hasn't been moved, tighten the bolts. If the switch requires adjustment, leave the bolts loose and follow the adjustment procedure below. The remainder of installation is the reverse of removal.

Adjustment

Refer to illustration 4.11

11   Line up the outer notch on the switch with the inner notch on the shift shaft and tighten the switch bolts **(see illustration)**.

12   Connect the negative battery cable and verify that the engine will start only in Neutral or Park.

5   Automatic transaxle shift cable - replacement and adjustment

1   Disconnect the negative cable from the

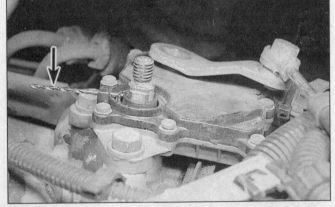

4.11 Align the outer notch of the Park/Neutral switch with the inner notch on the shaft using a drill bit (arrow)

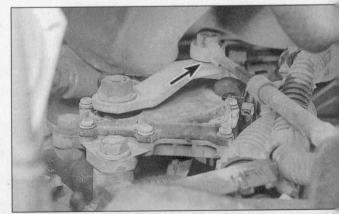

5.2 Disconnect the shift cable (arrow) at the shift lever and detach the cable housing from the bracket on the transaxle

5.3 Remove the retainer (arrow) from the shift handle

5.4 Location of the shift cable (1) and the park/lock cable (2)

battery. **Caution:** *On models equipped with the Theftlock audio system, be sure the lock-out feature is turned off before performing any procedure which requires disconnecting the battery.*

Replacement

Floor shift

Refer to illustrations 5.2, 5.3 and 5.4

2   Working in the engine compartment, disconnect the shift cable from the transaxle lever and bracket **(see illustration)**.

3   Remove the shift handle and the center console **(see illustration)**.

4   Disconnect the shift cable from the floor shift lever and bracket **(see illustration)**.

5   Remove the right and left side sound insulators from the under dash portion of the console, then pull back the carpet for access to the cable.

6   Trace the cable up to the firewall grommet.

7   Remove the screws from the grommet retainer, detach the grommet and retainer from the firewall and pull the cable assembly through the firewall.

8   Installation is the reverse of removal. After installation, adjust the cable as described in Step 14.

Column shift

Refer to illustration 5.11

9   Working in the engine compartment, remove the air cleaner, then disconnect the cable from the transaxle lever and bracket **(see illustration 5.2)**.

10   Working in the passenger compartment, remove the left sound insulator located under the dash.

11   Disconnect the cable bracket on the steering column and detach the cable from the column shift lever **(see illustration)**.

12   Dislodge the grommet in the firewall and withdraw the cable from the vehicle.

13   Installation is the reverse of removal. After installation, adjust the cable as described below.

5.11 Location of the column shift cable (arrow)

Adjustment

14   Place the shift lever and the transaxle lever in Neutral, then push the locking tab on the shift cable to automatically adjust the cable **(see illustration 3.8)**.

15   Reconnect the negative battery cable.

6   Automatic transaxle park/lock cable - removal and installation

Removal

1   Disconnect the negative cable from the battery. On V6 models, remove the air cleaner. **Caution:** *On models equipped with the Theftlock audio system, be sure the lock-out feature is turned off before performing any procedure which requires disconnecting the battery.*

2   Remove the console (Chapter 11).

3   Place the shift lever in Park and the ignition switch in the Run position.

4   Insert a screwdriver blade into the slot in the ignition switch inhibitor, depress the cable latch and detach the cable.

5   Push the cable connector lock button (located at the shift control base) to the Up position and detach the cable from the park lock lever pin. Depress the two cable connector latches and remove the cable from the

shift control base.

6   Remove the cable clips.

Installation

7   Make sure the cable lock button is in the Up position and the shift lever is in Park. Snap the cable connector into the shift control base.

8   With the ignition key in the Run position (this is very important), snap the cable into the inhibitor housing.

9   Turn the ignition key to the Lock position.

10   Snap the end of the cable onto the shifter park/lock pin.

11   Push the nose of the cable connector forward to remove the slack.

12   With no load on the connector nose, snap the cable connector lock button on.

13   Check the operation of the park/lock cable as follows.

a)   *With the shift lever in Park and the key in Lock, make sure the shift lever cannot be moved to another position and the key can be removed.*

b)   *With the key in Run and the shift lever in Neutral, make sure the key cannot be turned to Lock.*

14   If it operates as described above, the park/lock cable system is properly adjusted. Proceed to Step 16.

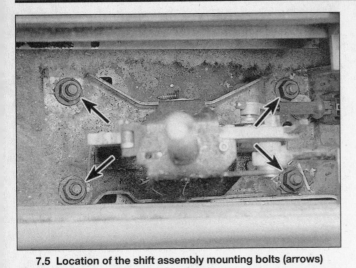

7.5 Location of the shift assembly mounting bolts (arrows)

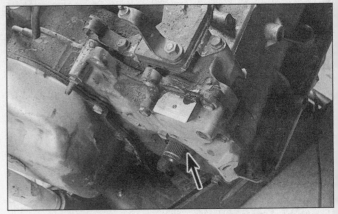

8.3 The rubber-type differential seal (arrow) used on earlier models can be pried out of the housing with a screwdriver - be careful not to damage the splines on the shaft (transaxle removed for clarity)

15   If the park/lock system doesn't operate as described, return the cable connector lock to the up position and repeat the adjustment procedure. Push the cable connector down and recheck the operation.

16   If the key cannot be removed in the Park position, snap the lock button to the up position and move the nose of the cable connector to the rear until the key can be removed from the ignition switch.

17   Install the cable in the retaining clips.

7   Automatic transaxle floor shift control assembly - removal and installation

Refer to illustration 7.5

1   Disconnect the negative cable from the battery. **Caution:** *On models equipped with the Theftlock audio system, be sure the lockout feature is turned off before performing any procedure which requires disconnecting the battery.*

2   Remove the console (see Chapter 11).

3   Disconnect the shift cable from the gear shift lever (see Section 5).

4   Disconnect the park/lock cable from the gear shift lever (see Section 6).

5   Remove the retaining nuts and lift the floor shift control assembly out of the vehicle **(see illustration)**.

6   Place the floor shift control assembly in position on the mounting studs and install the nuts. Tighten the nuts to the specified torque.

7   Connect the shift cables.

8   Install the console.

9   Reconnect the negative battery cable.

8   Transaxle differential seals - replacement

Refer to illustrations 8.3 and 8.4

1   Raise the vehicle and support it securely on jackstands.

2   Remove the driveaxle(s) (see Chapter 8).

3   If a rubber-type seal is involved, use a seal remover or a long screwdriver to pry it out of the transaxle. Be careful not to damage the splines on the output shaft **(see illustration)**.

4   If a metal-type seal is involved, use a hammer and chisel to pry up the outer lip of the seal to dislodge it so it can be pried out of the housing **(see illustration)**. The manufacturer recommends using a slide hammer to remove the metal-type seal.

5   Compare the new seal to the old one to make sure they're the same.

6   Coat the lips of the new seal with transmission fluid.

7   Place the new seal in position and tap it into the bore with a hammer and a large socket or a piece of pipe that's the same diameter as the outside edge of the seal.

8   Reinstall the various components in the reverse order of removal.

9   Transaxle mount - check and replacement

Refer to illustration 9.3

1   Insert a large screwdriver or prybar into the space between the transaxle bracket and the mount and try to pry the transaxle up slightly.

8.4 Dislodge the metal-type differential seal used on some later models by working around the outer edge with a chisel and hammer

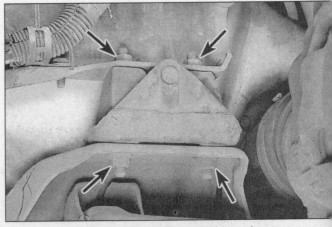

9.3 Transaxle mount nuts (arrows)

10.4 Torque converter inspection cover mounting bolt locations (arrows)

11.2 Vacuum modulator assembly

1   Vacuum connection   2   Plunger

2   The transaxle bracket should not move away from the insulator much at all.
3   To replace the mount, remove the nuts attaching the insulator to the crossmember and the nuts attaching the insulator to the transaxle (see illustration).
4   Raise the transaxle slightly with a jack and remove the insulator, noting which holes are used in the support for proper alignment during installation.
5   Installation is the reverse of the removal procedure. Be sure to tighten the nuts/bolts securely.

10   Automatic transaxle - removal and installation

Refer to illustration 10.4

Removal

1   Disconnect the negative cable from the battery. **Caution:** On models equipped with the Theftlock audio system, be sure the lockout feature is turned off before performing any procedure which requires disconnecting the battery.
2   Raise the vehicle and support it securely on jackstands.
3   Drain the transaxle fluid (Chapter 1).
4   Remove the torque converter cover (see illustration).
5   Mark the torque converter-to-driveplate relationship with white paint so they can be installed in the same position.
6   Remove the torque converter-to-driveplate bolts. Turn the crankshaft pulley bolt for access to each bolt.
7   Remove the starter motor (Chapter 5).
8   Disconnect the driveaxles from the transaxle (Chapter 8).
9   Disconnect the speedometer/speed sensor.
10   Disconnect the wire harness from the transaxle.
11   On models so equipped, disconnect the vacuum hose(s).
12   Remove any exhaust components which will interfere with transaxle removal (Chapter 4).

13   Disconnect the TV cable from the transaxle (Section 3).
14   Disconnect the shift linkage from the transaxle (Section 5).
15   Support the engine using a hoist from above or a jack and a block of wood under the oil pan to spread the load.
16   Support the transaxle with a jack - preferably a special jack made for this purpose. Safety chains will help steady the transaxle on the jack.
17   Remove any chassis or suspension components which will interfere with transaxle removal.
18   Remove the bolts securing the transaxle to the engine.
19   Remove the transaxle mount nuts and bolts.
20   Lower the transaxle slightly and disconnect and plug the transaxle cooler lines.
21   Remove the dipstick tube.
22   Move the transaxle back to disengage it from the engine block dowel pins and make sure the torque converter is detached from the driveplate. Secure the torque converter to the transaxle so it will not fall out during removal. Lower the transaxle from the vehicle.

Installation

23   Prior to installation, make sure that the torque converter hub is securely engaged in the pump. Lubricate the torque converter hub with multi-purpose grease.
24   With the transaxle secured to the jack, raise it into position. Be sure to keep it level so the torque converter does not slide out. Connect the fluid cooler lines.
25   Turn the torque converter to line up the bolt holes with the holes in the driveplate. The white paint mark on the torque converter and the driveplate made in Step 5 must line up.
26   Move the transaxle forward carefully until the dowel pins and the torque converter are engaged.
27   Install the transaxle housing-to-engine bolts. Tighten them securely.
28   Install the torque converter-to-driveplate bolts. Tighten the bolts to the specified torque.

29   Install the transaxle and any suspension and chassis components which were removed. Tighten the bolts and nuts to the specified torque.
30   Remove the jacks supporting the transaxle and the engine.
31   Install the dipstick tube.
32   Install the starter motor (Chapter 5).
33   Connect the vacuum hose(s) (if equipped).
34   Connect the shift and TV linkage.
35   Plug in the transaxle electrical connectors.
36   Install the torque converter cover.
37   Connect the driveaxles (Chapter 8).
38   Connect the speedometer/speed sensor cable.
39   Adjust the shift linkage (Section 5).
40   Install any exhaust system components that were removed or disconnected.
41   Lower the vehicle.
42   Fill the transaxle (Chapter 1), run the vehicle and check for fluid leaks.

11   Vacuum modulator (4T60-E) - check and replacement

Check

Refer to illustration 11.2

1   The vacuum modulator is connected to engine manifold vacuum to rapidly respond to changes in engine loading, and has an important effect on shift quality. If your vehicle exhibits shifting problems such as slipping or shifts that are either too soft or too harsh, check the vacuum supply to the modulator. With the engine running, connect a vacuum gauge to the modulator's vacuum line (at the modulator). Anything less than normal engine vacuum here of 13 to 17 inches (idling hot in Drive with the brakes applied) could cause shifting problems. If vacuum is too low, find the engine problem, hose kink or hose leak that is causing the low vacuum signal.
2   Connect a hand-held vacuum pump to the vacuum connection on the modulator (removed from vehicle) and apply 15 to 20

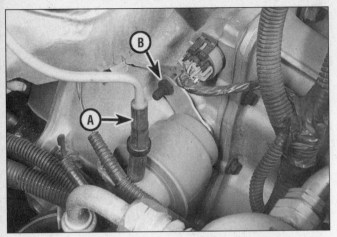

11.6 Disconnect the vacuum line (A) and remove the mounting bolt (B)

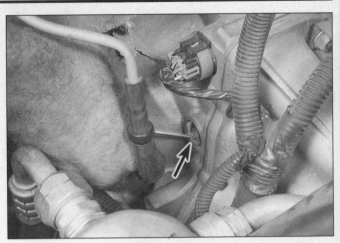

11.7 Remove the O-ring (arrow) from the transaxle case

inches of vacuum while watching the plunger **(see illustration)**. If the plunger isn't drawn in as vacuum is applied, the modulator should be replaced. The modulator should be able to hold this vacuum for at least half a minute.

3   When the modulator is withdrawn from the transaxle, turn it so the vacuum pipe is down. If any oil, water or other fluid drips out, the modulator should be replaced. **Note:** *A vehicle with a modulator whose diaphragm has a leak may exhibit excessive smoking at the tailpipe, due to transaxle fluid being drawn into the engine and burned.*

4   The body of the modulator can be checked for leaks by coating the outside with soapy water and blowing (by mouth, no more than 6 psi) into the vacuum connector, using a short length of vacuum hose. Bubbles on the outside of the modulator or along the seam indicate a leak.

Replacement

Refer to illustrations 11.6, 11.7 and 11.8

5   The modulator is located on the front (radiator) side of the transaxle, below the exhaust crossover pipe.

6   Disconnect the vacuum line, and remove the mounting bolt or stud, then withdraw the modulator **(see illustration)**.

7   Remove the O-ring from the modulator cavity, using a small screwdriver or hook **(see illustration)**.

8   Use a small magnet to remove the modulator valve from the transaxle **(see illustration)**. Inspect the valve for signs of abnormal wear or scoring.

9   Installation is the reverse of the removal procedure. Always use a new O-ring and make sure the modulator valve is installed the same way it came out.

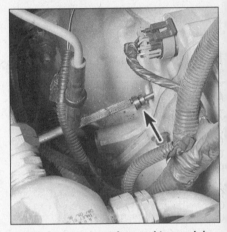

11.8 A magnet can be used to reach in and extract the modulator valve (arrow)

Chapter 8
Clutch and driveaxles

Contents

Specifications

Clutch fluid type	See Chapter 1
CV joint boot dimension (collapsed)	
1997 and earlier	5 1/16 in
1998 and later	4 29/32 in

Torque specifications — Ft-lbs

Clutch pressure plate-to-flywheel bolts	
Pull-type clutches	21
Push-type clutches	16
Clutch release cylinder mounting nuts	18
Clutch release cylinder canister (pull-type clutches)	28
Driveaxle hub nut	
1995 and earlier	184
1996 and 1997	151
1998 and later	118
Intermediate shaft mounting bolts	37
Intermediate shaft-to-bracket bolts	37
Intermediate shaft bracket-to-engine bolts	18
Wheel lug nuts	See Chapter 1

1   General information

The information in this Chapter deals with the components from the rear of the engine to the drive wheels, except for the transaxle, which is covered in the previous Chapter. For the purposes of this Chapter, these components are grouped into two categories: Clutch and driveaxles. Separate Sections within this Chapter cover components in both groups. **Warning:** *Since many of the procedures covered in this Chapter involve working under the vehicle, make sure it's securely supported on sturdy jackstands or on a hoist where the vehicle can easily be raised and lowered.*

2   Clutch - description and check

Refer to illustration 2.1

1   All vehicles with a manual transaxle use a single dry plate, diaphragm spring-type clutch **(see illustration)**. The clutch disc has a splined hub which allows it to slide along the splines of the transaxle input shaft. The clutch and pressure plate are held in contact by spring pressure exerted by the diaphragm in the pressure plate.

2   The clutch release system is operated by hydraulic pressure. The hydraulic release system consists of the clutch pedal, a master cylinder, the hydraulic line, a slave cylinder which actuates the clutch release lever and the clutch release (or throwout) bearing.

3   When pressure is applied to the clutch pedal to release the clutch, hydraulic pressure is exerted against the outer end of the clutch release lever. On 1990 and earlier models, the clutch uses a conventional "push-type" release mechanism. As the release lever pivots, the release bearing pushes against the fingers of the diaphragm spring in the pressure plate assembly, which in turn releases the clutch. On 1991 and later models, the clutch uses a "pull-type" mechanism. As the release lever pivots, the release bearing pulls the diaphragm spring, which in turn releases the clutch.

4   Terminology can be a problem when discussing the clutch components because common names are in some cases different from those used by the manufacturer. For example, the driven plate is also called the clutch plate or disc, the clutch release bearing is sometimes called a throwout bearing and the slave cylinder is often called an operating or release cylinder.

5   Other than to replace components with obvious damage, some preliminary checks should be performed to diagnose clutch problems.

a) *The first check should be of the fluid level in the clutch master cylinder. If the fluid level is low, add fluid as necessary and inspect the hydraulic system for leaks. If the master cylinder reservoir has run dry, bleed the system as described in Section 4 and recheck the clutch operation.*

b) *To check "clutch spin down time," run the engine at normal idle speed with the transaxle in Neutral (clutch pedal up - engaged). Disengage the clutch (pedal down), wait nine seconds and shift the transaxle into Reverse. No grinding noise should be heard. A grinding noise would most likely indicate a problem in the pressure plate or the clutch disc.*

c) *To check for complete clutch release, run the engine (with the parking brake on to prevent vehicle movement) and hold the clutch pedal approximately 1/2-inch from the floor. Shift the transaxle between First gear and Reverse several times. If the shift isn't smooth, compo-*

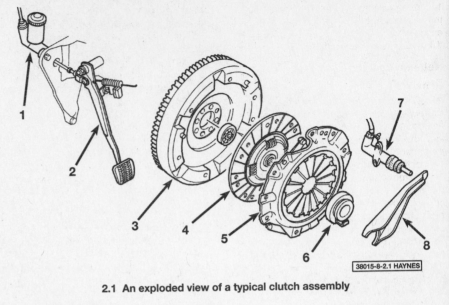

2.1 An exploded view of a typical clutch assembly

1	Clutch master cylinder	4	Pilot bearing	7	Clutch release bearing
2	Clutch pedal	5	Clutch disc	8	Clutch slave cylinder
3	Flywheel	6	Clutch cover	9	Clutch release lever

nent failure is indicated. Measure the slave cylinder pushrod travel. With the clutch pedal depressed completely the pushrod should extend about two inches. If it doesn't, check the fluid level in the clutch master cylinder.

d) *Visually inspect the clutch pedal bushing at the top of the clutch pedal to make sure it's not sticking or worn excessively.*

3   Hydraulic clutch components - removal and installation

1988 and 1991 through 1992

Note: *The hydraulic clutch release system on these models consists of two assemblies: the master cylinder and line and the release cylinder and line. The two assemblies are joined by a quick connect fitting. Neither the master cylinder nor release cylinder are rebuildable. Replacement assemblies are filled with fluid and have been bled (to remove air) at the factory. Individual components (i.e., the master cylinder, release cylinder or lines) are not available separately. Other than replacing the cylinder/line assemblies, bleeding the system is the only service procedure that may be necessary. There are no provisions for adjustment of clutch pedal height or freeplay.*

Master cylinder

Removal

1   Disconnect the cable from the negative battery terminal. **Caution:** *On models equipped with the Theftlock audio system, be sure the lockout feature is turned off before performing any procedure which requires disconnecting the battery.*

2   Remove the left side under-dash hush panel.

3   Remove the master cylinder pushrod clip (near the top of the clutch pedal arm) and slide the pushrod off the pedal pin.

4   Using a special hydraulic line disconnect tool (available at most auto parts stores), depress the white plastic sleeve on the clutch hydraulic line to separate the quick-connect fittings.

5   Remove the mounting screws for the remote reservoir of the clutch master cylinder. Keep the reservoir upright to avoid spilling fluid.

6   Remove the anti-rotation screw located next to the master cylinder flange at the pedal support plate.

7   To release the master cylinder, use the wrench flats on the front portion of the master cylinder and rotate it 45-degrees counterclockwise.

8   Remove the master cylinder from the vehicle.

Installation

9   Place the master cylinder into the firewall opening. Using the wrench flats on the front portion of the master cylinder, rotate it 45-degrees clockwise to engage it.

10   Install the anti-rotation screw for the master cylinder.

11   Install the reservoir and tighten the mounting screws. Connect the quick-connect fittings by pressing them together until you hear a snap. Check to be sure they're secure.

12   Lubricate and install a new bushing into the master cylinder pushrod.

13   Connect the pushrod to the clutch pedal (be sure the bushing tangs snap into the pedal pin groove). Reconnect the negative battery cable.

14 If the vehicle is equipped with cruise control, make sure the cruise control system disengages when the clutch pedal is depressed. If it doesn't, check the clutch switch.

Release cylinder

Removal

15 Disconnect the cable from the negative battery terminal. **Caution:** *On models equipped with the Theftlock audio system, be sure the lockout feature is turned off before performing any procedure which requires disconnecting the battery.*

16 Remove the left side under-dash hush panel.

17 Remove the clutch master cylinder pushrod retaining clip and slide the pushrod off the pedal pin.

18 Using a special hydraulic line disconnect tool, depress the white plastic sleeve on the clutch hydraulic line to separate the quick-connect fittings.

19 On 1991 and later models, remove the nuts holding the canister on the transaxle.

20 Remove the release cylinder mounting nuts at the transaxle.

21 Remove the release cylinder.

Installation

22 Connect the hydraulic lines by inserting the release cylinder fitting into the master cylinder fitting until they snap together.

23 While guiding the pushrod bushing into the release lever pocket, attach the release cylinder to the transaxle. Tighten the nuts securely. **Caution:** *Don't remove the plastic pushrod retainer from the release cylinder pushrod. It must remain in place during installation. The straps will break during the first clutch pedal application.*

24 Lubricate and install a new bushing into the master cylinder pushrod.

25 Connect the master cylinder pushrod to the clutch pedal (be sure the bushing tangs snap into the pedal pin groove). Attach the cable to the negative battery terminal.

26 The remainder of installation is the reverse of removal.

27 Press the clutch pedal down several times. This will break the plastic retaining strap on the release cylinder pushrod. Be sure the effort is normal.

28 If the vehicle is equipped with cruise control, make sure the cruise control system disengages when the clutch pedal is depressed. If it doesn't, check the clutch switch.

1989 and 1990 (assembly replacement)

Note: *The hydraulic clutch release system on these models is serviced as a complete unit. New assemblies have been bled (to remove air) at the factory. Adjustment of clutch pedal height or freeplay and bleeding of the hydraulic system are therefore usually not necessary.*

Removal

29 Disconnect the cable from the negative battery terminal. **Caution:** *On models equipped with the Theftlock audio system, be sure the lockout feature is turned off before performing any procedure which requires disconnecting the battery.*

30 On V6 models, remove the air cleaner and air intake duct as an assembly (see Chapter 4).

31 On all models, remove the left side under-dash hush panel.

32 Remove the clutch master cylinder pushrod retaining clip and slide the pushrod off the pedal pin.

33 Remove the mounting screws for the remote reservoir of the clutch master cylinder. Keep the reservoir upright to avoid spilling fluid.

34 Remove the anti-rotation screw located next to the master cylinder flange at the pedal support plate.

35 To release the master cylinder, use the wrench flats on the front portion of the master cylinder and rotate it 45-degrees counter-clockwise.

36 Remove the release cylinder mounting nuts at the transaxle.

37 Remove the hydraulic system as a unit from the vehicle.

Installation

38 While guiding the pushrod bushing into the pocket in the clutch release lever, attach the release cylinder to the transaxle. Tighten the nuts securely. **Caution:** *Don't remove the plastic pushrod retainer from the release cylinder pushrod. The straps will break during the first clutch pedal application.*

39 Place the master cylinder into the firewall opening. Using the wrench flats on the front portion of the master cylinder, rotate it 45-degrees clockwise to engage it.

40 Install the anti-rotation screw for the master cylinder.

41 Install the reservoir and tighten the mounting screws.

42 Lubricate and install a new bushing into the master cylinder pushrod.

43 Connect the pushrod to the clutch pedal, with the bushing tangs snapped into the pedal pin groove.

44 Press the clutch pedal down several times. This will break the plastic retaining straps on the release cylinder pushrod. Be sure the effort is normal.

45 The remainder of installation is the reverse of removal.

46 If the vehicle is equipped with cruise control, make sure the cruise control system disengages when the clutch pedal is depressed. If it doesn't, check the clutch switch.

---

4 Hydraulic clutch system - bleeding (1988 and 1991 through 1992 models only)

---

1 Using a special hydraulic line disconnect tool, depress the white plastic sleeve and disconnect the quick connect fittings.

2 Remove the cap and diaphragm and fill the reservoir with DOT 3 brake fluid.

3 Remove the upper secondary cowl on the left side of the engine compartment.

4 Remove the air from the supply hose by squeezing it until no air bubbles are seen in the reservoir.

5 Pump the clutch pedal slowly by hand until slight resistance is felt.

6 While holding pedal pressure, bleed air from the system by depressing the internal valve at the quick connect fitting. Don't use a sharp object. **Note:** *Be sure to maintain the fluid level in the master cylinder.*

7 Repeat steps 5 and 6.

8 Use a maximum of 50 pounds of force to check pedal firmness.

9 Reconnect the hydraulic line.

10 Replace the reservoir cap and diaphragm.

---

5 Clutch release bearing and lever - removal and installation

---

Removal

1988 through 1990 models

Refer to illustration 5.5

1 Disconnect the negative cable from the battery. **Caution:** *On models equipped with the Theftlock audio system, be sure the lockout feature is turned off before performing any procedure which requires disconnecting the battery.*

2 Remove the under-dash panel.

3 Disconnect the clutch master cylinder pushrod from the clutch pedal pin.

4 Remove the transaxle (see Chapter 7, Part A).

5 Remove the clutch release bearing from the lever shaft **(see illustration)**.

6 Hold the center of the bearing and turn the outer race. If the bearing doesn't turn smoothly or if it's noisy, replace it with a new one. Wipe the bearing with a clean rag and inspect it for damage, wear and cracks. Don't

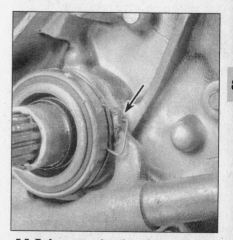

5.5 Before removing the release bearing from the transaxle, index the bearing pad to the clutch release fork (arrow)

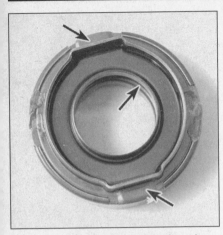

5.12 Lubricate the release bearing where the release lever contacts the release bearing - also, on 1988 through 1990 models, fill the inner groove of the bearing with grease

6.3 A clutch alignment tool in position to hold the disc during removal or center the disc during installation

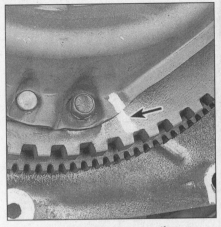

6.4 If you're going to re-use the same pressure plate, mark its relationship to the flywheel (arrow)

immerse the bearing in solvent - it's sealed and would be ruined by the solvent. It's common practice to replace the release bearing whenever the clutch components are replaced.

1991 and later models

7   Remove the release cylinder and canister from the transaxle.
8   Remove the access plug from the transaxle clutch housing.
9   Disconnect the release lever from the release bearing by pulling the lever. **Note:** *Pull the lever perpendicular to the transaxle center line until it reaches its stop.*
10   Remove the transaxle (see Chapter 7).
11   Remove the release bearing from the pressure plate as follows: Push the bearing toward the engine, fit a large flat-blade screwdriver between the clutch wedge collar and the release bearing, then twist the screwdriver. The bearing should pop out of the wedge collar.

Installation - all models

Refer to illustration 5.12

12   Lightly lubricate the release lever ends, where they contact the bearing, with white lithium-base grease. On 1988 to 1990 models, pack the inner groove of the bearing with grease as well **(see illustration)**.
13   Install the release bearing on the lever (be sure the levers fit between the large ears and small tangs on the bearing).
14   Be sure the lever moves freely with the bearing in place.
15   Install the transaxle (see Chapter 7).
16   On models with a pull-type clutch, seat the release bearing in the wedge collar by pushing the release cylinder end of the release lever toward the engine. The release bearing should pop into the wedge collar and become locked to the diaphragm spring.
17   On models with a pull-type clutch, install the access cover in the clutch housing.

18   Install the release cylinder and, if necessary, the canister, on the transaxle.
19   Reconnect the clutch master cylinder pushrod.
20   Install the under-dash cover.
21   Attach the cable to the negative battery terminal.
22   Check the clutch operation.

6   Clutch components - removal, inspection and installation

Refer to illustrations 6.3, 6.4, 6.9 and 6.11
Warning: *Dust produced by clutch wear and deposited on clutch components contains asbestos, which is hazardous to your health. DO NOT blow it out with compressed air and DO NOT inhale it. DO NOT use gasoline or petroleum-based solvents to remove the dust. Brake system cleaner should be used to flush the dust into a drain pan. After the clutch components are wiped clean with a rag, dispose of the contaminated rags and cleaner in a covered, marked container.*

Removal

1   Access to the clutch components is normally accomplished by removing the transaxle, leaving the engine in the vehicle. If the engine is being removed for major overhaul, then check the clutch for wear and replace worn components as necessary.
2   Referring to Chapter 7, Part A, remove the transaxle from the vehicle. Remove the release bearing (see Section 5).
3   To support the clutch disc during removal, install a clutch alignment tool through the splined hole **(see illustration)**.
4   Check the pressure plate and flywheel for indexing marks. They're usually an X, an O or some other white letter. If no marks are visible, make some to ensure installation of the components in the same relationship to each other **(see illustration)**.
5   Loosen the pressure plate-to-flywheel

bolts in 1/4-turn increments until they can be removed by hand. Follow a criss-cross pattern to avoid warping the pressure plate assembly. Support the pressure plate and completely remove the bolts, then detach the pressure plate and clutch disc.
6   On models with pull-type clutch only, remove the wedge collar and retainer ring from the pressure plate diaphragm spring by holding the collar against the diaphragm and working the retainer ring off the wedge collar legs with a small screwdriver. When the retainer has been removed, pry the wedge collar out of the diaphragm.

Inspection

7   Ordinarily, when a problem develops with the clutch, it can be attributed to wear of the clutch driven plate assembly (clutch disc). However, all components should be inspected at this time. **Note:** *If the clutch components are contaminated with oil, there will be shiny, black glazed spots on the clutch disc lining, which will cause the clutch to slip. Replacing clutch components won't completely cure the problem - be sure to check the rear crankshaft oil seal and the transaxle input shaft seal for leaks. If it looks like a seal is leaking, be sure to install a new one to avoid the same problem with a new clutch.*
8   Inspect the flywheel for cracks, heat checking, grooves and other obvious defects. If the imperfections are slight, a machine shop can machine the surface flat and smooth, which is highly recommended regardless of the surface appearance. Refer to Chapter 2 for the flywheel removal and installation procedure.
9   Inspect the lining on the clutch disc. There should be at least 1/16-inch of lining above the rivet heads. Check for loose rivets, distortion, cracks, broken springs and other obvious damage **(see illustration)**. As mentioned above, ordinarily the clutch disc is routinely replaced, so if in doubt about its condition, replace it with a new one.
10   Ordinarily, the release bearing is also

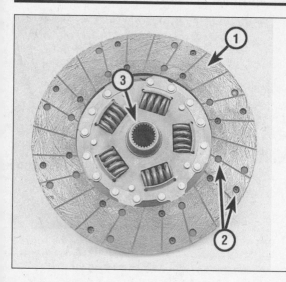

6.9 The clutch disc

1 **Lining** - *this will wear down in use*
2 **Rivets** - *these secure the lining and will damage the flywheel or pressure plate if allowed to contact the surfaces*
3 **Index marks** - *"Flywheel side" or something similar*

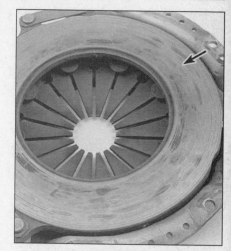

6.11 Inspect the pressure plate friction surface for score marks, cracks and signs of overheating

replaced along with the clutch disc (see Section 5).

11   Check the diaphragm spring fingers and the machined surfaces of the pressure plate **(see illustration)**. Light glazing can be removed with medium grit emery cloth. If a new pressure plate is required, new and factory-rebuilt units are available.

12   Check the condition of the wedge collar and retainer ring. If any of the wedge collar legs are bent or broken, or if the retainer ring is bent or deformed in any way, they should be replaced.

Installation

13   Before installation, clean the flywheel and pressure plate machined surfaces with lacquer thinner or acetone. It's important that no oil or grease is on these surfaces or the lining of the clutch disc. Handle the parts only with clean hands.

14   On models with pull-type clutch, install the wedge collar into the diaphragm spring from the transaxle side of the pressure plate. From the disc side of the pressure plate, install the retainer ring onto the "V" bend of the wedge collar legs. **Caution:** *Be sure you don't deform or otherwise damage the retainer ring or wedge collar legs or the release bearing will pull out of the wedge collar during operation.*

15   Position the clutch disc and pressure plate against the flywheel with the clutch plate held in place with an alignment tool. Make sure it's installed properly (most replacement clutch plates will be marked "flywheel side" or something similar - if it's not marked, install it with the damper springs toward the transaxle).

16   Tighten the pressure plate-to-flywheel bolts only finger tight, working around the pressure plate.

17   Center the clutch disc by inserting the alignment tool through the splined hub and into the bore in the crankshaft. Wiggle the alignment tool up, down or from side-to-side as needed to center the clutch. Tighten the pressure plate-to-flywheel bolts a little at a time, working in a criss-cross pattern to prevent distorting the cover. After all the bolts are snug, tighten them to the torque listed in this Chapter's Specifications. Remove the alignment tool.

18   Put some high-temperature grease on the release lever fingers. On push-type clutches, lubricate the inner groove of the release bearing also (refer to Section 5).

19   Install the clutch release bearing as described in Section 5.

20   Install the transaxle, slave cylinder and all components that were removed previously. **Caution:** *On models with pull-type clutch, be especially careful when installing the transaxle to avoid damaging the wedge collar or retainer ring with the transaxle input shaft.*

21   Adjust the shift linkage as outlined in Chapter 7.

7   Starter/clutch interlock switch - check and replacement

1   The starter/clutch interlock switch is mounted on the clutch pedal support and allows the engine to be started only with the clutch pedal fully depressed.

Check

2   Disconnect the electrical connector from the switch. The switch is located near the top of the clutch pedal arm.

3   Connect an ohmmeter or self-powered test light (continuity tester) to the two switch terminals (if there are three terminals on the switch, one is for the cruise control. You'll probably have to check a wiring diagram to determine which two terminals to connect to).

4   With the clutch pedal in it's normal position there should be no continuity (the test light should not come on or the ohmmeter should read infinity).

5   Depress the clutch pedal. There should now be continuity (the test light should come on or the ohmmeter should indicate close to zero ohms).

6   If the switch fails to operate as described, check to make sure the plunger on the switch is contacting the clutch pedal. If it is, and the pedal depresses the plunger when the pedal is released, the switch is probably faulty.

Replacement

7   Disconnect the negative cable from the battery. **Caution:** *On models equipped with the Theftlock audio system, be sure the lock-out feature is turned off before performing any procedure which requires disconnecting the battery.*

8   Disconnect the electrical connector and remove the switch.

9   Reference mark the amount of the switch protruding through the mounting bracket.

10   Remove the switch from the mount.

11   Install the new switch by reversing the removal procedure. Be sure the same amount of the switch protrudes through the mounting bracket as the old switch did.

12   Connect the negative battery cable.

13   Check to be sure the engine can be started only when the clutch pedal is fully depressed. Be sure to perform this test with the transaxle in Neutral.

8   Driveaxles - general information

Power is transmitted from the transaxle to the front wheels by two driveaxles, which consist of splined solid axles with constant velocity (CV) joints at each end.

There are two types of CV joints used. On the outer joint a double-offset design using ball bearings with an inner and outer race is used to allow angular movement. On the inner CV joint is a tri-pot design, with a spider bearing assembly and tri-pot housing to allow angular movement and permit the

9.2 A screwdriver inserted through the caliper and into a disc
cooling vane will hold the hub stationary while loosening
the hub nut

9.5 A two-jaw puller works well for pushing the axle
out of the hub

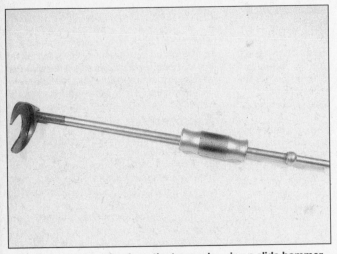

9.6 Remove the axles from the transaxle using a slide hammer
and adapter

9.9 A large punch or screwdriver, positioned in the groove on the
CV joint housing, can be used to seat the CV joint in the transaxle

driveaxle to slide in and out.

The CV joints are protected by rubber boots, which are retained by clamps so the joints are protected from water and dirt. The boots should be inspected periodically (see Chapter 1). The inner boots have very small breather holes which may leak a small amount of lubricant under some circumstances, such as when the joint is compressed during removal. Damaged CV joint boots must be replaced immediately or the joints can be damaged. Boot replacement involves removing the driveaxles. It's a good idea to disassemble, clean, inspect and repack the CV joint whenever replacing a CV joint boot to make sure the joint isn't contaminated with moisture or dirt, which would cause premature failure of the CV joint.

The most common symptom of worn or damaged CV joints, besides lubricant leaks, are a clicking noise in turns, a clunk when accelerating from a coasting condition or vibration at highway speeds.

9   Driveaxles - removal and installation

Refer to illustrations 9.2, 9.5, 9.6 and 9.9

Removal

1   Remove the wheel cover and loosen the hub nut. Loosen the wheel lug nuts, raise the front of the vehicle and support it securely on jackstands. Apply the parking brake and block the rear wheels to keep the vehicle from rolling off the jackstands. Remove the front wheel. Place a drain pan under the transaxle.
2   Remove the driveaxle hub nut. To prevent the hub from turning, insert a screwdriver through the caliper and into a rotor cooling vane, then remove the nut **(see illustration)**.
3   Remove the brake caliper and disc and support the caliper out of the way with a piece of wire (see Chapter 9).

4   Remove the control arm-to-steering knuckle balljoint stud nut and separate the lower arm from the steering knuckle (see Chapter 10 if necessary). On models equipped with anti-lock brakes (ABS), remove the ABS wheel speed sensor and place it to one side.
5   Push the driveaxle out of the hub with a puller **(see illustration)**, then support the outer end of the driveaxle with a piece of wire to prevent damage to the inner CV joint.
6   On automatic transaxle equipped models, use a slide hammer and adapter to separate the driveaxle from the transaxle **(see illustration)**.
7   On manual transaxle equipped models, the right driveaxle connects to an intermediate shaft. Use a slide hammer and adapter to separate the right axle from the intermediate shaft and the left axle from the transaxle.
8   On all models, support the CV joints and carefully remove the driveaxle from the vehicle.

10.4 Snap-ring pliers should be used to remove both the inner and outer retaining rings

10.8 Before installing the CV joint boot, wrap the axle splines with tape to prevent damage to the boot

a lot of time and work. Whatever is decided, check on the cost and availability of parts before disassembling the vehicle.

1   Remove the driveaxle (see Section 9).
2   Place the driveaxle in a vise lined with rags to avoid damage to the shaft.

Inner CV joint

Refer to illustrations 10.4, 10.8, 10.9, 10.10 and 10.12

3   Cut off the boot seal retaining clamps and slide the boot towards the center of the driveaxle. Mark the tri-pot housing and driveaxle so they can be reinstalled in the same relative positions, then slide the housing off the spider assembly.
4   Remove the spider assembly from the axle by first removing the inner retaining ring and sliding the spider assembly back to expose the front retaining ring. Remove the front retaining ring and slide the joint off the driveaxle **(see illustration)**.
5   Use tape or a cloth wrapped around the spider bearing assembly to retain the bearings during removal and installation.
6   Slide the boot off the axle.
7   Clean all of the old grease out of the housing and spider assembly. Carefully disassemble each section of the spider assembly, one at a time, and clean the needle bearings with solvent. Inspect the rollers, spider cross, bearings and housing for scoring, pitting and other signs of abnormal wear. Apply a coat of CV joint grease to the inner bearing surfaces to hold the needle bearings in place when reassembling the spider assembly.
8   Wrap the driveaxle splines with tape to avoid damaging the boot, then slide the boot onto the axle **(see illustration)**.
9   Install the spider bearing with the recess in the counterbore facing the end of the driveaxle **(see illustration)**.
10   Pack the housing with half of the grease furnished with the new boot and place the remainder in the boot **(see illustration)**.
11   Install the tri-pot housing.

Installation

9   Lubricate the differential seal with multi-purpose grease, raise the driveaxle into position while supporting the CV joints and insert the splined end of the inner CV joint into the differential side gear. Seat the shaft in the side gear by positioning the end of a screwdriver in the groove in the CV joint and tapping it into position with a hammer **(see illustration)**. On models with an intermediate shaft, slide the driveaxle inner CV joint onto the end of the intermediate shaft.
10   On all models, grasp the inner CV joint housing (not the driveaxle) and pull out to make sure the axle has seated securely.
11   Apply a light coat of multi-purpose grease to the outer CV joint splines, pull out on the strut/steering knuckle assembly and install the stub axle in the hub.
12   Insert the control arm balljoint stud into the steering knuckle and tighten the nut. Be sure to use a new cotter pin (refer to Chapter 10). If equipped with ABS, install the ABS

wheel speed sensor.
13   Install the brake disc and caliper (see Chapter 9 if necessary).
14   Install the hub nut. Lock the disc so it can't turn, using a screwdriver or punch inserted through the caliper into a disc cooling vane, and tighten the hub nut to the torque listed in this Chapter's Specifications.
15   Check the transaxle lubricant level and add some, if necessary, to bring it to the appropriate level (see Chapter 1).
16   Install the wheel and lower the vehicle. Install the wheel cover.

---

10   Driveaxle boot replacement

Note: *If the CV joints exhibit wear indicating the need for an overhaul (usually due to torn boots), explore all options before beginning the job. Complete rebuilt driveaxles are available on an exchange basis, which eliminates*

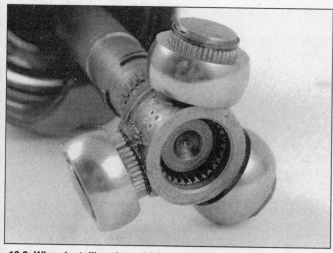

10.9 When installing the spider assembly on the driveaxle, make sure the recess in the counterbore (arrow) is facing the end of the driveaxle

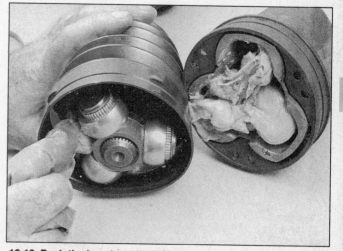

10.10 Pack the housing with half of the grease furnished with the new boot and place the remainder in the boot

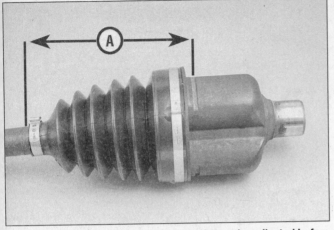

10.12 The length of all inner CV joints (A) must be adjusted before the large boot clamp is tightened

10.14 Carefully tap around the circumference of the seal retainer to remove it from the housing

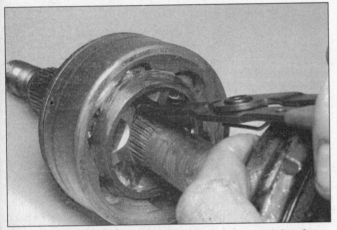

10.16 Use snap-ring pliers to remove the inner retaining ring

10.19 Gently tap the inner race with a brass punch to tilt it enough to allow ball bearing removal

12   Seat the boot in the housing and axle seal grooves, then adjust the length of the joint **(see illustration)**.
13   Install the retaining clamps, then install the driveaxle as described in Section 9.

Outer CV joint

Refer to illustrations 10.14, 10.16, 10.19, 10.20, 10.21, 10.22, 10.23a, 10.23b, 10.26, 10.27 and 10.31

14   Tap lightly around the outer circumference of the seal retainer with a hammer and punch to dislodge and remove it. Be very careful not to deform the retainer, or it won't seal properly **(see illustration)**.
15   Cut off the band retaining the boot to the shaft.
16   Remove the snap-ring and slide the joint assembly off **(see illustration)**.

10.20 Using a dull screwdriver, carefully pry the balls out of the cage

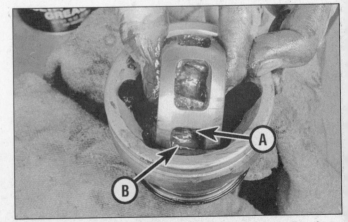

10.21 Tilt the inner race and cage 90-degrees, then align the windows in the cage (A) with the lands (B) and rotate the inner race up and out of the outer race

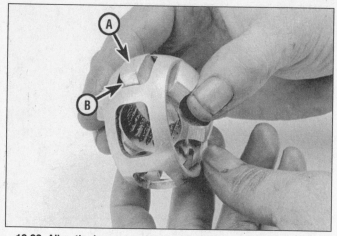

10.22 Align the inner race lands (A) with the cage windows (B) and rotate the inner race out of the cage

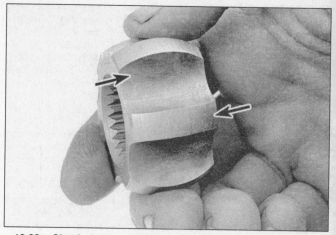

10.23a Check the inner race lands and grooves for pitting and score marks

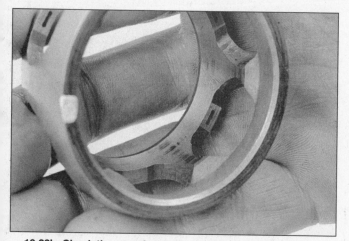

10.23b Check the cage for cracks, pitting and score marks - shiny spots are normal and don't affect operation

10.26 Align the cage windows and the inner and outer race grooves, then tilt the cage and inner race to insert the balls

17   Slide the old boot off the driveaxle.
18   Place marks on the inner race and cage so they both can be installed facing out when reassembling the joint.
19   Press down on the inner race far enough to allow a ball bearing to be removed. If it's difficult to tilt, tap the inner race with a brass punch and hammer (see illustration).
20   Pry the balls out of the cage, one at a time, with a blunt screwdriver or wooden tool (see illustration).
21   With all of the balls removed from the cage and the cage/inner race assembly tilted 90-degrees, align the cage windows with the outer race lands and remove the assembly from the outer race (see illustration).
22   Remove the inner race from the cage by turning the inner race 90-degrees in the cage, aligning the inner lands with the cage windows and rotating the inner race out of the cage (see illustration).
23   Clean the components with solvent to remove all traces of grease. Inspect the cage and races for pitting, score marks, cracks and other signs of wear and damage. Shiny, polished spots are normal and won't adversely

affect CV joint operation (see illustrations).
24   Install the inner race in the cage by reversing the technique described in Step 21.
25   Install the inner race and cage assembly in the outer race by reversing the procedure in Step 20. The marks that were previously applied to the inner race and cage must both be visible after the assembly is installed in the outer race.
26   Press the balls into the cage windows (see illustration).
27   Pack the CV joint assembly with lubricant through the inner splined hole. Force the grease into the bearing by inserting a wooden dowel through the splined hole and pushing it to the bottom of the joint. Repeat this procedure until the bearing is completely packed (see illustration).
28   Install the boot on the driveaxle as described in Step 10. Apply a liberal amount of grease to the inside of the axle boot.
29   Position the CV joint assembly on the driveaxle, aligning the splines. Using a soft-face hammer, drive the CV joint onto the driveaxle until the retaining ring is seated in the groove.

10.27 Apply grease through the splined hole, then insert a wooden dowel (approximately 15/16-inch diameter) through the splined hole and push down - the dowel will force the grease into the joint

30   Seat the inner end of the boot in the seal groove and install the retaining clamp.

31   Install the seal retainer securely by tapping evenly around the outer circumference with a hammer and punch **(see illustration)**. **Note:** *Earlier models use a seal retainer while later models use boot retaining clamps on the outboard housing. Always follow the boot manufacturers instructions when installing boot clamps since several different types are available.*

32   Install the driveaxle as described in Section 9.

11   Intermediate axleshaft - removal and installation

Removal

1   Loosen the right front wheel lug nuts, raise the front of the vehicle and support it securely on jackstands. Apply the parking brake and block the rear wheels to keep the vehicle from rolling off the jackstands. Remove the right wheel.

2   Drain the transaxle.

3   Remove the right driveaxle (see Section 9).

4   Remove the bracket bolts retaining the intermediate shaft assembly.

5   Remove the bolts connecting the shaft assembly-to-transaxle.

6   Pull the shaft out of the transaxle.

Installation

7   Lubricate the lips of the differential seal with multi-purpose grease and slide the intermediate shaft into the transaxle.

8   The remainder of installation is the reverse of removal. Refill the transaxle with the special lubricant (see Chapter 1).

10.31 Carefully tap around the circumference of the seal retainer to install it on the housing (early model shown)

Chapter 9 Brakes

Contents

Specifications

General
Brake fluid type ... See Chapter 1

Disc brakes
Brake pad lining minimum thickness ... See Chapter 1
Front and rear
Minimum thickness ... Refer to dimension stamped on disc
Runout (maximum) ... 0.003 in
Thickness variation limit ... 0.0005 in

Drum brakes
Brake shoe lining minimum thickness ... See Chapter 1
Maximum allowable diameter ... Cast into drum

Specifications
Ft-lbs (unless otherwise indicated)

Caliper mounting (slide) bolts	
Front	
1996 and earlier	80
1997 and later	63
Rear	
1996 and earlier	20
1997 and later	32
Caliper mounting bracket bolts	
Front	
1996 and earlier	148
1997 and later	137
Rear	
1996 and earlier	81
1997 and later	92
Brake hose-to-caliper bolt	32
Wheel cylinder mounting bolts	15
Wheel lug nuts	See Chapter 1

1 General information

Conventional (non-ABS) brake system

The vehicles covered by this manual are equipped with hydraulically operated front and rear brake systems. All front brake systems are disc type, while the rear brakes are either disc or drum. All brakes are self adjusting. Front and rear disc brakes automatically compensate for pad wear. Rear drum brakes incorporate an adjusting mechanism which is activated any time the service brakes are applied.

Hydraulic system

The hydraulic system consists of two separate circuits. The master cylinder has separate reservoirs for the two circuits and in the event of a leak or failure in one hydraulic circuit, the other circuit will remain operative. A visual warning of circuit failure or air in the

2.5a Before disassembling the brake, wash it thoroughly with brake system cleaner and allow it to dry - position a drain pan under the brake to catch the residue - DO NOT use compressed air to blow off brake dust!

2.5b To make room for the new pads, use a C-clamp to depress the piston into the caliper before removing the caliper and pads - do this a little at a time, keeping an eye on the fluid level in the master cylinder to make sure it doesn't overflow

system is given by a warning light activated on the dash.

Proportioner valves

The proportioner valves are designed to provide better front to rear braking balance with heavy brake application. These valves allow more pressure to be applied to the front brakes (under certain braking operations) due to the fact the rear of the vehicle is lighter and does not require as much braking force.

Power brake booster

The power brake booster, utilizing engine manifold vacuum and atmospheric pressure to provide assistance to the hydraulically operated brakes, is mounted on the firewall in the engine compartment.

Anti-lock Brake System (ABS)

This system is available as an option. It is designed to reduce lost traction while braking. The system is similar to the non-ABS system except for the controller (computer) and related wiring, speed sensors and the hydraulic pump which replaces the master cylinder and power brake booster.

Anti-lock braking occurs only when a wheel is about to lock up (lose traction). Input signals from the wheel speed sensors to the computer are used to determine when a wheel is about to lose traction during braking. Hydraulic pressure will be reduced for the wheel about to loose traction. Diagnosis of this system is beyond the scope of the home mechanic.

Parking brake

The parking brake operates the rear brakes only, through cable actuation. It's activated by a pedal mounted on the left side kick panel.

Service

After completing any operation involving disassembly of any part of the brake system, always test drive the vehicle to check for

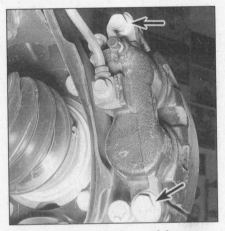

2.6a On 1996 and earlier models remove the caliper mounting bolts (arrows) . . .

2.6b . . . then slide the caliper out of the mounting bracket . . .

proper braking performance before resuming normal driving. When testing the brakes, perform the tests on a clean, dry flat surface. Conditions other than these can lead to inaccurate test results.

Test the brakes at various speeds with both light and heavy pedal pressure. The vehicle should stop evenly without pulling to one side or the other. Avoid locking the brakes because this slides the tires and diminishes braking efficiency and control of the vehicle.

Tires, vehicle load and front-end alignment are factors which also affect braking performance.

2 Disc brake pads (front and rear) - replacement

Warning: *Disc brake pads must be replaced on both front or rear wheels at the same time - never replace the pads on only one wheel. Also, the dust created by the brake system may contain asbestos, which is harmful to your health. Never blow it out with compressed air*

and don't inhale any of it. An approved filtering mask should be worn when working on the brakes. Do not, under any circumstances, use petroleum-based solvents to clean brake parts. Use brake system clean-er only!

Front

Removal

Refer to illustrations 2.5a, 2.5b, 2.6a, 2.6b, 2.6c, 2.6d, 2.6e, 2.7a, 2.7b, 2.8a, 2.8b and 2.8c

1   Remove and discard two-thirds of the brake fluid from the master cylinder.
2   Loosen the wheel lug nuts.
3   Raise the vehicle and support it securely on jackstands.
4   Remove the wheel and reinstall two lug nuts to hold the disc in position.
5   Clean the caliper, disc and other components with brake system cleaner before beginning any brake work **(see illustration)**. This should remove traces of potentially hazardous brake dust. Collect the drippings in a plastic container. Use a large C-clamp over the caliper to push the pistons back into the body of the caliper **(see illustration)**.

2.6c ... and suspend the caliper with a piece of wire - DO NOT let the caliper hang by the hose

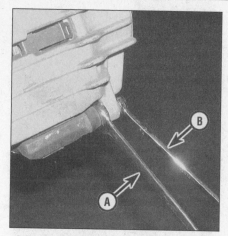

2.6d On 1997 and later models, hold the caliper slide pin with an open-end wrench (A) and loosen the lower mounting bolt with another wrench (B) ...

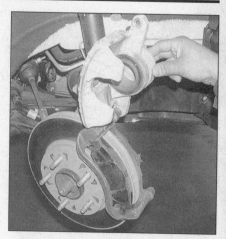

2.6e ... then pivot the caliper up and support it in this position for access to the brake pads

2.7a Use a screwdriver to pry the spring of the outside pad away from the caliper while removing the pad

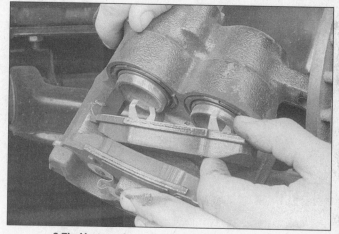

2.7b Unsnap the pad retainers from the pistons

6   On 1996 and earlier models remove the caliper mounting bolts and pull the caliper straight up and off, but do not allow it to hang by the brake hose **(see illustrations)**. On 1997 and later models remove the lower caliper bolt and swing the caliper upward (see illustrations).

7   On 1996 and earlier models remove the outer pad from the caliper **(see illustration)**. If necessary, use a screwdriver to pry up on the retaining spring of the outer pad while sliding it out of the caliper. Unclip the inner pad from the pistons **(see illustrations)**.

8   On 1997 and later models remove the inner and outer pads from the caliper mount then remove the pad retaining clips from the caliper mount **(see illustrations)**.

2.8a On 1997 and later models, remove the inner brake pad ...

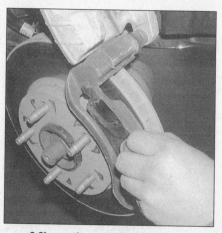

2.8b ... the outer brake pad ...

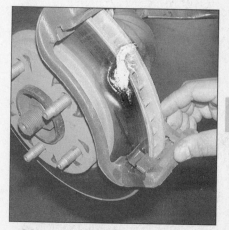

2.8c ... and the upper and lower pad retainers from the caliper mounting bracket

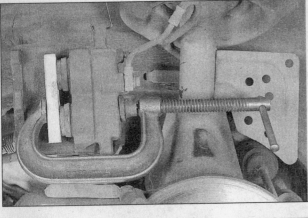

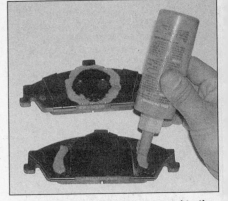

2.9a Using a "C" clamp and flat piece of wood or metal, push the pistons into the caliper

2.9b Apply anti-squeal compound to the back of both pads (let the compound "set up" a few minutes before installing them)

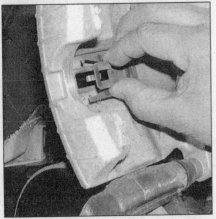

2.11 On some 1997 and later models, it will be necessary to check the condition of the anti-rattle spring in the center of the caliper, replacing it if necessary

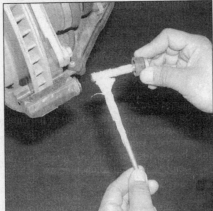

2.12 Clean the caliper slide pin and inspect it for scoring and corrosion; coat the pin with high-temperature grease

Installation

Refer to illustrations 2.9a, 2.9b, 2.11 and 2.12

9   Compress the piston(s) back into the caliper bore using a "C" clamp and a block of wood **(see illustration)** then coat the back of the new brake pads with an anti-squeal compound **(see illustration)**.

10   If the boots for the caliper mounting bolts are torn or damaged, pry them out with a small screwdriver and replace them with new ones.

11   To install the new inner and outer pads reverse the removal procedure. **Note:** *On 1997 and later models make sure the anti rattle clip (if equipped) is securely fastened in place* **(see illustration)**.

12   Install the caliper over the disc and align the bolt holes. The rubber boots should be between the caliper and the mounting bracket. Lubricate the entire length of the mounting bolts with a thin film of silicone grease **(see illustration)**. Be sure to tighten the caliper mounting bolts to the torque listed in this Chapter's Specifications.

13   Lower the vehicle. Fill the reservoir with the recommended fluid (see Chapter 1) and pump the brake pedal several times to bring the pads into contact with discs.

14   Road test the vehicle carefully before returning it to normal service.

Rear

Removal

Refer to illustrations 2.24, 2.31 and 2.32

15   Remove and discard two-thirds of the brake fluid from the reservoir.

16   Loosen the wheel lug nuts, raise the vehicle and support it securely on jackstands. Remove the wheel.

17   Install two lug nuts to retain the disc when the caliper is removed.

1994 and later models (except Lumina)

18   On 1996 and earlier models remove the bolt and washer attaching the cable support bracket to the caliper body.

19   On 1996 and earlier models remove the sleeve bolt from the bottom of the caliper, then pivot the caliper up. On 1997 and later models remove the upper caliper bolt and pivot the caliper downward.

20   Remove the pads and retainers from the caliper.

Others

21   Remove the bolts retaining the shield assembly, then remove the shield.

22   Loosen the tension on the parking brake cables at the equalizer (see Section 9).

23   Disconnect the cable and return spring from the parking brake lever.

24   Hold the caliper lever and remove the locknut **(see illustration)**. Remove the lever and the lever seal.

25   Compress the piston into the caliper bore. Using two pairs of adjustable pliers, squeeze the inner pad and the outer flanges on the caliper housing together as much as possible.

26   Reinstall the lever seal with the rubber sealing bead against the caliper housing.

27   Reinstall the lever and it's retaining nut.

28   Remove the caliper mounting bolts.

29   Detach the brake hose from the suspension bracket.

30   Remove the caliper and support it with a piece of wire.

31   Using a screwdriver, disengage the buttons on the outer pad from the holes in the caliper housing **(see illustration)**. Remove the outer pad.

32   From the open side of the caliper, press in on the edge of the inner pad and tilt it out to release the pad from the retainer **(see illustration)**. Remove the inner pad. Also remove the check valve from the center of the caliper piston.

Installation

Refer to illustrations 2.35, 2.38a, 2.38b, 2.43 and 2.44

33   Check the bolt boots for damage and

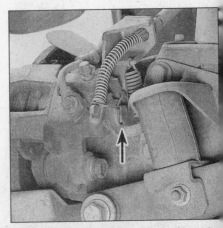

2.24 Parking brake lever mounting nut

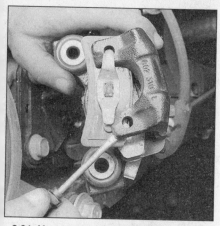

2.31 Use a screwdriver to pry the outer pad from the caliper

2.32 Unclip the inner pad from the retainer

2.35 Using a pair of needle nose pliers, engage the tips of the pliers into the cutouts of the piston face, then turn the piston into the cylinder bore - after the piston bottoms out, align the cutouts in the piston 90 degrees from the centerline of the caliper mounting bolts (1994 through 1996 models)

replace them if necessary.

34   Check the mounting bolts and sleeves for corrosion, replacing them if necessary. Note that the sleeves on 1994 through 1996 models are installed in the caliper and that the sleeves on 1997 and later models are installed in the caliper mount.

1994 and later models (except Lumina)

35   On 1994 through 1996 models bottom the piston in the caliper by threading it in using a suitable spanner-like tool **(see illustration)**. Turn the piston as necessary so that a line drawn through the piston slots would be exactly perpendicular (90 degrees) to a line drawn through the caliper mounting bolt holes.

36   On 1997 and later models compress the piston back into the caliper bore using a "C" clamp and a block of wood **(see illustration 2.9a)**.

37   Coat the back of the new brake pads with an anti-squeal compound **(see illustration 2.9b)**.

38   On all 1994 and later models install the pad retainers on the caliper mount **(see illustrations)**. Install the inner and outer pads in the caliper mount.

39   Pivot the caliper assembly over the disc and align the bolt holes. The rubber boots should be between the caliper and the mounting bracket. Lubricate the entire length of the mounting bolts with a thin film of silicone grease **(see illustration 2.12)**.

40   The remainder of reassembly is the opposite of disassembly. Tighten fasteners to the torques listed at the beginning of this chapter. Road test the vehicle carefully before returning it to normal service. **Warning:** *Before driving the car, pump the brakes several times to seat the pads against the disc.*

Others

41   Push the sleeves into the caliper, away from the inner pad.

42   Apply a thin film of silicone grease to a new piston check valve and install it in the face of the piston.

43   With the inner pad retainer properly positioned **(see illustration)**, place the new inner pad on the piston with the tab fitting into the "D" notch in the piston face. If the tab does not line up with the notch, turn the piston with a pair of needle-nose pliers until it is in proper alignment.

2.38a Install the upper anti-rattle clip . . .

44   Install the outer pad so the spring retainer fits into the slots in the caliper housing **(see illustration)**.

2.38b . . . and the lower anti-rattle clip in the caliper mounting bracket - make sure they're both fully seated

2.43 Position the tangs of the pad retainer as shown, then clip the pad into the retainer - make sure the tab on the retainer fits into the "D" notch in the piston face

2.44 Slide the outer pad into the caliper, making sure the ends of the retainer spring fit into the slots

45   Slide caliper over the disc. Install the mounting bolts and tighten them to the torque listed in this Chapter's Specifications.

46   Pump the brake pedal several times to seat the pads against the disc.

47   Install the bracket and retaining bolt.

48   Clean the area around the lever seal.

49   Lubricate the lever seal and install it with the sealing bead against the caliper housing.

50   Install the lever on the actuator screw hex with the lever pointing down.

51   To prevent parking brake actuation, rotate the lever back against the stop and tighten the nut.

52   Install the return spring.

53   Connect and adjust the parking brake cable (see Section 8).

54   Remove the two lug nuts used to retain the disc. Install the wheel and tighten the lug nuts to the torque listed in the Chapter 1 Specifications.

---

3   Disc brake caliper - removal, overhaul and installation

Warning: *Dust created by the brake system may contain asbestos, which is harmful to your health. Never blow it out with compressed air and don't inhale any of it. An approved filtering mask should be worn when working on the brakes. Do not, under any circumstances, use petroleum-based solvents to clean brake parts. Use brake system cleaner only!*

Note: *If an overhaul is indicated (usually because of fluid leakage) explore all options before beginning the job. New and factory rebuilt calipers are available on an exchange basis, which makes this job quite easy. If it's decided to rebuild the calipers, make sure a rebuild kit is available before proceeding. Always rebuild the calipers in pairs - never rebuild just one of them.*

Removal

1   The removal procedure is the same as in Section 2 with the exception of disconnecting the hydraulic line from the caliper and, on

3.4 Place a block of wood or rags in front of the pistons, then apply compressed air to the hydraulic line inlet to force the pistons out

1994 through 1996 models (except Lumina), the parking brake cable (rear disc only) and upper caliper bolts. On 1997 and later models it will be necessary to remove the hydraulic brake line and lower caliper bolt. **Warning:** *Before servicing any ABS component, it is mandatory that hydraulic pressure in the system be relieved. With ignition off, pump the brake pedal a MINIMUM of 40 times, using full pedal strokes.*

Overhaul

Front (1996 and earlier models)

Refer to illustrations 3.4, 3.6, 3.7a, 3.7b, 3.13 and 3.15

Note: *Due to the unavailability of overhaul kits for these models the calipers on 1997 and later models should be replaced with a new or rebuilt unit.*

2   Refer to Section 2 and remove the brake pads from the caliper.

3   Clean the exterior of the caliper with brake system cleaner. Never use gasoline, kerosene or petroleum-based cleaning solvents. Place the caliper on a clean workbench.

3.6 Use a screwdriver to pry the boots from the caliper

4   Position a wood block or several shop rags in the caliper as a cushion, then use compressed air to remove the pistons from the caliper **(see illustration)**. Use only enough air pressure to ease the pistons out of the bore. If a piston is blown out, even with the cushion in place, it may be damaged. **Warning:** *Never place your fingers in front of the pistons in an attempt to catch or protect it when applying compressed air, as serious injury could occur.*

5   If only one piston comes out it will have to be slightly reinserted. Wedge a spacer between the reinserted piston and the caliper housing. Again apply air pressure to release the frozen piston.

6   Carefully pry the dust boots out of the caliper bores **(see illustration)**.

7   Using a wood or plastic tool, remove the piston seal from the groove in the caliper bores **(see illustrations)**. Metal tools may cause bore damage.

8   Remove the caliper bleeder screw. Discard all rubber parts.

9   Clean the remaining parts with brake system cleaner and let them dry.

10   Carefully examine the pistons for nick

3.7b The piston seal should be removed with a plastic or wooden tool to avoid damage to the bore and seal groove - a pencil will do the job

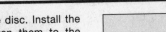

3.7a Exploded view of a typical front brake caliper

A   *Caliper body*
B   *Piston seals*
C   *Pistons*
D   *Dust boots*
E   *Caliper bleeder screw*
F   *Bleeder screw dust cap*

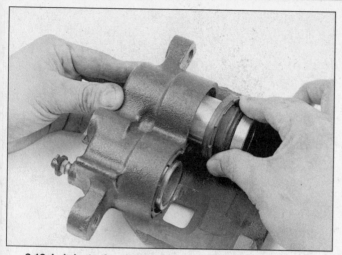

3.13 Lubricate the piston and insert it (with boot attached) carefully into the bore - do not twist the seal

3.15 Use the proper size driver to install the boot to the caliper

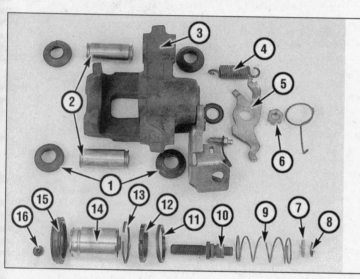

3.18 Exploded view of the rear brake caliper

1	Bolt boot	9	Balance spring and retainer
2	Sleeve	10	Actuator screw
3	Caliper body	11	Piston seal
4	Return spring	12	Piston locator
5	Parking brake lever	13	Snap-ring
6	Nut	14	Piston
7	Thrust washer	15	Caliper boot
8	Shaft seal	16	Check valve

and burrs and loss of plating. If surface defects are present, the parts must be replaced.

11   Check the caliper bores in a similar way. Light polishing with crocus cloth is permissible to remove light corrosion and stains. Discard the mounting bolts if they're corroded or damaged.

12   When assembling, lubricate the piston bores and seals with clean brake fluid. Position the seals in the caliper bore grooves.

13   Lubricate the pistons with clean brake fluid, then install new boots in the piston grooves with the fold toward the open end of the piston **(see illustration)**.

14   Insert the pistons squarely into the caliper bores, then apply force to bottom them.

15   Position the dust boots in the caliper counterbores, then use a driver to install them into position **(see illustration)**. If a seal driver isn't available, carefully tap around the outer circumference of the seal using a blunt punch. Make sure the boot is recessed evenly below the caliper face.

16   Install the bleeder valve.

17   Remove the bushings, support bushings and boots from the caliper mounting bracket and install new ones **(see illustration 3.7a)**.

Rear (1993 and earlier and 1994 Lumina)

Refer to illustrations 3.18, 3.22, 3.26, 3.46 and 3.49

Note: *Later model calipers require special tools for overhaul and should be replaced with a new or rebuilt unit.*

18   Remove the sleeves and bolt boots **(see illustration)**.

19   Remove the pad retainer from the end of the piston by rotating the retainer until the inside tabs line up with notches in the piston.

20   Remove the lever nut, lever and lever seal.

21   Pad the interior of the caliper with rags.

22   Using a wrench to rotate the actuator screw **(see illustration)**, work the piston out of the caliper bore (in the direction the parking brake is applied).

23   Remove the balance spring.

24   Remove the actuator screw by pressing

on the threaded end.

25   Remove the shaft seal and thrust washer from the actuator screw.

26   While being careful not to scratch the

3.22 Rotate the actuator screw (arrow) to remove the piston

3.26 Carefully pry the dust boot out of the housing, taking care not to scratch the surface of the bore

3.46 Use a seal driver to seat the boot in the caliper housing

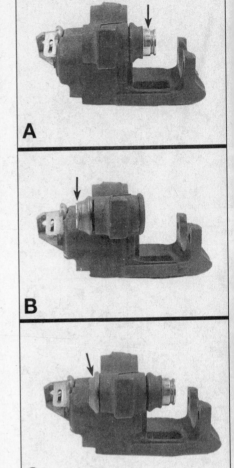

3.49 Sleeve boot installation procedure

A   *Install the sleeve boot into the groove in the sleeve bore, then push the sleeve into the boot*
B   *Install the other boot into the groove in the other side of the bore*
C   *Push the sleeve back and seat both boots into the grooves in the ends of the sleeve*

housing bore, remove the boot **(see illustration)**.
27   Using snap-ring pliers, remove the snap-ring.
28   Remove the piston locator.
29   Using a plastic or wooden tool (so as not to damage the bore), remove the seal from the caliper bore **(see illustration 3.7b)**.
30   Remove the bleeder valve.
31   Clean all parts not included in the repair kit with brake cleaner or denatured alcohol.
32   Dry all parts with filtered unlubricated compressed air.
33   Blow out all passages in the housing and bleeder valve.
34   Inspect all parts for wear, damage and corrosion. Replace any parts as needed. **Note:** *It is okay to remove minor corrosion from the caliper bore using crocus cloth.*
35   Lubricate all of the internal parts with clean brake fluid **(see illustration 3.18)**.
36   Install a new piston seal into the caliper bore groove. Be sure not to twist the seal.
37   Install a new piston locator on the piston.
38   Install the thrust washer on the actuator screw with the grayish side towards the caliper housing and the copper side towards the piston assembly.
39   Install the shaft seal on the actuator screw.
40   Install the actuator screw onto the piston assembly.
41   Install the balance spring and retainer into the piston recess.
42   Install the piston assembly with the actuator screws and balance spring into the lubricated bore of the caliper.
43   Push the piston into the caliper bore so the locator is past the retainer groove in the caliper bore. **Note:** *A pair of adjustable pliers might be necessary to press the piston into the caliper bore.*
44   Using snap-ring pliers, install the snap-ring.
45   With the inside lip of the boot in the piston groove and boot fold toward the end of the piston that contacts the brake pad, install

the caliper boot onto the piston **(see illustration 3.13)**.
46   Use a seal driver to seat the caliper boot into the cylinder counterbore **(see illustration)**. If a seal driver isn't available, use a blunt punch and carefully tap around the outer circumference of the seal to seat it.
47   Install the pad retainer on the piston.
48   Lubricate outside diameter of the mounting bolt sleeves and the caliper sleeve cavities with silicone grease.
49   Install one sleeve boot into a groove of the sleeve cavity **(see illustration)**.
50   Install a sleeve through the opposite side of the cavity and continue pushing sleeve until the boot lip seats in the sleeve groove.
51   Install boot in the opposite side of the sleeve cavity groove and push the sleeve through the cavity far enough for that boot lip to slip into the remaining groove of the sleeve.
52   Repeat the sleeve boot installation procedure to the remaining sleeve.

Installation

53   The installation procedure is the same as in Section 2.
54   When attaching the hydraulic line to the caliper, use new copper washers at the line-to-caliper connection.
55   Bleed the hydraulic system as described in Section 7.

4   Brake disc - inspection, removal and installation

Refer to illustrations 4.3, 4.4a, 4.4b, 4.5a and 4.5b

Inspection

1   Loosen the wheel lug nuts, raise the vehicle and support it securely on jackstands. Apply the parking brake and block the wheels to keep the vehicle from rolling off the jackstands. Remove the wheel and install three lug nuts to hold the disc in place.
2   Remove the brake caliper as outlined in

Section 2 (it's part of the pad replacement procedure). You don't have to disconnect the brake hose. After removing the caliper bolts, suspend the caliper out of the way with a piece of wire - DO NOT let it hang by the hose.
3   Visually inspect the disc surface for score marks and other damage. Light scratches and shallow grooves are normal and may not be detrimental to brake operation, but deep score marks - over 0.015-inch (0.38 mm) deep - require disc removal and refinishing by an automotive machine shop. Be sure to check both sides of the disc **(see illustration)**. If pulsating has been felt during application of the brakes, suspect excessive disc runout.
4   To check disc runout, mount a dial indicator with the stem resting at a point about 1/2-inch from the outer edge of the disc **(see illustration)**. Set the indicator to zero and

4.3 The brake pads on this vehicle were obviously neglected, as they wore down to the rivets and cut deep grooves into the disc - wear this severe will require replacement of the disc

4.4a Check for runout with a dial indicator - mount it with the indicator needle about 1/2-inch from the outer edge of the disc

4.4b If you don't have the discs machined, at the very least be sure to remove the glaze from the disc surface with sandpaper or emery cloth (use a swirling motion as shown here)

4.5a The minimum wear (or discard) thickness (arrow) is cast into the disc

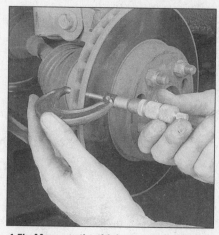

4.5b Measure the thickness of the disc at several points with a micrometer

rn the disc. The indicator reading should ot exceed the specified allowable runout mit. If it does, the disc should be refinished / an automotive machine shop. **Note:** *The scs should be resurfaced, regardless of the al indicator reading, to impart a smooth fin- h and ensure perfectly flat brake pad sur- ces which will eliminate pedal pulsations. At e very least, if you don't have the discs surfaced, remove the glaze with sandpaper emery cloth using a swirling motion* **(see ustration)**.

Never machine the disc to a thickness ss than the specified minimum allowable finish thickness. The minimum wear (or dis- rd) thickness is cast into the disc **(see ustration)**. The disc thickness can be ecked with a micrometer **(see illustration)**.

emoval

fer to illustration 4.6

Remove the caliper mounting bracket if uipped **(see illustration)**.

Installation

7   Place the disc in position over the threaded studs.

8   Install the caliper and brake pad assem- bly (refer to Section 2 for the caliper installa- tion procedure, if necessary). Tighten the caliper bolts to the torque listed in this Chap- ter's Specifications.

9   Install the wheel, then lower the vehicle to the ground. Tighten the lug nuts to the torque listed in the Chapter 1 Specifications. Depress the brake pedal a few times to bring the brake pads into contact with the disc. Bleeding of the system won't be necessary unless the brake hose was disconnected from the caliper. Check the operation of the brakes carefully before driving the vehicle in traffic.

5   Rear brake shoes - replacement

Warning: *Drum brake shoes must be replaced on both rear wheels at the same time - never replace the shoes on only one wheel. Also, the dust created by the brake*

system contains asbestos, which is harmful to your health. Never blow it out with com- pressed air and don't inhale any of it. An approved filtering mask should be worn when working on the brakes. Do not, under any cir- cumstances, use petroleum-based solvents

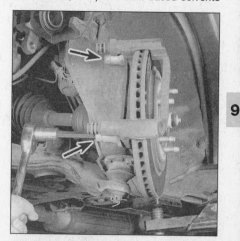

4.6 Remove the bolts (arrows) and detach the caliper mounting bracket

9

5.5a Drum brake components

1   Actuator spring
2   Trailing shoe
3   Retractor spring
4   Leading shoe
5   Adjusting screw
6   Actuator lever
7   Wheel cylinder
8   Backing plate

5.5b Use a pair of needle-nose pliers to remove the actuator spring

5.5c Wedge a flat-bladed screwdriver under the spring and pry it out of the leading brake shoe, then remove the shoe, adjusting screw and actuator lever

5.5d Lift the retractor spring from the trailing brake shoe and swing the shoe out from the hub area to gain access to the parking brake cable

to clean brake parts. Use brake cleaner or denatured alcohol only!

Caution: *Whenever the brake shoes are replaced, the return and hold-down springs should also be replaced. Due to the continuous heating/cooling cycle the springs are* subjected to, they lose tension over a period of time and may allow the shoes to drag on the drum and wear at a much faster rate than normal. When replacing the rear brake shoes, use only high-quality, nationally-recognized brand name parts.

Removal

Refer to illustrations 5.5a, 5.5b, 5.5c, 5.5d, 5.5e and 5.5f

1   Loosen the wheel lug nuts, raise the rea of the vehicle and support it securely on jac stands. Block the front wheels to keep th vehicle from rolling off the jackstands.

2   Release the parking brake and remov the wheel. **Note:** *All four rear shoes must b replaced at the same time, but to avoid mi ing up parts, work on only one brake asser bly at a time.*

3   Remove the brake drum. If it's difficult remove, back off the parking brake cabl remove the access hole plug from the bac ing plate, insert a screwdriver through t hole and press in to push the parking bra lever off its stop. This will allow the bra shoes to retract slightly. Insert a pun through the hole at the bottom of the spla shield and tap gently on the punch to loos the drum. Use a rubber mallet to tap ge on the outer rim of the drum and/or arou the inner drum diameter by the spindle. Avc using excessive force.

4   Clean the brake assembly with bra

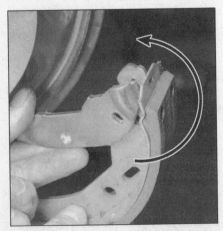

5.5e Rotate the brake shoe to release the parking brake lever from the shoe

5.5f Use a screwdriver to pry the retractor spring over the alignment peg (arrow)

5.6a Lubricate the contact surfaces of the backing plate with high-temperature grease

5.6b Lubricate the adjuster screw with high-temperature grease prior to installation

system cleaner - DO NOT use compressed air to blow the dust out of the brake assembly.

5   Refer to the accompanying illustrations and perform the brake show removal procedure. Be sure to stay in order and to read the caption under each illustration.

Installation

Refer to illustrations 5.6a, 5.6b and 5.8

6   Installation of the shoes is the reverse of removal. Lubricate components as shown **(see illustrations)**.

7   Set the preliminary shoe adjustment by turning the star wheel on the adjuster so that the diameter of the linings is 0.05 inch less than the inner diameter of the drum.

8   Before reinstalling the drum, check it for cracks, score marks, deep scratches and hard spots, which will appear as blue discolored areas. If the hard spots can't be removed with fine emery cloth or if any of the other conditions listed above exist, the drum must be taken to an automotive machine shop to have it turned. **Note:** *The drums should be resurfaced, regardless of the surface appearance, to impart a smooth finish and ensure a perfectly round drum (which will eliminate brake pedal pulsations related to out-of-round drums). At the very least, if you don't have the drums resurfaced, remove the glaze from the surface with medium-grit emery cloth using a swirling motion. If the drum won't "clean up" before the maximum service limit is reached in the machining operation, install a new one. The maximum wear diameter is cast into each brake drum* **(see illustration)**.

9   Install the brake drum on the axle flange.

10   Mount the wheel, install the lug nuts, then lower the vehicle.

11   Apply and release the brake pedal 30 to 35 times using normal pedal force. Pause about one second between pedal applications. After adjustment, make sure that both wheels turn freely.

6   Rear wheel cylinder - removal, overhaul and installation

Refer to illustrations 6.4, 6.5 and 6.7

Note: *If an overhaul is indicated (usually because of fluid leakage or sticking brakes) explore all options before beginning the job. New wheel cylinders are available, which makes this job quite easy. If you do rebuild the wheel cylinder, make sure rebuild kits are available before proceeding.*

Removal

1   Raise the rear of the vehicle and support it securely on jackstands. Block the front wheels to keep the vehicle from rolling off the jackstands.

2   Remove the brake shoe assembly (see Section 5).

3   Carefully clean the area around the wheel cylinder on both sides of the backing plate.

4   Unscrew the brake line fitting **(see illustration)**, but don't pull the line away from the wheel cylinder.

5   Remove the wheel cylinder retaining

5.8 The drum has a maximum permissible diameter cast into it which must not be exceeded when removing scoring or other imperfections in the friction surface

bolts **(see illustration)**.

6   Remove the wheel cylinder from the brake backing plate and place it on a clean workbench. Immediately plug the brake line to prevent fluid loss and contamination.

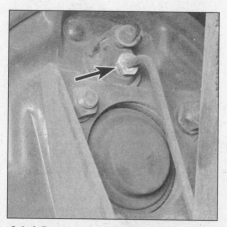

6.4 A flare-nut wrench should be used to disconnect the brake line

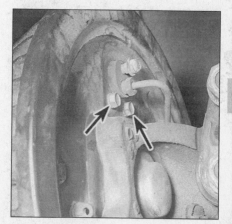

6.5 To remove the wheel cylinder, remove these two bolts (arrows)

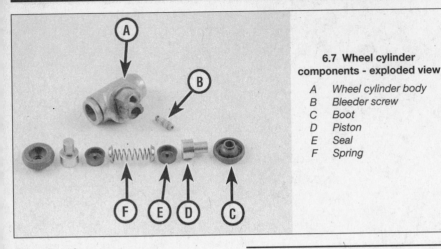

6.7 **Wheel cylinder components - exploded view**

A   Wheel cylinder body
B   Bleeder screw
C   Boot
D   Piston
E   Seal
F   Spring

7.2 Unplug the fluid level sensor connector (arrow) and unscrew the brake line fittings (arrows)

Overhaul

7   Remove the bleeder valve, seals, pistons, boots and spring assembly from the wheel cylinder body **(see illustration)**.
8   Clean the wheel cylinder with brake system cleaner. **Warning:** *Do not, under any circumstances, use petroleum-based solvents to clean brake parts.*
9   Check the bore for corrosion and score marks. Crocus cloth may be used to remove light corrosion and stains, but the cylinder must be replaced with a new one if the defects can't be removed easily, or if the bore is scored.
10   Lubricate the new seals with brake fluid.
11   Assemble the brake cylinder components, making sure the boots are properly seated.

Installation

12   Place the wheel cylinder in position.
13   Connect the brake line loosely.
14   Install the wheel cylinder bolts and tighten them securely.
15   Tighten the brake line fittings.
16   Install the brake shoes (see Section 5).
17   Bleed the brakes (see Section 9).

7   Master cylinder - removal, overhaul and installation

Note: *Before deciding to overhaul the master cylinder, check on the availability and cost of a new or factory-rebuilt unit and the availability of a rebuild kit. This procedure does not apply to vehicles equipped with ABS.*

Removal

Refer to illustrations 7.2 and 7.6

1   Detach the cable from the negative battery terminal. **Caution:** *On models equipped with the Theftlock audio system, be sure the lockout feature is turned off before performing any procedure which requires disconnecting the battery.*
2   Unplug the electrical connector from the fluid level sensor switch **(see illustration)**.
3   Place rags under the line fittings and prepare caps or plastic bags to cover the ends of the lines once they're disconnected. **Caution:** *Brake fluid will damage paint. Cover all painted parts and be careful not to spill fluid during this procedure.*
4   Loosen the fittings at the ends of the brake lines where they enter the master cylin-

der. To prevent rounding off the flats on the fittings, use a flare-nut wrench, which wraps around the hex.
5   Pull the brake lines away from the master cylinder and plug the ends to prevent contamination.
6   Remove the two mounting nuts **(see illustration)** and detach the master cylinder from the vehicle.
7   Remove the reservoir cover and reservoir diaphragm, then discard any remaining fluid.

Overhaul

Refer to illustrations 7.9, 7.10, 7.11, 7.12, 7.16, 7.18, 7.19a, 7.19b, 7.19c, 7.19d, 7.19e, 7.19f, 7.20 and 7.32

8   Mount the master cylinder in a vise. Be sure to line the vise jaws with rags or blocks of wood to prevent damage to the cylinder body.
9   Drive out the roll pins **(see illustration)** with a 1/8-inch punch. Pull straight up on the

7.6 Remove the master cylinder mounting nuts (arrows)

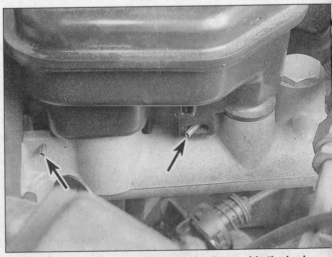

7.9 Location of the retaining roll pins (arrows) in the brake master cylinder

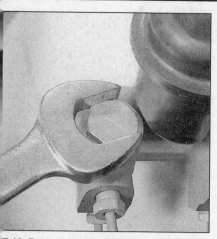

7.10 Remove the proportioner valve caps

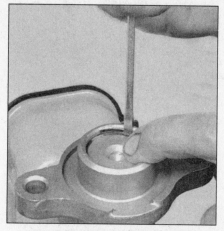

7.11 Press down on the piston and remove the primary piston lock ring

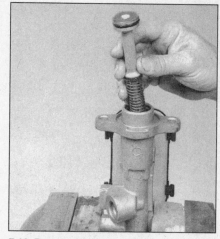

7.12 Remove the primary piston assembly

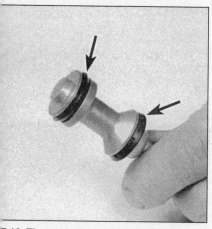

7.16 The secondary piston seals must be installed with the lips facing out as shown

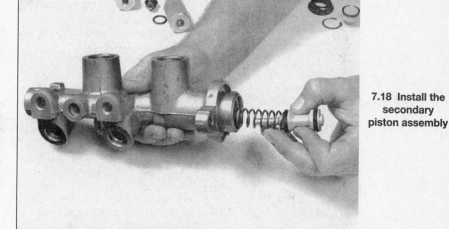

7.18 Install the secondary piston assembly

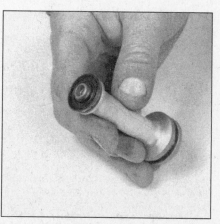

7.19a The primary piston seal must be installed with the lip facing away from the piston

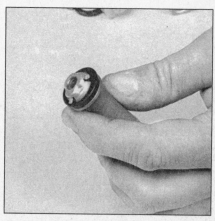

7.19b Install the seal guard over the seal

reservoir assembly and separate it from the master cylinder body. Remove and discard the two O-rings.

10   Remove the proportioner valve caps **(see illustration)**. Using needle-nose pliers, carefully remove the O-rings, the springs, the proportioner valve pistons and seals. Make sure you don't scratch or otherwise damage the piston stems or the bores. Set each proportioner valve assembly aside.

11   Remove the primary piston lock ring by depressing the piston and prying the ring out with a screwdriver **(see illustration)**.

12   Remove the primary piston assembly from the bore **(see illustration)**.

13   Remove the secondary piston assembly from the bore. It may be necessary to remove the master cylinder from the vise and invert it, carefully tapping it against a block of wood to expel the piston.

14   Clean the master cylinder body, the primary and secondary piston assemblies, the proportioner valve assemblies and the reservoir with brake system cleaner. **Warning:** *DO NOT, under any circumstances, use petroleum-based solvents to clean brake parts.*

15   Inspect the master cylinder piston bore for corrosion and score marks. If any corro-

sion or damage in the bore is evident, replace the master cylinder body - don't use abrasives to try to clean it up.

16   Remove the old seals from the secondary piston assembly and install the new seals with the cup lips facing out **(see illustration)**.

17   Attach the spring retainer to the sec-

ondary piston assembly.

18   Lubricate the cylinder bore with clean brake fluid and install the spring and secondary piston assembly **(see illustration)**.

19   Disassemble the primary piston assembly, noting the locations of the parts, then lubricate the new seals with clean brake fluid and install them on the piston **(see illustrations)**.

9

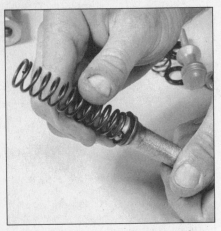

7.19c   Place the primary piston spring in position

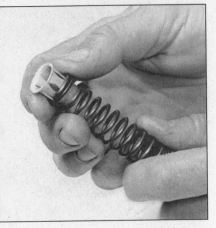

7.19d   Insert the spring retainer into the spring

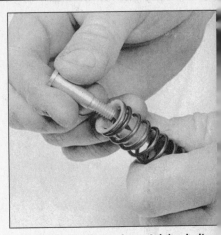

7.19e   Insert the spring retaining bolt through the retainer and spring and thread it into the piston

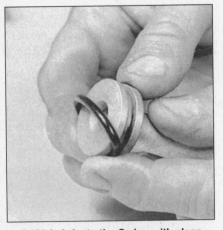

7.19f   Lubricate the O-ring with clean brake fluid, then install it on the piston

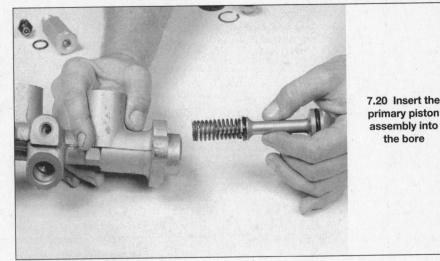

7.20   Insert the primary piston assembly into the bore

20   Install the primary piston assembly in the cylinder bore **(see illustration)**, depress it and install the lock ring.

21   Inspect the proportioner valves for corrosion and score marks. Replace them if necessary.

22   Lubricate the new O-rings and proportioner valve seals with the silicone grease supplied with the rebuild kit. Also lubricate the stem of the proportioner valve pistons.

23   Install the new seals on the proportioner valve pistons with the seal lips facing toward the cap assembly.

24   Install the proportioner valve pistons and seals in the master cylinder body.

25   Install the springs in the master cylinder body.

26   Install the new O-rings in their respective grooves in the proportioner valve cap assemblies.

27   Install the proportioner valve caps in the master cylinder and tighten them to the torque listed in this Chapter's Specifications.

28   Inspect the reservoir for cracks and distortion. If any damage is evident, replace it.

29   Lubricate the new reservoir O-rings with clean brake fluid and press them into their respective grooves in the master cylinder body. Make sure they're properly seated.

30   Lubricate the reservoir fittings with clean brake fluid and install the reservoir on the master cylinder body by pressing it straight down.

31   Drive in new reservoir retaining (roll) pins. Make sure you don't damage the reservoir or master cylinder body.

32   Inspect the reservoir diaphragm and cover for cracks and deformation. Replace any damaged parts with new ones and attach the diaphragm to the cover **(see illustration)**.
Note: *Whenever the master cylinder is removed, the complete hydraulic system must be bled. The time required to bleed the system can be reduced if the master cylinder is filled with fluid and bench bled (refer to Steps 33 through 36) before it's installed on the vehicle.*

33   Insert threaded plugs of the correct size into the brake line outlet holes and fill the reservoirs with brake fluid. The master cylinder should be supported so brake fluid won't spill during the bench bleeding procedure.

34   Loosen one plug at a time and push the piston assembly into the bore to force air from the master cylinder. To prevent air from being drawn back in, the appropriate plug must be replaced before allowing the piston to return to its original position.

35   Stroke the piston three or four times for each outlet to ensure that all air has been expelled.

7.32   Install the reservoir diaphragm in the cover

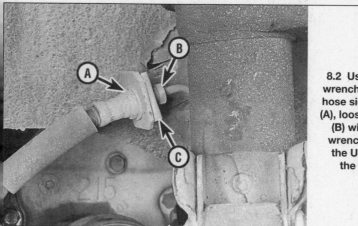

8.2 Using a back-up wrench on the flexible hose side of the fitting (A), loosen the tube nut (B) with a flare-nut wrench and remove the U-clip (C) from the hose fitting

36   Since high pressure isn't involved in the bench bleeding procedure, there is an alternative to the removal and replacement of the plugs with each stroke of the piston assembly. Before pushing in on the piston assembly, remove one of the plugs completely. Before releasing the piston, however, instead of replacing the plug, simply put your finger tightly over the hole to keep air from being drawn back into the master cylinder. Wait several seconds for the brake fluid to be drawn from the reservoir into the piston bore, then repeat the procedure. When you push down on the piston it'll force your finger off the hole, allowing the air inside to be expelled. When only brake fluid is being ejected from the hole, replace the plug and go on to the other port.

37   Refill the master cylinder reservoirs and install the diaphragm and cover assembly. **Note:** *The reservoirs should only be filled to the top of the reservoir divider to prevent overflowing when the cover is installed.*

Installation

38   Carefully install the master cylinder by reversing the removal steps, fill the reservoir with the recommended fluid (see Chapter 1), then bleed the brakes at each wheel (see Section 9).

8   Brake hoses and lines - inspection and replacement

Refer to illustration 8.2
Warning: *Before servicing any ABS component it is mandatory that hydraulic pressure in the system be relieved. With the ignition off, pump the brake pedal a MINIMUM of 40 times using full pedal strokes.*

About every six months, raise the vehicle and support it securely on jackstands, then check the flexible hoses that connect the steel brake lines to the front and rear brake assemblies. Look for cracks, chafing of the outer cover, leaks, blisters and other damage. The hoses are important and vulnerable parts of the brake system and the inspection should be thorough. A light and mirror will be helpful to see into restricted

areas. If a hose exhibits any of the above conditions, replace it with a new one.

Front brake hose

2   Using a back-up wrench, disconnect the brake line from the hose fitting, being careful not to bend the frame bracket or brake line **(see illustration).**
3   Use pliers to remove the U-clip from the female fitting at the bracket, then remove the hose from the bracket.
4   At the caliper end of the hose, remove the bolt from the fitting block, then remove the hose and the copper washers on either side of the fitting block.
5   When installing the hose, always use new copper washers on either side of the fitting block and lubricate all bolt threads with clean brake fluid.
6   With the fitting flange engaged with the caliper locating ledge, attach the hose to the caliper.
7   Without twisting the hose, install the female fitting in the hose bracket. It'll fit the bracket in only one position.
8   Install the U-clip retaining the female fitting to the frame bracket.
9   Using a back-up wrench, attach the brake line to the hose fitting.
10   When the brake hose installation is complete, there shouldn't be any kinks in the hose. Make sure the hose doesn't contact any part of the suspension. Check it by turning the wheels to the extreme left and right positions. If the hose makes contact, remove the hose and correct the installation as necessary.

Rear brake hose

11   Using a back-up wrench, disconnect the hose at both ends, being careful not to bend the bracket or steel lines.
12   Remove the U-clip with pliers and separate the female fittings from the brackets.
13   Unbolt the hose retaining clip and remove the hose.
14   Without twisting the hose, install the female ends in the frame bracket. It'll fit the bracket in only one position.
15   Install the U-clip retaining the female end to the bracket.

16   Using a back-up wrench, attach the steel line fittings to the female fittings. Again, be careful not to bend the bracket or steel line.
17   Make sure the hose installation didn't loosen the frame bracket. Tighten the bracket if necessary.
18   Fill the master cylinder reservoir and bleed the system (refer to Section 9).

Metal brake lines

19   When replacing brake lines, be sure to buy the correct replacement parts. Don't use copper or any other tubing for brake lines.
20   Prefabricated brake lines, with the ends already flared and fittings installed, are available at auto parts stores and dealer service departments. The lines are also bent to the proper shapes if necessary.
21   If prefabricated lines aren't available, obtain the recommended steel tubing and fittings to match the line to be replaced. Determine the correct length by measuring the old brake line (a piece of string can usually be used for this) and cut the new tubing to length, allowing about 1/2-inch extra for flaring the ends.
22   Install the fittings on the cut tubing and flare the ends of the line with an ISO flaring tool.
23   If necessary, carefully bend the line to the proper shape. A tube bender is recommended for this. **Caution:** *Don't crimp or damage the line.*
24   When installing the new line, make sure it's securely supported in the brackets with plenty of clearance between moving or hot components.
25   After installation, check the master cylinder fluid level and add fluid as necessary. Bleed the brake system as outlined in the next Section and test the brakes carefully before driving the vehicle in traffic.

9   Brake system bleeding

Refer to illustrations 9.8 and 9.38
Warning: *Wear eye protection when bleeding the brake system. If you get fluid in your eyes, rinse them immediately with water and seek medical attention.*
Note: *Bleeding the brakes is necessary to remove air that manages to find its way into the system when its been opened during removal and installation of a hose, line, caliper or master cylinder.*

Conventional brake system (non-ABS equipped vehicles)

1   It'll probably be necessary to bleed the system at all four brakes if air has entered the system due to low fluid level, or if the brake lines have been disconnected at the master cylinder.
2   If a brake line was disconnected at only one wheel, then only that caliper or wheel cylinder must be bled.
3   If a brake line is disconnected at a fitting

9.8 When bleeding the brakes, a hose is connected to the bleeder valve at the caliper and then submerged in brake fluid - air will be seen as bubbles in the container and in the tube (all air must be expelled before continuing to the next wheel)

9.38 Later model ABS components

1   ABS control assembly
2   ABS modulator rear bleeder valve
3   ABS modulator front bleeder valve

located between the master cylinder and any of the brakes, that part of the system served by the disconnected line must be bled.

4   Remove any residual vacuum from the power brake booster by applying the brake several times with the engine off.

5   Remove the master cylinder reservoir cover and fill the reservoir with brake fluid. Reinstall the cover. **Note:** *Check the fluid level often during the bleeding procedure and add fluid as necessary to prevent the level from falling low enough to allow air bubbles into the master cylinder.*

6   Have an assistant on hand, as well as a supply of new brake fluid, an empty, clear plastic container, a length of 3/16-inch plastic, rubber or vinyl tubing to fit over the bleeder valve and a wrench to open and close the bleeder valve.

7   Beginning at the right front wheel, loosen the bleeder valve slightly, then tighten it to a point where it's snug but can still be loosened quickly and easily.

8   Place one end of the tubing over the bleeder valve and submerge the other end in brake fluid in the container **(see illustration)**.

9   Have your assistant pump the brakes slowly a few times to get pressure in the system, then hold the pedal down firmly.

10   While the pedal is held down, open the bleeder valve just enough to allow fluid to flow out of the valve. Watch for air bubbles to exit the submerged end of the tube. When the fluid slows after a couple of seconds, close the valve and have your assistant release the pedal.

11   Repeat Steps 9 and 10 until no more air is seen leaving the tube, then tighten the bleeder valve and proceed to the right rear wheel, the left front wheel and the left rear wheel, in that order, and perform the same procedure. Be sure to check the fluid in the master cylinder reservoir frequently.

12   Never use old brake fluid. It contains moisture which can boil, rendering the brakes useless.

13   Fill the master cylinder with fluid at the end of the operation.

14   Check the operation of the brakes. The pedal should feel firm when depressed. If necessary, repeat the entire procedure. **Warning:** *Don't operate the vehicle if you're in doubt about the effectiveness of the brake system.*

Anti-lock Brake System (ABS) equipped vehicles

Warning: *Before servicing any ABS component it is mandatory that hydraulic pressure in the system be relieved. With the ignition off, pump the brake pedal a MINIMUM of 40 times using full pedal strokes.*

1991 and earlier models

15   Relieve the system pressure.

16   With the ignition off, be sure the fluid reservoir is full.

17   Connect a clear hose to the bleeder valve on the right front brake caliper and submerge the other end of the hose in a container of brake fluid **(see illustration 9.8)**.

18   Open the bleeder valve.

19   Have an assistant slowly depress the brake pedal. As the fluid flows from the caliper, tap on the caliper housing to free trapped air.

20   Close the bleeder valve and have the assistant release the brake pedal.

21   Repeat this step until air is no longer seen in the hose.

22   Repeat this procedure for the left front caliper.

23   Before bleeding the rear brakes, turn the ignition On and allow the pump motor to charge the system. The pump motor should stop running within one minute.

24   With the reservoir full, connect the hose to the bleeder valve on the right rear caliper and submerge the other end of the hose in a container of clean brake fluid.

25   Open the bleeder valve.

26   Have an assistant slowly depress the brake pedal, part way only. Allow the fluid to flow for 15 seconds, while lightly tapping on the caliper housing to free trapped air.

27   Close the bleeder valve and have the assistant release the brake pedal.

28   Repeat this step until air is no longer seen in the hose.

29   Check the brake fluid level, then repeat this procedure for the left rear caliper.

30   Once all the calipers have been bled, the isolation valves on the ABS unit should be bled. Connect the bleeder hose to the inner side of the ABS unit and submerge the other end in the container of clean brake fluid. With the ignition still On, have an assistant apply light pressure to the brake pedal. Slowly open the bleeder valve and allow the fluid to flow until no more air bubbles are seen.

31   Repeat this procedure to the bleeder valve on the outer side of the unit.

32   Depressurize the system and check the brake fluid, adding as necessary.

33   Before turning the key On, sharply apply the brake pedal three times, using full force.

34   Turn the ignition key On and allow the pump motor to run. When the motor stops, turn the key Off.

35   Depress the brake pedal, using moderate pressure, then turn the ignition key On for three seconds. Repeat this procedure ten times.

36   Depressurize the system and recheck the fluid level.

1992 and later models

37   Clean the reservoir and cap then check that the reservoir is full of fluid.

38   Prime the ABS modulator/master cylinder assembly. **Note:** *Priming is only necessary if the assembly has been replaced or is suspected of having air trapped inside it.* First attach a bleeder hose to the rear bleeder valve **(see illustration)**. Submerge the other end of the hose in clean brake fluid. Open the bleeder valve 1/2 to 3/4 turns and have an a

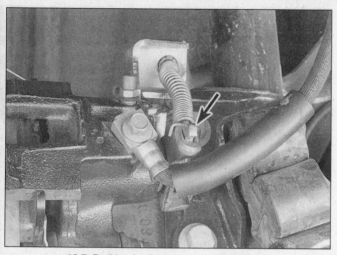

10.7 Parking brake lever location (arrow)

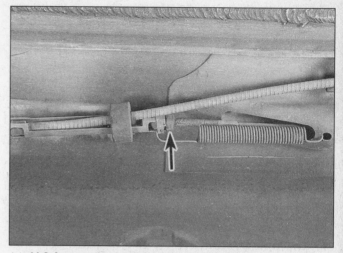

11.2 Loosen the parking brake cable adjusting nut (arrow) located at the equalizer assembly

assistant depress the brake pedal until fluid begins to flow, then close the valve. Repeat this on the front bleeder valve.

39   Check the fluid level and fill the reservoir as necessary.

40   Raise the car and support it securely on jackstands.

41   Bleed the brakes as described in steps 5 through 12 of this section. **Note:** *The recommended bleeding sequence for ABS-equipped models is right rear, left rear, right front, left front, in that order.* While bleeding each brake, be sure to allow at least five seconds to elapse between closing and re-opening the bleeder valve, and be sure to check the fluid level in the reservoir often.

42   Lower the car.

43   Attach a bleeder hose to the rear bleeder valve of the ABS modulator/master cylinder assembly **(see illustration 9.38)**. Submerge the other end of the hose in clean brake fluid.

44   Open the bleeder valve 1/2 to 3/4 turns, have an assistant depress the brake pedal until fluid begins to flow, then close the valve and allow at least five seconds to elapse.

45   Repeat step 44 until all air is purged from the system.

46   Repeat steps 43 through 45 on the front bleeder valve of the ABS modulator/master cylinder assembly.

47   Fill the reservoir as required.

48   Turn the ignition on, then apply and hold the brake pedal with moderate force.

49   If the pedal feels firm and does not have too much travel, proceed to step 50. If the pedal feels soft or has too much travel, remove your foot from the pedal, start the engine and let it run for at least 10 seconds to "initialize" the ABS. **Warning:** *Do not drive the car.* After 10 seconds, turn the ignition off. Repeat this procedure 5 times to ensure that trapped air has been dislodged, then repeat the bleeding procedure.

50   Start the engine and recheck the feel and travel of the pedal. If it continues to feel firm, road test the car and make several normal

from moderate speeds stops (i.e., stops that do not require the ABS to operate) to be sure the brake system is functioning normally.

---

10   Parking brake - adjustment

Rear disc brakes (1996 and earlier)

Refer to illustration 10.7

1   Using heavy pedal pressure, depress the brake pedal three times.

2   Apply and release the parking brake three times.

3   Raise the rear of the vehicle and support it securely on jackstands.

4   Make sure the parking brake is fully released.

5   Turn the ignition key to the On position.

6   If the brake warning light is on, operate the manual brake release and pull down on the front parking brake cable to remove the slack from the pedal assembly.

7   At the rear caliper housings, the two parking brake levers should be against the lever stops **(see illustration)**. If they are not, check for binding in the rear cables and/or loosen the cables at the adjuster until both the left and the right levers are against the stops.

8   Tighten the parking brake cable at the equalizer **(see illustration 11.2)** until either the left or right lever just begins to move off the stop.

9   Operate the parking brake to check for proper adjustment. Lower the vehicle.

Rear drum brakes and 1997 and later rear disc brakes

10   Adjust the rear brake shoes on drum brake equipped models (see Section 5). On 1997 and later (rear disc equipped) models adjust the rear parking brake shoes (see Section 12).

11   Apply the parking brake 10 clicks, then release it. Repeat this 5 times.

12   Make sure the parking brake is fully released. Turn the ignition on. If the brake warning light is on, operate the manual brake release and pull down on the front parking brake cable to remove slack.

13   Raise the car and support it securely on jackstands.

14   Adjust the parking brake by turning the nut on the equalizer **(see illustration 11.2)** while spinning both rear wheels. When either wheel begins to drag, stop adjusting and back off the equalizer nut one full turn.

15   Apply the parking brake 4 clicks and check the rear wheel rotation. You should not be able to turn the wheel by hand in the direction of forward movement; the wheel should drag or not turn at all when spun in the direction or rearward rotation.

16   Release the parking brake and be sure the rear wheels spin freely, then lower the car.

---

11   Parking brake cables - replacement

Refer to illustrations 11.2 and 11.13

Front cable

1   Raise and support the vehicle.

2   Loosen the adjuster nut at the equalizer **(see illustration)**.

3   Detach the front cable from the retainer.

4   Remove the nut at the underbody bracket.

5   Detach the clip from the underbody.

6   Detach the cable from the parking brake lever assembly.

7   Installation is the reverse of removal.

Left rear cable

8   Raise the vehicle and support it.

9   Disconnect the spring from the equalizer, if necessary.

10   Detach the equalizer from the cable.

11   On models with rear drum brakes,

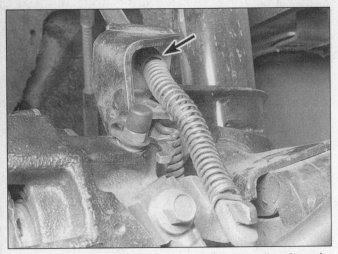

11.13 Detach the cable from the bracket on the caliper (arrow)

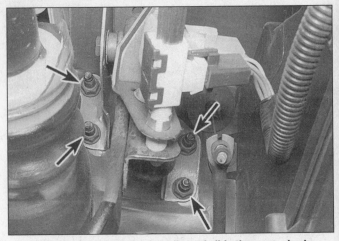

13.10 Remove the retaining clip and slide the power brake pushrod off the brake pedal pin, then remove the booster-to-firewall nuts (arrows)

detach the cable from the backing plate and parking brake lever.

12   Detach any remaining cable-to-frame brackets.

13   On 1996 and earlier rear disc models, detach the cable from the parking brake lever at the caliper. Detach the cable from the caliper bracket **(see illustration)**.

14   On 1997 and later rear disc models, detach the cable from the brake cable support bracket and from the cable actuator lever.

15   Installation is the reverse of removal.

Right rear cable

16   Raise and support the vehicle.

17   Detach the spring from the equalizer, if necessary.

18   Detach the equalizer from the cable.

19   Detach the cable from the underbody bracket.

20   Remove the bolts from the clips above the fuel tank.

21   On models with rear drum brakes, detach the cable from the backing plate and parking brake lever. On 1996 and earlier rear disc models, detach the cable from the parking brake lever at the caliper. Detach the cable from the caliper bracket. On 1997 and later rear disc models detach the cable from the brake cable support bracket and from the cable actuator lever.

22   Installation is the reverse of removal.

12   Parking brake shoes - replacement

Note: *This procedure applies only to 1997 and later models with rear disc brakes.*

1   Loosen the rear wheel lug nuts, raise the rear of the vehicle and support it securely on jackstands. Remove the wheels.

2   Remove the brake caliper (see Section 3) and the brake disc (see Section 4).

3   Disconnect the parking brake cable

from the bracket and the actuator lever.

4   Unbolt the parking brake cable bracket.

5   Remove the rear hub and bearing assembly (see Chapter 10).

6   Remove the parking brake actuator and the parking brake shoe.

7   When installing the new shoe and lining assembly, turn the adjuster screw until the shoe lining just drags on the braking surface inside the disc. Then remove the disc and back-off the adjuster screw until the shoe lining doesn't drag when the disc is installed and turned.

8   Installation is otherwise the reverse of the removal procedure. Be sure to tighten the hub and bearing assembly bolts to the torque listed in the Chapter 10 Specifications, the caliper mounting bolts to the torque listed in the Chapter 9 Specifications, and the wheel lug nuts to the torque listed in the Chapter 1 Specifications.

13   Power brake booster - check, removal and installation

1   The power brake booster unit requires no special maintenance apart from periodic inspection of the vacuum hose and case.

Operating check

2   Depress the brake pedal several times with the engine off and make sure there is no change in the pedal reserve distance (the minimum distance to the floor).

3   Depress the pedal and start the engine. If the pedal goes down slightly, operation is normal.

Airtightness check

4   Start the engine and turn it off after one or two minutes. Depress the pedal several times slowly. If the pedal goes down farther the first time but gradually rises after the second or third depression, the booster is airtight.

5   Depress the brake pedal while the engine is running, then stop the engine with the brake pedal depressed. If there is no change in the pedal reserve travel after holding the pedal for 30 seconds, the booster is airtight.

Removal

1988 through 1990 models

Refer to illustration 13.10

6   Dismantling of the power unit requires special tools and is not ordinarily done by the home mechanic. If a problem develops, install a new or factory rebuilt unit.

7   Remove the nuts attaching the master cylinder to the booster and carefully pull the master cylinder forward until it clears the mounting studs. Be careful to avoid bending or kinking the brake lines.

8   Disconnect the vacuum hose where it attaches to the power brake booster.

9   From the passenger compartment, disconnect the power brake pushrod from the top of the brake pedal.

10   Also from this location, remove the nuts attaching the booster to the firewall **(see illustration)**.

11   Carefully lift the booster away from the firewall and out of the engine compartment.

1991 and later models

Refer to illustration 13.16

12   Dismantling of the power unit requires special tools and is not ordinarily done by the home mechanic. If a problem develops, install a new or factory rebuilt unit.

13   Remove the nuts attaching the master cylinder to the booster and carefully pull the master cylinder forward until it clears the mounting studs. Be careful to avoid bending or kinking the brake lines. **Warning:** *On ABS-equipped models, do not disconnect the master cylinder from the hydraulic modulator assembly.*

14   Disconnect the vacuum hose where it attaches to the power brake booster. On 3.4L

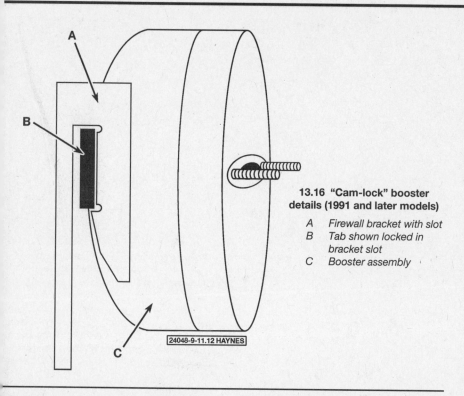

13.16 "Cam-lock" booster
details (1991 and later models)

A Firewall bracket with slot
B Tab shown locked in
   bracket slot
C Booster assembly

1991 and later models, make sure the locking tab is completely seated in the firewall bracket slot.
20   Carefully test the operation of the brakes before driving the car.

14   Brake light switch - removal, installation and adjustment

Removal

1   The brake light switch is located on a bracket at the top of the brake pedal. The switch activates the brake lights at the rear of the vehicle when the pedal is depressed.
2   Remove the under dash cover.
3   If necessary, slide the steering shaft protective sleeve towards the cowl.
4   If necessary, disconnect the vacuum hose at the cruise control cut-off switch.
5   If necessary, remove the retainer clip holding the brake light switch to the steering column bracket.
6   Detach the retainer from the four wire connector.
7   Detach the two wire connector (if equipped with cruise control).
8   If necessary, detach the switch arm from the pedal by pushing the arm to the left and towards the cowl. Release the switch by pulling down and releasing the top snap clip.
9   On later models, remove the mounting screw.
10   Detach the four wire connector and remove the switch.

Installation and adjustment

11   Installation is the reverse of removal.
12   With the brake pedal fully depressed, use a wire with a hooked end and pull on the switch set lever while listening for a click. **Note:** *If there is no click the switch may be faulty. Release the brake pedal and repeat this step.*
13   After adjustment, be sure the brake lights are functioning properly.

engines it may be necessary to remove the upper intake plenum (see Chapter 4). On 1996 and later 3800 engines remove the acoustic engine cover (see Chapter 2, Part F). On some models it may be necessary to remove the transaxle filler tube to allow removal of the brake booster.
15   From the passenger compartment, disconnect the brake pushrod from the top of the brake pedal. **Caution:** *Keep the pedal from moving while doing this, to avoid damage to the brake switch.*
16   The booster is mounted to the firewall with a "cam-lock" bracket. On the right side of the bracket, use a screwdriver to pry the booster's locking tab out of the notch on the bracket **(see illustration)**. Insert the screwdriver between the tab and the notch.
17   If you do not have access to the factory tool that holds the front of the booster for removal/installation, there are two other ways to do the job. If you have a large strap wrench, wrap it around the body of the booster. You can also install two short lengths of rubber fuel hose over the mastercylinder mounting studs at the front of the booster, and use a prybar between the two studs to turn the booster as described in the next Step.
18   While holding the tab out of the notch (toward the firewall), rotate the booster counterclockwise until it comes free of the bracket. **Caution:** *Be careful not to tear the boot (interior side of the firewall) around the booster pushrod as you remove the booster.*

Installation

19   Installation is the reverse of removal. On

Notes

Chapter 10
Suspension and steering systems

Contents

Specifications

Torque specifications

Front suspension

	Ft-lbs
Control arm pivot bolts	
1988 through 1989	52
1990 through 1993	56
1994 through 1996	52
1997 and later	83
Subframe-to-body bolts	
1996 and earlier	103
1997 and later	133
Lower balljoint nut	
1988 and 1989	
Initial	84 in-lbs
Final	Tighten an additional 120-degrees
1990 and 1991	
Initial	15
Final	Tighten an additional 90-degrees
1992 through 1996	63
1997 and later	40

Torque specifications (continued)

	Ft-lbs
Front suspension	
Stabilizer bar clamp nuts	35
Front hub and wheel bearing assembly bolts	
1996 and earlier	52
1997 and later	96
Strut mount cover nuts	
1994 and earlier	17
1995 and later	24
Strut damper shaft nut	
1994 and earlier	72
1995	82
1996	59
1997 and later	63
Strut cartridge nut	82
Driveaxle hub nut	See Chapter 8
Rear suspension	
Rear hub and wheel bearing assembly bolts	
1996 and earlier	52
1997 and later	55
Strut-to-body bolt	34
Strut-to-knuckle nut	
1994 and earlier	133
1995 through 1996	122
1997 through 1998	82
1999	90
Trailing arm-to-body nut/bolt	
1994 and earlier	48
1995 through 1996	44 plus an additional 90 degrees
1997	37 plus an additional 60 degrees
1998	77
1999	40
Trailing arm-to-knuckle nut/bolt	
1994 and earlier	192
1995	66 plus an additional 90 degrees
1996 through 1997	66 plus an additional 75 degrees
1998	192
1999	52 plus an additional 65 degrees
Suspension crossmember-to-body bolt	
1996 and earlier	85
1997 through 1998	77
1999	81
Stabilizer bar link bolt	40
Stabilizer bar clamp bolt	18
Jack pad bolt	18
Spring retention plate bolt	
1994 and earlier	15
1995 through 1996	22
Lateral link rod	
1988 and 1989	
To-knuckle	66 plus an additional 120-degrees (two flats)
To-crossmember (rear link rod)	66 plus an additional 120-degrees (two flats)
To-toe adjuster (front link rod)	81 plus an additional 120-degrees (two flats)
1990 through 1995	
To-knuckle	66 plus an additional 90-degrees rotation
To-crossmember	81 plus an additional 90-degrees rotation
1996	
To-knuckle	177
To-crossmember	111
1997 and later	
To-knuckle	110
To-crossmember	103
Airbag system	
Airbag module mounting bolts	25 in-lbs

Torque specifications (continued)

Ft-lbs

Steering system

Steering gear mounting bolts	59
Tie-rod end nut	
1996 and earlier	40
1997 through 1998	18 plus an additional 180-degrees rotation
1999	22 plus an additional 120-degrees rotation
Tie-rod jam nut	50
Intermediate shaft pinch-bolt	35
Steering wheel hub nut	30 to 35
Wheel lug nuts	See Chapter 1

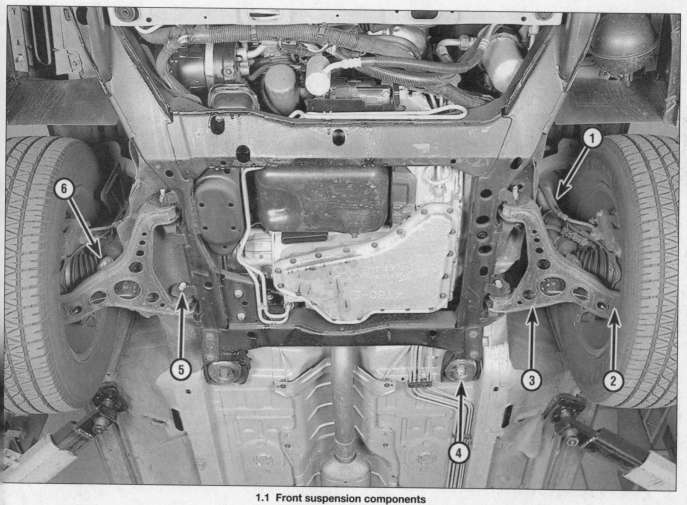

1.1 Front suspension components

1	Strut/knuckle assembly	3	Control arm	5	Control arm pivot bolt
2	Balljoint	4	Subframe bolt	6	Stabilizer bar

1   General information

Refer to illustrations 1.1 and 1.2

Warning: *Whenever any of the suspension or steering fasteners are loosened or removed, they must be inspected and, if necessary, replaced with new ones of the same part number or of original equipment quality and design. Torque specifications must be followed for proper reassembly and component*

retention. Never attempt to heat or straighten any suspension or steering components. Instead, replace any bent or damaged part with a new one.

The front suspension is a strut design which is made up of a strut welded to a knuckle and supported by a coil spring. The strut/knuckle bearing is located under the lower spring seat, with the upper spring seat attaching to the chassis. Because of the location of the bearing, the strut cartridge on

1996 and earlier models can be removed from the engine compartment without disassembling the spring and strut/knuckle. The strut/knuckle assemblies are connected by balljoints to the lower control arms, which are mounted to the frame. The control arms are connected by a stabilizer bar, which reduces body lean during cornering **(see illustration)**.

The rear suspension is independent, with earlier models having a transversely mounted fiberglass leaf spring connected to

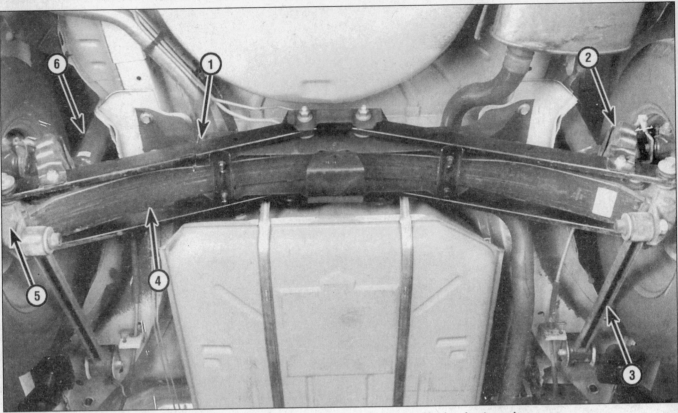

1.2 Rear suspension components - transversely mounted leaf spring type shown

1	Lateral link	3	Trailing arm	5	Knuckle
2	Auxiliary damper	4	Fiberglass spring	6	Strut

the knuckle assemblies and two shock struts (not coil over spring struts) with an auxiliary damper **(see illustration)**. While later models use a traditional coil over spring strut connected to knuckle assemblies. Both types of suspensions are located by trailing arms and parallel lateral link rods, and an optional stabilizer bar.

The rack-and-pinion steering gear is located behind the engine/transaxle assembly on the firewall and actuates the steering arms which connect to the strut housings. All vehicles are equipped with power steering. The steering column is connected to the steering gear through an insulated coupler. The steering column is designed to collapse in the event of an accident.

Note: *These vehicles have a combination of standard and metric fasteners on the various suspension and steering components, so it would be a good idea to have both types of tools available when beginning work.*

2   Front stabilizer bar and bushings - removal and installation

Refer to illustrations 2.3a, 2.3b, 2.4a and 2.4b

Removal

1   Loosen the lug nuts on both front wheels, raise the vehicle and support it securely on jackstands. Remove the front wheels.

2.3a Loosen the stabilizer bar bolts (arrows) on the control arm (early models)

2.3b On later models use a wrench to hold the bolt head while loosening the stabilizer link nut (arrow)

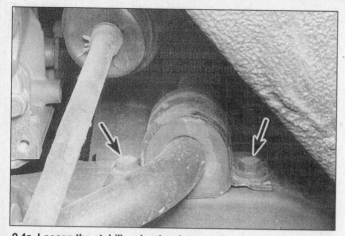

2.4a Loosen the stabilizer bar insulator-to-frame bolts (arrows) . . .

2.4b . . . and nuts (arrows), which are accessible from underneath

2   Remove the steering shaft pinch bolt.
3   Loosen the stabilizer bar-to-control arm nuts/bolts **(see illustration)**.
4   Loosen the stabilizer bar insulator-to-frame nuts/bolts **(see illustrations)**.
5   Place a jack under the rear of the subframe crossmember, then loosen the two front subframe-to-body bolts four turns. Remove the two rear subframe-to-body bolts. Slowly lower the jack and allow the rear of the subframe to drop down.
6   Remove the stabilizer clamps and insulators.
7   Pull the stabilizer bar to the rear, swing it down then remove it through the left side wheel well.
8   Inspect the bushings for wear and damage and replace them if necessary. To remove them, pry the bushing clamp off with a screwdriver and pull the bushings off the bar. To ease installation, spray the inside and outside of the bushings with a silicone-based lubricant. Do not use petroleum-based lubricants on any rubber suspension part!

Installation

9   Assemble the shaft bushings and clamps on the bar, guide the bar through the wheel well, over the frame and into position.
10   Install the clamps loosely to the frame and control arm and install the nuts finger tight. Tighten the bolts to the torque listed in this Chapter's Specifications.
11   Raise the frame into place while guiding the steering shaft into position in the steering gear and install the rear bolts. Tighten the frame bolts to the torque figures listed in this Chapter's Specifications.
12   Install the pinch bolt.
13   Install the wheels and lower the vehicle. Tighten the lug nuts to the torque listed in the Chapter 1 Specifications.

3   Balljoint - check and replacement

Refer to illustrations 3.3a and 3.3b

Check

1   Raise the front of the vehicle and support it securely on jackstands. Apply the parking brake and block the rear wheels to keep the vehicle from rolling off the jackstands.
2   Visually inspect the rubber seal for damage, deterioration and leaking grease. If any of these conditions are noticed, the balljoint should be replaced.
3   Place a large pry bar under the balljoint and attempt to push the balljoint up. Next, position the pry bar between the steering knuckle and control arm and pry down **(see illustrations)**. If any movement is seen or felt during either of these checks, a worn out balljoint is indicated.
4   Have an assistant grasp the tire at the top and bottom and move the top of the tire in-and-out.
5   Separate the control arm from the steering knuckle (see Section 4). Using your fingers (don't use pliers), try to twist the stud in the socket. If the stud turns, replace the balljoint.

Replacement

6   Loosen the wheel lug nuts, raise the front of the vehicle and support it securely on jackstands. Apply the parking brake and block the rear wheels to keep the vehicle from rolling off the jackstands. Remove the wheel.
7   Separate the control arm from the steering knuckle (see Section 4). Temporarily insert the balljoint stud back into the control arm (loosely). This will ease balljoint removal

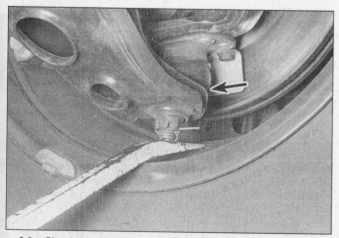

3.3a Check for movement between the balljoint and steering knuckle (arrow) when prying up

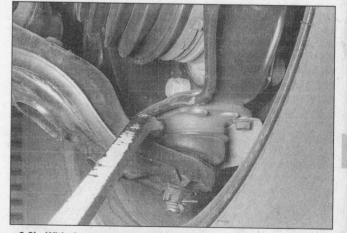

3.3b With the pry bar positioned between the steering knuckle boss and the balljoint, pry down and check for play in the balljoint - if there is any play, replace the balljoint

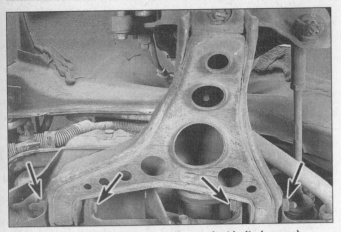

4.5a Location of the control arm pivot bolts (arrows)

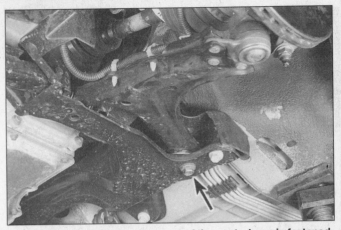

4.5b On later models the rear part of the control arm is fastened by a vertically mounted bushing bolt (arrow)

after Step 9 has been performed, as well as hold the assembly stationary while drilling out the rivets.

8   Using a 1/8-inch drill bit, drill a pilot hole into the center of each balljoint-to-steering knuckle rivet.

9   Using a 1/2-inch drill bit, drill the head off each rivet. Work slowly and carefully to avoid deforming the holes in the control arm.

10   Loosen (but don't remove) the stabilizer bar-to-control arm nut. Pull the control arm down to remove the balljoint stud from the hole in the arm, then dislodge the balljoint from the steering knuckle.

11   Position the new balljoint on the steering knuckle and install the bolts (supplied in the balljoint kit) from the top of the steering knuckle. Tighten the bolts to the torque specified in the new balljoint instruction sheet.

12   Insert the balljoint into the control arm, install the castellated nut, tighten it to the torque listed in this Chapter's Specifications and install a new cotter pin. It may be necessary to tighten the nut some to align the cotter pin hole with an opening in the nut, which is acceptable. Never loosen the castellated nut to allow cotter pin insertion.

13   Tighten the stabilizer bar-to-control arm nut to the torque listed in this Chapter's Specifications.

14   Install the wheel, lower the vehicle and tighten the lug nuts to the torque listed in the Chapter 1 Specifications. It's a good idea to take the vehicle to a dealer service department or alignment shop to have the front end alignment checked and, if necessary, adjusted.

4   Control arm - removal and installation

Refer to illustration 4.5a and 4.5b

Removal

1   Loosen the wheel lug nuts, raise the front of the vehicle and support it securely on jackstands. Apply the parking brake and block the rear wheels to keep the vehicle from rolling off the jackstands. Remove the wheel.

2   If only one control arm is being removed, disconnect only that end of the stabilizer bar. If both control arms are being removed, disconnect both ends (see Section 2 if necessary).

3   Remove the balljoint stud-to-control arm castellated nut and cotter pin.

4   Separate the balljoint from the control arm with a separation tool. If a tool is not available a large pry bar positioned between the control arm and steering knuckle can be used to "pop" the balljoint out of the control arm. **Caution:** *Be careful not to overextend the inner CV joint or it may be damaged.*

5   Remove the two control arm pivot bolts and detach the control arm **(see illustrations)**.

6   The control arm bushings are replaceable, but special tools and expertise are necessary to do the job. Carefully inspect the bushings for hardening, excessive wear and cracks. If they appear to be worn or deteriorated, take the control arm to a dealer service department or repair shop.

Installation

7   Position the control arm in the suspension support and install the pivot bolts. Don't tighten the bolts completely yet.

8   Insert the balljoint stud into the hole in the control arm, install the castellated nut and tighten it to the torque listed in this Chapter's Specifications. If necessary, tighten the nut a little more (but not more than one flat of the nut) if the cotter pin hole doesn't line up with an opening on the nut. Install a new cotter pin.

9   Place a floor jack under the balljoint and raise it to simulate normal ride height. Tighten the control arm pivot bolts to the torque listed in this Chapter's Specifications.

10   Install the stabilizer bar-to-control arm bolt, spacer, bushings and washers and tighten the nut to the torque listed in this Chapter's Specifications.

11   Install the wheel and lower the vehicle. Tighten the lug nuts to the torque specified in Chapter 1.

12   Drive the vehicle to a dealer service department or an alignment shop to have the

front wheel alignment checked and, if necessary, adjusted.

5   Front strut and spring assembly - removal, inspection and installation

Refer to illustration 5.2

Note 1: *On 1996 and earlier models the strut cartridge can be replaced with the strut and spring assembly installed in the vehicle (see Section 6). Because it isn't necessary to remove and disassemble the strut and spring assembly for strut cartridge replacement, removal of the assembly should only be required to repair damage to the spring, seats and strut/knuckle components.*

Note 2: *On 1997 and later models, there is no serviceable strut cartridge. If wear or damage has occurred on these models, the entire strut assembly will have to be removed and replaced.*

Removal

1   Loosen the wheel lug nuts, raise the front of the vehicle and support it securely on

5.2 Mark the position of the strut mount cover before removing the nuts

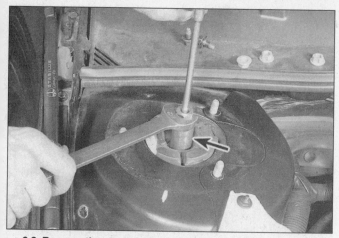

6.2 Remove the strut shaft nut using this tool (arrow) with a wrench on it, and an extension with a Torx bit through the top to hold the strut shaft from turning

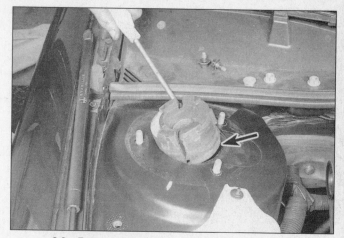

6.3a Remove the strut mount bushing (arrow) . . .

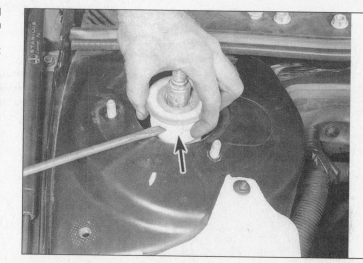

6.3b . . . and the upper strut bumper (arrow)

jackstands. Apply the parking brake and block the rear wheels to keep the vehicle from rolling off the jackstands. Remove the wheel.

2   Mark the position of the strut mount cover to the strut tower **(see illustration)**.

3   On 1997 and later models remove the strut to knuckle bolts **(see illustration 9.11)** and remove the strut. Be extremely careful not to over extend the CV joint or stretch the brake hose. If necessary support the control arm with a jack. On 1996 and earlier models separate the tie-rod end from the steering arm as described in Section 14.

4   Remove the brake caliper, bracket and disc. Hang the caliper out of the way. On models so equipped, remove the bolt and secure the ABS sensor out of the way.

5   Remove the hub and wheel bearing bolts (see Section 7).

6   Separate the balljoint from the lower control arm (see Section 4). Be careful not to overextend the inner CV joint or stretch the brake hose.

7   Remove the driveaxle (see Chapter 8).

8   Support the strut and spring assembly with one hand and remove the three strut mounting nuts. Remove the assembly out through the fenderwell.

Inspection

9   Check the strut body for leaking fluid, dents, cracks and other obvious damage which would warrant repair or replacement.

10   Check the coil spring for chips and cracks in the spring coating (this will cause premature spring failure due to corrosion). Inspect the spring seats for hardening, cracks and general deterioration.

11   Inspect the knuckle/strut for corrosion, bending or twisting.

12   If wear or damage is evident which would require removing the coil spring, take the strut and spring assembly to a dealer or properly equipped shop for repair.

Installation

13   Slide the strut assembly up into the fenderwell, align the marks and install the

strut cover, then insert the three upper mounting studs through the holes in the shock tower. Once the three studs protrude from the shock tower, install the strut cover to the marked position, then the nuts so the strut won't fall back through. This may require an assistant, since the strut is quite heavy and awkward. On 1997 and later models, install the strut-to-knuckle bolts, then tighten all of the strut mounting bolts to the torque listed in this Chapter's Specifications.

14   On 1996 and earlier models, install the driveaxle.

15   Connect the lower balljoint to the control arm and install the nut. Tighten the nut to the torque listed in this Chapter's Specifications and install a new cotter pin. If the cotter pin won't pass through, tighten the nut a little more, but just enough to align the hole in the stud with a castellation on the nut (don't loosen the nut).

16   Install the hub and brake components.

17   Install the tie-rod end in the steering arm and tighten the castellated nut to the torque listed in this Chapter's Specifications. Install a new cotter pin. If the cotter pin won't pass through, tighten the nut a little more, but just enough to align the hole in the stud with a

castellation on the nut (don't loosen the nut).

18   Install the wheel, lower the vehicle and tighten the lug nuts to the torque listed in the Chapter 1 Specifications.

19   Tighten the three upper mounting nuts to the torque listed in this Chapter's Specification.

6   Strut cartridge (1996 and earlier models) - replacement

Refer to illustrations 6.2, 6.3a, 6.3b, 6.4a, 6.4b and 6.5

1   Mark the position of the strut mount cover **(see illustration 5.2)** and remove the nuts and the cover. **Note:** *The vehicle should be on the ground with the full weight on the suspension.*

2   Remove the strut shaft nut, using a special tool that grips the nut while a Torx bit is installed into the shaft to prevent the shaft from turning **(see illustration)**.

3   Pry out the strut mount bushing and the upper strut bumper **(see illustrations)**. You may have to compress the strut shaft down into the cartridge using a length of pipe that just fits over the strut shaft.

1

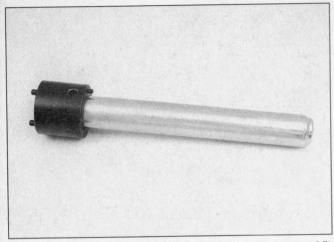

6.4a Strut cartridge nut removal tool - the lugs at the large end fit into the slots in the nut

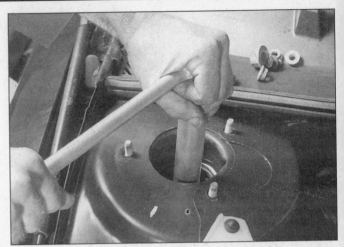

6.4b In use, the tool fits down over the strut shaft and a breaker bar can be used to turn it

4   Remove the strut cartridge nut using a special tool which engages with the slots in the nut (available at most auto parts stores) **(see illustrations)**.

5   Grasp the strut cartridge and lift it out **(see illustration)**.

6   If the old cartridge had been leaking oil, use a suction pump to remove the damper fluid from the strut body and pour it into an approved oil container.

7   Insert the replacement strut cartridge, which is a self-contained unit, into position and install the nut. Tighten the cartridge nut to the torque listed in this Chapter's Specifications.

8   Install the bumper.

9   Temporarily install the strut shaft nut enough to grip it with locking pliers to raise the shaft if it does not come up by itself, then remove the nut and install the bushing.

10   Install the strut shaft nut. Tighten the nut to the torque listed in this Chapter's Specifications.

11   Install the strut mount cover and nuts. Tighten the nuts to the torque listed in this Chapter's Specifications.

7   Front hub and wheel bearing assembly - removal and installation

Refer to illustration 7.6

Note: *The front hub and wheel bearing assembly is sealed-for-life and must be replaced as a unit.*

Removal

1   Loosen the driveaxle nut one turn. Loosen the wheel lug nuts, raise the front of the vehicle and support it securely on jackstands. Apply the parking brake and block the rear wheels to keep the vehicle from rolling off the jackstands. Remove the wheel.

2   Disconnect the stabilizer bar from the control arm (see Section 2 if necessary).

3   Remove the caliper and bracket from the steering knuckle and hang it out of the way with a piece of wire (see Chapter 9). On ABS-equipped models, remove the bolt and secure the sensor out of the way.

4   Pull the disc off the hub.

5   Remove the driveaxle nut.

6   Remove the hub/bearing bolts **(see illustration)**.

7   Use a puller, pull the hub/bearing assembly off the driveaxle splines and detach it from the knuckle. **Note:** *It may be necessary to tap hub and bearing assembly out of the knuckle.*

8   Wrap a rag around the end of the driveaxle to avoid damaging the splines.

Installation

9   Apply a coat of multi-purpose grease to the driveaxle splines. Carefully guide the hub/bearing assembly over the end of the driveaxle, making sure the splines engage properly. Push the hub/bearing assembly into position and install the bolts. Tighten the bolts to the torque listed in this Chapter's Specifications. Install the hub and tighten it securely.

10   Install the brake disc, caliper and (if equipped) the ABS sensor (see Chapter 9).

11   Install the wheel, lower the vehicle and tighten the lug nuts to the torque listed in the Chapter 1 Specifications.

12   Tighten the hub nut to the torque specified in Chapter 8.

6.5 Lift the cartridge out

7.6 Hub mounting bolt locations (arrows)

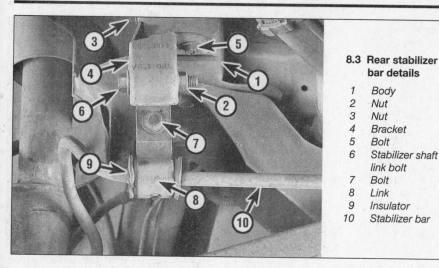

8.3 Rear stabilizer bar details

1 Body
2 Nut
3 Nut
4 Bracket
5 Bolt
6 Stabilizer shaft link bolt
7 Bolt
8 Link
9 Insulator
10 Stabilizer bar

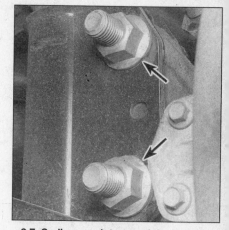

9.7 Scribe or paint around the knuckle nuts (arrow)

8 Rear stabilizer bar - removal and installation

Refer to illustration 8.3

1 Raise the rear of the vehicle and support it securely on jackstands. Block the front wheels to keep the vehicle from rolling off the jackstands.

2 On 1997 and later models remove the stabilizer shaft link from each end of the stabilizer bar, then detach the insulator brackets at the rear frame support and remove the shaft from the vehicle. On 1996 and earlier models mark the strut-to-knuckle position.

3 Remove the four nuts and detach the stabilizer shaft mounting bracket **(see illustration)**.

4 Remove the two stabilizer link mounting brackets.

5 Support the left knuckle with a jack and remove the lower left knuckle mounting bolts.

6 Remove the stabilizer bar through the left side, under the knuckle.

7 Inspect the bushings for cracks, hardening and wear. Replace them if necessary.

8 Installation is the reverse of the removal procedure. Make sure the marks made in

Step 2 line up before tightening the knuckle bolts.

9 Remove the strut-to-knuckle nuts and knock the bolts out with a brass, lead or plastic hammer.

9 Rear strut - removal and installation

Transversely mounted leaf spring

Refer to illustrations 9.7, 9.10 and 9.11

Note: *This procedure requires special tools to compress the auxiliary damper and the rear leaf spring.. These tools are available at specialty tool suppliers. There are two different versions of the auxiliary damper compression tool, an early and late version.*

1 Loosen the wheel lug nuts, raise the rear of the vehicle and support it securely on jackstands. Block the front wheels to keep the vehicle from rolling off the jackstands. Remove the wheel.

2 Compress and remove the auxiliary damper (if equipped) using an auxiliary

damper compression tool.

3 On 1991 through 1995 models (except 1995 Chevrolet) equipped with dual exhaust, remove the exhaust system as necessary.

4 On 1991 through 1995 models (except 1995 Chevrolet) remove the jack pad and retention plates from the rear transverse spring. Install a special rear leaf spring compression tool and compress, but do not remove, the leaf spring.

5 Detach the brake hose bracket.

6 Remove the brake caliper and hang it out of the way on a piece of wire.

7 Mark the strut-to-knuckle relationship and outline the knuckle nuts **(see illustration)**.

8 Detach the stabilizer bar from the knuckle.

9 Support the knuckle with a jack.

10 Remove the strut-to-body bolts **(see illustration)**.

11 Remove the strut-to-knuckle nuts and knock the bolts out with a brass, lead or plastic hammer **(see illustration)**.

12 Separate the strut from the knuckle and remove the strut from the vehicle.

13 Installation is the reverse of the removal procedure. Make sure the strut-to-knuckle

9.10 Location of the strut-to-body bolts (arrows)

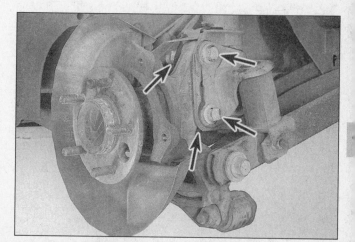

9.11 Strut-to-knuckle mounting bolts

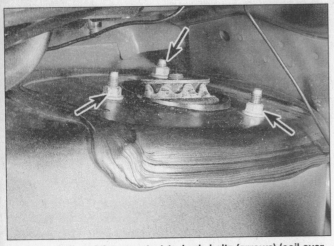

9.14 Location of the rear strut-to-body bolts (arrows) (coil over strut type rear suspensions)

10.2 Remove the jack pad mounting bolts (arrows)

10.3 Remove the bolts and detach the spring retention plate (arrows)

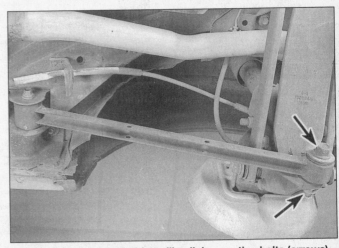

10.4 Detach the right side trailing link mounting bolts (arrows)

alignment marks made during Step 7 line up before tightening the bolts to the torque listed in this Chapter's Specifications.

Coil over strut type suspension

Refer to illustration 9.14

Note: *This procedure applies to 1997 and later models and early Chevrolet models with coil over strut type suspensions.*

14   Loosen the wheel lug nuts, raise the rear of the vehicle and support it securely on jackstands. Block the front wheels to keep the vehicle from rolling off the jackstands. Remove the wheel.

15   Support the bottom of the knuckle with a floor jack.

16   Open the trunk and loosen the three strut-to-body nuts **(see illustration)**.

17   Detach the brake hose bracket from the strut housing if equipped.

18   Mark the relationship of the strut to the knuckle and outline the knuckle bolts/nuts.

19   Detach the stabilizer bar link from the strut (see Section 9).

20   Remove the strut-to-body nuts **(see**

illustration 9.14), and lower the strut assembly until there is no more spring pressure on the knuckle.

21   Remove the strut-to-knuckle nuts and knock the bolts out with a brass or plastic hammer **(see illustration 9.11)**.

22   Separate the strut from the knuckle and remove the strut (with coil spring assembly in place) from the vehicle.

23   If the strut is being replaced, install the coil spring and upper strut components from the old assembly to the new strut. **Note 1:** *Replacement of the coil springs on the rear struts requires a number of special tools and a procedure that is typically beyond the scope of the home mechanic. If the coil springs are in need of replacement, take the car or the strut assembly to a dealer service department or other qualified shop.* **Note 2:** *There is no serviceable cartridge on the rear struts. The strut itself must be replaced.*

24   Installation is the reverse of the removal procedure. Make sure the strut-to-knuckle alignment marks made during Step 18 line up before tightening the bolts to the torque listed in this Chapter's Specifications.

10   Rear transverse spring - removal and installation

Refer to illustrations 10.2, 10.3 and 10.4

Warning: *Wear safety goggles and don't place any part of your body under the ends of the spring while performing this procedure.*

Note: *This procedure either requires a special tool to compress the rear leaf spring (all models except 1997 and later and early Chevrolet with coil over strut type suspensions) or two floor jacks. The tool is available at most specialty tool suppliers.*

Removal

Using special tool

1   Loosen the wheel lug nuts, raise the rear of the vehicle and support it securely on jackstands. Block the front wheels to keep the vehicle from rolling off the jackstands. Remove both rear wheels.

2   Remove the jack pad **(see illustration)**.

3   Remove the spring retention plates **(see illustration)**.

4   Detach the right trailing link at the

knuckle **(see illustration)**.

5   Unplug the ABS electrical connectors, if equipped.

6   Attach the spring compressor tool to the center of the spring, make sure the rollers are centered, then compress the rear leaf spring.

7   Detach the spring from the right knuckle and slide it toward the left side of the vehicle. If it is necessary to pry the spring to the left using a prybar against the right knuckle, pry only on the rubber cover on the end of the spring and be very careful not to gouge or scratch the spring surface.

8   Relax the spring with the tool sufficiently to allow the spring and tool assembly to be removed from the right side of the vehicle.

Without special tool

Note: *This procedure requires the use of two floor jacks.*

9   Perform Steps 1 through 3 above.

10   Unplug the ABS electrical connectors, if equipped.

11   Loosen, but do not remove, the front and rear trailing link bolts on both sides. Remove the right side lateral link-to-knuckle bolt.

12   Disconnect the brake hoses at the rear calipers (see Chapter 9 if necessary). Plug the ends of the hoses to prevent excessive fluid loss and contamination.

13   Place a floor jack under each knuckle and raise the jacks just enough to support the knuckles.

14   Unbolt the struts from the body (see Section 9).

15   Slowly lower the jacks until the spring is no longer under tension. Remove the spring. If the ends of the spring are stuck to the knuckles, carefully pry on the rubber covers on the ends of the spring only. Be very careful not to nick or gouge the spring surface.

Installation

Using special tool

16   Compress the spring, raise it into position, then insert it into the left knuckle.

17   Compress the spring further and insert the right end into the right knuckle.

18   Center the spring and install the retention plate and bolts. Make sure the retention plate tabs line up with support so they don't contact the fuel tank. Tighten the retention plate bolts to the torque listed in this Chapter's Specifications.

19   Connect the trailing link to the right knuckle and install the bolt, but don't tighten the bolt yet. Install the jack pad.

20   If equipped, reconnect the ABS electrical harness.

21   Install the wheels and lower the vehicle. Tighten the trailing link bolt to the torque listed in this Chapter's Specifications.

Without special tool

22   Position the ends of the spring into the knuckles.

23   Slowly raise the knuckles high enough for the right side lateral links to be connected

11.3 The rear hub and bearing assembly is secured to the knuckle with four bolts

to the knuckle. Install the bolt, but don't tighten it yet. It would be helpful to have an assistant support the strut while doing this.

24   Continue to slowly raise the jacks, guide the upper mounts of the struts into position and install the bolts, tightening them to the torque listed in this Chapter's Specifications.

25   Install the spring retention plates. Make sure the retention plate tabs line up with the supports so they don't contact the fuel tank, then tighten the bolts to the torque listed in this Chapter's Specifications.

26   Connect the brake hoses to the calipers, using new copper washers, and tighten the inlet fitting bolt to the torque listed in the Chapter 9 Specifications. Bleed the brakes following the procedure in Chapter 9.

27   Reconnect the ABS electrical connectors, if equipped.

28   Install the wheels. Lower the vehicle and tighten the trailing link bolts and the lateral link-to-knuckle bolt to the torque listed in this Chapter's Specifications. Tighten the wheel lug nuts to the torque listed in the Chapter 1 Specifications.

11   Rear hub and wheel bearing assembly - removal and installation

Refer to illustration 11.3
Note: *The rear hub and wheel bearing assembly is sealed-for-life and must be replaced as a unit.*

Removal

1   Loosen the wheel lug nuts, raise the rear of the vehicle and support it securely on jackstands. Block the front wheels to keep the vehicle from rolling off the jackstands. Remove the wheel.

2   Remove the brake drum or the brake caliper and disc (see Chapter 9). Support the caliper with a piece of wire. If equipped, unplug the ABS electrical connector.

3   Remove the four hub-to-knuckle bolts **(see illustration)**.

4   Remove the hub and bearing assembly, maneuvering it out through the brake caliper bracket as necessary.

Installation

5   Position the hub and bearing assembly on the knuckle and align the holes. Install the bolts. After all four bolts have been installed, tighten them to the torque listed in this Chapter's Specifications.

6   Install the brake drum or the brake caliper and disc, tightening the caliper bolts to the torque listed in the Chapter 9 Specifications. Install the wheel. Lower the vehicle and tighten the wheel lug nuts to the torque listed in Chapter 1 Specifications.

12   Rear suspension components - removal and installation

1   Loosen the wheel lug nuts, raise the rear of the vehicle and support it securely on jackstands. Block the front wheels to keep the vehicle from rolling off the jackstands. Remove the wheels.

2   Remove the stabilizer bar as outlined in Section 8.

3   Remove the brake drums or calipers and discs from the hubs. See Chapter 9 if difficulty is encountered. If equipped, unplug the ABS electrical connectors.

Trailing arms

Refer to illustrations 12.4 and 12.5

4   Remove the trailing arm-to-knuckle nut and bolt **(see illustration)**.

12.4 Remove the trailing arm-to-knuckle bolt and nut

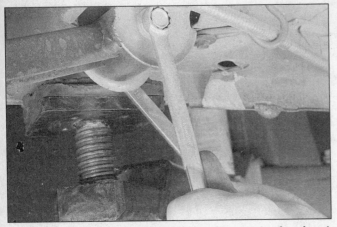

12.5 Use two wrenches to remove the trailing arm-to-chassis nut and bolt

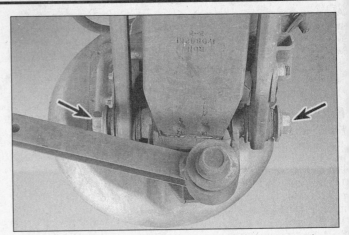

12.9 Lateral link rod-to-knuckle mounting bolts (arrows)

5   Remove the trailing arm-to-chassis nut and bolt **(see illustration)**.
6   Detach the trailing arm and lower it from the vehicle.
7   Installation is the reverse of removal. Do not tighten the nuts and bolts to the torque listed in this Chapters Specifications until the vehicle weight has been lowered onto the suspension.

Lateral link rods

Refer to illustrations 12.9, 12.10 and 12.11

8   Compress and detach the auxiliary spring if equipped (see Section 9).
9   Remove the nuts and bolts connecting the rods to the knuckle **(see illustration)**.
10   Remove the rod inner bolts at the cross-member **(see illustration)**.
11   On 1988 and 1989 models mark the position of the toe adjusting cam to the crossmember. Remove the toe adjusting cam and push the bolt through far enough to allow

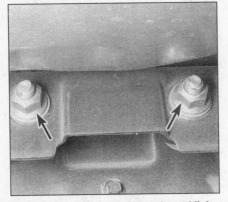

12.10 Locations of the inner lateral link rod nuts (arrows)

12.11 Mark the positions of the toe adjusting cams to the crossmember

the rod to be removed **(see illustration)**. (Later models don't have a toe adjusting cam.) On some models it may be necessary

to lower the fuel tank for access to the front rod bolt.
12   Installation is the reverse of removal.

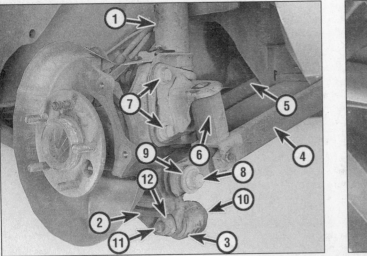

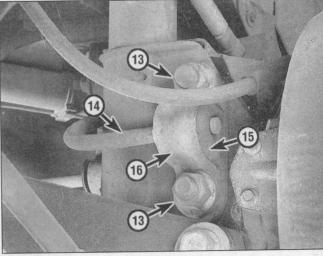

12.16 Knuckle installation details

1 Rear strut	5 Front lateral link rod	9 Washer	13 Nut
2 Knuckle	6 Auxiliary damper	10 Washer	14 Stabilizer bar
3 Trailing arm	7 Bolt	11 Bolt	15 Insulator
4 Rear lateral link rod	8 Bolt	12 Nut	16 Stabilizer bar bracket

Apply thread locking compound to all bolts and don't tighten the bolts to the torque listed in this Chapter's Specifications until the vehicle weight has been lowered onto the suspension. After installation have the rear toe checked by a dealer service department or alignment shop.

Knuckle

Refer to illustration 12.16

13 Mark the strut-to-knuckle relationship and make a line around the knuckle nuts **(see illustration 9.7)**.

14 Remove the rear brake drum or caliper and disc (see Chapter 9). On ABS equipped models, unplug the ABS electrical harness.

15 On cars with rear drum brakes, remove the brake shoe assembly (see Chapter 9) and the hub and bearing assembly (see Section 11). Remove the rear transverse spring, if equipped (see Section 10).

16 Remove the bolts and detach the trailing arm, lateral link rods and, if equipped, auxiliary damper and stabilizer bar from the knuckle **(see illustration)**.

17 Remove the strut-to-knuckle nuts and knock the bolts out with a brass, lead or plastic hammer.

18 Separate the knuckle from the strut and lower it from the vehicle.

19 To install the knuckle, slide it into the strut flange. If equipped, install the stabilizer bar bracket and auxiliary damper and insert the two bolts. Apply thread locking compound to the threads and install the nuts finger tight. Connect the lateral link rods and trailing arm to the knuckle. Align the marks made in Step 13 and tighten the strut-to-knuckle bolts to the torque listed in this Chapter's Specifications. The remainder of installation is the reverse of removal.

13 Steering system - general information

Warning: *Whenever any of the steering fasteners are removed, they must be inspected and, if necessary, replaced with new ones of the same part number or of original equipment quality and design. Torque specifications must be followed for proper reassembly and component retention. Never attempt to heat or straighten any suspension or steering components. Instead, replace any bent or damaged part with a new one.*

All vehicles covered by this manual have power rack-and-pinion steering systems. The components making up the system are the steering wheel, steering column, rack-and-pinion steering gear, tie-rods and tie-rod ends. The power steering system has a belt-driven pump to provide hydraulic pressure.

In the power steering system, the motion of turning the steering wheel is transferred through the column to the pinion shaft of the rack-and-pinion assembly. Teeth on the pinion shaft are meshed with teeth on the rack, so when the shaft is turned, the rack is

moved left or right in the housing. A rotary control valve in the rack-and-pinion unit directs hydraulic fluid under pressure from the power steering pump to either side of the integral rack piston, which is connected to the rack, thereby reducing steering force. Depending on which side of the piston this hydraulic pressure is applied to, the rack will be forced either left or right, which moves the tie-rods, etc. If the power steering system loses hydraulic pressure it will still function manually, though with increased effort.

Some models are also equipped with Variable Effort Steering (VES). With VES, power assist is maximum when the vehicle is moving at low speeds, such as when parking, and minimum when the vehicle is moving at high speeds, such as highway cruising. This allows the driver to turn the steering wheel easily when it's most necessary, while still maintaining "road feel." The VES uses signals from the engine control module, the ABS control module and a steering wheel rotation sensor to control the opening of a power steering solenoid actuator. The actuator, which is located on the power steering pump on VES-equipped cars, responds to these signals by varying the amount of power steering fluid flow. Thus, as vehicle speed decreases power steering fluid flow (and steering assist) increases, and as vehicle speed increases, fluid flow (and steering assist) decreases.

The steering column is a collapsible, energy-absorbing type, designed to compress in the event of a front end collision to minimize injury to the driver. The column also houses the ignition switch lock, key warning buzzer, turn signal controls, headlight dimmer control and windshield wiper controls. The ignition and steering wheel can both be locked while the vehicle is parked.

Due to the column's collapsible design, it's important that only the specified screws, bolts and nuts be used as designated and that they're tightened to the specified torque. Other precautions particular to this design are noted in appropriate Sections.

In addition to the standard steering column, optional tilt and key release versions are also offered. The tilt model can be set in five different positions, while with the key release model the ignition key is locked in the column until a lever is depressed to extract it.

Because disassembly of the steering column is more often performed to repair a switch or other electrical part than to correct a problem in the steering, the upper steering column disassembly and reassembly procedure is included in Chapter 12.

14 Tie-rod ends - removal and installation

Refer to illustration 14.3

Removal

1 Loosen the wheel lug nuts, raise the

14.3 Using white paint, mark the relationship of the tie-rod end and the threaded adjuster

front of the vehicle and support it securely on jackstands. Apply the parking brake and block the rear wheels to keep the vehicle from rolling off the jackstands. Remove the wheel.

2 Loosen the tie-rod end jam nut.

3 Mark the relationship of the tie-rod end to the threaded portion of the tie-rod **(see illustration)**. This will ensure the toe-in setting is restored when reassembled.

4 Remove the cotter pin and disconnect the tie-rod from the steering knuckle arm with a puller.

5 Unscrew the tie-rod end from the tie-rod.

Installation

6 Thread the tie-rod end onto the tie-rod to the marked position and connect the tie-rod end to the steering arm. Install the castellated nut and tighten it to the torque listed in this Chapter's Specifications. Install a new cotter pin.

7 Tighten the jam nut securely and install the wheel. Lower the vehicle and tighten the lug nuts to the torque listed in the Chapter 1 Specifications.

8 Have the front end alignment checked by a dealer service department or an alignment shop.

15 Steering gear - removal and installation

Removal

1 Disconnect the cable from the negative battery terminal. **Caution:** *On models equipped with the Theftlock audio system, be sure the lockout feature is turned off before performing any procedure which requires disconnecting the battery.*

2 On models equipped with the 3.4L V6 engine, remove the air cleaner and duct assembly.

3 Loosen the front wheel lug nuts, raise the front of the vehicle and support it

securely on jackstands. Apply the parking brake and block the rear wheels to keep the vehicle from rolling off the jackstands. Remove both front wheels.

4   On models equipped with the 3.4L V6 engine, remove the right side engine splash shield.

5   On cars equipped with airbags, set the steering wheel straight ahead, check that the wide spline or "block tooth" on the upper steering shaft assembly is at the 12 o'clock position and that the ignition switch is set to "lock." **Caution:** *Failure to do this can result in damage to the airbag system's coil unit in the proceeding steps.*

6   Roll back the boot at the bottom of the steering column to expose the flange and steering coupler assembly. Mark the coupler and steering column shaft, remove the pinch bolt and separate the steering column from the power steering input shaft.

7   Unplug the electrical connector from the power steering idle speed switch.

8   On models equipped with the 3.4L V6 engine, remove the exhaust pipe and catalytic converter assembly.

9   Separate the tie-rod ends from the steering arms (see Section 14).

10   Support the rear of the subframe assembly with a jack and remove the rear bolts. Loosen the front bolts of the subframe and lower the rear approximately five inches (three inches on models equipped with the 3.4L V6 engine).

11   Remove the heat shield from the steering gear.

12   Remove the fluid pipe clip from the rack.

13   Place a drain pan or tray under the vehicle, positioned beneath the steering gear. Using a flare-nut wrench, disconnect the pressure and return lines from the steering gear. Plug the lines to prevent excessive fluid loss.

14   Remove the steering gear through-bolts and nuts.

15   Lift the steering gear out of the mounts them move it forward and remove the pinch bolt from the coupler.

16   Support the steering gear and carefully maneuver the entire assembly out through the left side wheel opening.

Installation

17   Pass the steering gear assembly through the left wheel opening and place it in position in the mounts. Install the through-bolts and nuts, tightening them to the torque listed in this Chapter's Specifications.

18   Attach the pressure and return lines to the steering gear. Connect the line retainer.

19   Install the heat shield.

20   Raise the subframe into position and install the bolts. Tighten the frame bolts to the torque listed in this Chapter's Specifications.

21   Connect the tie-rod ends to the steering arms and tighten the nuts to the torque listed in this Chapter's Specifications. Install new cotter pins.

22   Install the exhaust system, if removed.

23   Center the steering gear and have an

assistant guide the coupler onto the steering column shaft, with the previously applied marks aligned. Install and tighten the pinch bolt securely. **Warning:** *Be sure the shaft is seated before you install the pinch bolt or the shafts may disengage.*

24   Install the front wheels, lower the vehicle and tighten the lug nuts to the torque listed in the Chapter 1 Specifications.

25   Reconnect the negative battery cable.

26   Fill the power steering pump with the recommended fluid, bleed the system (see Section 18) and recheck the fluid level. Check for leaks.

27   Have the front end alignment checked by a dealer service department or an alignment shop.

16   Steering gear boots - replacement

1   Remove the steering gear from the vehicle (see Section 15).

2   Remove the tie-rods ends from the steering gear (see Section 14).

3   Remove the jam nuts and outer boot clamps.

4   Cut off both inner boot clamps and discard them.

5   Mark the location of the breather tube, then remove the boots and the tube.

6   Install a new clamp on the inner end of the boot.

7   Apply multi-purpose grease to the tie-rod and the mounting groove on the steering gear.

8   Line up the breather tube with the marks made during removal and slide the boot onto the steering gear housing.

9   Make sure the boot isn't twisted, then tighten the clamp.

10   Install the outer clamps and tie-rod end jam nuts.

11   Install the tie-rod ends and tighten the nuts securely (see Section 14).

12   Install the steering gear assembly.

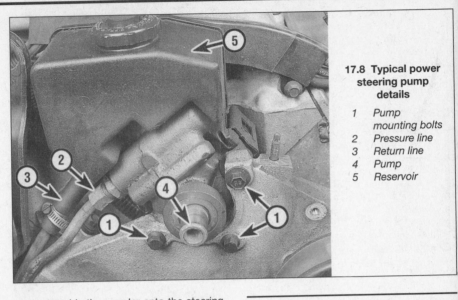

17.8   Typical power steering pump details

1   *Pump mounting bolts*
2   *Pressure line*
3   *Return line*
4   *Pump*
5   *Reservoir*

17   Power steering pump - removal and installation

Refer to illustration 17.8

Removal

1   Disconnect the cable from the negative battery terminal. **Caution:** *On models equipped with the Theftlock audio system, be sure the lockout feature is turned off before performing any procedure which requires disconnecting the battery.*

2   Remove the coolant recovery reservoir if necessary.

3   On models equipped with the 3.4L V6 engine, remove the air cleaner and duct assembly.

4   On models equipped with the 3800 engine, remove the ECM cover.

5   Remove the pump drivebelt (see Chapter 1).

6   Position a drain pan under the vehicle. Remove as much fluid as possible with a suction pump, then remove the return line from the pump.

7   Using a flare-nut wrench and a back-up wrench, disconnect the pressure hose from the pump. **Note:** *On Quad-4 engines it may be necessary to remove the left side torque strut from the engine, disconnect the accelerator cable bracket and move it out of the way, then remove the torque strut bracket for access.*

8   Remove the pump mounting bolts and detach the pump from the engine, being careful not to spill the remaining fluid (see illustration).

Installation

9   Position the pump on the mounting bracket and install the bolts.

10   Connect the pressure and return lines to the pump.

11   Fill the reservoir with the recommended fluid and bleed the system, following the procedure described in the next Section.

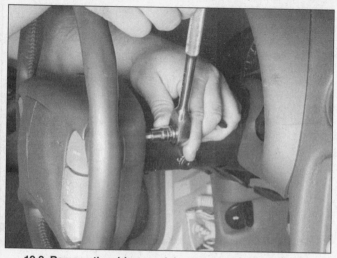

19.3 Remove the airbag module retaining screws from the backside of the steering wheel

19.4 Remove the airbag module and unplug the two electrical connectors (arrows)

18 Power steering system - bleeding

1   Following any operation in which the power steering fluid lines have been disconnected, the power steering system must be bled to remove air and obtain proper steering performance.
2   With the front wheels turned all the way to the left, check the power steering fluid level and, if low, add fluid until it reaches the Cold mark on the dipstick.
3   Start the engine and allow it to run at fast idle. Recheck the fluid level and add more if necessary to reach the Cold mark on the dipstick.
4   Bleed the system by turning the wheels from side-to-side, without hitting the stops. This will work the air out of the system. Don't allow the reservoir to run out of fluid.
5   When the air is worked out of the system, return the wheels to the straight ahead position and leave the engine running for several minutes before shutting it off. Recheck the fluid level.
6   Road test the vehicle to be sure the

steering system is functioning normally with no noise.
7   Recheck the fluid level to be sure it's up to the Hot mark on the dipstick while the engine is at normal operating temperature. Add fluid if necessary.

19 Steering wheel - removal and installation

Refer to illustrations 19.3, 19.4, 19.5, 19.6 and 19.7

Warning: *Some models covered by this manual are equipped with Supplemental Inflatable Restraint (SIR) systems, more commonly known as airbags. Always disable the airbag system before working in the vicinity of any airbag system components to avoid the possibility of accidental deployment of the airbag(s) which could cause personal injury (see Chapter 12, Section 22). The yellow wiring harnesses and connectors routed through the console and instrument panel are for this system. Do not use electrical test*

equipment on any of the airbag system wiring or tamper with it in any way.
Note: *If your vehicle is not equipped with an airbag, disregard the steps which do not apply.*
1   Disconnect the cable from the negative battery terminal. **Caution:** *On models equipped with the Theftlock audio system, be sure the lockout feature is turned off before performing any procedure which requires disconnecting the battery.*
2   Disable the airbag system (see Chapter 12).
3   Remove the airbag module retaining screws **(see illustration)**.
4   Remove the airbag module and unplug the electrical connectors **(see illustration)**. Carry the airbag module with the trim side facing away from your body, and set it in an isolated area with the trim side facing up.
5   If equipped, remove the safety clip from the steering shaft **(see illustration)**.
6   Remove the steering wheel retaining nut. Mark the relationship of the steering wheel to the steering shaft **(see illustration)**.
7   Remove the steering wheel with a puller

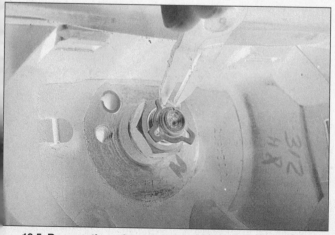

19.5 Remove the safety clip with a pair of snap-ring pliers

19.6 Make alignment marks (arrows) on the steering wheel and shaft, if none exist

19.7 Use a steering wheel puller to separate the steering wheel from the shaft - DO NOT attempt to remove the wheel with a hammer!

20.2 Use a special tool to push the stud out of the flange

(see illustration). **Warning:** *Don't allow the steering shaft to turn with the steering wheel removed. If for some reason the steering shaft does turn, refer to Chapter 12, Section 8, for the airbag coil centering procedure.*
8   Installation is the reverse of removal. Be sure to tighten the steering wheel nut to the torque listed in this Chapter's Specifications.
9   To enable the airbag system, refer to Chapter 12.

20   Wheel studs - replacement

Refer to illustrations 20.2 and 20.3
Note: *This procedure applies to both the front and rear wheel studs.*
1   Install a lug nut part way onto the stud being replaced.
2   Push the stud out of the hub flange using a special tool (see illustration).
3   Insert the new stud into the hub flange from the back side and install four flat washers and a lug nut on the stud (see illustration).

4   Tighten the lug nut until the stud is seated in the flange.
5   Reinstall the hub and wheel bearing assembly.

21   Wheels and tires - general information

Refer to illustration 21.1
All vehicles covered by this manual are equipped with metric-size fiberglass or steel belted radial tires (see illustration). The use of other size or type tires may affect the ride and handling of the vehicle. Don't mix different types of tires, such as radials and bias belted, on the same vehicle, since handling may be seriously affected. Tires should be replaced in pairs on the same axle, but if only one tire is being replaced, be sure it's the same size, structure and tread design as the other.
Because tire pressure affects handling and wear, the tire pressures should be checked at least once a month or before any

20.3 Install the four washers and a lug nut on the stud, then tighten the nut to draw the stud into place

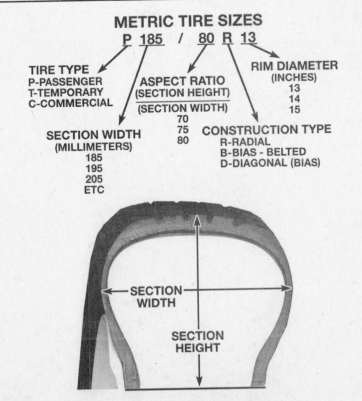

21.1 Metric tire size code

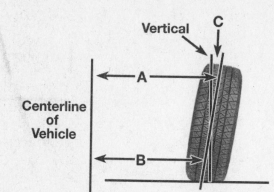

CAMBER ANGLE (FRONT VIEW)

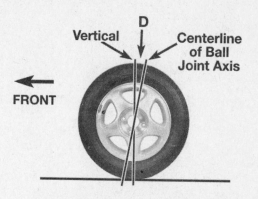

CASTER ANGLE (SIDE VIEW)

TOE-IN (TOP VIEW)

22.1 Front end alignment details

A minus B = C (degrees camber)
D = caster (expressed in degrees)
E minus F = toe-in (measured in inches)
G = toe-in (expressed in degrees)

extended trips (see Chapter 1).

Wheels must be replaced if they're bent, ented, leak air, have elongated bolt holes, re heavily rusted, out of vertical symmetry or the lug nuts won't stay tight. Wheel repairs y welding or peening aren't recommended.

Tire and wheel balance is important to e overall handling, braking and perfor- ance of the vehicle. Unbalanced wheels an adversely affect handling and ride char- cteristics as well as tire life. Whenever a tire installed on a wheel, the tire and wheel ould be balanced by a shop with the roper equipment.

2 Wheel alignment - general information

efer to illustration 22.1

A wheel alignment refers to the adjust- ents made to the wheels so they're in oper angular relationship to the suspension d the ground. Wheels that are out of proper ignment not only affect steering control, but also increase tire wear. The adjustment most commonly required is the toe-in adjustment **(see illustration)**, but camber adjustment is also possible.

Getting the proper wheel alignment is a very exacting process, one in which compli- cated and expensive machines are necessary to perform the job properly. Because of this, you should have a technician with the proper equipment perform these tasks. We will, however, attempt to give you a basic idea of what's involved with wheel alignment so you can better understand the process and deal intelligently with the shop that does the work.

Toe-in is the turning in of the wheels. The purpose of a toe specification is to ensure parallel rolling of the wheels. In a vehi- cle with zero toe-in, the distance between the front edges of the wheels will be the same as the distance between the rear edges of the wheels. The actual amount of toe-in is nor- mally only a fraction of an inch. Incorrect toe- in will cause the tires to wear improperly by making them scrub against the road surface. On the front end, toe adjustment is controlled by the tie-rod end position on the tie-rod. On the rear on some models, toe is adjusted by turning a cam bolt on the inner end of the front link rod. On other models, a special tool is required to reposition the front link rod at its inner mount.

Camber is the tilting of the wheels from the vertical when viewed from the front or rear of the vehicle. When the wheels tilt out at the top, the camber is said to be positive (+). When the wheels tilt in at the top the camber is negative (-). The amount of tilt is measured in degrees from the vertical and this mea- surement is called the camber angle. This angle affects the amount of tire tread which contacts the road and compensates for changes in the suspension geometry when the vehicle is cornering or travelling over an undulating surface. On the front end, camber can be adjusted after the strut mounting holes have been elongated with a file. On the rear, camber can be adjusted by changing the position of the rear knuckle on the strut.

Caster is not adjustable. If the caster angle is not to specification, the suspension components must be checked for damage.

10

Notes

Chapter 11 Body

Contents

General information

These models are available in two-door coupe and four-door sedan body styles. The vehicle is a "unibody" type, which means the body is designed to provide vehicle rigidity so a separate frame isn't necessary. Body maintenance is an important part of the retention of the vehicle's market value. It's far less costly to handle small problems before they grow into larger ones. Major body components which are particularly vulnerable in accidents are removable. These include the hood, front fenders, grille, doors, trunk lid and tail light assembly. It's often cheaper and less time consuming to replace an entire panel than it is to attempt a restoration of the old one. However, this must be decided on a case-by-case basis.

2 Maintenance - body

1 The condition of the body is very important, because the value of the vehicle is dependent on it. It's much more difficult to repair a neglected or damaged body than it is to repair mechanical components. The hidden areas of the body, such as the fender wells and the engine compartment, are equally important, although they obviously don't require as frequent attention as the rest of the body.
2 Once a year, or every 12,000 miles, it's a good idea to have the underside of the body steam cleaned. All traces of dirt and oil will be removed and the underside can then be inspected carefully for rust, damaged brake lines, frayed electrical wiring, damaged cables and other problems.

3 At the same time, clean the engine and the engine compartment with a water soluble degreaser.
4 The fender wells should be given particular attention, as undercoating can peel away and stones and dirt thrown up by the tires can cause the paint to chip and flake, allowing rust to set in. If rust is found, clean down to the bare metal and apply an anti-rust paint.
5 The body should be washed as needed. Wet the vehicle thoroughly to soften the dirt, then wash it down with a soft sponge and plenty of clean soapy water. If the surplus dirt isn't washed off very carefully, it will in time wear down the paint.
6 Spots of tar or asphalt coating thrown up from the road should be removed with a cloth soaked in solvent.
7 Once every six months, wax the body thoroughly. If a chrome cleaner is used to

1

remove rust from any of the vehicle's plated parts, remember that the cleaner also removes part of the chrome, so use it sparingly.

3   Maintenance - upholstery and carpets

1   Every three months remove the carpets or mats and clean the interior of the vehicle (more frequently if necessary). Vacuum the upholstery and carpets to remove loose dirt and dust.
2   If the upholstery is soiled, apply upholstery cleaner with a damp sponge and wipe it off with a clean, dry cloth.

4   Vinyl trim - maintenance

Vinyl trim should not be cleaned with detergents, caustic soaps or petroleum-based cleaners. Plain soap and water or a mild vinyl cleaner is best for stains. Test a small area for color fastness. Bubbles under the vinyl can be eliminated by piercing them with a pin and then working the air out.

5   Body repair - minor damage

See photo sequence

Repair of minor scratches

1   If the scratch is superficial and does not penetrate to the metal of the body, repair is very simple. Lightly rub the scratched area with a fine rubbing compound to remove loose paint and built up wax. Rinse the area with clean water.
2   Apply touch-up paint to the scratch, using a small brush. Continue to apply thin layers of paint until the surface of the paint in the scratch is level with the surrounding paint. Allow the new paint at least two weeks to harden, then blend it into the surrounding paint by rubbing with a very fine rubbing compound. Finally, apply a coat of wax to the scratch area.
3   If the scratch has penetrated the paint and exposed the metal of the body, causing the metal to rust, a different repair technique is required. Remove all loose rust from the bottom of the scratch with a pocket knife, then apply rust inhibiting paint to prevent the formation of rust in the future. Using a rubber or nylon applicator, coat the scratched area with glaze-type filler. If required, the filler can be mixed with thinner to provide a very thin paste, which is ideal for filling narrow scratches. Before the glaze filler in the scratch hardens, wrap a piece of smooth cotton cloth around the tip of a finger. Dip the cloth in thinner and then quickly wipe it along the surface of the scratch. This will ensure that the surface of the filler is slightly hollow.

The scratch can now be painted over as described earlier in this section.

Repair of dents

4   When repairing dents, the first job is to pull the dent out until the affected area is as close as possible to its original shape. There is no point in trying to restore the original shape completely as the metal in the damaged area will have stretched on impact and cannot be restored to its original contours. It is better to bring the level of the dent up to a point which is about 1/8-inch below the level of the surrounding metal. In cases where the dent is very shallow, it is not worth trying to pull it out at all.
5   If the back side of the dent is accessible, it can be hammered out gently from behind using a soft-face hammer. While doing this, hold a block of wood firmly against the opposite side of the metal to absorb the hammer blows and prevent the metal from being stretched.
6   If the dent is in a section of the body which has double layers, or some other factor makes it inaccessible from behind, a different technique is required. Drill several small holes through the metal inside the damaged area, particularly in the deeper sections. Screw long, self tapping screws into the holes just enough for them to get a good grip in the metal. Now the dent can be pulled out by pulling on the protruding heads of the screws with locking pliers.
7   The next stage of repair is the removal of paint from the damaged area and from an inch or so of the surrounding metal. This is easily done with a wire brush or sanding disk in a drill motor, although it can be done just as effectively by hand with sandpaper. To complete the preparation for filling, score the surface of the bare metal with a screwdriver or the tang of a file or drill small holes in the affected area. This will provide a good grip for the filler material. To complete the repair, see the Section on filling and painting.

Repair of rust holes or gashes

8   Remove all paint from the affected area and from an inch or so of the surrounding metal using a sanding disk or wire brush mounted in a drill motor. If these are not available, a few sheets of sandpaper will do the job just as effectively.
9   With the paint removed, you will be able to determine the severity of the corrosion and decide whether to replace the whole panel, if possible, or repair the affected area. New body panels are not as expensive as most people think and it is often quicker to install a new panel than to repair large areas of rust.
10   Remove all trim pieces from the affected area except those which will act as a guide to the original shape of the damaged body, such as headlight shells, etc. Using metal snips or a hacksaw blade, remove all loose metal and any other metal that is badly affected by rust. Hammer the edges of the hole inward to create a slight depression for the filler material.

11   Wire brush the affected area to remove the powdery rust from the surface of the metal. If the back of the rusted area is accessible, treat it with rust inhibiting paint.
12   Before filling is done, block the hole in some way. This can be done with sheet metal riveted or screwed into place, or by stuffing the hole with wire mesh.
13   Once the hole is blocked off, the affected area can be filled and painted. See the following subsection on filling and painting.

Filling and painting

14   Many types of body fillers are available, but generally speaking, body repair kits which contain filler paste and a tube of resin hardener are best for this type of repair work. A wide, flexible plastic or nylon applicator will be necessary for imparting a smooth and contoured finish to the surface of the filler material. Mix up a small amount of filler on a clean piece of wood or cardboard (use the hardener sparingly). Follow the manufacturer's instructions on the package, otherwise the filler will set incorrectly.
15   Using the applicator, apply the filler paste to the prepared area. Draw the applicator across the surface of the filler to achieve the desired contour and to level the filler surface. As soon as a contour that approximates the original one is achieved, stop working the paste. If you continue, the paste will begin to stick to the applicator. Continue to add thin layers of paste at 20-minute intervals until the level of the filler is just above the surrounding metal.
16   Once the filler has hardened, the excess can be removed with a body file. From then on, progressively finer grades of sandpaper should be used, starting with a 180 grit paper and finishing with 600 grit wet-or-dry paper. Always wrap the sandpaper around a flat rubber or wood block, otherwise the surface of the filler will not be completely flat. During the sanding of the filler surface, the wet-or-dry paper should be periodically rinsed in water. This will ensure that a very smooth finish is produced in the final stage.
17   At this point, the repair area should be surrounded by a ring of bare metal, which in turn should be encircled by the finely feathered edge of good paint. Rinse the repair area with clean water until all of the dust produced by the sanding operation is gone.
18   Spray the entire area with a light coat of primer. This will reveal any imperfections in the surface of the filler. Repair the imperfections with fresh filler paste or glaze filler and once more smooth the surface with sandpaper. Repeat this spray-and-repair procedure until you are satisfied that the surface of the filler and the feathered edge of the paint are perfect. Rinse the area with clean water and allow it to dry completely.
19   The repair area is now ready for painting. Spray painting must be carried out in a warm, dry, windless and dust free atmosphere. These conditions can be created

9.2 Mark the hinge-to-hood bolt positions by drawing around the heads with a marking pen

9.4a Use a small screwdriver to pry the clip out . . .

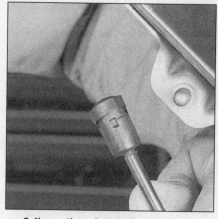

9.4b . . . then detach the strut from the hood

you have access to a large indoor work area, but if you are forced to work in the open, you will have to pick the day very carefully. If you are working indoors, dousing the floor in the work area with water will help settle the dust which would otherwise be in the air. If the repair area is confined to one body panel, mask off the surrounding panels. This will help minimize the effects of a slight mismatch in paint color. Trim pieces such as chrome strips, door handles, etc., will also need to be masked off or removed. Use masking tape and several thicknesses of newspaper for the masking operations.

0   Before spraying, shake the paint can thoroughly, then spray a test area until the spray painting technique is mastered. Cover the repair area with a thick coat of primer. The thickness should be built up using several thin layers of primer rather than one thick one. Using 600-grit wet-or-dry sandpaper, rub down the surface of the primer until it is very smooth. While doing this, the work area should be thoroughly rinsed with water and the wet-or-dry sandpaper periodically rinsed as well. Allow the primer to dry before spraying additional coats.

1   Spray on the top coat, again building up the thickness by using several thin layers of paint. Begin spraying in the center of the repair area and then, using a circular motion, work out until the whole repair area and about two inches of the surrounding original paint is covered. Remove all masking material 10 to 15 minutes after spraying on the final coat of paint. Allow the new paint at least two weeks to harden, then use a very fine rubbing compound to blend the edges of the new paint into the existing paint. Finally, apply a coat of wax.

Body repair - major damage

Major damage must be repaired by an auto body/frame repair shop with the necessary welding and hydraulic straightening equipment.

2   If the damage has been serious, it is vital that the structure be checked for proper alignment or the vehicle's handling characteristics may be adversely affected. Other problems, such as excessive tire wear and wear in the driveline and steering may occur.

3   Due to the fact that all of the major body components (hood, fenders, etc.) are separate and replaceable units, any seriously damaged components should be replaced rather than repaired. Sometimes these components can be found in a wrecking yard that specializes in used vehicle components, often at considerable savings over the cost of new parts.

7   Maintenance - hinges and locks

Every 3000 miles or three months, the door, hood and trunk lid hinges should be lubricated with a few drops of oil. The door striker plates should also be given a thin coat of white lithium-base grease to reduce wear and ensure free movement.

8   Windshield and fixed glass - replacement

1   Replacement of the windshield and fixed glass requires the use of special fast setting adhesive/caulk materials. These operations should be left to a dealer or a shop specializing in glass work.

2   Windshield-mounted rear view mirror support removal is also best left to experts, as the bond to the glass also requires special tools and adhesives.

9   Hood - removal, installation and adjustment

Refer to illustrations 9.2, 9.4a, 9,4b and 9.10
Note: *The hood is heavy and somewhat awkward to remove and install - at least two people should perform this procedure.*

Removal and installation

1   Use blankets or pads to cover the cowl area of the body and the fenders. This will protect the body and paint as the hood is lifted off.

2   Scribe or paint alignment marks around the bolt heads to insure proper alignment during installation **(see illustration)**.

3   Disconnect any cables or wire harnesses which will interfere with removal.

4   Have an assistant support the weight of the hood. Detach the upper ends of the hood support struts **(see illustrations)**.

5   Remove the hinge-to-hood bolts and lift off the hood.

6   Installation is the reverse of removal.

Adjustment

7   Fore-and-aft and side-to-side adjustment of the hood is done by moving the hood in relation to the hinge plate after loosening the bolts or nuts.

8   Scribe a line around the entire hinge plate so you can judge the amount of movement **(see illustration 9.2)**.

9   Loosen the bolts or nuts and move the hood into correct alignment. Move it only a little at a time. Tighten the hinge bolts and carefully lower the hood to check the alignment.

10   Finally, adjust the hood bumpers on the radiator support so the hood, when closed, is

9.10 Screw the hood bumpers in or out to adjust the hood flush with the fenders

1

These photos illustrate a method of repairing simple dents. They are intended to supplement *Body repair - minor damage* in this Chapter and should not be used as the sole instructions for body repair on these vehicles.

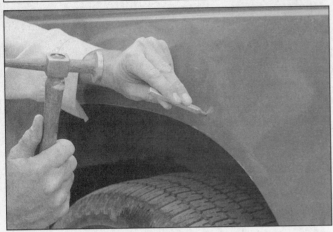

1   If you can't access the backside of the body panel to hammer out the dent, pull it out with a slide-hammer-type dent puller. In the deepest portion of the dent or along the crease line, drill or punch hole(s) at least one inch apart . . .

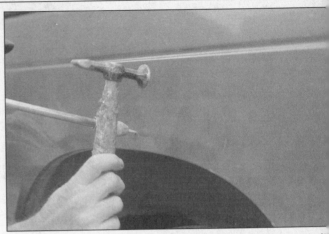

2   . . . then screw the slide-hammer into the hole and operate it Tap with a hammer near the edge of the dent to help 'pop' the metal back to its original shape. When you're finished, the dent area should be close to its original contour and about 1/8-inch below the surface of the surrounding metal

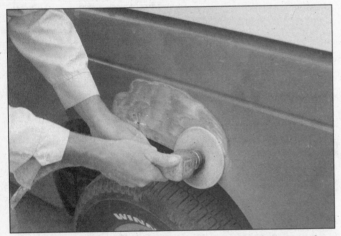

3   Using coarse-grit sandpaper, remove the paint down to the bare metal. Hand sanding works fine, but the disc sander shown here makes the job faster. Use finer (about 320-grit) sandpaper to feather-edge the paint at least one inch around the dent area

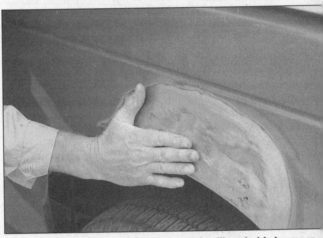

4   When the paint is removed, touch will probably be more helpful than sight for telling if the metal is straight. Hammer down the high spots or raise the low spots as necessary. Clean the repair area with wax/silicone remover

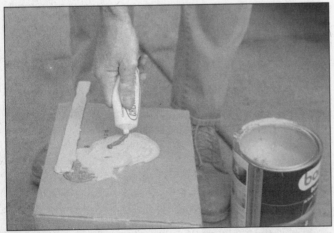

5   Following label instructions, mix up a batch of plastic filler and hardener. The ratio of filler to hardener is critical, and, if you mix it incorrectly, it will either not cure properly or cure too quickly (you won't have time to file and sand it into shape)

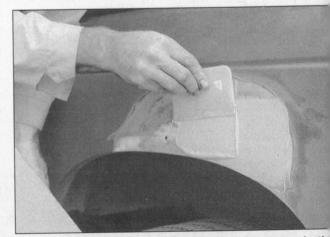

6   Working quickly so the filler doesn't harden, use a plastic applicator to press the body filler firmly into the metal, assuring bonds completely. Work the filler until it matches the original contour and is slightly above the surrounding metal

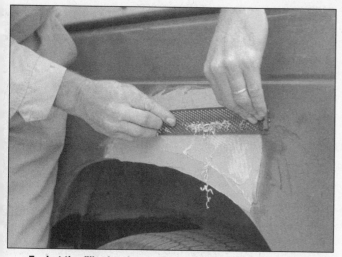

7 Let the filler harden until you can just dent it with your fingernail. Use a body file or Surform tool (shown here) to rough-shape the filler

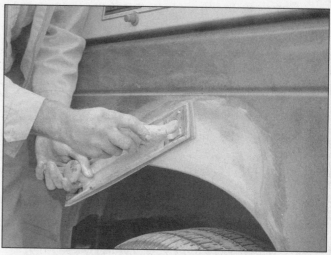

8 Use coarse-grit sandpaper and a sanding board or block to work the filler down until it's smooth and even. Work down to finer grits of sandpaper - always using a board or block - ending up with 360 or 400 grit

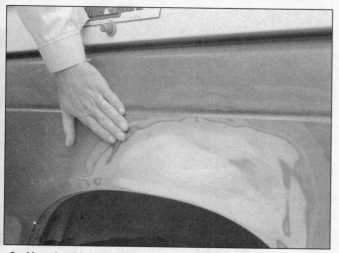

9 You shouldn't be able to feel any ridge at the transition from the filler to the bare metal or from the bare metal to the old paint. As soon as the repair is flat and uniform, remove the dust and mask off the adjacent panels or trim pieces

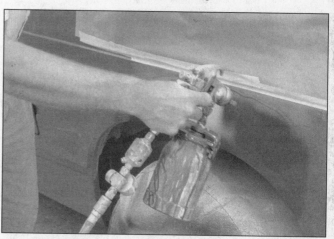

10 Apply several layers of primer to the area. Don't spray the primer on too heavy, so it sags or runs, and make sure each coat is dry before you spray on the next one. A professional-type spray gun is being used here, but aerosol spray primer is available inexpensively from auto parts stores

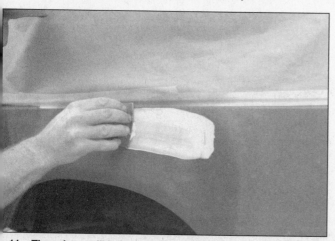

11 The primer will help reveal imperfections or scratches. Fill these with glazing compound. Follow the label instructions and sand it with 360 or 400-grit sandpaper until it's smooth. Repeat the glazing, sanding and respraying until the primer reveals a perfectly smooth surface

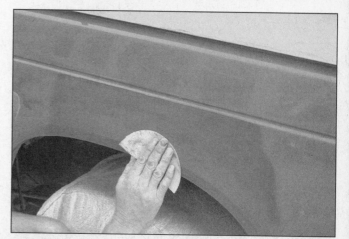

12 Finish sand the primer with very fine sandpaper (400 or 600-grit) to remove the primer overspray. Clean the area with water and allow it to dry. Use a tack rag to remove any dust, then apply the finish coat. Don't attempt to rub out or wax the repair area until the paint has dried completely (at least two weeks)

10.2a Typical hood latch release handle trim cover mounting screws (arrows) (floor mounted type)

10.2b Hood latch release handle mounting bolt (arrow) (floor mounted type)

10.2c Typical hood latch release handle mounting details (door pillar mounted)

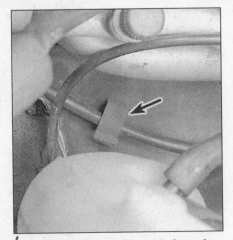

10.3 Hood latch cable guide (arrow)

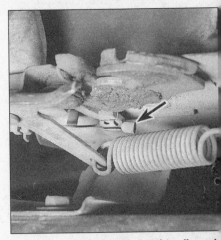

10.4 Hood latch cable mounting clip and cable end (arrow)

flush with the fenders (see illustration).

11   The hood latch assembly, as well as the hinges, should be periodically lubricated with white lithium-base grease to prevent sticking and wear.

10   Hood latch cable - replacement

Refer to illustrations 10.2a, 10.2b, 10.2c, 10.3 and 10.4

1   In the passenger compartment, remove the trim panel and, if necessary, pull back the carpet.

2   Remove the latch handle retaining screws and detach the handle (see illustrations).

3   In the engine compartment, detach the cable guides and grommet (see illustration). Depending on the year and model of your car, you may have to remove the air cleaner and duct assembly, the headlamp access panel and/or the battery.

4   Detach the end of the cable from the latch (see illustration).

5   Connect a piece of string or thin wire of suitable length to the end of the cable and

pull the cable through into the passenger compartment.

6   Connect the string or wire to the new cable and pull it back into the engine compartment.

7   Connect the cable and install the latch screws and trim panel.

11   Front fender liner - removal and installation

Refer to illustration 11.2

1   Raise the vehicle, support it securely on jackstands and remove the front wheel.

11.2 Inner front fender mounting screws (arrows)

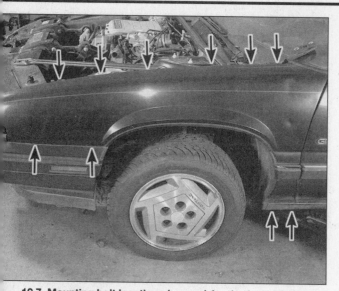

12.7 Mounting bolt locations (arrows) for the front fender

14.3 After removing the cover, the door lock rod linkage is easily accessible on coupe models

Remove the liner retaining screws and the nut **(see illustration)**.

Once all of the retainers are removed, detach the liner and remove it from the vehicle.

To install, place the liner in position and install the screws and nut.

2 Front fender - removal and installation

Refer to illustration 12.7

Loosen the wheel lug nuts, raise the front of the vehicle and support it securely on jackstands. Remove the wheel.

Remove the hood (see Section 9).

If removing the left fender, remove the following components as required:

Left strut tower brace
Windshield washer reservoir
(some models)
Battery cover
Vacuum tank

If removing the right fender, remove the following components as required:

Coolant recovery reservoir (some models)
Convenience center (some models)
Windshield washer reservoir (some models)

For either fender, remove the following components as required:

Headlamp access panel
Turn signal lamp
Rocker panel and/or rocker panel finish molding
Fender liner (see Section 11)
Fender-to-fascia bolts
Fender-to-headlamp panel bolts

Disconnect any wiring harness connectors and other components that would interfere with fender removal.

Remove the fender mounting bolts **(see illustration)**.

8   Detach the fender. It is a good idea to have an assistant support the fender while it's being moved away from the vehicle to prevent damage to the surrounding body panels.

9   Installation is the reverse of removal.

10   Tighten all nuts, bolts and screws securely.

13 Door trim panel - removal and installation

Removal

1   Disconnect the negative cable from the battery. **Caution:** *On models equipped with the Theftlock audio system, be sure the lock-out feature is turned off before performing any procedure which requires disconnecting the battery.*

2   Remove all door trim panel retaining screws and door pull/armrest assemblies. On some models it will be necessary to pry out the trim panel plugs to access the trim panel retaining screws.

3   On manual window regulator equipped models, remove the window crank. On power regulator models, pry out the control switch assembly and unplug it.

4   On models with power mirrors, remove the mirror remote control (driver's side only).

5   Insert a putty knife between the trim panel and the door and disengage the retaining clips. Work around the outer edge until the panel is free.

6   Once all of the clips are disengaged, detach the trim panel, unplug any wire harness connectors and remove the trim panel from the vehicle.

7   Remove the door courtesy light or reflector, as necessary.

8   For access to the inner door, carefully peel back the plastic watershield and remove the energy absorber pad (if equipped).

Installation

9   Prior to installation of the door panel, be sure to reinstall any clips in the panel which may have come out during the removal procedure and remain in the door itself.

10   Plug in the wire harness connectors and place the panel in position in the door. Press the door panel into place until the clips are seated and install the armrest/door pulls. Install the manual regulator window crank or power window switch assembly.

14 Door lock assembly - removal and installation

Refer to illustrations 14.3 and 14.7

Removal

1   With the window glass in the full up position, remove the door trim panel and water shield (see Section 13).

2   Remove the energy absorber pad, if equipped.

3   Disconnect spring clips from the remote control connecting rods at the lock assembly **(see illustration)**. Note: *Some models are equipped with non-reusable locking rod clips. On these models it will be necessary to use side cutters to cut off the clips that attach to the latch and purchase new clips.*

4   Remove any electrical connectors.

5   If necessary, remove the door lock module and separate it from the lock by bending the retaining clips or removing the rivets, as applicable.

6   As necessary, remove the inner and outer door handles.

11

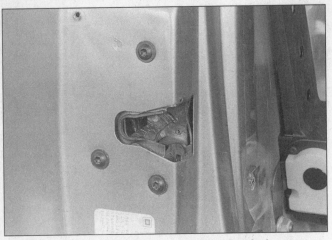

14.7 Use a Torx-head tool to remove the door lock screws

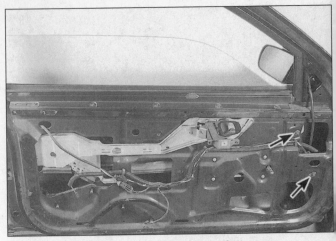

15.4 Forward run channel mounting screws

15.5a Front door glass removal

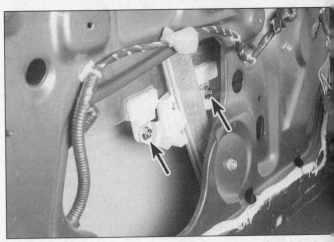

15.5b On later models raise the window just enough to access
the glass retaining bolts (arrows) through the hole in the
door frame

7   Remove the lock assembly-to-door screws and lift the assembly from the door **(see illustration)**.

Installation

8   Installation is the reverse of removal.

---

15   Door window glass - removal, installation and adjustment

1   With the window glass in the full up position, remove the door trim panel and water shield (Section 13).

Front door glass

Refer to illustrations 15.4, 15.5a and 15.5b

2   Remove the energy absorber pad, if equipped.

3   Pry the inner (and, if necessary, outer) belt line sealing strip out of the glass opening, then remove the run channel screws.

4   If necessary, remove front run channel **(see illustration)**.

5   Raise the glass half way up, then disengage the rear run channel, if necessary, by

pushing on the rear guide retainer. Tilt the glass inboard of the door frame to remove it **(see illustration)**. **Note:** *On some later models it will only be necessary to raise the window half way up to access and remove the glass retaining bolts, then tilt the glass forward and remove the glass from the door* **(see illustration)**.

6   Install the glass by inserting it into the door, then engaging the regulator arm roller to the sash channel.

7   Lower the glass to the half way up position and pull the glass rearward to engage the rear guide retainer to the run channel.

8   Raise the glass to about three inches above the door belt line and install the front run channel, if removed.

9   Engage the front clip on the glass in the front run channel and tighten the bolts securely.

10 The remainder of installation is the reverse of removal.

11 If adjustment is necessary, loosen the front run channel bolts, then lower the glass fully. Tighten the top bolt finger tight and the bottom bolt securely. Raise the glass to the Up position and tighten the top bolt securely.

Rear door glass

12 Remove the inner and outer door bel sealing strips.

13 Remove the energy absorber pad, i equipped.

14 If necessary, remove the run channe from the door frame at the front and rear o the glass division channel.

15 If necessary, remove the front weather strip from the door.

16 Remove the sash nuts or drill out the riv ets, as necessary, and lift the glass from th door. **Note:** *On some later models it will onl be necessary to raise the window half way up to access and remove the glass retaining bolts then tilt the glass forward and remove th glass from the door* **(see illustration 5.15b)**.

17 Insert the glass and sash into the doc and install the nuts. Tighten the nut securely. Connect the weatherstrip and div sion channel, making sure to engage th glass guide securely to the channel.

18 Install the sealing strips.

19 The remainder of installation is th reverse of removal.

20 The rear door glass is not adjustable.

17.2a Door handle cover mounting screw locations (arrows)

16 Window regulator - removal and installation

Removal

1   On power window equipped models, disconnect the negative cable at the battery. **Caution:** *On models equipped with the Theft-lock audio system, be sure the lockout feature is turned off before performing any procedure which requires disconnecting the battery.*

2   On manual window glass regulator equipped models, remove the handle by pressing the bearing plate and door trim panel in and, with a piece of hooked wire, pulling off the spring clip. A special tool is available for this purpose but its use is not essential. With the clip removed, take off the handle and the bearing plate.

3   With the window glass in the full up position, remove the door trim panel and water shield (see Section 13).

4   Secure the window glass in the Up position with strong adhesive tape fastened to the glass and wrapped over the door frame place a rag over the top of the door so the

tape doesn't contact the weatherstrip).

5   Punch out the center pins of the rivets that secure the window regulator and drill the rivets out with a 1/4-inch drill bit (if equipped).

6   On power window equipped models, unplug the electrical connector.

7   Remove the retaining bolts and move the regulator until it is disengaged from the sash channel. Lift the regulator from the door.

Installation

8   Place the regulator in position in the door and engage it in the sash channel.

9   Secure the regulator to the door using 3/16-inch rivets and a rivet tool (if equipped).

10   Install the bolts and tighten them securely.

11   Plug in the electrical connector (if equipped).

12   Install the water shield, door trim panel and window regulator handle. Connect the negative battery cable.

17 Door handles - removal and installation

Refer to illustrations 17.2a, 17.2b and 17.9

1   With the window glass in the full up position, remove the door trim panel, the energy absorber pad, if equipped, and the water shield (see Section 13).

Outside handle

Coupe models

2   On early models remove the cover assembly, if equipped **(see illustrations)**. Remove the retaining nut or drill out the rivets, as required, then detach the lock rods and remove the handle from the door.

3   On later models remove the lock rods from the latch and remove the handle retaining nuts through the opening in the door frame. Rotate the lock cylinder pawl for clearance and remove the door handle

4   Installation is the reverse or removal.

Sedan models

5   If you're working on a rear door handles, remove the window glass (see Section 15).

6   Pry the remote rod out of the handle with a small screwdriver, remove the nuts or drill out the rivets, as required, and lift the handle off.

7   Installation is the reverse of removal.

Inside handle

8   Disconnect the rod from the handle.

9   Remove the bolt or punch out the center pins of the rivets that secure the window regulator and drill the rivets out with a 3/16-inch drill bit **(see illustration)**.

10   Lift the handle from the door.

11   To install, place the handle in position and secure it to the door, using 3/16-inch rivets and a rivet tool.

12   The remainder of installation is the reverse of removal.

18 Door lock cylinder - removal and installation

1   With the window glass in the full up position, remove the door trim panel, the energy absorber pad, if equipped, and water shield (see Section 13).

2   If necessary, disconnect the rod from the lock cylinder and remove the anti-theft shield, if equipped.

3   Disconnect the electrical connector, if necessary.

4   Use a screwdriver to pry the retainer off and withdraw the lock cylinder from the door **Note:** *On some models it will be necessary to remove the outside door handle first (see Section 17), then pry the lock cylinder retaining clip off to remove it from the door handle..*

5   Installation is the reverse of removal.

19 Door lock striker - removal and installation

1   Mark the position of the striker bolt on

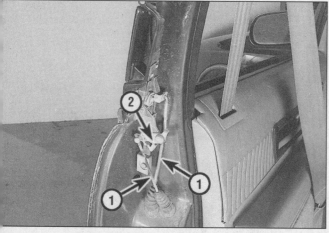

17.2b Door handle linkage details

1   *Lock rod*   2   *Linkage*

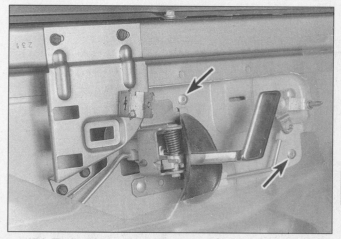

17.9 The inside handle assembly is retained to the door by rivets (arrows)

20.3a Typical door check link details

1	Check link assembly	3	Door
2	Bolt	4	Sealing grommet

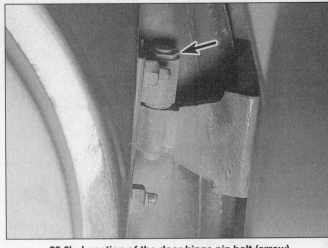

20.3b Location of the door hinge pin bolt (arrow)

**21.2 Typical trunk lid mounting bolt
locations (arrows)**

the door pillar with a pencil or marking pen.
2   On most models, it will be necessary to
use a special tool to fit the star-shaped
recess in the striker bolt head. Unscrew the
bolt and remove it.
3   To install, screw the lock striker bolt into
the tapped cage plate in the door pillar and

tighten it finger tight at the marked position.
Tighten the bolt securely.

20 Door - removal and installation

Refer to illustrations 20.3a and 20.3b
1   Unplug any electrical connectors that
connect the electrical accessories and lights
in the door to the dash wiring harness. Some
of these may be accessible by removing the
door panel - others may be tucked away
behind the kick panels under the dash or
behind the center pillar trim panels.
2   Open the door all the way and support it
on jacks or blocks covered with cloth or pads
to prevent damage to the paint.
3   On earlier models remove the door
check link bolt and the hinge pin bolts, then
with the help of an assistant, lift the door
away **(see illustrations)**.
4   On later models remove the hinge-to-
door bolts and remove the door.
5   Install the door by reversing the removal
procedure. On early models apply thread
locking compound to the threads, install the
hinge pin bolts and tighten the hinge pin bolts
securely.

21 Trunk lid - removal and installation

Refer to illustration 21.2
1   Open the trunk lid and unplug any elec
trical connectors, then disconnect the
solenoid (if equipped).
2   With an assistant supporting the trunk lid
detach the upper ends of the trunk suppor
struts if equipped **(see illustrations 9.4a and
9.4b)**, then remove retaining bolts and lift the
trunk lid from the vehicle **(see illustration)**.
3   Installation is the reverse of removal.

22 Trunk lock cylinder - removal and
installation

Refer to illustrations 22.3a and 22.3b
1   Open the trunk lid.
2   If necessary, remove the trunk lid rea
garnish by removing the bolts from inside th
trunk lid. Remove the back-up lamp socket
too, if applicable.
3   Pry the retaining clip off and withdraw
the lock cylinder from the vehicle **(see illus**

**22.3a Pry the retainer clip from the trunk
lock cylinder**

**22.3b Typical trunk
latch and lock cylinder
mounting details**

1	Screw or rivet
2	Lock cylinder retaining clip
3	Solenoid mounting bolts
4	Solenoid
5	Latch mounting bolts
6	Latch

23.4 Typical trunk latch striker mounting nuts (arrows)

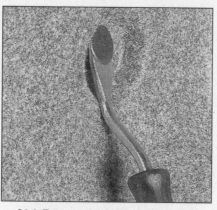

24.4 Remove the fasteners from the trim panel using a special tool to prevent breakage

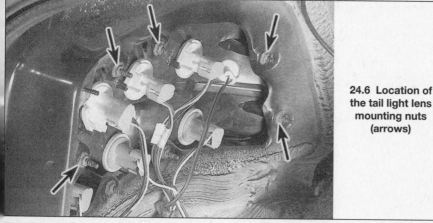

24.6 Location of the tail light lens mounting nuts (arrows)

tration). Disconnect the actuator cable (if equipped). Some retaining clips will be secured with a rivet which must be drilled out with a 5/32-inch drill bit **(see illustration)**.

4   To install, place the lock cylinder in place and secure it with the retaining clip.

23   Trunk latch and striker - removal and installation

Refer to illustration 23.4

1   Disconnect the negative cable from the battery. **Caution:** *On models equipped with the Theftlock audio system, be sure the lockout feature is turned off before performing any procedure which requires disconnecting the battery.*

2   On models so equipped, disconnect the trunk lock cylinder cable by inserting a small screwdriver into the connector to hold the release tab down and then pull the cable out of the solenoid.

3   Remove the electronic solenoid (if equipped) and unbolt and remove the latch and (if equipped) the ajar switch.

4   Remove the retaining nuts and lift off the striker **(see illustration)**.

5   Installation is the reverse of removal.

24   Rear lens assembly - removal and installation

Refer to illustrations 24.4 and 24.6

1   Disconnect the negative cable at the battery. **Caution:** *On models equipped with the Theftlock audio system, be sure the lockout feature is turned off before performing any procedure which requires disconnecting the battery.*

2   Open the trunk lid.

3   Remove the convenience net, if equipped.

4   Remove the rear compartment trim panel **(see illustration)**.

5   Remove the bulb and socket assemblies, if necessary.

6   Unscrew the plastic wing nuts, pull the lens assembly out and lean it back **(see illustration)**. If not already removed, disconnect the bulb holders (see Chapter 12) and lift the assembly from the vehicle.

7   Installation is the reverse of removal.

25   Console - removal and installation

Refer to illustrations 25.2 and 25.4

1   Disconnect the negative cable at the battery. **Caution:** *On models equipped with the Theftlock audio system, be sure the lockout feature is turned off before performing any procedure which requires disconnecting the battery.*

2   Remove the shift handle trim, the coin holder, storage compartment, armrest and shift knob and any other components which would block access to the console bolts **(see illustration)**.

3   Remove the retaining screw or pry the clip out of the automatic transaxle shift handle and manual transaxle shift handles. Pull the handle or knob off the shift lever.

4   Remove the bolts, then pull the console

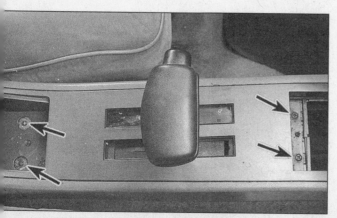

25.2 Console trim mounting screws (arrows)

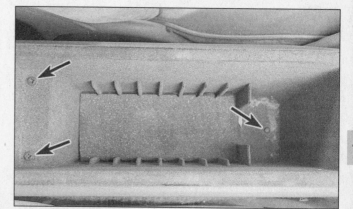

25.4 Typical console mounting screws (arrows)

up **(see illustration)**.

5   Lift the console up for access and unplug the electrical connectors.

6   Installation is the reverse of removal.

26   Seats - removal and installation

Front seat

1   Move the seat all the way forward.

2   Remove the seat track covers and pull the carpet away from the adjuster and retaining nuts.

3   Remove the seat adjuster-to-floor panel retaining nuts.

4   Move the seat all the way to the rear.

5   Remove the front retaining nuts. On power seats, unplug the electrical connector. Lift the seat from the vehicle.

6   Installation is the reverse of removal.

Rear seat

7   Remove the seat cushion retaining bolts, detach the seat cushion and remove it from the vehicle.

8   Remove the nuts securing the bottom of the seatback, then swing the seatback up to disengage it from its upper retainers.

9   Installation is the reverse of removal.

27   Outside mirror - removal and installation

Refer to illustration 27.3

1   Remove the door trim panel (see Section 13) and the mirror reinforcement panel.

2   On power mirrors, unplug the electrical connector.

3   Remove the three nuts and lift off the mirror assembly **(see illustration)**.

4   Installation is the reverse of removal.

28   Seat belt check

1   Check the seat belts, buckles and guide loops for obvious damage and signs of wear.

2   Passive restraint type seat belts on some models are designed to lock up during a sudden stop or impact, yet allow free movement during normal driving. Verify that

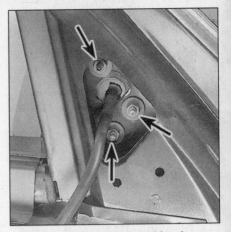

27.3 Location of the outside mirror mounting nuts (arrows)

the retractors return the belt against your chest while driving and rewind the belt fully when the buckle is unlatched.

3   If any of the above checks reveal problems with the seat belt system, replace parts as necessary.

Chapter 12
Chassis electrical system

Contents

General information

The electrical system is a 12-volt, negative ground type. Power for the lights and all electrical accessories is supplied by a lead/acid-type battery which is charged by the alternator.

This Chapter covers repair and service procedures for the various electrical components not associated with the engine. Information on the battery, alternator, ignition system and starter motor can be found in Chapter 5.

It should be noted that when portions of the electrical system are serviced, the negative battery cable should be disconnected from the battery to prevent electrical shorts and/or fires.

2 Electrical troubleshooting - general information

A typical electrical circuit consists of an electrical component, any switches, relays, motors, fuses, fusible links or circuit breakers related to that component and the wiring and connectors that link the component to both the battery and the chassis. To help you pinpoint an electrical circuit problem, wiring diagrams are included at the end of this book.

Before tackling any troublesome electrical circuit, first study the appropriate wiring diagrams to get a complete understanding of what makes up that individual circuit. Trouble spots, for instance, can often be narrowed down by noting if other components related to the circuit are operating properly. If several components or circuits fail at one time, chances are the problem is in a fuse or ground connection, because several circuits are often routed through the same fuse and ground connections.

Electrical problems usually stem from simple causes, such as loose or corroded connections, a blown fuse, a melted fusible link or a bad relay. Visually inspect the condition of all fuses, wires and connections in a problem circuit before troubleshooting it.

If testing instruments are going to be utilized, use the diagrams to plan ahead of time where you will make the necessary connections in order to accurately pinpoint the trouble spot.

The basic tools needed for electrical troubleshooting include a circuit tester or voltmeter (a 12-volt bulb with a set of test

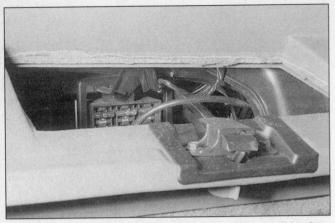

3.1a The passenger compartment fuse block is accessible after lifting out the glove box insert

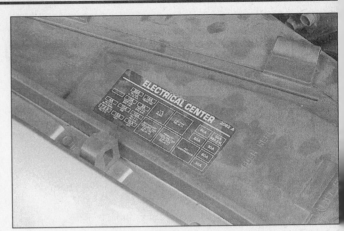

3.1b This fuse block is located under a cover in the right front corner of the engine compartment

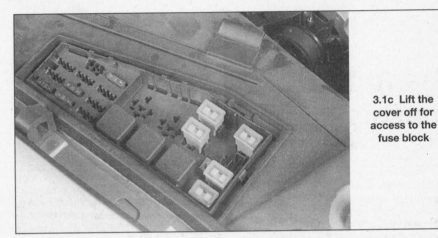

3.1c Lift the cover off for access to the fuse block

leads can also be used), a continuity tester, which includes a bulb, battery and set of test leads, and a jumper wire, preferably with a circuit breaker incorporated, which can be used to bypass electrical components. Before attempting to locate a problem with test instruments, use the wiring diagram(s) to decide where to make the connections.

Voltage checks

Voltage checks should be performed if a circuit is not functioning properly. Connect one lead of a circuit tester to either the negative battery terminal or a known good ground. Connect the other lead to a connector in the circuit being tested, preferably nearest to the battery or fuse. If the bulb of the tester lights, voltage is present, which means that the part of the circuit between the connector and the battery is problem free. Continue checking the rest of the circuit in the same fashion. When you reach a point at which no voltage is present, the problem lies between that point and the last test point with voltage. Most of the time the problem can be traced to a loose connection. **Note:** *Keep in mind that some circuits receive voltage only when the ignition key is in the Accessory or Run position.*

Finding a short

One method of finding shorts in a circuit is to remove the fuse and connect a test light or voltmeter in its place to the fuse terminals. There should be no voltage present in the circuit. Move the wiring harness from side–to–side while watching the test light.

If the bulb goes on, there is a short to ground somewhere in that area, probably where the insulation has rubbed through. The same test can be performed on each component in the circuit, even a switch.

Ground check

Perform a ground test to check whether a component is properly grounded. Disconnect the battery and connect one lead of a selfpowered test light, known as a continuity tester, to a known good ground. Connect the

other lead to the wire or ground connection being tested. If the bulb goes on, the ground is good. If the bulb does not go on, the ground is not good.

Continuity check

A continuity check is done to determine if there are any breaks in a circuit - if it is passing electricity properly. With the circuit off (no power in the circuit), a self–powered continuity tester can be used to check the circuit. Connect the test leads to both ends of the circuit (or to the "power" end and a good ground), and if the test light comes on the circuit is passing current properly. If the light doesn't come on, there is a break somewhere in the circuit. The same procedure can be used to test a switch, by connecting the continuity tester to the switch terminals. With the switch turned On, the test light should come on.

Finding an open circuit

When diagnosing for possible open circuits, it is often difficult to locate them by sight because oxidation or terminal misalignment are hidden by the connectors. Merely wiggling a connector on a sensor or in the wiring harness may correct the open circuit condition. Remember this when an open circuit is indicated when troubleshooting a cir-

cuit. Intermittent problems may also be caused by oxidized or loose connections.

Electrical troubleshooting is simple if you keep in mind that all electrical circuits are basically electricity running from the battery through the wires, switches, relays, fuses and fusible links to each electrical component (light bulb, motor, etc.) and to ground, from which it is passed back to the battery. Any electrical problem is an interruption in the flow of electricity to and from the battery.

3   Fuses - general information

Refer to illustrations 3.1a, 3.1b, 3.1c and 3.3

The electrical circuits of the vehicle are protected by a combination of fuses, circuit breakers and fusible links. The passenger compartment fuse block is located under the instrument panel on the right side of the dashboard. It is accessible after lifting out the glove box **(see illustration)**. A second fuse block is located at the right front corner of the engine compartment, under a cover **(see illustrations)**. Some models may also have other fuse blocks in the engine compartment.

Each of the fuses is designed to protect a specific circuit, and the various circuits are identified on the fuse panel itself.

Miniaturized fuses are employed in the

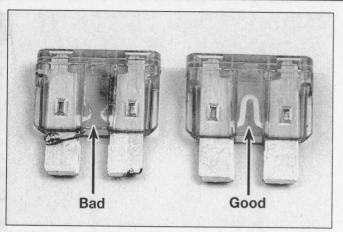

3.3 When a fuse blows, the element between the terminals melts - the fuse on the left is blown, the fuse on the right is good

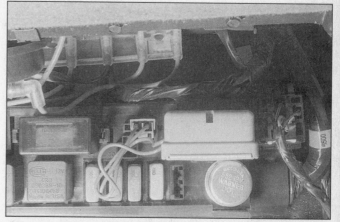

6.2 The convenience center is located under the right side of the dashboard and houses relays and other components

fuse block. These compact fuses, with blade terminal design, allow removal and replacement by hand. If an electrical component fails, always check the fuse first. The easiest way to check fuses is with a test light. Check for power at the exposed terminal tips of each fuse. If power is available on one side of the fuse but not the other, the fuse is blown. A blown fuse can also be confirmed by visually inspecting it **(see illustration)**.

Be sure to replace blown fuses with the correct type. Fuses of different ratings are physically interchangeable, but only fuses of the proper rating should be used. Replacing a fuse with one of a higher or lower value than specified is not recommended. Each electrical circuit needs a specific amount of protection. The amperage value of each fuse is molded into the fuse body.

If the replacement fuse immediately fails, don't replace it again until the cause of the problem is isolated and corrected. In most cases, the cause will be a short circuit in the wiring caused by a broken or deteriorated wire.

4 Fusible links - general information

Some circuits are protected by fusible links or fusible elements. The fusible elements used on some models are similar to a fuse and are located in the engine compartment fuse box. They are replaced in the same manner as a fuse. The fusible links are used in circuits which are not ordinarily fused, such as the ignition circuit.

Although the fusible links appear to be a heavier gauge than the wire they are protecting, the appearance is due to the thick insulation. All fusible links are several wire gauges smaller than the wire they are designed to protect.

Fusible links cannot be repaired, but a new link of the same size wire can be put in its place. The procedure is as follows:

a) *Disconnect the negative cable from the*

battery. **Caution:** *On models equipped with the Theftlock audio system, be sure the lockout feature is turned off before performing any procedure which requires disconnecting the battery.*
b) *Disconnect the fusible link from the wiring harness.*
c) *Cut the damaged fusible link out of the wiring just behind the connector.*
d) *Strip the insulation back approximately 1/2-inch.*
e) *Position the connector on the new fusible link and crimp it into place.*
f) *Use rosin core solder at each end of the new link to obtain a good solder joint.*
g) *Use plenty of electrical tape around the soldered joint. No wires should be exposed.*
h) *Connect the battery ground cable. Test the circuit for proper operation.*

5 Circuit breakers - general information

Circuit breakers protect components such as power windows, power door locks and headlights. Some circuit breakers are located in the fuse block (see Section 3) and the convenience center (see Section 6).

On some models the circuit breaker resets itself automatically, so an electrical overload in a circuit breaker protected system will cause the circuit to fail momentarily, then come back on. If the circuit doesn't come back on, check it immediately. Once the condition is corrected, the circuit breaker will resume its normal function. Some circuit breakers must be reset manually.

6 Relays - general information

Refer to illustration 6.2

Several electrical accessories in the vehicle use relays to transmit the electrical signal to the component. If the relay is defective, that component will not operate properly.

The various relays are grouped together in several locations. Some relays are grouped together in the convenience center which is located under the right side of the dashboard behind the sound insulator panel **(see illustration)**.

If a faulty relay is suspected, it can be removed and tested by a dealer service department or a repair shop. Defective relays must be replaced as a unit.

7 Turn signal and hazard flashers - check and replacement

Refer to illustration 7.1

Turn signal flasher

1   The turn signal flasher is a small canister–shaped or rectangular unit located under the dash, either in a clip adjacent to the steering column **(see illustration)**, or in the convenience center **(see illustration 6.2)**. **Note:** *Later models are equipped with a single combination turn signal/hazard flasher that is located behind the left kick panel.*.
2   When the flasher unit is functioning properly, an audible click can be heard during

7.1 The turn signal flasher on some models is clipped to the steering column bracket

8.3a Remove the snap-ring and lift the airbag coil off

8.3b Use a special tool (available at most auto parts stores) to compress the lock plate for access to the retaining ring

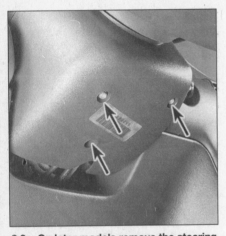

8.3c On later models remove the steering column cover screws (arrows) and detach the covers (lower cover shown)

8.4 Remove the cancel cam assembly (arrow)

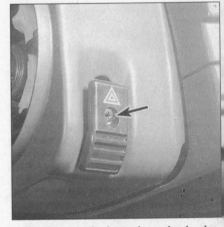

8.5 Remove the hazard warning knob screw (arrow) and the knob

its operation. If the turn signals fail on one side or the other and the flasher unit does not make its characteristic clicking sound, a faulty turn signal bulb is indicated.

3   If both turn signals fail to blink, the problem may be due to a blown fuse, a faulty flasher unit, a broken switch or a loose or open connection. If a quick check of the fuse box indicates that the turn signal fuse has blown, check the wiring for a short before installing a new fuse.

4   To replace the flasher, simply unplug it and pull it out of the clip.

5   Make sure that the replacement unit is identical to the original. Compare the old one to the new one before installing it.

6   Installation is the reverse of removal.

Hazard flasher

7   The hazard flasher, a small canister–shaped or rectangular unit located under the dash, either in a clip adjacent to the steering **(see illustration 7.1)** or in the convenience center **(see illustration 6.2)**, flashes all four turn signals simultaneously when activated.

8   The hazard flasher is checked in a fashion similar to the turn signal flasher (see

Steps 2 and 3).

9   To replace the hazard flasher, pull it from it's retainer or the convenience center. **Note:** *Later models are equipped with a single combination turn signal/hazard flasher that is located behind the left kick panel.*

10   Make sure the replacement unit is identical to the one it replaces. Compare the old one to the new one before installing it.

11   Installation is the reverse of removal.

8   Turn signal switch assembly - removal and installation

Warning: *Some models covered by this manual are equipped with Supplemental Inflatable Restraint (SIR) systems, more commonly known as airbags. Always disable the airbag system before working in the vicinity of any airbag system components to avoid the possibility of accidental deployment of the airbag(s) which could cause personal injury (see Section 22). The yellow wiring harnesses and connectors routed through the console and instrument panel are for this system. Do not use electrical test equipment on any of the airbag system wiring or tamper with it in any way.*

Removal

Refer to illustrations 8.3a, 8.3b, 8.3c, 8.4, 8.5, 8.8, 8.12 and 8.13

1   Detach the cable from the negative battery terminal and disable the airbag system, if equipped (see Section 22). **Caution:** *On models equipped with the Theftlock audio system, be sure the lockout feature is turned off before performing any procedure which requires disconnecting the battery.*

2   Remove the steering wheel (see Chapter 10).

3   Remove the airbag coil retaining snapring and remove the coil assembly. Let the coil hang by the wiring harness. Remove the wave washer. Using a lock plate removal tool, depress the lock plate for access to the retaining ring **(see illustrations)**. Use a small screwdriver to pry the retaining ring out of the groove in the steering column and remove the lock plate. **Note:** *On later models it is not necessary to remove the airbag coil assembly, To remove the switch on these models simply remove the retaining screws from the steering column upper and lower covers and remove the covers* **(see illustration)**. Then detach the switch electrical connectors and remove the switch mounting screws.

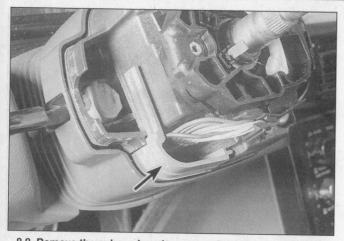

8.8 Remove the column housing cover plastic spacer (arrow) - don't forget this piece when reassembling the column housing

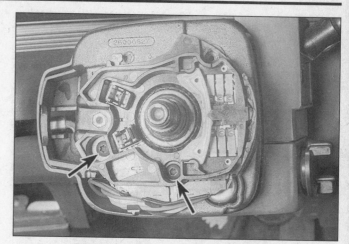

8.12 Remove the two screws (arrows) . . .

4   On earlier models remove the cancel cam assembly **(see illustration)**.

5   Remove the hazard warning knob **(see illustration)**.

6   Remove the column housing cover screw (which is longer and blue in color). Unplug the electrical connector behind the turn signal lever, grasp the lever and pull it straight out of the housing.

7   Detach the wiring protector from under the steering column, separate the wiper pulse switch wires, then unplug the connector. Remove the pulse switch pivot pin, then pull the switch wires up through the column and remove switch and wires.

8   Remove the plastic spacer in the lower left corner of the housing **(see illustration)**. Don't forget this spacer during reassembly.

9   Remove the under–dash panel below the steering column.

10   Remove the steering column bracket bolts and lower the steering column from the dash.

11   Locate the turn signal switch electrical connector. Detach the wires to the buzzer switch assembly (tan and black wire and light green wire) from the back of the connector.

12   Remove the turn signal switch mounting screws **(see illustration)**.

13   Pull the wiring harness and electrical connector up through the steering column and remove the switch assembly **(see illustration)**.

Installation

Refer to illustrations 8.18 and 8.19

14   On later models installation is the reverse of removal. Make sure the airbag coil is centered as described in Step 18 before installing the steering wheel. On earlier models feed the turn signal switch connector and wiring harness down through the column. Use a section of mechanics wire to pull it through, if necessary.

15   Plug in the connector and replace the wiring protector.

16   Seat the turn signal switch on the column and install the turn signal switch mount-

8.13 . . . and detach the turn signal switch assembly

ing screws and lever arm. Install the hazard knob and multi-function lever. Press the multi-function lever straight in until it snaps in place.

17   Install the cancel cam and the lock plate. Depress the lock plate and install the retaining ring.

18   If necessary center the airbag coil as follows (it will only become uncentered if the spring lock is depressed and the hub rotated with the coil off the column) **(see illustration)**:

a)   Turn the coil over and depress the spring lock.

b)   Rotate the hub in the direction of the arrow until it stops.

c)   Rotate the hub in the opposite direction 2-1/2 turns and release the spring lock.

19   Install the wave washer and the airbag coil **(see illustration)**. Pull the slack out of airbag coil lower wiring harness to keep it tight through the steering column, or it may be cut when the steering wheel is turned. Install the airbag coil retaining snap-ring.

20   The remainder of installation is the reverse of removal.

24017-12-8.14 HAYNES

8.18 To center the airbag coil, depress the spring lock (arrow), rotate the hub in the direction of the arrow until it stops, then back the hub off 2-1/2 turns and release the spring lock

8.19 When properly installed, the airbag coil will be centered with the marks aligned (circle) and the tab (arrow) fitted between the projections on the top of the steering column

1

10.3 Location of the headlight mounting bolts (arrows)

10.4 Unplug the electrical connector

9   Ignition switch key lock cylinder - replacement

Warning: *Some models covered by this manual are equipped with Supplemental Inflatable Restraint (SIR) systems, more commonly known as airbags. Always disable the airbag system before working in the vicinity of any airbag system components to avoid the possibility of accidental deployment of the airbag(s) which could cause personal injury (see Section 22). The yellow wiring harnesses and connectors routed through the console and instrument panel are for this system. Do not use electrical test equipment on any of the airbag system wiring or tamper with it in any way.*

1996 and earlier models

Fixed steering column

1   Detach the cable from the negative battery terminal. **Caution:** *On models equipped with the Theftlock audio system, be sure the lockout feature is turned off before performing any procedure which requires disconnecting the battery.*
2   Remove the steering wheel (see Chapter 10).
3   Remove the turn signal switch assembly (see Section 8).
4   Place the lock cylinder in the Run position.
5   Remove the turn signal switch housing screws.
6   Remove the turn signal switch housing and steering shaft assembly as a complete unit.
7   Using a screwdriver, lift the switch tab, then pull gently on the buzzer switch wires and remove the buzzer switch.
8   Place the lock cylinder in the Accessory position.
9   Remove the lock cylinder retaining screw.
10   Remove the lock cylinder.
11   Installation is the reverse of removal.

10.5a Remove the headlight bulb holder by turning it clockwise . . .

Tilt steering column

12   Take the vehicle to a dealer service department or other repair shop. A special tool is needed to remove the pivot pins in the steering column housing on tilt steering columns, making key lock cylinder replacement impossible for the home mechanic.

1997 and later models

13   Detach the cable from the negative battery terminal. **Caution:** *On models equipped with the Theftlock audio system, be sure the lockout feature is turned off before performing any procedure which requires disconnecting the battery.*
14   Remove the steering wheel (see Chapter 10).
15   Remove the steering column covers **(see illustration 8.3c)**.
16   Unplug the turn signal switch connectors from the bulk head connector. Disconnect the key alarm connector if equipped.
17   Remove the ignition switch housing screws and detach the ignition switch/alarm module from the steering column.
18   To remove the lock cylinder turn the key to the RUN position. Insert a 1/16 allen

10.5b . . . and pull it out of the headlight assembly (don't try to pull the bulb out of the holder - they're sold as a single unit)

wrench into the hole at the top of the housing and depress the lock retaining pin. Then remove the key lock cylinder from the steering column.
19   Installation is the reverse of removal.

10   Headlight bulb - replacement

Refer to illustrations 10.3, 10.4, 10.5a and 10.5b

Warning: *Halogen gas filled bulbs are under pressure and may shatter if the surface is scratched or the bulb is dropped. Wear eye protection and handle the bulbs carefully, grasping only the base whenever possible. Do not touch the surface of the bulb with your fingers because the oil from your skin could cause it to overheat and fail prematurely. If you do touch the bulb surface, clean it with rubbing alcohol.*

1   Detach the cable from the negative terminal of the battery. **Caution:** *On models equipped with the Theftlock audio system, be sure the lockout feature is turned off before performing any procedure which requires disconnecting the battery.*

11.1 Headlight adjustment screw locations (arrows)

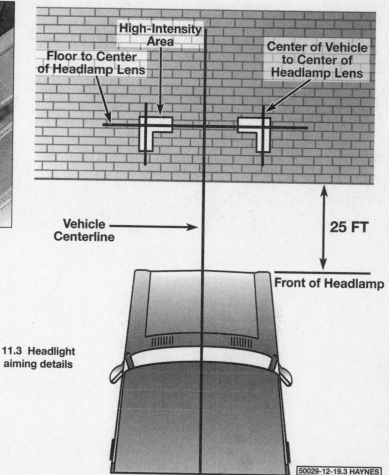

11.3 Headlight aiming details

2   If equipped, raise the flap or remove the plastic cover over the headlight for access to the bulbs.
3   On some models it may be necessary to remove the screws or nuts and pull the headlight assembly forward **(see illustration)**.
4   Unplug the electrical connector **(see illustration)**.
5   Twist the bulb lock ring counterclockwise **(see illustration)** and pull the bulb holder assembly out of the headlight **(see illustration)**.
6   Installation is the reverse of removal.

11   Headlights - adjustment

Refer to illustrations 11.1 and 11.3
Note: *The headlights must be aimed correctly. If adjusted incorrectly they could blind the driver of an oncoming vehicle and cause a serious accident or seriously reduce your ability to see the road. The headlights should be checked for proper aim every 12 months and any time a new headlight is installed or front end body work is performed. It should be emphasized that the following procedure is only an interim step which will provide temporary adjustment until the headlights can be adjusted by a properly equipped shop.*
1   Headlights have two spring loaded adjusting screws, one on the top controlling up-and-down movement and one on the side controlling left-and-right movement **(see illustration)**.
2   There are several methods of adjusting the headlights. The simplest method requires a blank wall 25–feet in front of the vehicle and a level floor.
3   Position masking tape vertically on the wall in reference to the vehicle centerline and the centerlines of both headlights **(see illustration)**.
4   Position a horizontal tape line in reference to the centerline of all the headlights.
Note: *It may be easier to position the tape on the wall with the vehicle parked only a few inches away.*
5   Adjustment should be made with the

vehicle sitting level, the gas tank half–full and no unusually heavy load in the vehicle.
6   Starting with the low beam adjustment, position the high intensity zone so it's two inches below the horizontal line and two inches to the side of the headlight vertical line, away from oncoming traffic. Adjustment is made by turning the top adjusting screw clockwise to raise the beam and counterclockwise to lower the beam. The adjusting screw on the side should be used in the same manner to move the beam left or right.
7   With the high beams on, the high intensity zone should be vertically centered with the exact center just below the horizontal line.
Note: *It may not be possible to position the*

headlight aim exactly for both high and low beams. If a compromise must be made, keep in mind that the low beams are the most used and have the greatest effect on driver safety.
8   Have the headlights adjusted by a dealer service department or service station at the earliest opportunity.

12   Bulb replacement

Refer to illustrations 12.1 and 12.2
1   The lenses of many lights are held in place by screws, which makes it a simple procedure to gain access to the bulbs **(see illustration)**.

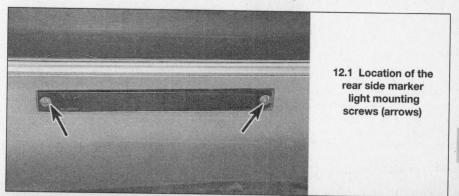

12.1 Location of the rear side marker light mounting screws (arrows)

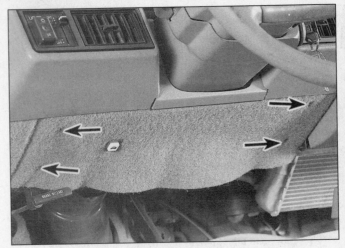

12.2 Press the two tabs (arrows) on the front sidemarker light assembly, then slide it forward for access to the bulb

13.6 Steering column trim panel mounting screws

2   On some lights, the lenses are held in place by tabs. Simply pop them off with your fingers or pry them off with a small screwdriver **(see illustration)**.

3   Several types of bulbs are used. Some are removed by pushing in and turning them counterclockwise; others can simply be pulled straight out of the socket.

4   To gain access to the instrument panel lights, the instrument cluster will have to be removed first (see Section 15).

13   Radio and speakers - removal and installation

1   Detach the cable from the negative terminal of the battery prior to performing any of the following procedures. **Caution:** *On models equipped with the Theftlock audio system, be sure the lockout feature is turned off before performing any procedure which requires disconnecting the battery.*

Radio

Early models

Refer to illustrations 13.6

2   On these models, the radio receiver is mounted under the right side of the dashboard and is operated remotely by the radio control panel.

3   Remove the right side sound insulator or, on some models, the glove compartment.

4   Remove the radio receiver mounting nuts/screws. Lower the radio receiver from the dash

5   Reach up behind the receiver and unplug the antenna, speaker, power and ground connectors from the radio.

6   To remove the radio control panel it will be necessary to remove the steering column trim panel **(see illustration)**, the radio control knobs, instrument panel trim plate and radio control panel and on some models the heater and air conditioner control assembly as a single unit.

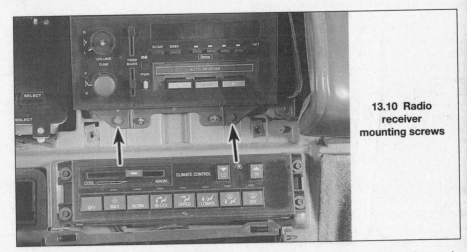

13.10 Radio receiver mounting screws

7   Once the assembly is removed detach the radio from the heater and air conditioner control assembly.

8   Installation is the reverse of removal.

Late models

Refer to illustrations 13.10

9   On these models remove the accessory trim panel surrounding the radio and on some models the instrument cluster bezel.

10   Remove the mounting screws and the electrical connectors at the rear of the radio and the antenna lead **(see illustration)**. detach the radio from the instrument panel.

11   Installation is the reverse of removal.

Speakers

Dash mounted speaker

12   Remove the instrument panel pad (see Section 15).

13   Remove the speaker mounting screws.

14   Lift the speaker out of its enclosure, unplug the electrical connector and remove the speaker.

15   Installation is the reverse of removal.

Door mounted speaker

16   Remove the door trim panel (see Chapter 11).

17   Remove the screws securing the speakers to the door.

18   Disconnect the electrical connector and remove the speaker.

19   Installation is the reverse of removal.

Rear speaker

20   Remove the rear seat back and the seat back–to–window trim panel.

21   Remove the screws securing the speaker housing (coupe models) or the speaker itself (sedan models). Remove the speaker and detach the electrical connector. On coupe models, remove the screws and separate the speaker from the housing.

22   Installation is the reverse of removal.

14   Radio antenna - removal and installation

Refer to illustration 14.2

1   Disconnect the negative battery cable. **Caution:** *On models equipped with the Theftlock audio system, be sure the lockout feature is turned off before performing any procedure which requires disconnecting the battery.*

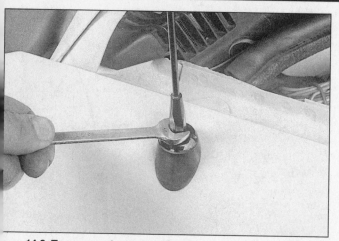

14.2 To remove the standard antenna, simply unscrew it

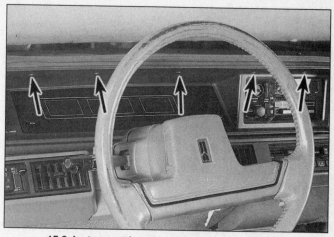

15.2 Instrument bezel mounting screws (arrows)

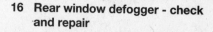

15.4 Instrument cluster mounting screws (arrows)

Standard antenna

Simply use a small wrench to unscrew the antenna from the base and screw on a new antenna **(see illustration)**.

Power antenna

Open the trunk and remove the trim panel.

Unscrew the escutcheon from the fender.

Unplug the electrical connector, disconnect the ground and relay cables, drain tube and antenna lead, then remove the antenna and motor assembly.

Installation is the reverse or removal.

15 Instrument panel cluster - removal and installation

Refer to illustrations 15.2 and 15.4

Warning: Some models covered by this manual are equipped with Supplemental Inflatable Restraint (SIR) systems, more commonly known as airbags. Always disable the airbag system before working in the vicinity of any airbag system components to avoid the possibility of accidental deployment of the airbag(s) which could cause personal injury. The yellow wiring harnesses and connectors routed through the console and instrument panel are for this system. Do not use electrical test equipment on any of the airbag system wiring or tamper with it in any way.

1   Detach the cable from the negative battery terminal. **Caution:** On models equipped with the Theftlock audio system, be sure the lockout feature is turned off before performing any procedure which requires disconnecting the battery.

2   On early models remove the instrument panel pad and bezel **(see illustration)**. On later models remove the instrument panel trim plate. On some later models it will be necessary to remove the floor console before removing the instrument panel trim plate.

3   On early models remove the steering column bracket and lower the steering column. On later models lower the tilt wheel to its lowest position and on column shift models place the gear selector lever in the L1 position.

4   Remove the instrument panel cluster **(see illustration)**.

5   Remove any switches or controls which will interfere with removal.

6   On some early models it may be necessary to remove the parking brake handle and (if equipped) center console (see Chapter 11).

7   Remove the radio, controls and speakers as necessary (see Section 13).

8   Remove the bolts and pull the instrument panel from its cavity in the dashboard, unplug the electrical connectors and remove it.

9   Installation is the reverse of removal.

16 Rear window defogger - check and repair

Refer to illustrations 16.5a and 16.5b

1   This option consists of a rear window with a number of horizontal elements baked into the glass surface during the glass forming operation.

2   Small breaks in the element can be successfully repaired without removing the rear window.

Check

3   To test the grids for proper operation, start the engine and turn on the system.

4   Ground one lead of a test light and carefully touch the other lead to each element line.

5   The brilliance of the test light should increase as the lead is moved across the element **(see illustrations)**. If the test light

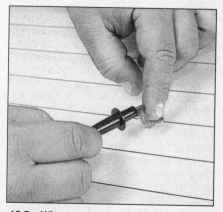

16.5a When measuring the voltage at the rear window defogger grid, wrap a piece of aluminum foil around the probe of the voltmeter and press the foil against the wire with your finger

12

16.5b To determine if the heating element has broken, check the voltage at the center of each element - if the voltage is 6-volts, the element is unbroken

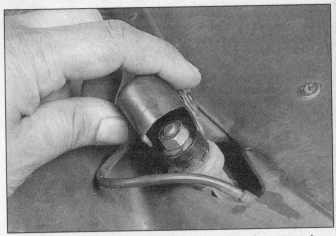

17.4 Lift the protective cap and remove the wiper arm nut

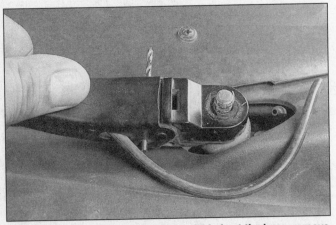

17.5 With a pin or drill bit inserted in the hole at the base, remove the wiper arm by grasping it securely and rocking it back-and-forth while lifting it off the shaft

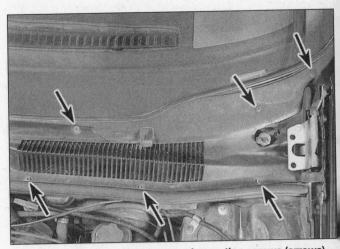

17.7 Air inlet panel and glass cowl mounting screws (arrows)

glows brightly at both ends of the lines, check for a loose ground wire. All of the lines should be checked in at least two places.

Repair

6   To repair a break in a line, it is recommended that a repair kit specifically for this purpose be purchased from a dealer parts department. Included in the repair kit will be a decal, a container of silver plastic and hardener, a mixing stick and instructions.

7   To repair a break, first turn off the system and allow it to de–energize for a few minutes.

8   Lightly buff the element area with fine steel wool, then clean it thoroughly with alcohol.

9   Use the decal supplied in the repair kit or apply strips of electrician's tape above and below the area to be repaired. The space between the pieces of tape should be the same width as the existing lines. This can be checked from outside the vehicle. Press the tape tightly against the glass to prevent seepage.

10   Mix the hardener and silver plastic thoroughly.

11   Using the wood spatula, apply the silver plastic mixture between the pieces of tape, overlapping the undamaged area slightly on either end.

12   Carefully remove the decal or tape and apply a constant stream of hot air directly to the repaired area. A heat gun set at 500 to 700–degrees F is recommended. Hold the gun one inch from the glass for two minutes.

13   If the new element appears off color, tincture of iodine can be used to clean the repair and bring it back to the proper color. This mixture should not remain on the repair for more than 30 seconds.

14   Although the defogger is now fully operational, the repaired area should not be disturbed for at least 24 hours.

17 Windshield wiper module and motor - removal and installation

Refer to illustrations 17.4, 17.5, 17.7, 17.8a, 17.8b, 17.9, 17.10a, 17.10b and 17.11

1   Detach the cable from the negative terminal of the battery. **Caution:** *On models equipped with the Theftlock audio system, be sure the lockout feature is turned off before performing any procedure which requires disconnecting the battery.*

2   Raise the hood.

3   Disconnect the connector for the windshield washer fluid line at the wiper arms.

4   Pop off the windshield wiper blade cap over the nut and remove the nut **(see illustration)**.

5   Insert a pop rivet or a pin through the holes next to the wiper arm pivots, then grasp the wiper arms and remove them by rocking them back–and–forth, while lifting up **(see illustration)**.

6   Remove the screws from the lower edge of the reveal molding, then lower the hood and remove the molding by lifting it toward the windshield.

7   Raise the hood and remove the air inlet panel screws and (if equipped) the under-hood lamp switch, then remove the panel **(see illustration)**.

8   If the wiper motor will run, rotate the crank arm to the inner Wipe position **(see illustration)**. If the motor will not run, us-

17.8a The wiper arm must be in the inner Wipe position when wiper module assembly is removed

17.8b If the wiper motor won't run, use pliers to rotate the arm to the inner Wipe position

17.9 Unplug the electrical connectors from the wiper motor (arrows)

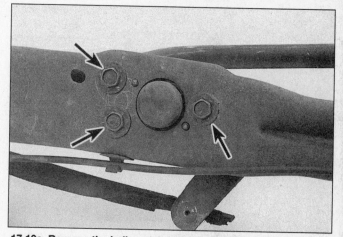

17.10a Remove the bellcrank housing screws (arrows) and lower the wiper transmission . . .

adjustable pliers to move the arm to the inner Wipe position by engaging the upper jaw against the top edge of the crank arm and the lower jaw against crank arm nut, using it as a pivot **(see illustration)**.

9   Unplug the electrical connectors **(see**

illustration).

10   Remove the three screws from the bell-crank housing, then lower the wiper transmission and lift the module from the vehicle **(see illustrations)**.

11   Remove the motor mounting bolts, then

remove the nut that fastens the transmission linkage arm to the motor shaft (crank arm nut), detach the linkage and remove the motor from the module **(see illustration)** .

12   Installation is the reverse of removal.

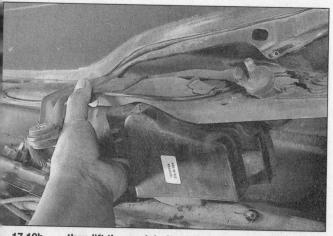

17.10b . . . then lift the module from the engine compartment

17.11 Remove the wiper motor bolts and the crank arm nut, detach the linkage and remove the motor from the module

1

18 Horn - replacement

1   Detach the cable from the negative terminal of the battery. **Caution:** *On models equipped with the Theftlock audio system, be sure the lockout feature is turned off before performing any procedure which requires disconnecting the battery.*
2   Open the hood.
3   The horns are located in a variety of places, some are located in the engine compartment in front of the left strut tower behind the air cleaner and battery , some are located at the front of the engine compartment attached to the radiator support below the hood latch and others are located behind the right headlight housing.
4   Locate the horns on particular model.
5   On vehicles with the horns mounted in front of the left strut tower remove the air cleaner assembly, remove the two bolts from the left side engine compartment cross brace, then loosen the remaining bolt and move the brace out of the way. remove the windshield washer reservoir and the battery.
6   On vehicles with the horns mounted at the front of the engine compartment by the hood latch, remove the headlight housing panel or the air deflector panel above the hood latch to access the horns.
7   On vehicles with the horns located behind the right headlight housing, remove the headlight housing (see Section 10).
8   Remove the horn mounting bracket bolt, lower the horn, unplug the electrical connector and remove the horn.
9   Installation is the reverse of removal.

19 Cruise control system - description and check

The cruise control system maintains vehicle speed with a vacuum actuated servo motor located in the engine compartment, which is connected to the throttle linkage by a cable. The system consists of the servo motor, clutch switch, brake switch, control switches, a relay and associated vacuum hoses.

Diagnosis can usually be limited to simple checks of the wiring and vacuum connections for minor faults which can be easily repaired. These include:

a) *Inspect the cruise control actuating switches for broken wires and loose connections.*
b) *Check the cruise control fuse.*
c) *The cruise control system is operated by vacuum so it's critical that all vacuum switches, hoses and connections are secure. Check the hoses in the engine compartment for tight connections, cracks and obvious vacuum leaks.*

20 Power window system - description and check

The power window system operates the electric motors mounted in the doors which lower and raise the windows. The system consists of the control switches, the motors (regulators), glass mechanisms and associated wiring.

Diagnosis can usually be limited to simple checks of the wiring connections and motors for minor faults which can be easily repaired. These include:

a) *Inspect the power window actuating switches for broken wires and loose connections.*
b) *Check the power window fuse/and or circuit breaker.*
c) *Remove the door panel(s) and check the power window motor wires to see if they're loose or damaged. Inspect the glass mechanisms for damage which could cause binding.*

21 Power door lock system - description and check

The power door lock system operates the door lock actuators mounted in each door. The system consists of the switches, actuators and associated wiring. Diagnosis can usually be limited to simple checks of the wiring connections and actuators for minor faults which can be easily repaired. These include:

a) *Check the system fuse and/or circuit breaker.*
b) *Check the switch wires for damage and loose connections. Check the switches for continuity.*
c) *Remove the door panel(s) and check the actuator wiring connections to see if they're loose or damaged. Inspect the actuator rods (if equipped) to make sure they aren't bent or damaged. Inspect the actuator wiring for damaged or loose connections. The actuator can be checked by applying battery power momentarily. A discernible click indicates that the solenoid is operating properly.*

22 Airbags - general information and precautions

Warning: *Some models covered by this manual are equipped with Supplemental Inflatable Restraint (SIR) systems, more commonly known as airbags. Always disable the airbag system before working in the vicinity of any airbag system components to avoid the possibility of accidental deployment of the airbag(s) which could cause personal injury. The yellow wiring harnesses and connectors routed through the console and instrument panel are for this system. Do not use electrical test equipment on any of the airbag system wiring or tamper with it in any way.*

Description

1   Some models covered by this manual are equipped with a Supplemental Inflatable Restraint (SIR) system, more commonly known as an airbag system. The SIR system is designed to protect the driver and passenger from serious injury in the event of a head-on or frontal collision.
2   The SIR system consists of an airbag located in the center of the steering wheel, and on some models, another located in the top of the dashboard, above the glove box. 1994 models utilize two impact sensors; one located in the instrument panel and another located just in front of the radiator; an arming sensor located under the center console; and a diagnostic/energy reserve module located at the right end of the instrument panel. On 1996 and later models, the sensors and the diagnostic/energy reserve module have been incorporated into one unit and is located under the center console.

Sensors

3   The 1994 system has three separate sensors; two impact sensors and an arming sensor. The sensors are basically pressure sensitive switches that complete an electrical circuit during an impact of sufficient G force. The electrical signal from the crash sensors is sent to the diagnostic module, that then completes circuit and inflates the airbags.
4   On the 1995 and later systems, the sensing circuitry is contained in the sensing diagnostic/energy reserve module. If a frontal crash of sufficient force is detected, the circuitry allows current to flow to the airbags, inflating them.

Diagnostic/energy reserve module

5   The diagnostic/energy reserve module contains an on-board microprocessor which monitors the operation of the system. It performs a diagnostic check of the system every time the vehicle is started. If the system is operating properly, the AIRBAG warning light will blink on and off seven times. If there is a fault in the system, the light will remain on and the airbag control module will store fault codes indicating the nature of the fault. If the AIRBAG warning light remains on after staring, or comes on while driving, the vehicle should be taken to your dealer immediately for service. The diagnostic/energy reserve module also contains a back-up power supply to deploy the airbags in the event battery power is lost during a collision.

Operation

6   For the airbag(s) to deploy, an impact of sufficient G force must occur within 30 degrees of the vehicle centerline. When this condition occurs, the circuit to the airbag inflator is closed and the airbag inflates. If the battery is destroyed by the impact, or is too low to power the inflators, a back-up power supply inside the diagnostic/energy reserve module supplies current to the airbags.

Self-diagnosis system

7   A self-diagnosis circuit in the module displays a light when the ignition switch is turned to the On position. If the system

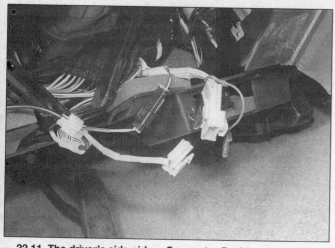

22.11 The driver's side airbag Connector Position Assurance (CPA) connector is found at the base of the steering column

22.13 The passenger airbag Connector Position Assurance (CPA) connector is located under the instrument panel and is accessible through the glovebox opening

operating normally, the light should go out after seven flashes. If the light doesn't come on, or doesn't go out after seven flashes, or if it comes on while you're driving the vehicle, there's a malfunction in the SIR system. Have it inspected and repaired as soon as possible. Do not attempt to troubleshoot or service the SIR system yourself. Even a small mistake could cause the SIR system to malfunction when you need it.

Servicing components near the SIR system

3   Nevertheless, there are times when you need to remove the steering wheel, radio or service other components on or near the instrument panel. At these times, you'll be working around components and wiring harnesses for the SIR system. SIR system wiring is easy to identify; they're all covered by a bright yellow conduit. Do not unplug the connectors for the SIR system wiring, except to disable the system. And do not use electrical test equipment on the SIR system wiring. *ALWAYS DISABLE THE SIR SYSTEM BEFORE WORKING NEAR THE SIR SYSTEM COMPONENTS OR RELATED WIRING.*

Disabling the SIR system

Refer to illustrations 22.11 and 22.13

Turn the steering wheel to the straight ahead position, place the ignition switch in lock and remove the key. Remove the airbag fuse from the fuse block (see Section 3). *Note: Some models designate the airbag fuse as ARBG#1 while others are designated ARBAG#1 or SIR.*

Unplug the yellow Connector Position

Assurance (CPA) connectors at the base of the steering column and under the right side of the instrument panel as described in the following steps.

Driver's side airbag

11   Remove the knee bolster and sound insulator panel below the instrument panel (see Chapter 11) and unplug the yellow Connector Position Assurance (CPA) steering column harness connector **(see illustration)**.

Passenger's side airbag

12   Remove the glove box (see Chapter 11).
13   Unplug the CPA electrical connector from the passenger inflator module under the right side of the dash **(see illustration)**.

Enabling the SIR system

14   After you've disabled the airbag and performed the necessary service, plug in the steering column (driver's side) and passenger side CPA connectors. Reinstall the knee bolster, sound insulator panel and the glove box.
15   Install the airbag fuse.

23   Head–up display (HUD) - general information

Some models use a Head–up Display (HUD) to project information such as vehicle speed and turn signal indicators, as well as battery, engine temperature, high beam and low fuel indicators, toward the front of the car (as seen from the driver's seat).

The system is made up of the HUD unit, which is built into the top of the dash, the HUD dimmer switch on the left side of the

instrument panel, and a HUD–specific windshield. On some models a HUD English/metric switch is also used. Troubleshooting of the HUD system is best left to a dealer or other qualified service facility. However, if the HUD display seems dim or unclear, try cleaning the HUD projector window with a soft, clean cloth moistened with glass cleaner. **Caution:** *Do not spray glass cleaner on the HUD unit's projector window as this may cause damage to the HUD components inside. Wipe the HUD window gently, then dry it. Also check to be sure that the windshield area directly in front of the HUD unit is clear and unobstructed.*

If the windshield on a HUD–equipped car ever needs to be replaced, be sure to replace it with a HUD–specific windshield for optimum HUD performance.

24   Wiring diagrams - general information

Since it isn't possible to include all wiring diagrams for every year covered by this manual, the following diagrams are those that are typical and most commonly needed.

Prior to troubleshooting any circuit, check the fuse and circuit breakers (if equipped) to make sure they're in good condition. Make sure the battery is properly charged and check the cable connections (see Chapter 1).

When checking a circuit, make sure that all connectors are clean, with no broken or loose terminals. When unplugging a connector, do not pull on the wires. Pull only on the connector housings themselves.

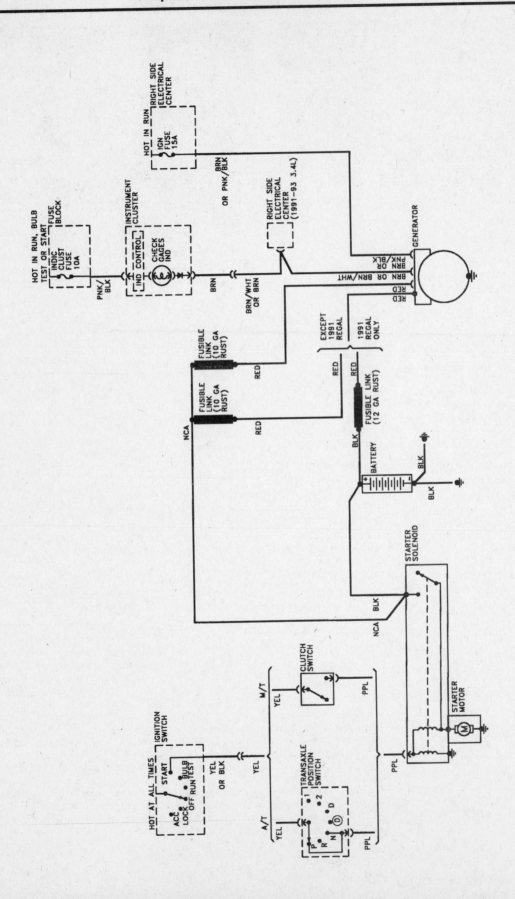

1993 and earlier starting and charging systems

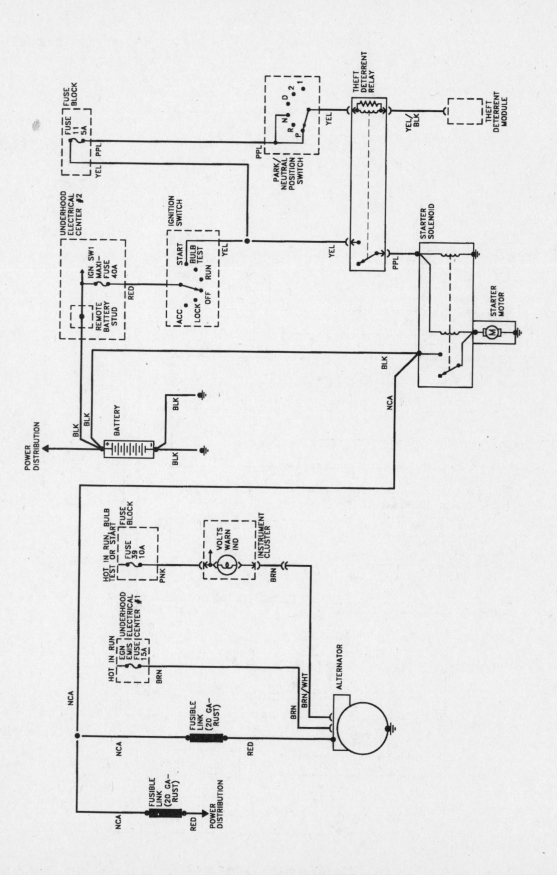

1994 and 1995 starting and charging systems

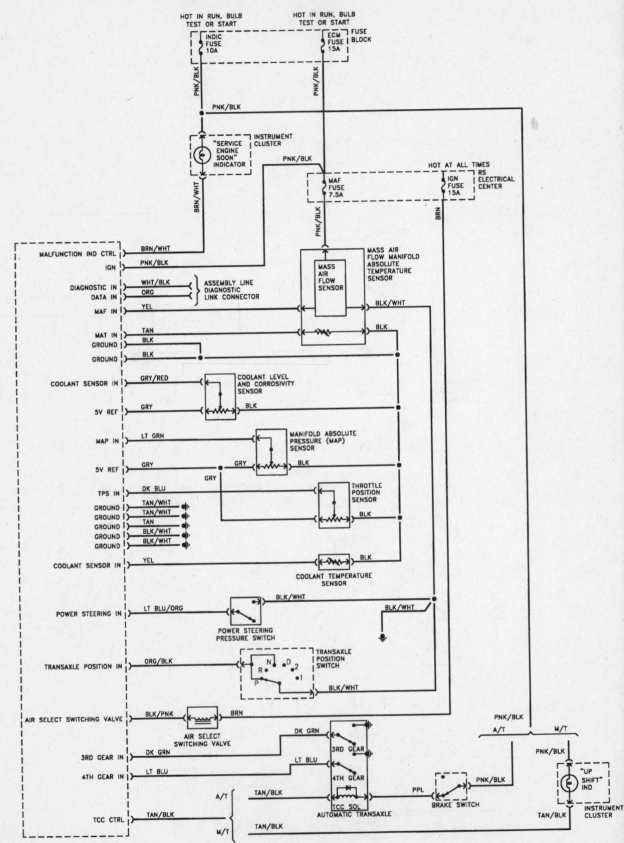

2.8L engine control system (1 of 2)

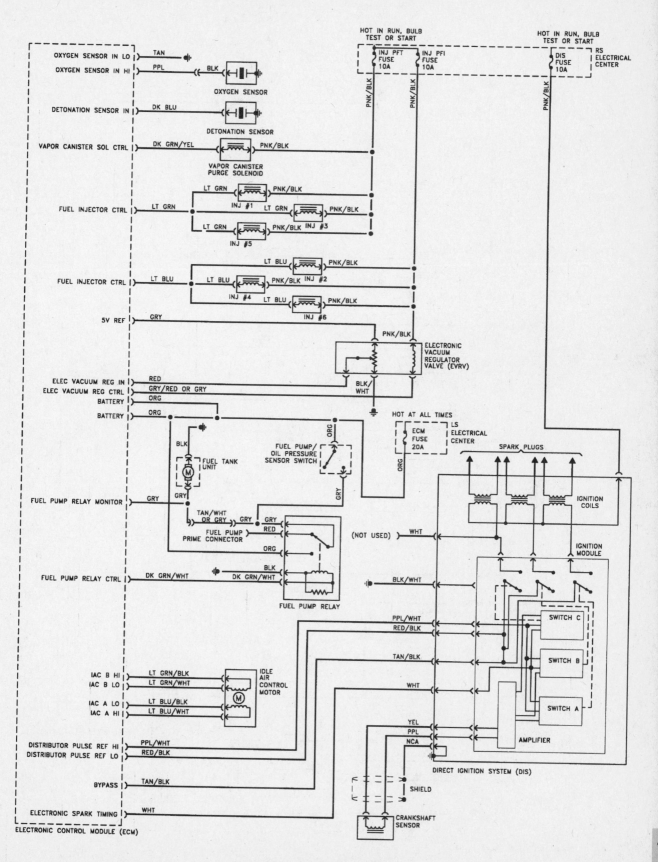

2.8L engine control system (2 of 2)

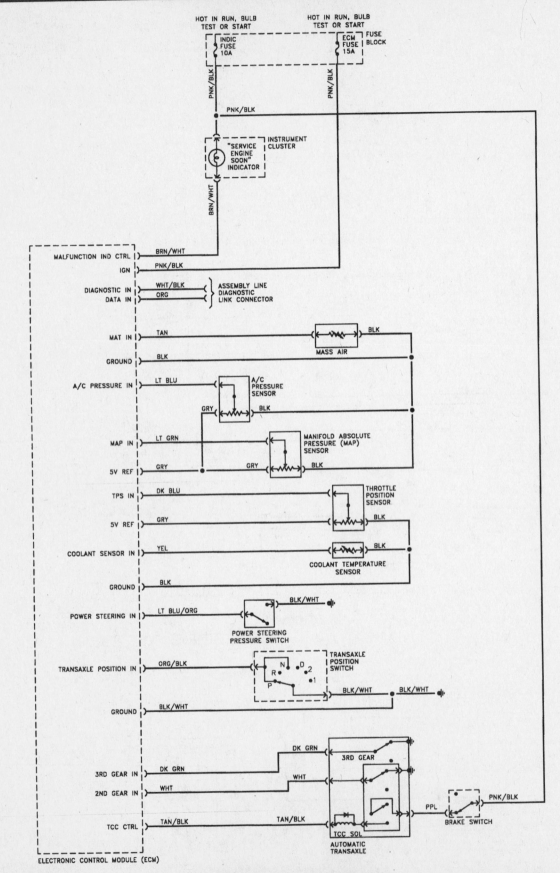

2.3L engine control system (1 of 2)

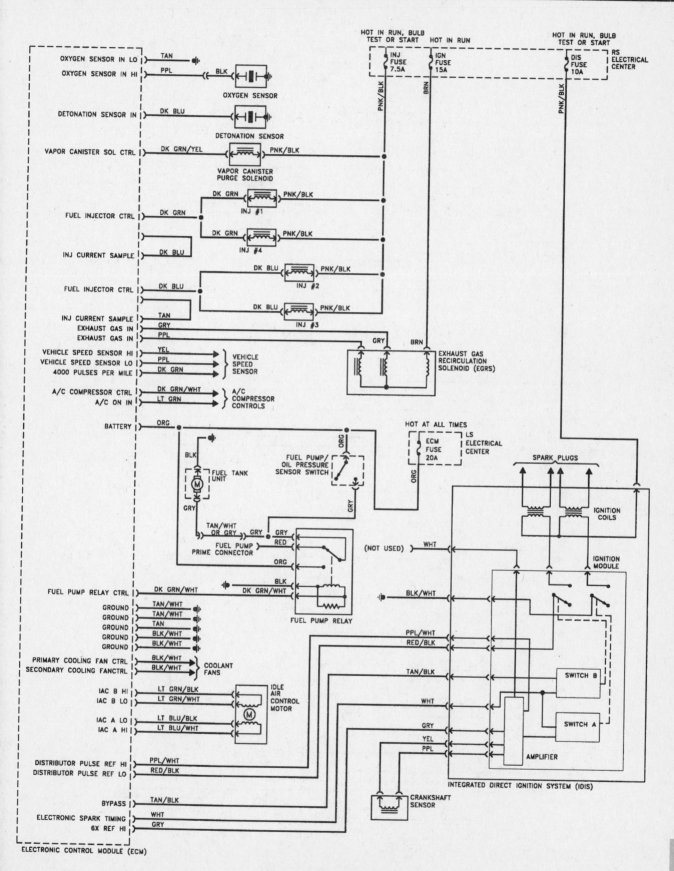

2.3L engine control system (2 of 2)

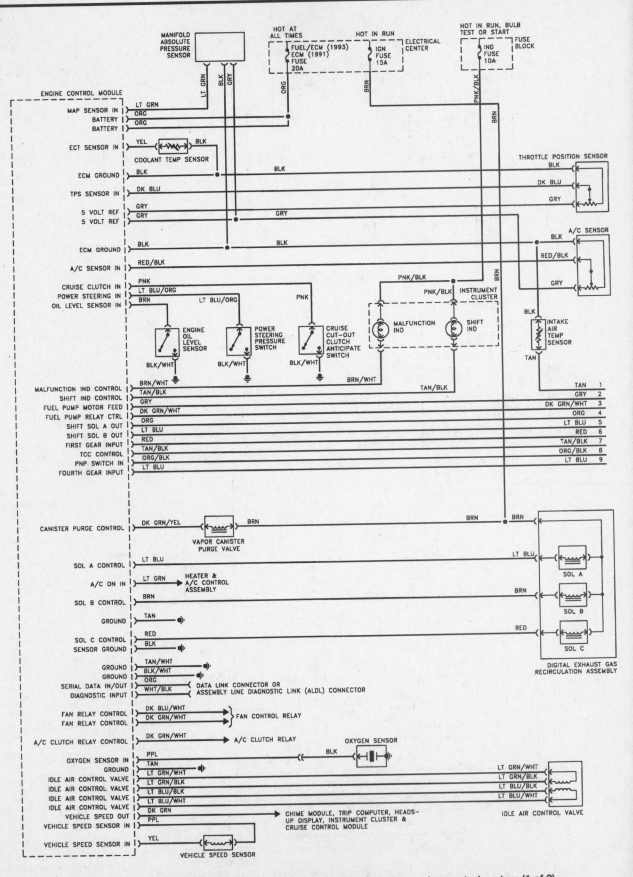

1993 and earlier 3.1L (except Generation II) and 3.4L Cutlass Supreme engine control system (1 of 2)

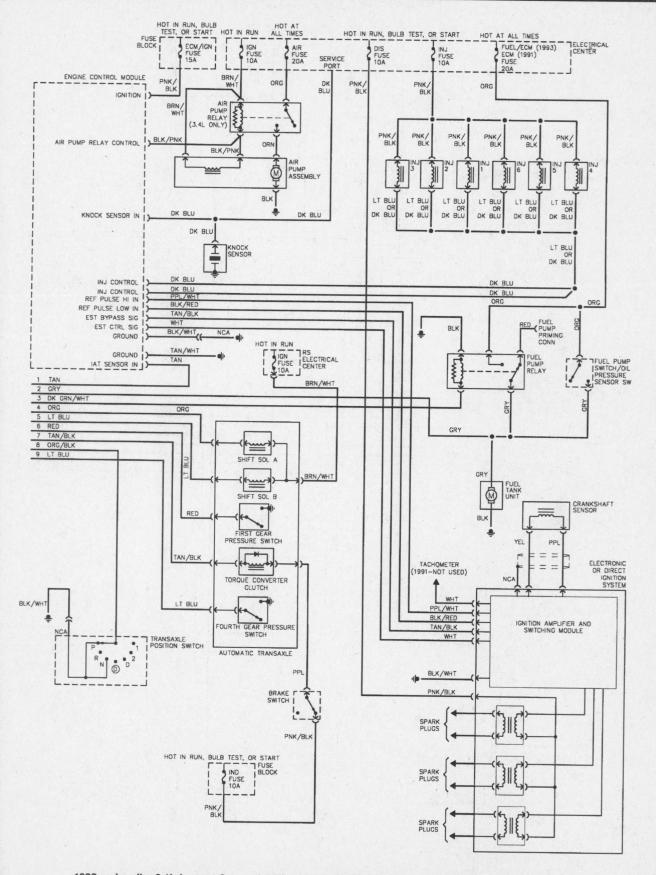

1993 and earlier 3.1L (except Generation II) and 3.4L Cutlass Supreme engine control system (2 of 2)

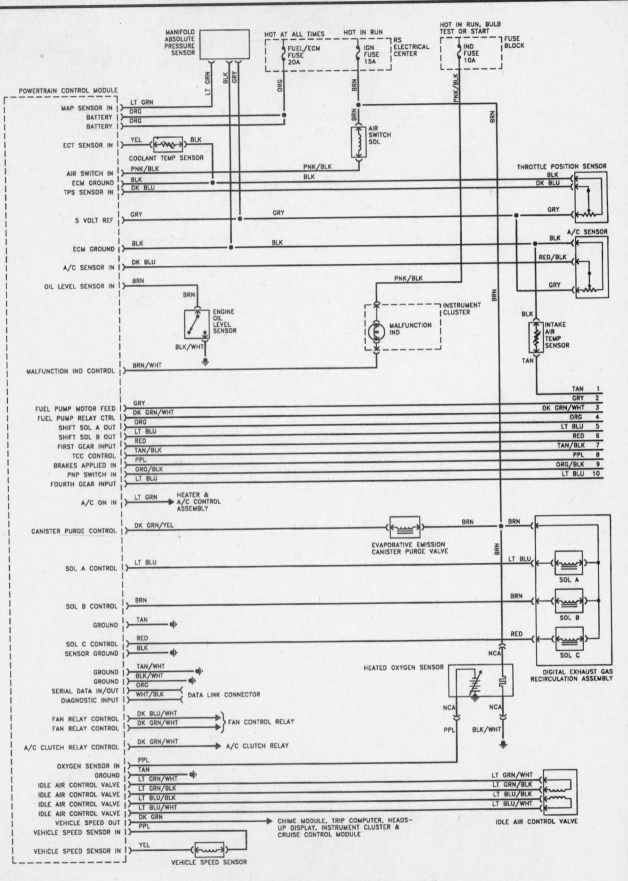

1993 3.1L Generation II engine control system (1 of 2)

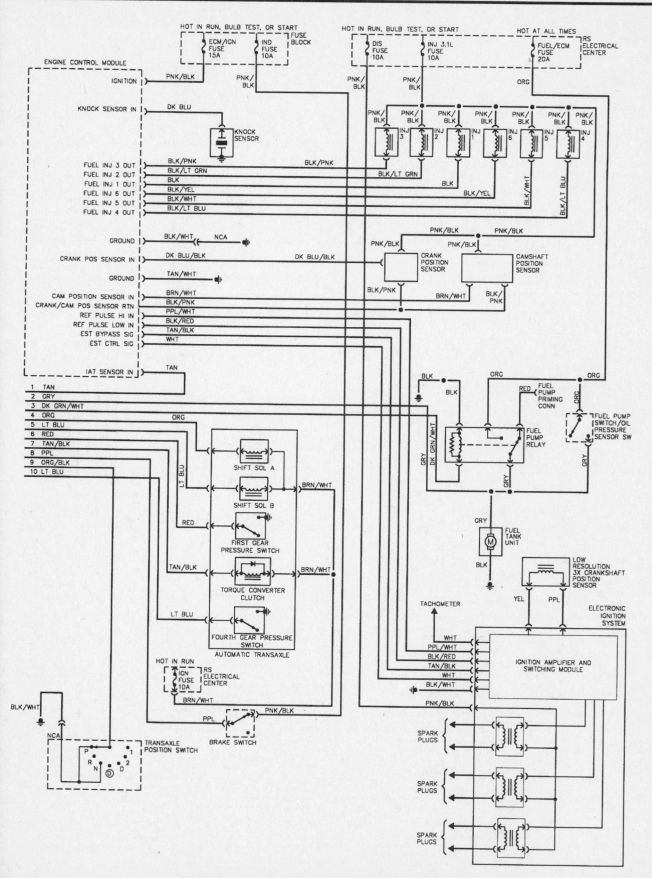

1993 3.1L Generation II engine control system (2 of 2)

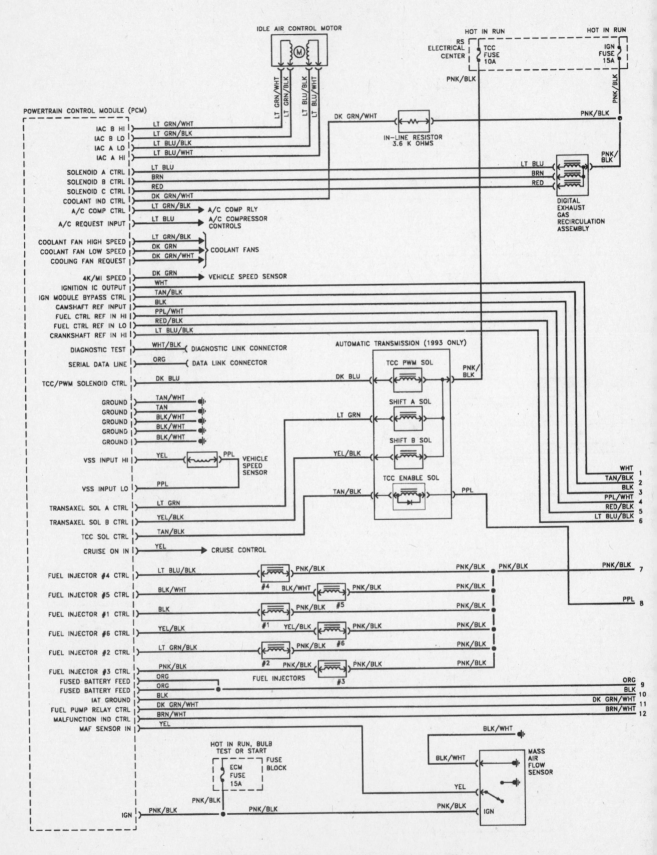

1993 and earlier 3.8L engine control system (1 of 3)

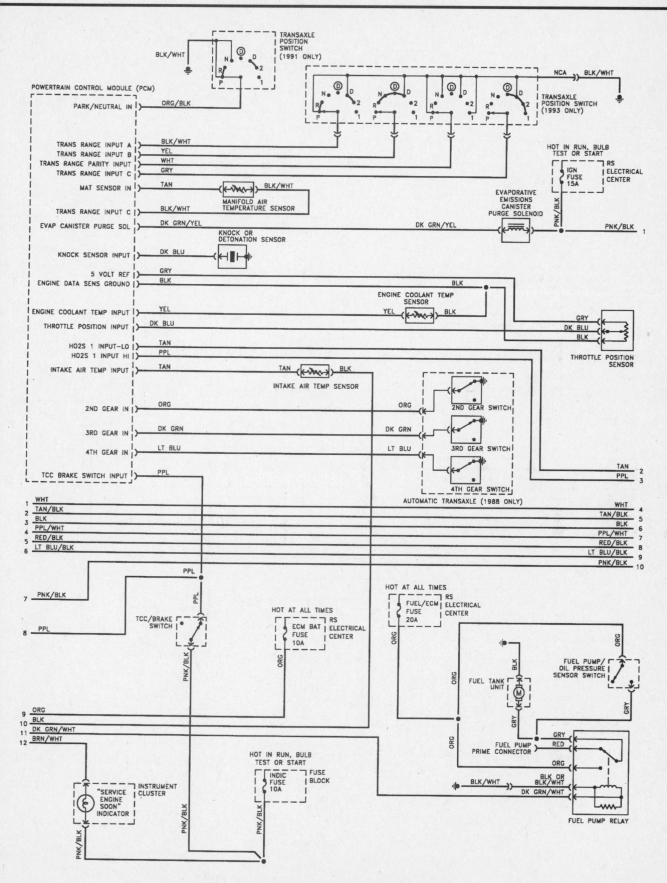

1993 and earlier 3.8L engine control system (2 of 3)

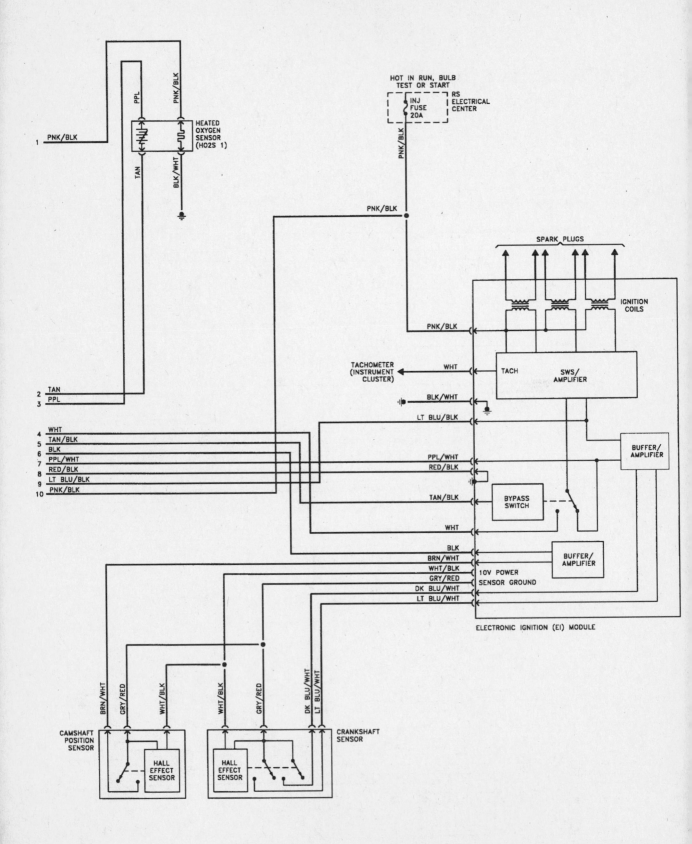

1993 and earlier 3.8L engine control system (3 of 3)

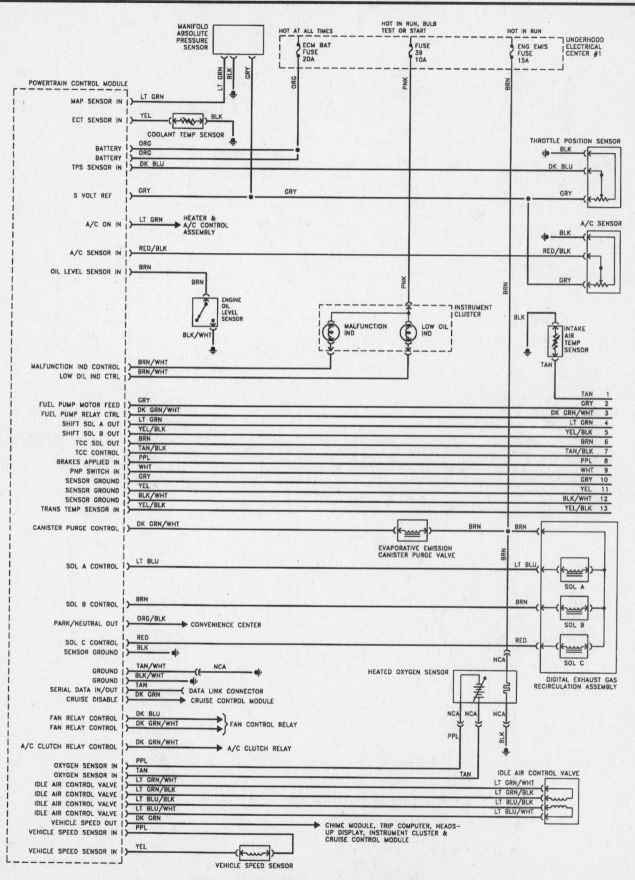

1994 and 1995 3100 engine control system (1 of 2)

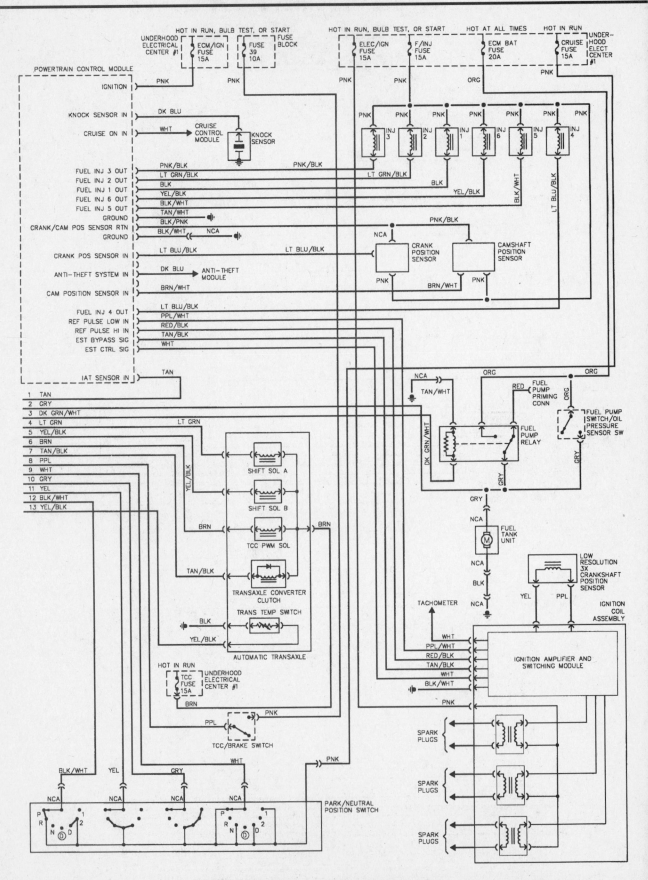

1994 and 1995 3100 engine control system (2 of 2)

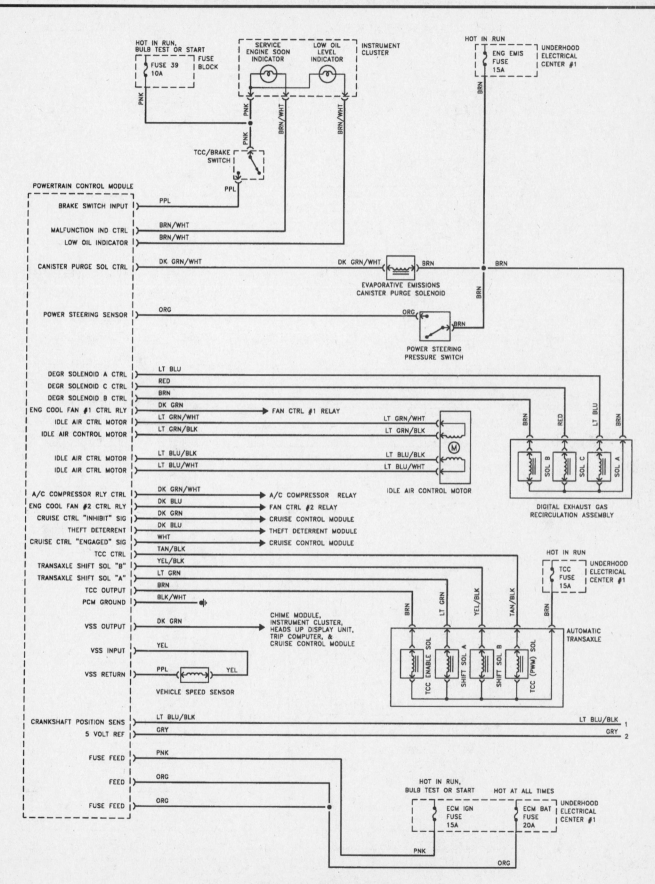

1994 and 1995 3.4L engine control system (1 of 3)

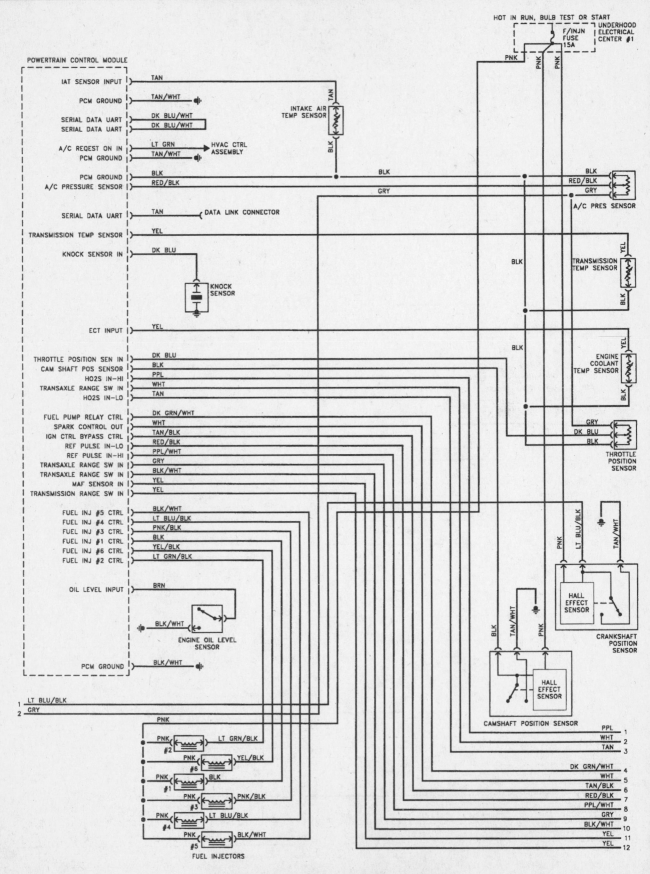

1994 and 1995 3.4L engine control system (2 of 3)

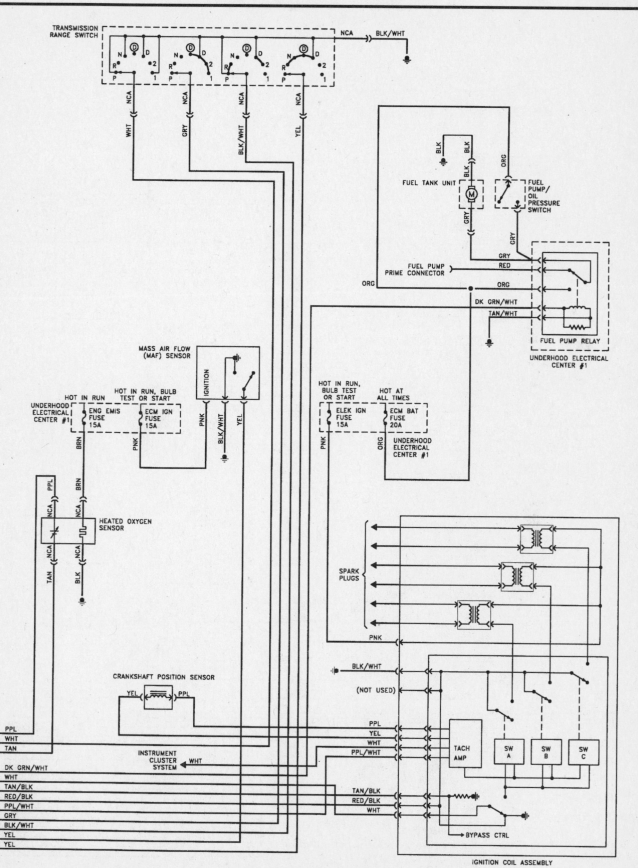

1994 and 1995 3.4L engine control system (3 of 3)

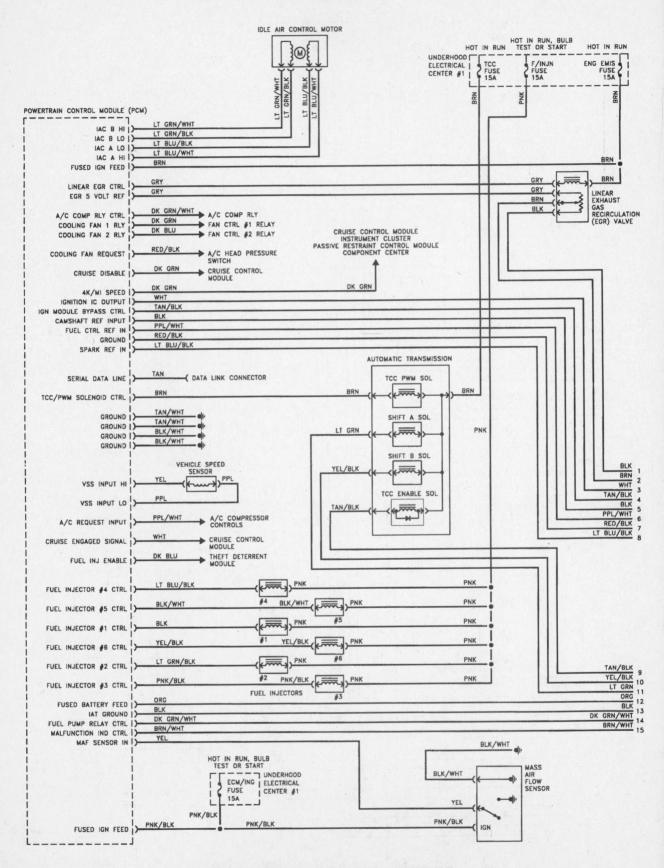

1994 and 1995 3.8L engine control system (1 of 3)

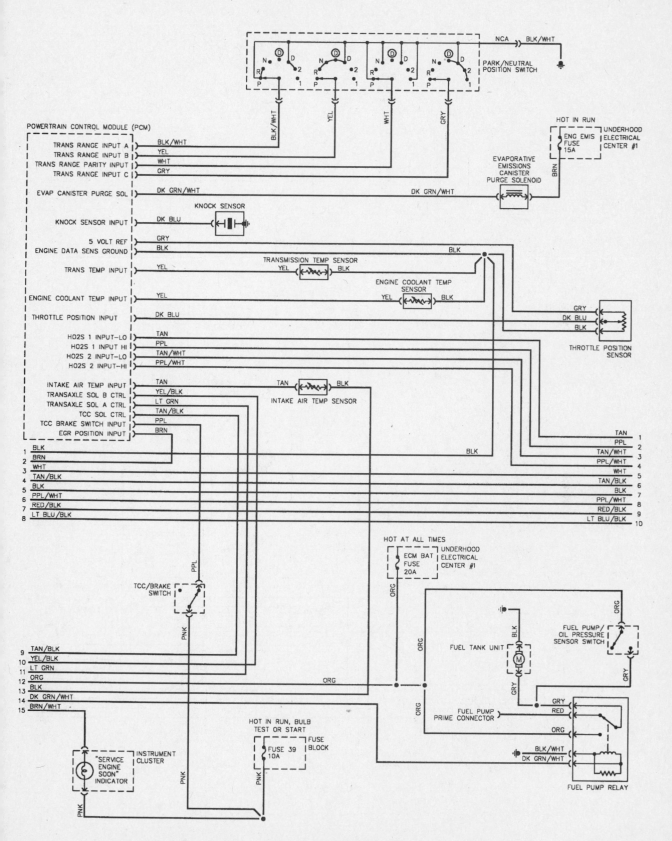

1994 and 1995 3.8L engine control system (2 of 3)

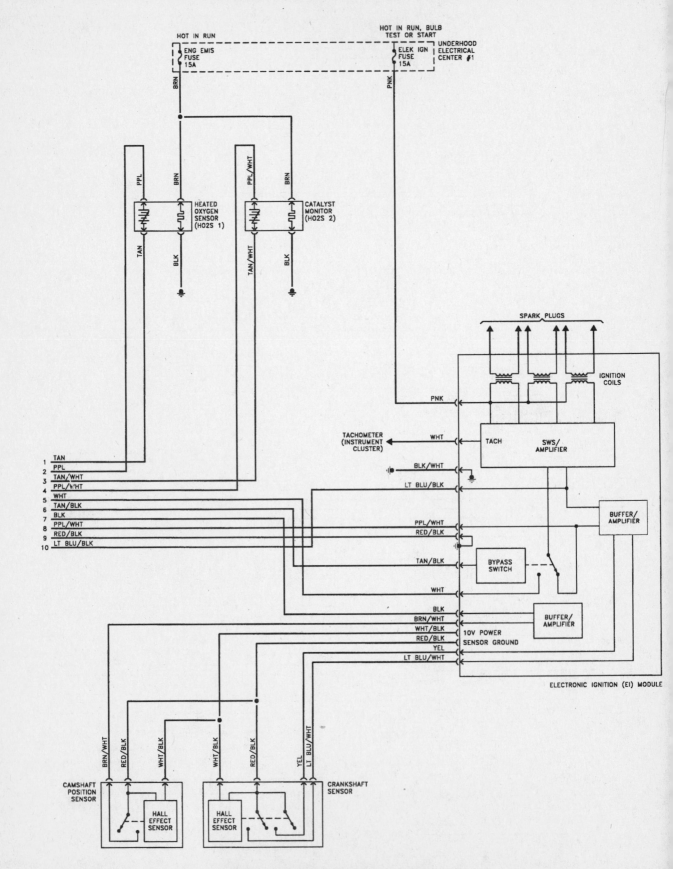

1994 and 1995 3.8L engine control system (3 of 3)

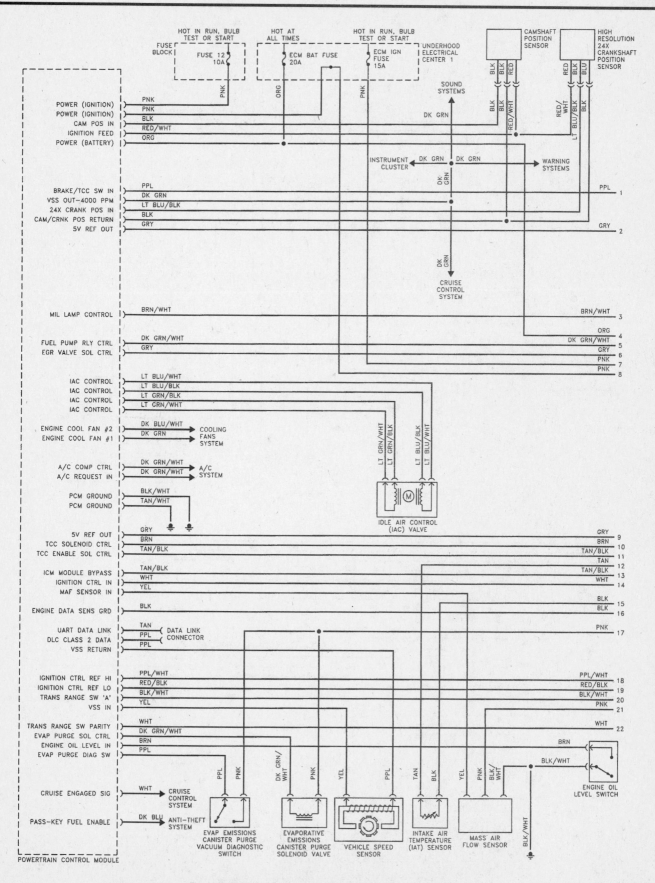

Typical 1996 3100 engine control system (part 1 of 3)

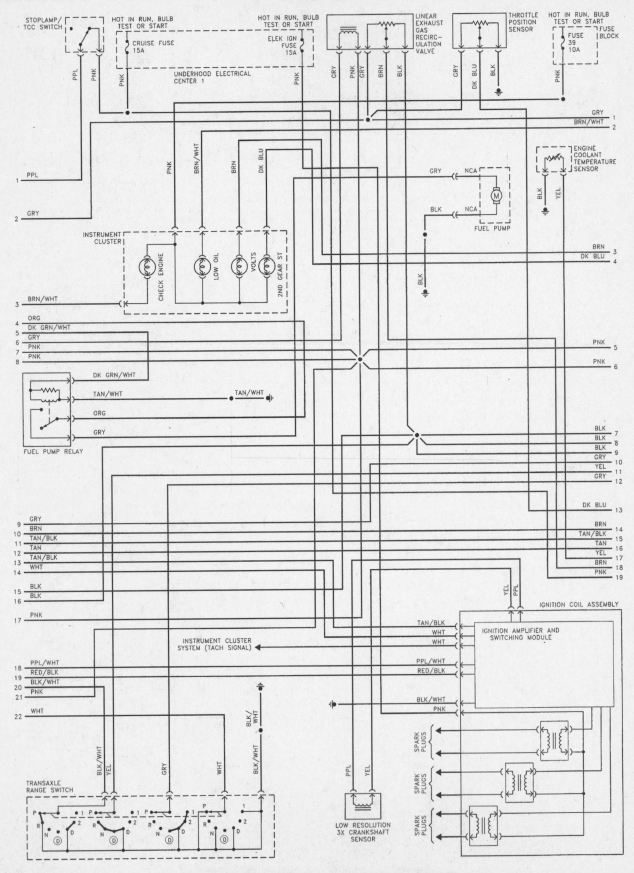

Typical 1996 3100 engine control system (part 2 of 3)

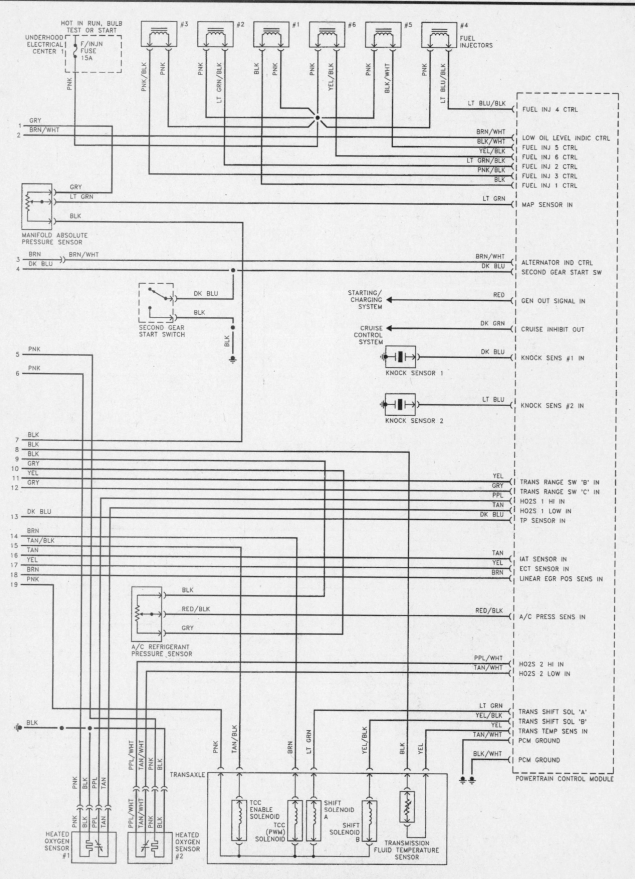

Typical 1996 3100 engine control system (part 3 of 3)

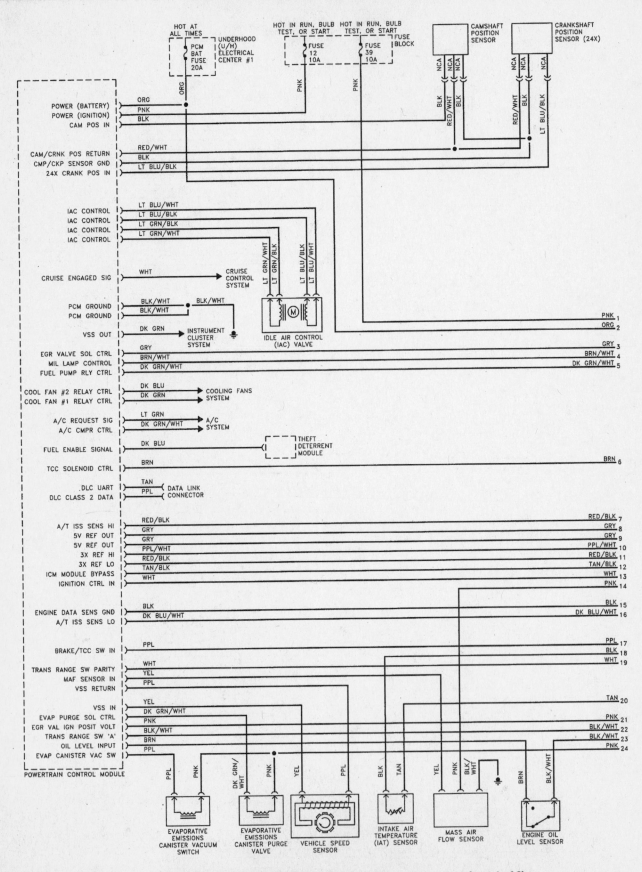

Typical 1996 3.4L and 1997 and later 3100 engine control system (part 1 of 3)

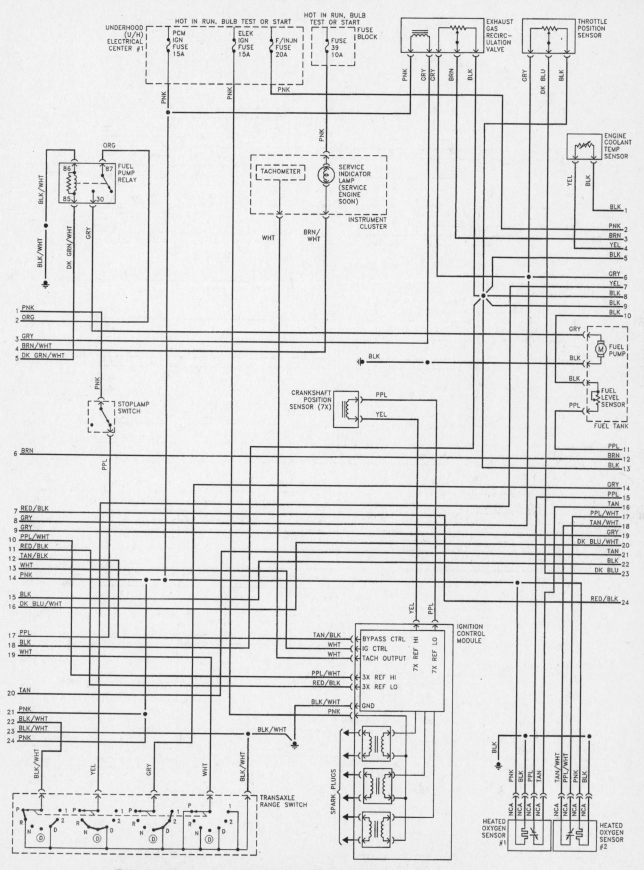

Typical 1996 3.4L and 1997 and later 3100 engine control system (part 2 of 3)

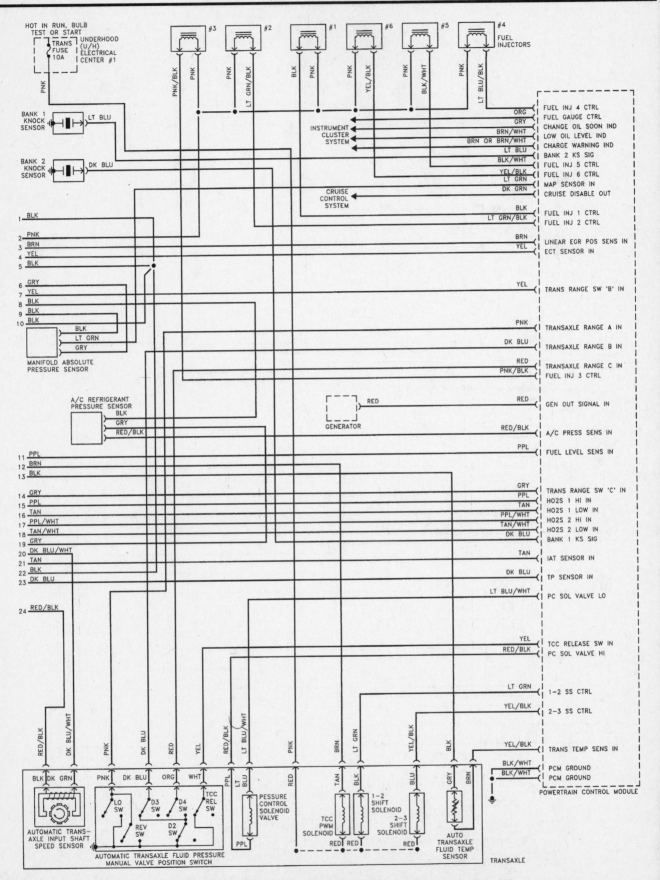

Typical 1996 3.4L and 1997 and later 3100 engine control system (part 3 of 3)

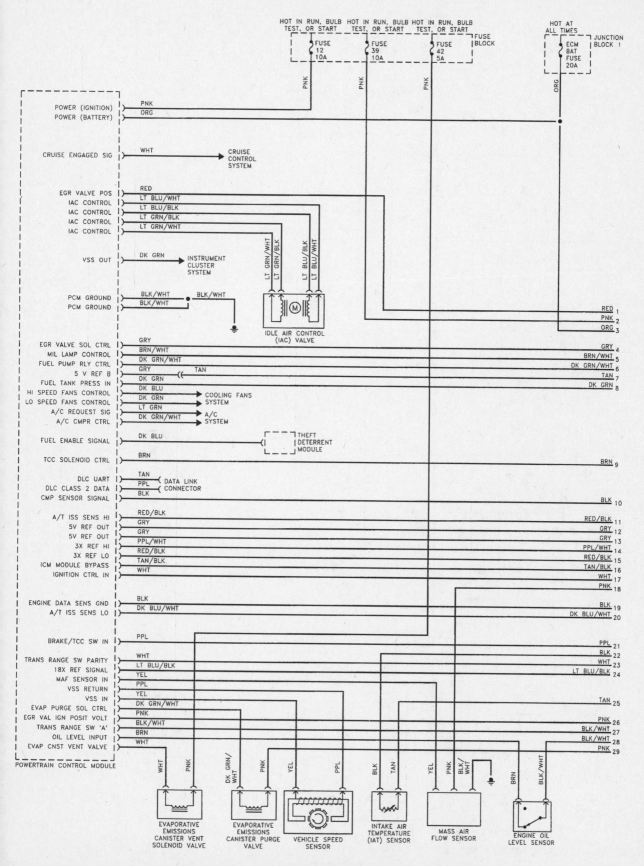

Typical 1996 and later 3800 engine control system (part 1 of 3)

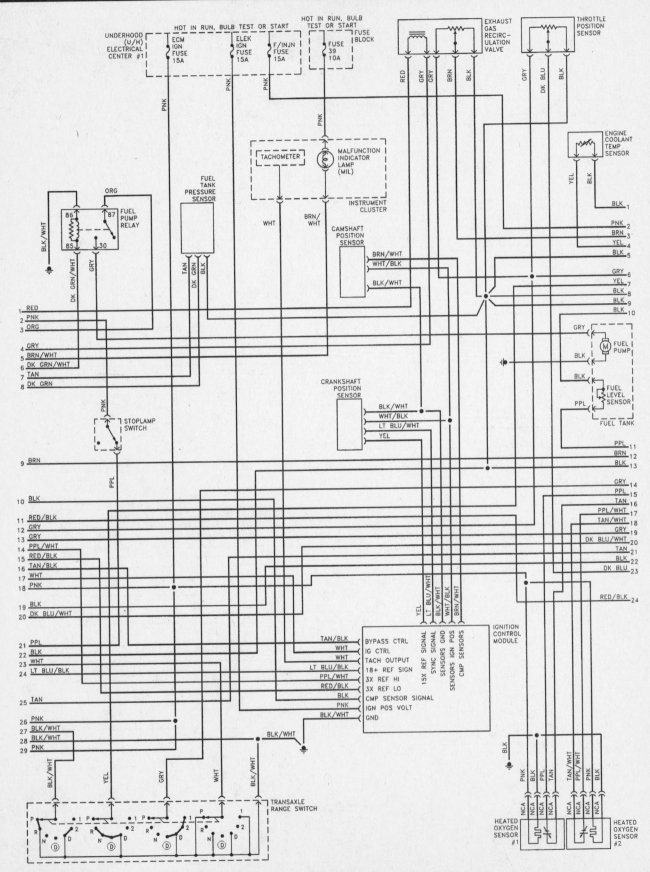

Typical 1996 and later 3800 engine control system (part 2 of 3)

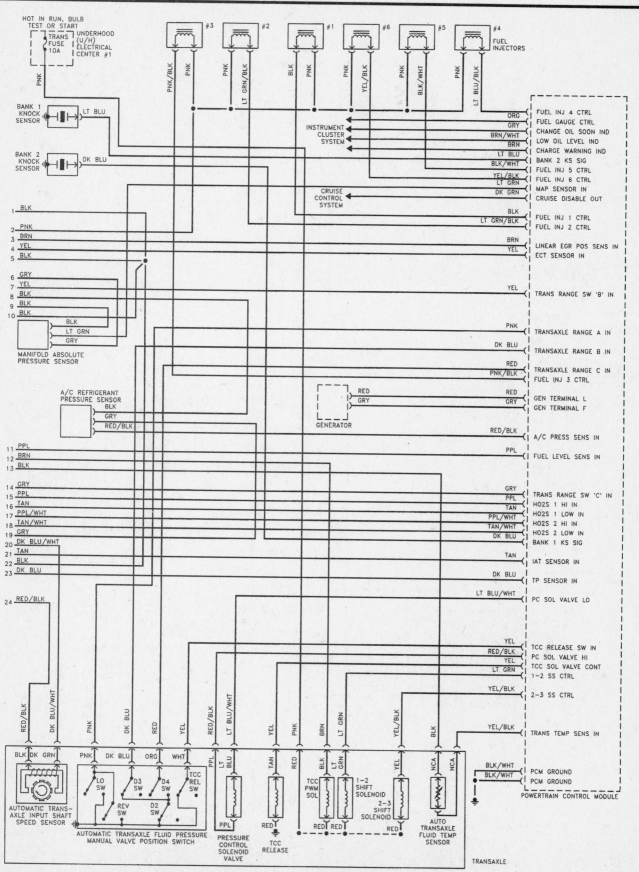

Typical 1996 and later 3800 engine control system (part 3 of 3)

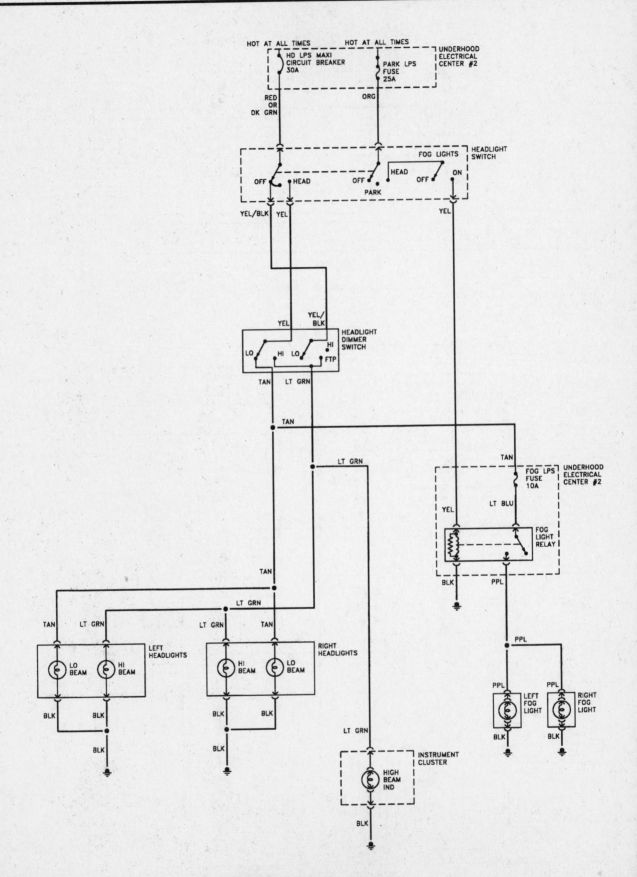

1988 through 1990 headlight circuit

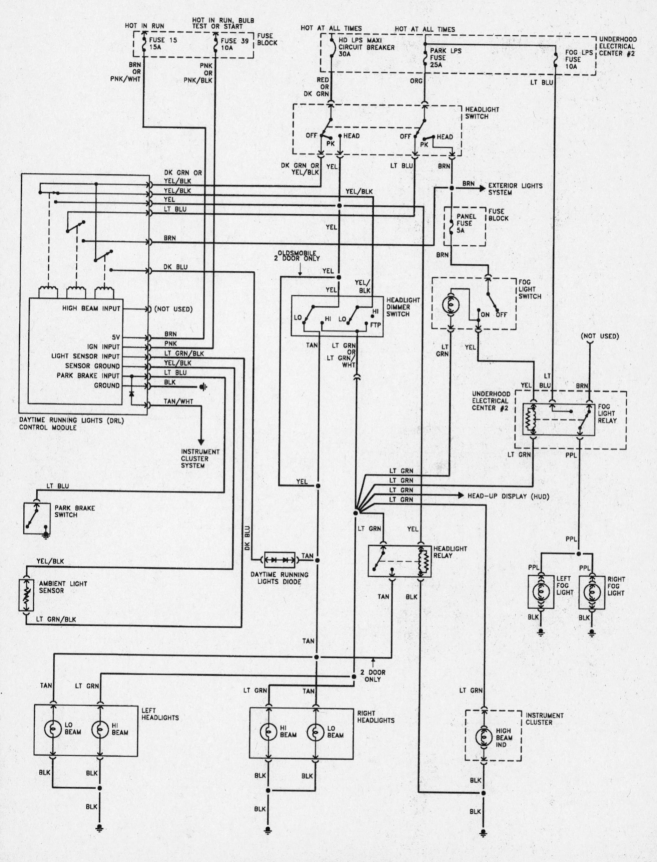

1991 through 1995 headlight circuit

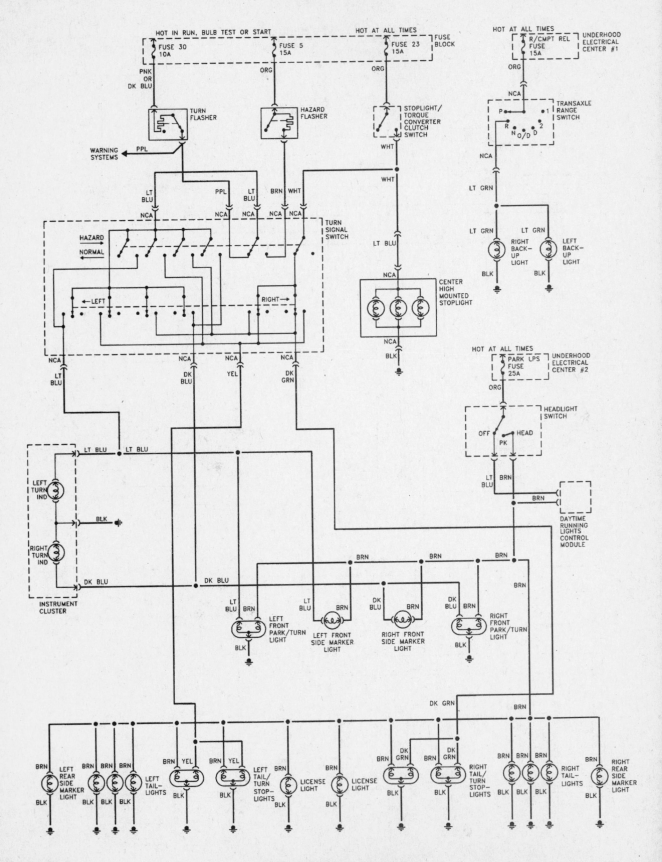

Exterior lights circuit (all models)

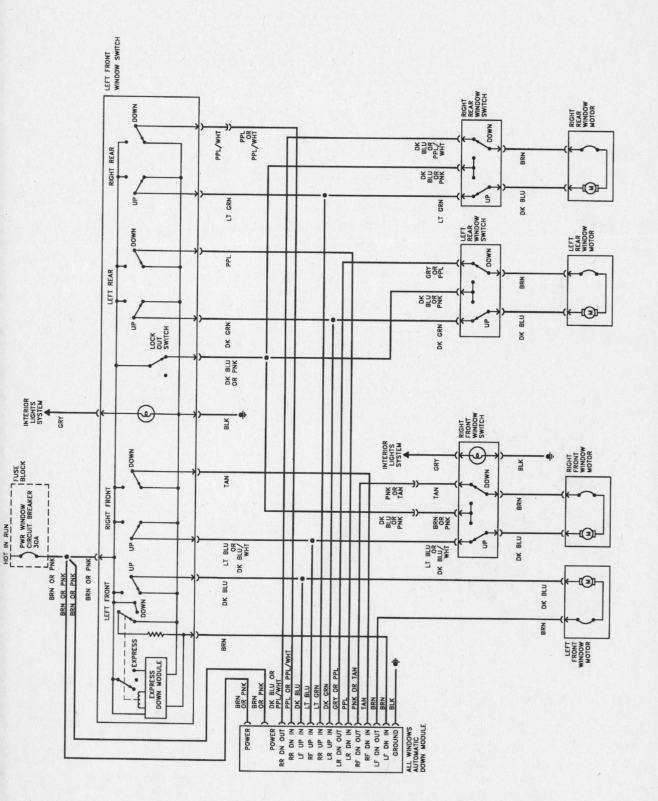

Power window circuit for 2-door models

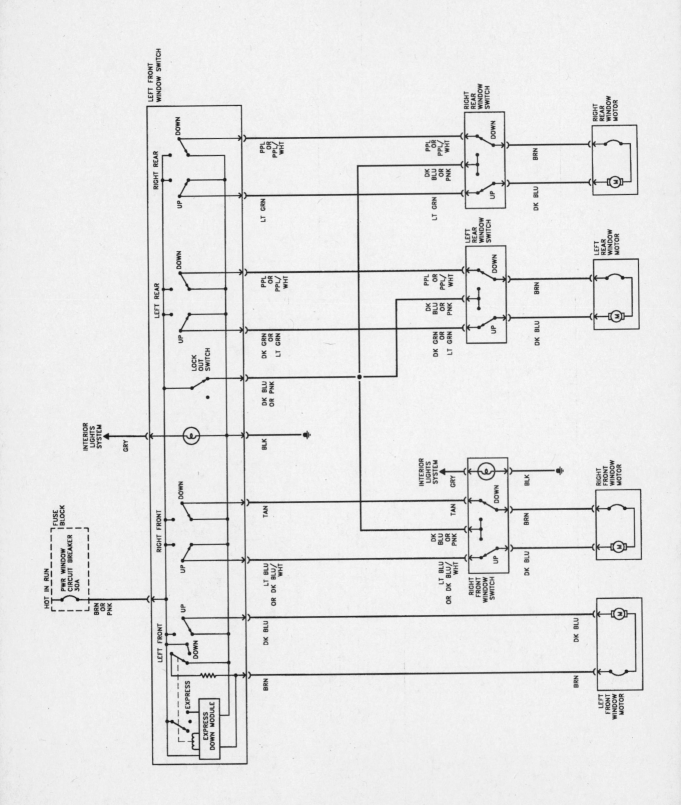

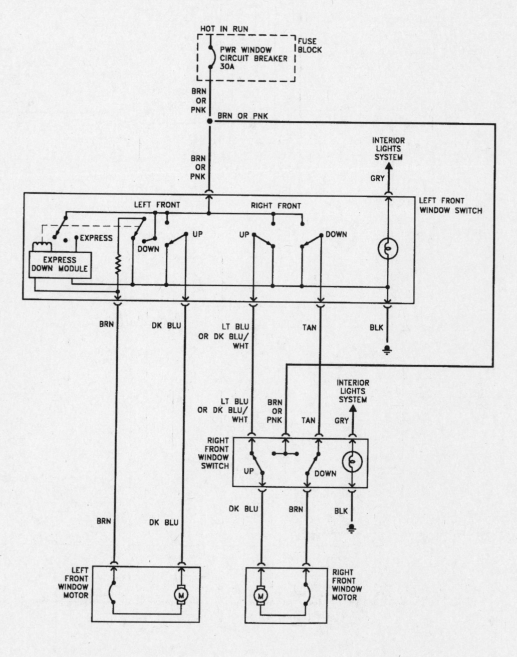

Power window circuit for convertible Cutlass Supreme

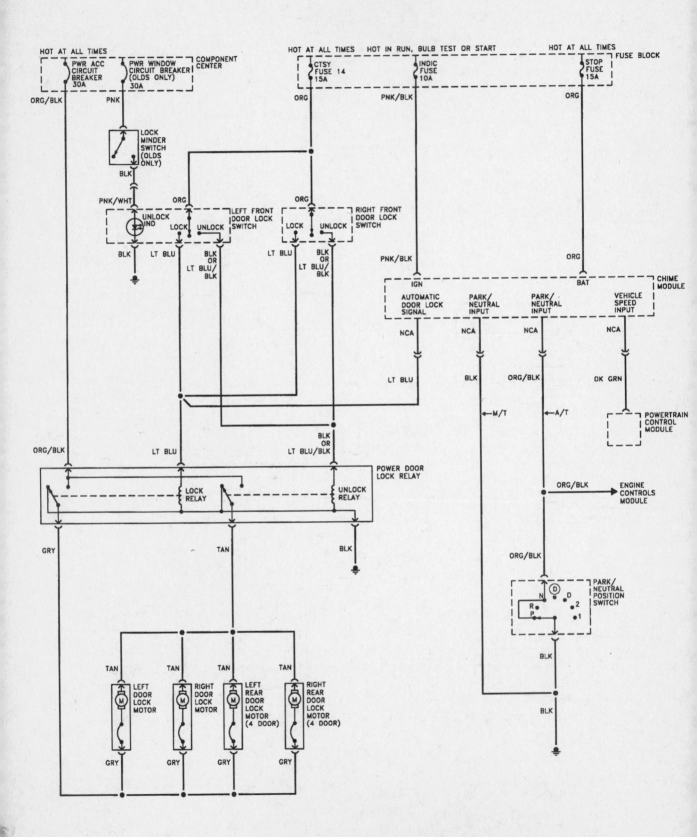

Power door lock circuit - 1988 through 1993

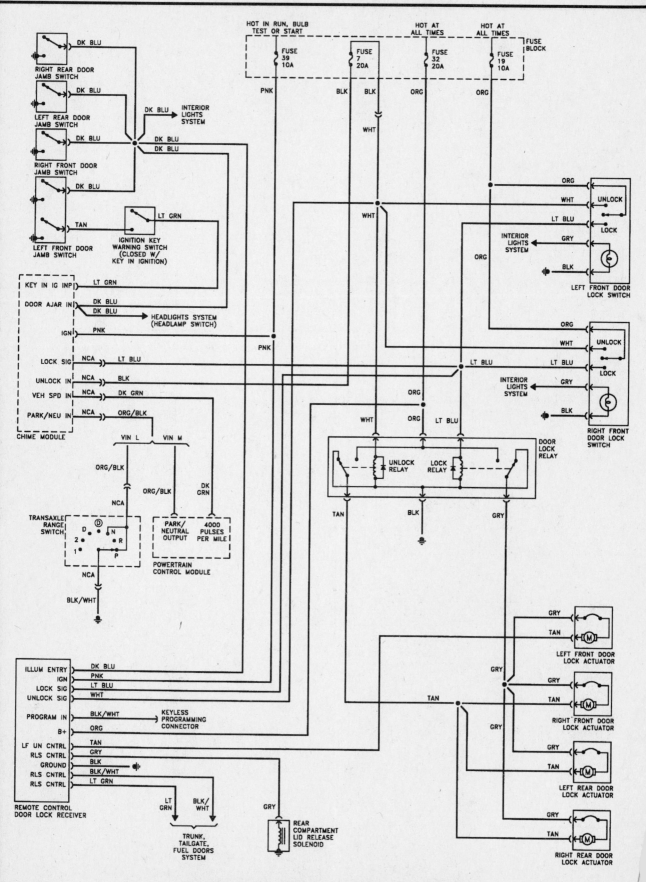

Power door lock and keyless entry circuit - 1994 and 1995

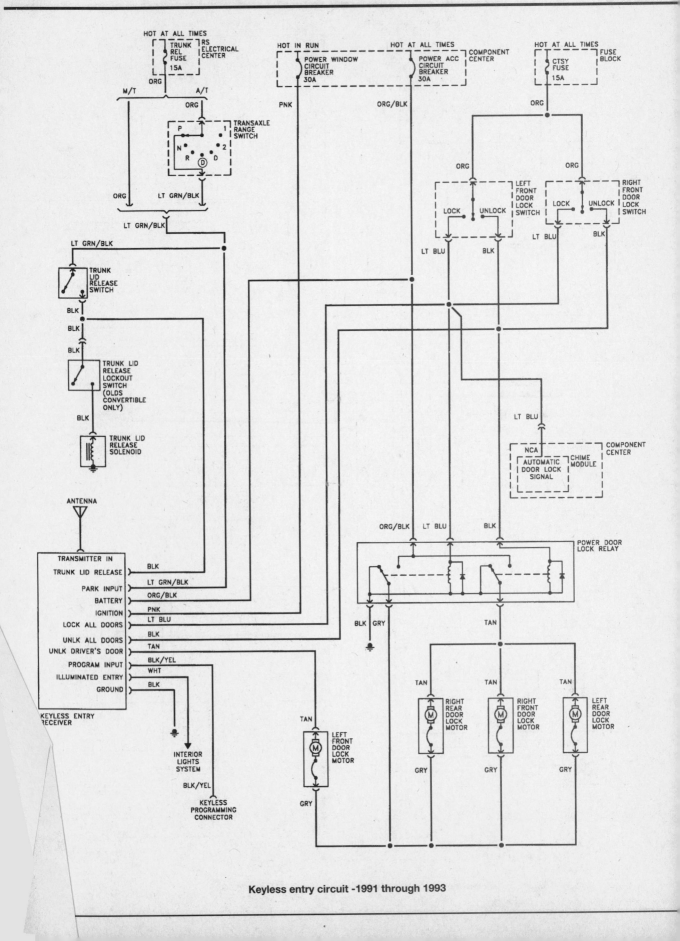

Keyless entry circuit -1991 through 1993

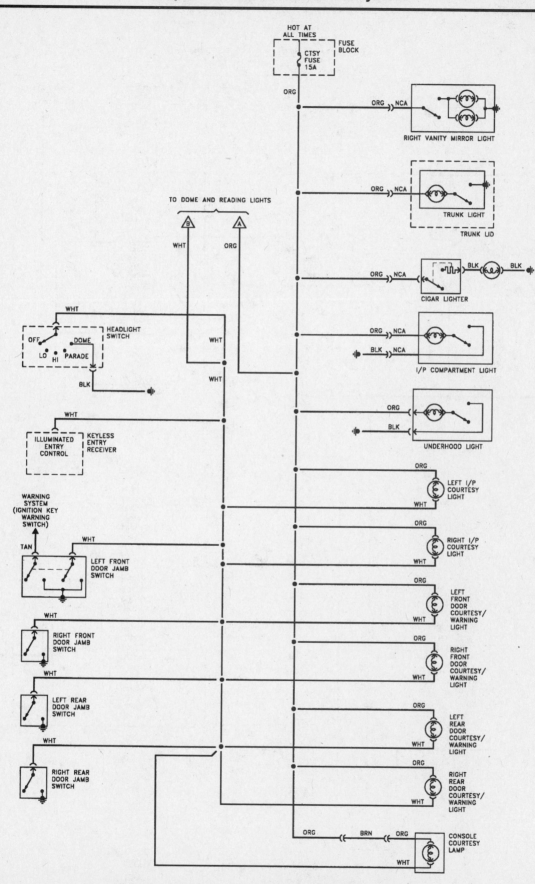

Courtesy lights circuit - 1988 through 1993 (1 of 2)

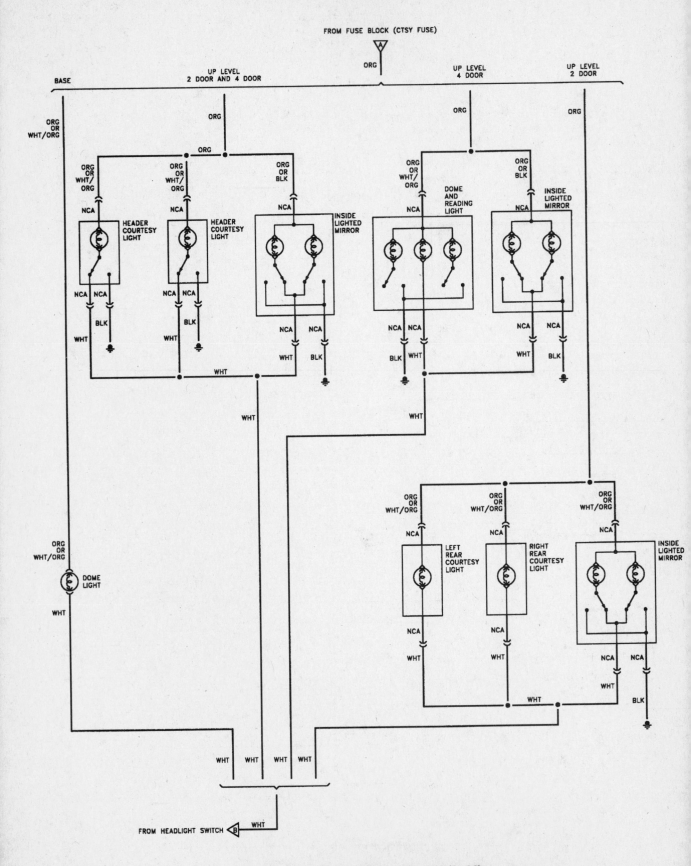

Courtesy lights circuit - 1988 through 1993 (2 of 2)

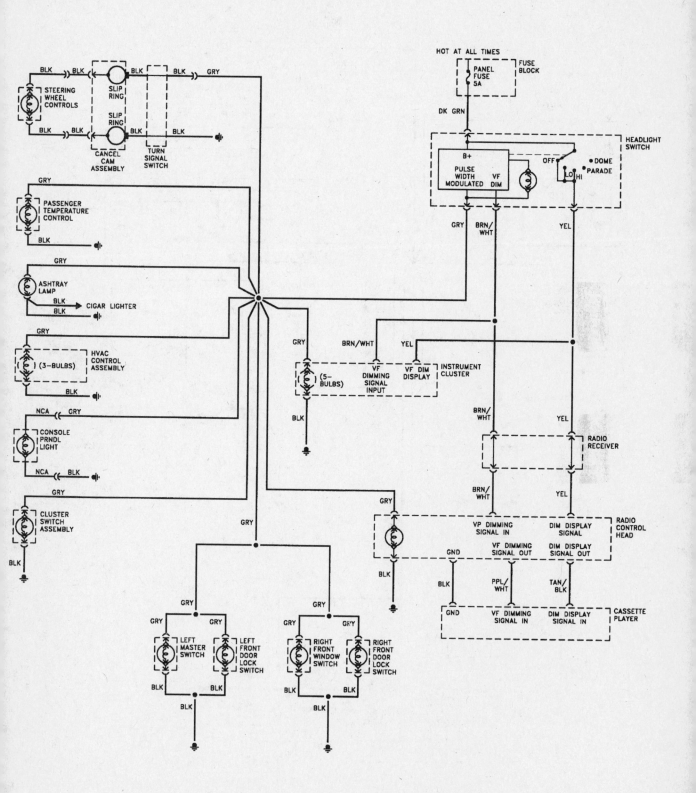

Interior lights circuit - 1988 through 1993 (Buick Regal)

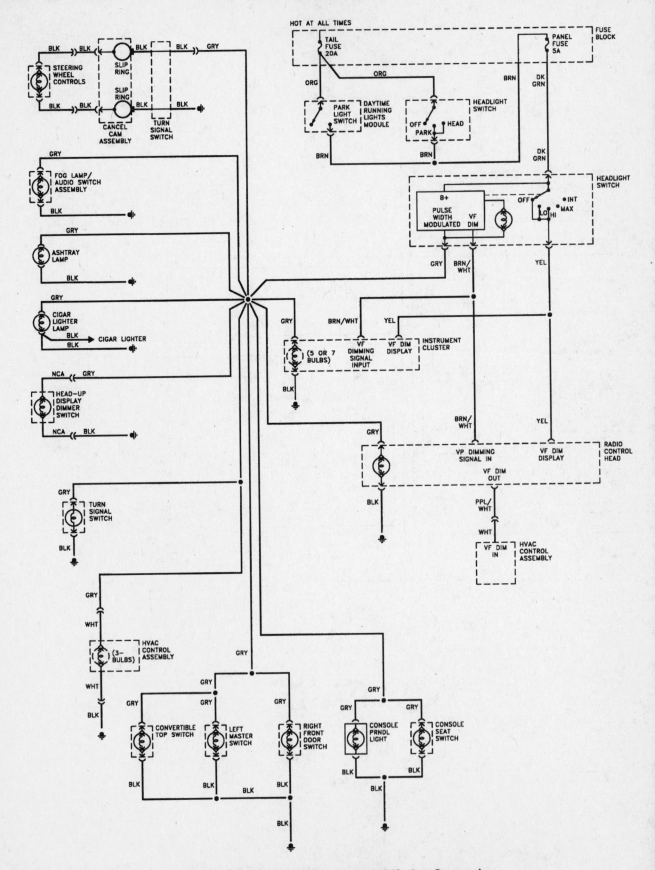

Interior lights circuit - 1988 through 1993 (Cutlass Supreme)

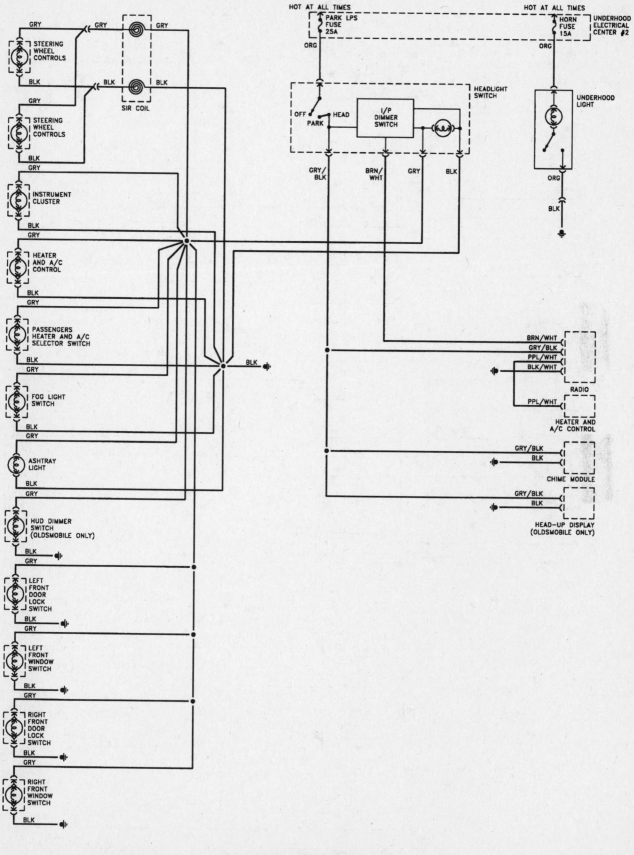

Interior lights circuit - 1994 and 1995

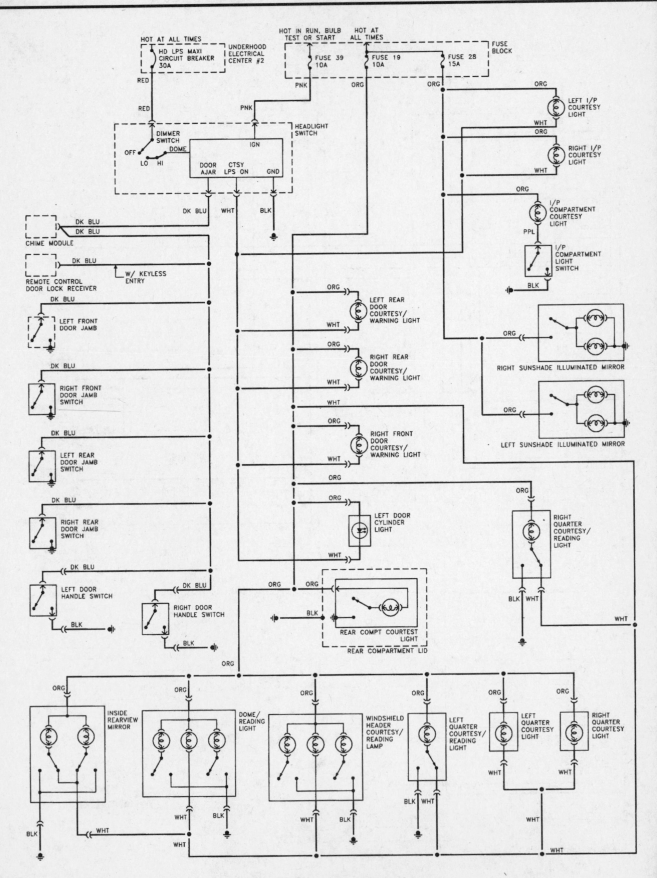

Courtesy lights circuit - 1994 and 1995

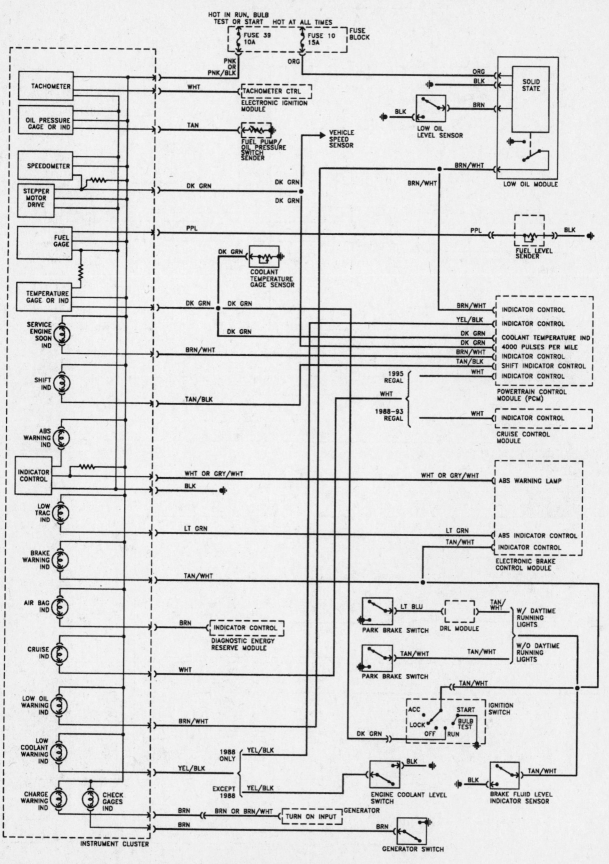

Typical warning lamp circuit

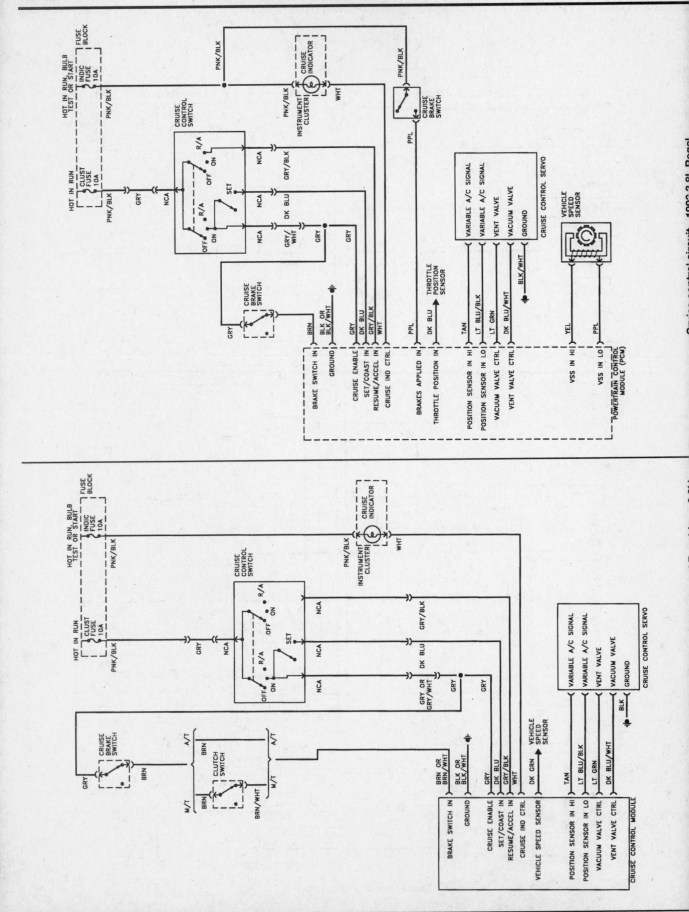

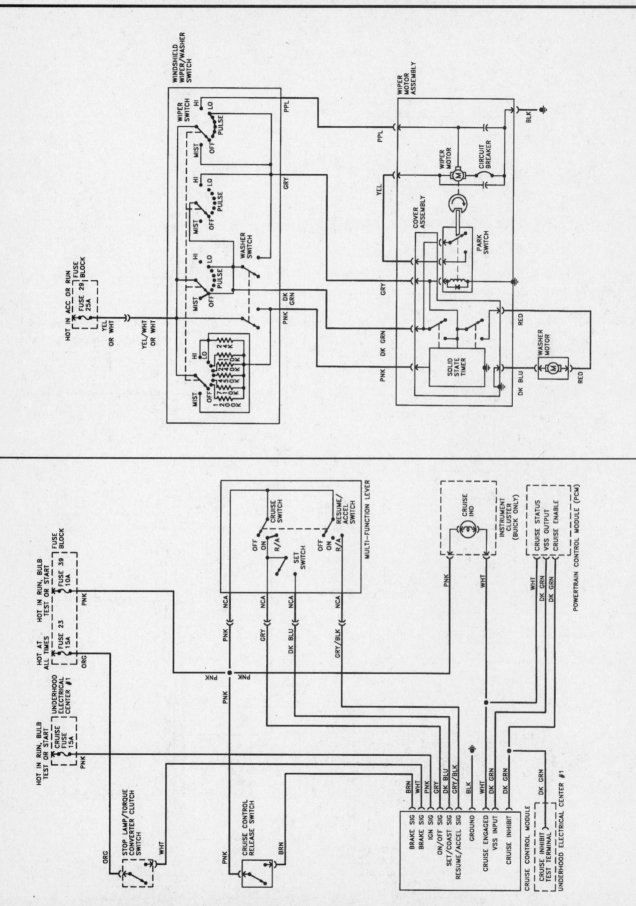

Typical windshield wiper/washer circuit

Cruise control circuit - 1994 and 1995

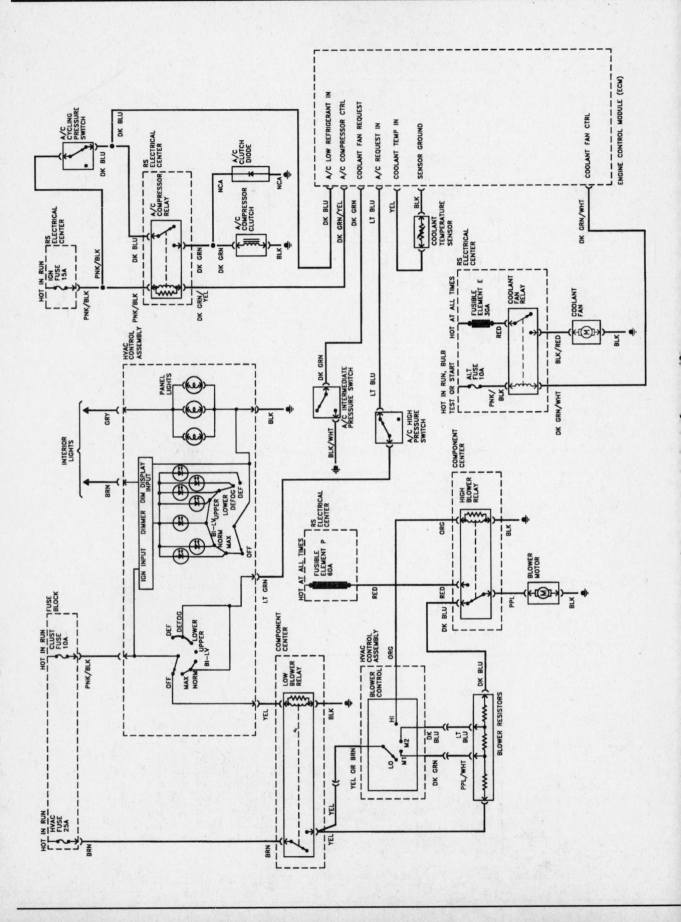

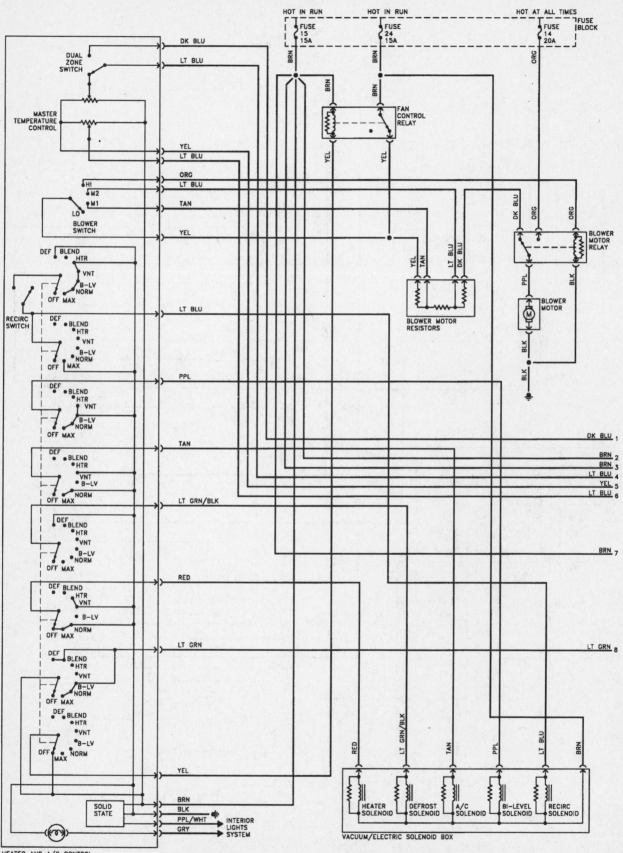

Air conditioning circuit - 1995 models with manual temperature control and 3.8L engines (1 of 2)

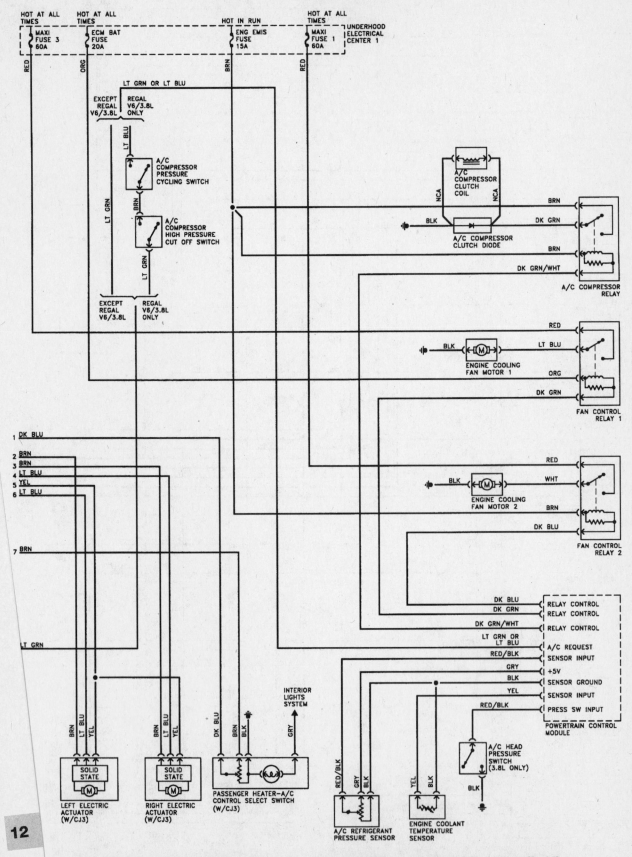

Air conditioning circuit - 1995 models with manual temperature control and 3.8L engines (2 of 2)

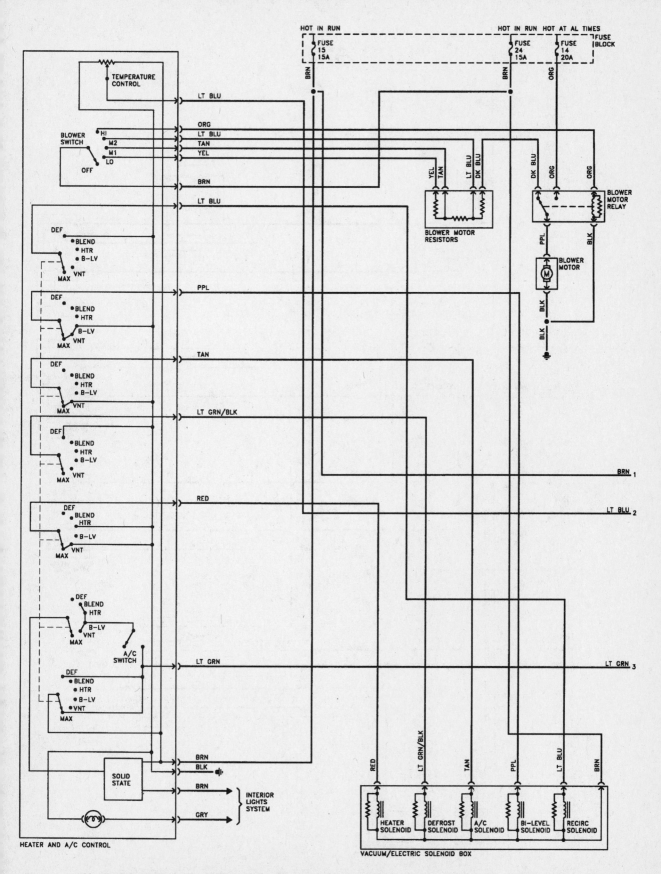

Air conditioning circuit - 1995 models with manual temperature control and 3.1L/3.4L engines (1 of 2)

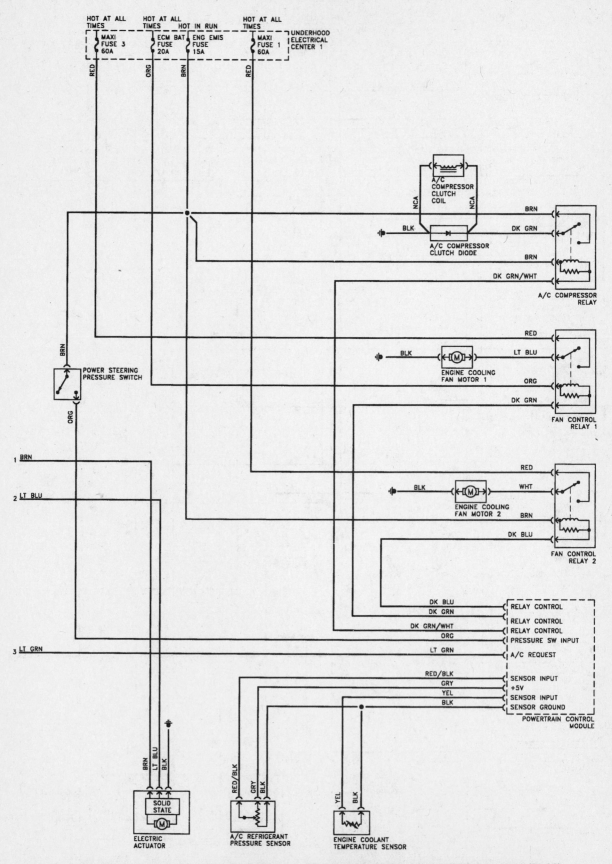

Air conditioning circuit - 1995 models with manual temperature control and 3.1L/3.4L engines (2 of 2)

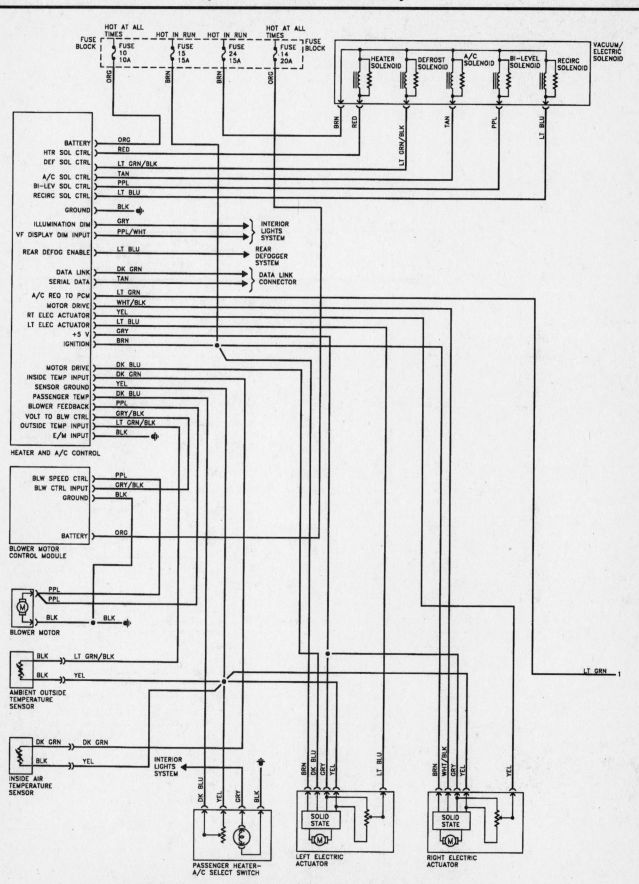

Air conditioning circuit - 1995 models with automatic temperature control (1 of 2)

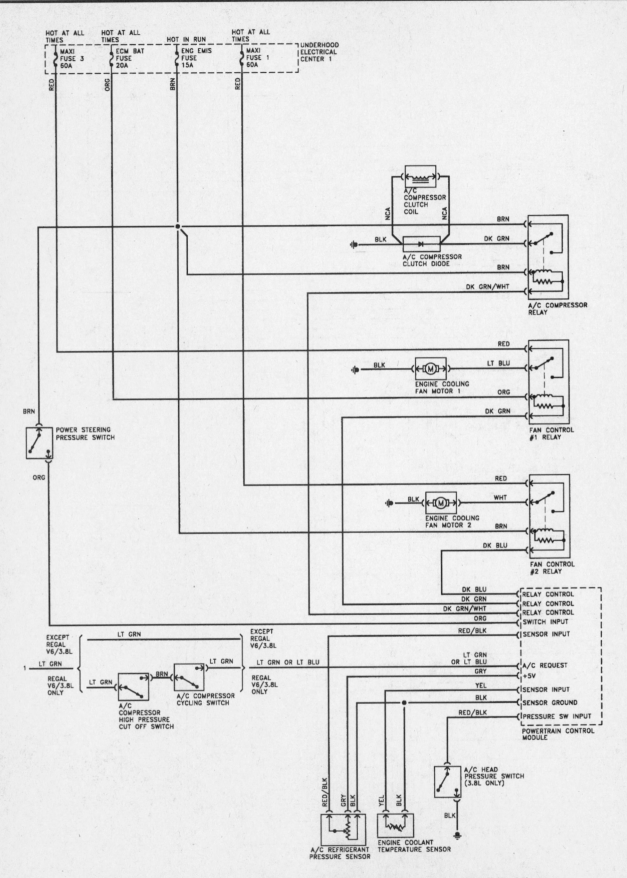

Air conditioning circuit - 1995 models with automatic temperature control (2 of 2)

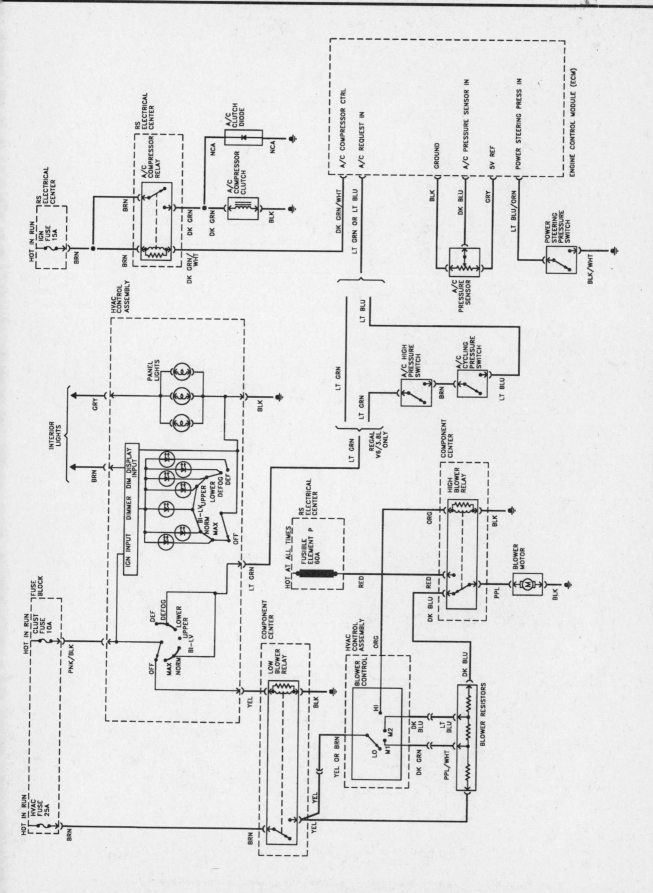

Air conditioning circuit - 1991 through 1993 models with manual temperature control

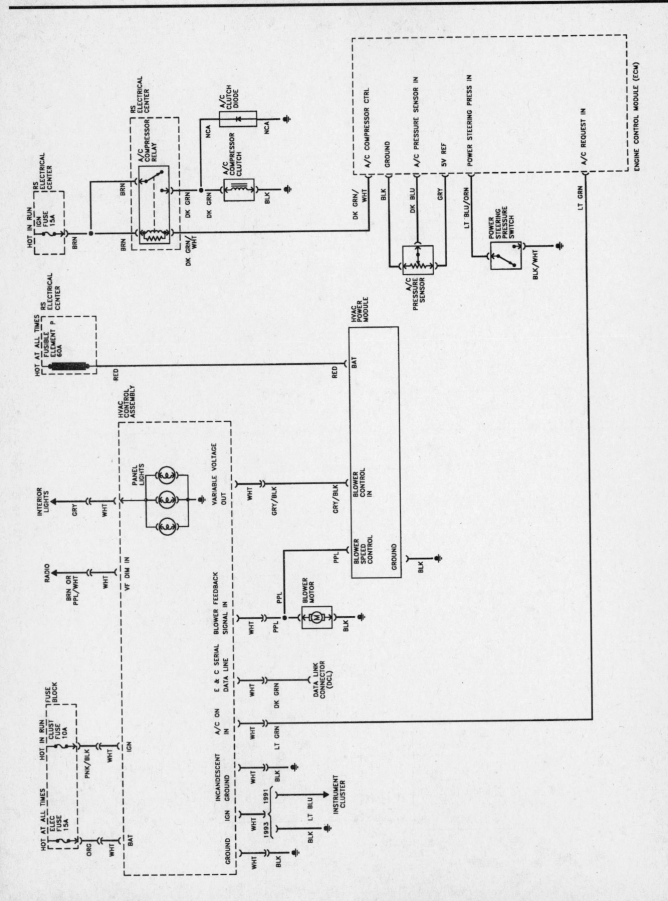

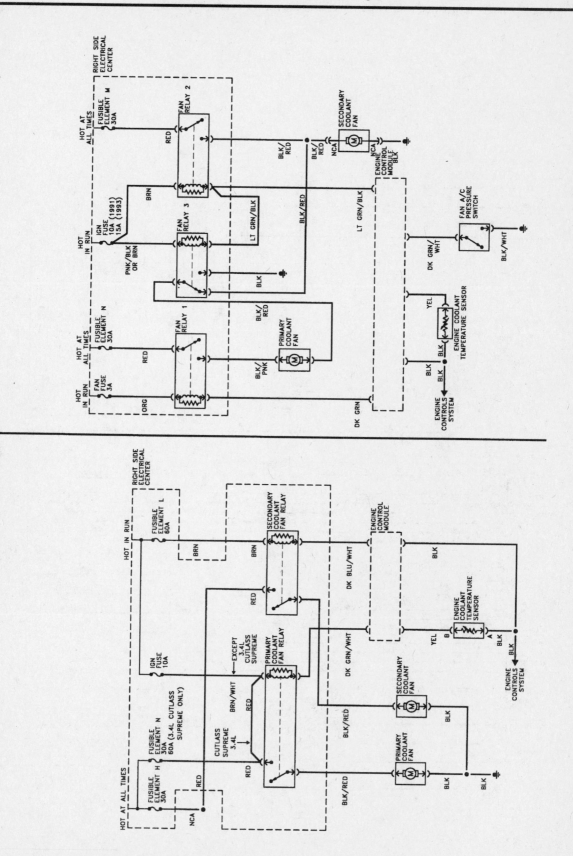

Typical cooling fans circuit

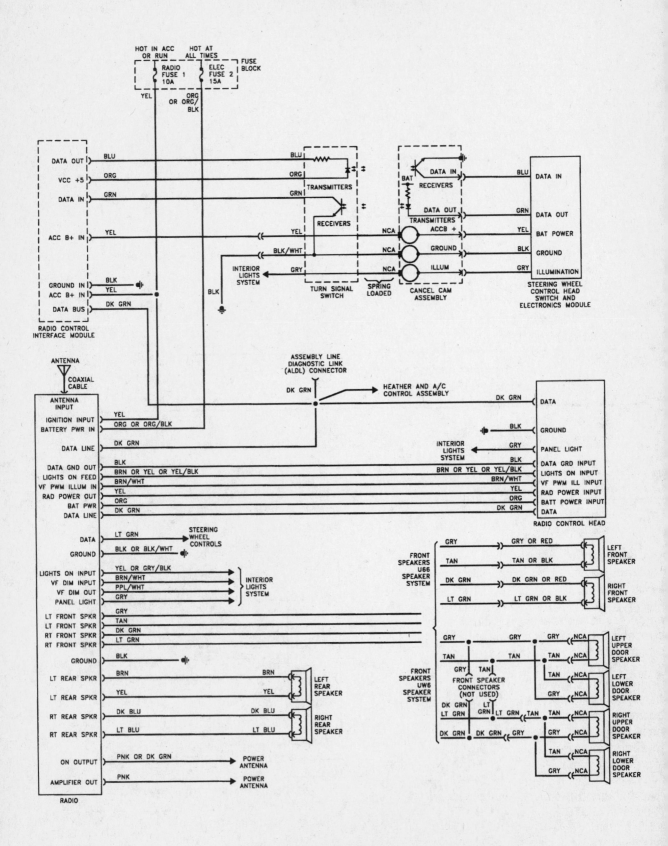

Typical radio circuit

ndex